Aspects of Statistical Inference

WILEY SERIES IN PROBABILITY AND STATISTICS

Established by WALTER A. SHEWHART and SAMUEL S. WILKS

A complete list of the titles in this series appears at the end of this volume

Aspects of Statistical Inference

A.H. WELSH
The Australian National University

A Wiley-Interscience Publication
JOHN WILEY & SONS, INC.
New York • Chichester • Brisbane • Toronto • Singapore • Weinheim

The cover design is based on Figure 3.6, which shows the relationship between a one sided p-value and the posterior probability of the null hypothesis under a Gaussian model.

This text is printed on acid-free paper.

Library of Congress Cataloging in Publication Data:

Welsh, A.H., 1960–
Aspects of statistical inference / A.H. Welsh.
p. cm.—(Wiley series in probability and statistics. Probability and statistics)
"A Wiley-Interscience publication."
Includes bibliographical references and indexes.
ISBN 0-471-11591-6 (cloth : alk. paper)
1. Mathematical statistics. I. Title. II. Series.
QA276.W4525 1996
519.5′4—dc20 96-5234
CIP

Printed in the United States of America

10 9 8 7 6 5 4 3 2 1

To Mary, Claire, and Laura

Contents

Preface

This book provides an introduction to the central ideas and methods of statistical inference by integrating abstract conceptual development with the analysis of data. Within the context of a data-based approach, the book adopts a balanced eclectic presentation in which the different approaches to inference are introduced positively (but without making them appear monolithic) and compared critically, and expands the often narrow focus of discussion about inference by including important topics like robustness, randomization, finite population inference, computational methods based on simulation and smoothing methods.

Data is used throughout the book to motivate the inference problem, explain the importance of context, explain the choice of models, illustrate the nature of inference in practice, make concrete the effect of model choice and emphasize the need to discuss robustness. The data sets are intentionally simple so that their analysis does not overwhelm the concepts they are used to develop. The data-based approach can make the presentation more complex because it precludes prolonged discussion of artificial problems. On the other hand, it is liberating in its clarification of the role of mathematics in statistics because mathematical techniques are introduced when required and in pursuit of a clear data driven objective. The use of data also helps make the abstract concepts concrete and ensures that the practical aspects of inference are not neglected in the theoretical discussion. Indeed, the data-based approach makes it possible to acquire simultaneously both a conceptual framework and practical tools for inference.

This text has grown from my experience in teaching a one semester course in statistical inference, first to graduate students at the University of Chicago and then over a number of years to advanced undergraduate and graduate students at The Australian National University. The course has been offered in the honours program in Statistics, the Graduate Diploma, and Master of Statistics programs, often simultaneously, so that the students have varied backgrounds. This is reflected to some extent in this book through the presentation of material at different levels and the effort to make the book relatively selfcontained. The prerequisites for a course based on this book are a knowledge of calculus (up to the level of Taylor's theorem), a course in distribution theory, and some exposure to the application of statistics.

Chapter 1 presents data on the incidence of severe visual loss due to diabetic retinopathy, the times to failure of pressure vessels subjected to stress, the effect of caffeine ingestion on stress hormones and the volume of urine voided, the velocities of stars, and the corrosion of enamel covered steel plates exposed to hydrochloric acid. Simple models for data are introduced and then important classes of models such as location–scale models, the exponential family, regression models, and generalized linear models are presented to facilitate general development. The inference problem is then introduced and discussed in some detail with attention to the interpretation and meaning of statistical inference. Informal inferences are presented and used to explore the nuisance aspects of the models introduced for the data.

Chapter 2 presents the Bayesian, fiducial, and likelihood approaches to inference in the context of the diabetic retinopathy and caffeine investigations. The use in inference of posterior distributions and credibility sets is discussed. de Finetti's justification for the introduction of prior distributions is explored and the problem of specifying the prior distribution is discussed in detail. The role of conjugate, hierarchical, dominated, robust, and Jeffreys' priors is described. The use of improper priors and some difficulties associated with their use are then discussed. The difficulties of testing sharp hypotheses in general and then using improper priors are presented. The likelihood and fiducial approaches are presented as attempts to capture aspects of the Bayesian approach. They provide interesting insights into the Bayesian paradigm by showing the difficulties of relaxing the requirements for Bayesian inference and help create a historical perspective.

The frequentist approach is developed in Chapter 3, mainly in the context of the caffeine experiment. Point estimation is introduced and discussed briefly. The Fisherian and Neyman–Pearson approaches to testing are then presented separately to explain the differences between them. The actual hybrid mixture of the two approaches which is in widespread use today is then presented. The relationship between likelihood ratio tests and the likelihood and Bayesian approaches to testing is then discussed. Confidence intervals are presented and derived first from pivotal quantities and then by inverting hypothesis tests. Difficulties with setting confidence intervals when the data are discrete are then discussed. The famous Behrens–Fisher and Fieller–Creasy problems are introduced to highlight the differences between the Bayesian/fiducial and frequentist approaches to interval estimation. This material leads to a discussion of conditional inference. Ancillarity, relevant subsets, and the work of Fisher and Pitman on finding exact conditional densities for transformation group models are discussed. The close relationship between the conditional, fiducial and Bayesian approaches to inference is emphasized. The chapter concludes with a discussion of the use of simulation to explore repeated sampling properties.

The pressure vessel failure data is used to illustrate the use of large sample approximations for carrying out approximate inferences in complicated problems in Chapter 4. Initially, an exponential model is adopted and direct approximation methods are used to approximate the distribution function of the sample mean. It is then shown that theorems from probability theory can be used to simplify the procedure for obtaining asymptotic expansions. A number of alternative approximations and inferences based on these are derived and compared. The more complicated problem of how to make approximate inferences if we adopt a gamma model for the data is then discussed. This discussion leads to the development of general results for estimators defined by estimating equations (including maximum likelihood estimators) in multi-parameter problems. The method of moments estimator is also obtained and used to motivate the problem of comparing procedures. The use of saddlepoint methods to obtain higher order approximations to sampling distributions is then discussed. Large sample theory for tests, focusing on the Wald, score, and likelihood ratio tests is then presented with a brief discussion of Bartlett and other adjustments to likelihood ratios. Finally, approximations to likelihoods and posteriors (with emphasis on the use of Laplace's method) are discussed.

The presentation of large sample theory is nonstandard. First, the emphasis is on approximating sampling or posterior distributions. This provides a conceptual framework which clarifies the relationship between nonprobabilistic methods and probabilistic methods like the central limit theorem, and motivates consideration of higher order expansions. More importantly, it integrates large sample theory and inference and thereby provides motivation for what is traditionally rather abstract material. Also, the main results are derived without assuming that the models hold exactly because they are needed in this form in Chapter 5 and because this helps to clarify what is special about the situation in which the model holds.

The stellar velocity data is used as a point of departure for the presentation of robustness in Chapter 5. Inference for this type of data is often based on the standard deviation under Gaussian model assumptions. The effect of departures from a Gaussian model is explored and used to motivate abstract concepts like influence functions and breakdown points. It is shown that standard deviations are sensitive to outlying points of the type found in the data. The important but often overlooked problem of departures from independence is also discussed. The objectives of robustness theory are then considered. Bounded influence estimators are developed for the location–scale problem and then applied to the stellar velocity data. An advantage of using real data here is that there is no possibility of treating the unrealistic problem of making inference about location with known scale or of exploiting the special features of the location problem. The steel plate data is then analyzed to show how robust methods can be used in the regression context. Tests based on M-estimators and some other approaches to robustness are then presented. Finally, the effects of outliers on posterior distributions and likelihoods are explored, the need for robustness in nonfrequentist analyses is discussed, and some of the robustness ideas which have been advanced for these approaches are considered.

Chapter 6 begins with discussion of randomized comparative experiments and the randomization approach to inference in the simple case of paired data (like the caffeine data). The properties of the sample mean and variance under randomization models are developed in detail and the fact that the randomization approach to analyzing data from designed experiments gives the same result as the Gaussian theory approach in large samples is explored. Finite population problems in which the model-based and design-based approaches often do not agree are then discussed. Finite population methods are relevant to data collected by sample surveys, so this section is practically as well as conceptually valuable. The development of nonparametric methods continues with the presentation of permutation tests which are compared with randomization tests. Simulation methods such as the bootstrap, sampling–importance–resampling, and the Gibbs sampler, which provide insight as well as practical tools for inference, are presented. Finally, a review of the scope of nonparametric methods is presented. The connection between kernel smoothing and randomization highlights the central role of randomization in nonparametric methods.

The final chapter discusses the principles of inference, their relationships, and their impact on statistical inference. Proofs are given because they deepen understanding of what the principles mean but they are given in the discrete case to avoid unnecessary complications. For example, Cornfield's proof that coherency and Bayesian inference are equivalent is presented in the discrete case. The likelihood principle is discussed and derived from the sufficiency and ancillary principles. The repeated sampling principle is then presented. The chapter concludes with a discussion of some of the vague and unstated but implicitly important principles which underly statistical inference.

The appendix contains a number of useful facts concerning expansions, matrices, integrals, and distribution theory which are used in the book and the exercises but would break the smooth development were they presented there.

The data-based approach, the inclusion of nonstandard topics (robustness, randomization, finite population inference, simulation, and computational methods based on simulation), and the effort to make the book relatively short but self-contained have meant that some topics which have been viewed as standard have had to be omitted. Optimality results are still mentioned and discussed, though perhaps more critically than usual. Indeed, the exact finite sample optimality of Pitman estimators is highlighted. The mathematical details of optimality results and the decision theoretic framework in which they are usually developed are not presented here. These topics are well presented elsewhere and are more appropriate in a more abstract, mathematical development than that intended here.

It is perhaps important to state that this book is neither a methodological cookbook nor a collection of a neat set of mathematical results. I believe that the data-based approach adopted in this book precludes either of these outcomes. Instead, it opens the possibility of a book which is genuinely about the ideas underlying inference and enforces a focus on the applicable ideas. This book is also not a history book, although it does follow a rough historical framework. The historical references are important for understanding the development and flow of ideas as well as their chronological relationships. They are necessarily incomplete but hopefully sufficient to provide the required framework. The history of statistics is accessibly presented in Stigler (1986).

In spite of the importance of the subject, there are only a limited number of texts which can be used for a senior undergraduate or graduate course on statistical inference. Of these, Bickel and Doksum (1977) and Casella and Berger (1990) are rather more mathematically oriented and adopt a strong frequentist perspective. Silvey (1970), Cox and Hinkley (1974), and Barnett (1982) try to be accessible and to include substantial conceptual material. None of these books adopts a data-based presentation, so they differ fundamentally in philosophy and approach from this book. This is most clearly seen in the presentation of the different approaches to inference, large sample theory, robustness, computational methods based on simulation, the positioning of the principles of inference at the end of the book, and in differences in topic selection. For example, while Bickel and Doksum and Cox and Hinkley discuss robustness in a limited way, none of the other texts mentions the topic. This of course means that related ideas like estimating equation theory are not presented. Furthermore, none of the books discusses randomization in detail and none discusses finite population inference, the bootstrap, or recent developments in Bayesian computation. To be fair, some of these developments have occurred since the publication of these books, but that simply underlines my perception of the need for a new book.

There is a huge volume of literature on the subject of statistical inference. Indeed, statistical inference is so fundamental to statistics that almost everything written about statistics can be interpreted as a contribution to statistical inference. It is impossible to give adequate credit to all the contributors to the subject and I apologize to the many whose views and contributions are not reflected in this book. In one sense, the fact that there are so many contributors and so many different perspectives, means that one has to be courageous or crazy to undertake the writing of a book on statistical inference. On the other hand, the subject is so important that one would be crazy not to.

A.H. Welsh

Canberra, Australia

Acknowledgments

Many people have helped, often inadvertently, in the preparation of this manuscript and I am grateful to them all. The book reflects many influences and I would like to acknowledge the impact of a number of colleagues and co-authors who have affected my thinking and provided indirect but essential encouragement at often crucial times.

Special thanks are due to D.L. Wallace who has not only been a profound influence on my thinking, but who combined with W. H. Wong to encourage me to teach a course in statistical inference and unselfishly assisted me in my first attempt. I am grateful to G.S. Watson and E. Ronchetti who acted generously and unselfishly in providing very detailed comments on early versions of this book and thereby influenced the final shape and content of this book. I am grateful for the suggestions, valuable advice and encouragement I have received at crucial moments from K.R.W. Brewer, R.J. Carroll, R.L. Chambers, R.B. Cunningham, C.A. Field, M.A. Martin, H.L. Morrison, S. Morgenthaler, D.S. Poskitt, S. Stern, and A. Weinberg. I am also grateful to G.A. Barnard, J.O. Berger, W.H. Jefferys, D.V. Lindley, and N.H. Welsh who kindly sent me information and references on specific topics. I am also grateful to the anonymous readers whose suggestions have improved this book. I owe a debt to the numerous students who suffered through early versions of the course and whose feedback encouraged the improvement of the course. I would like to acknowledge the technical help of S. Barry, T. Hesterberg, and P. Minogue in placing nonstandard labels on to my Splus graphics, of R. Boyce who helped greatly by locating obscure references, and of J. Gilpin who helped me obtain various addresses in the United States.

I would also like to thank Steve Quigley, Jessica Downey, Kate Roach, Erin Singletary, Angela Volan, and the editorial staff at John Wiley & Sons for their assistance.

Finally, I would like to thank the following sources of quotations and data for permission to reproduce them:

The Ophthalmic Publishing Company for the diabetic retinopathy data (Table 1.1 and Problem 1.2.3) from the Diabetic Retinopathy Study (1976)

The American Statistical Association for the pressure vessel data (Table 1.2) from Keating, Glaser and Ketchum (1990), the Kevlar 49/epoxy vessel data (Problem 1.5.4) from Schmoyer (1991), the air conditioning data (Problem 1.5.11) from Proschan (1964), and for the quotation from Pratt (1962)

W.B. Saunders and Co. for the caffeine data (Table 1.3) and the quotation from Bellet et al. (1969)

H.L. Morrison for the stellar velocity data (Table 1.4)

N.L. Johnson and F.C. Leone for the enamel covered steel plate data (Table 1.5 and Problem 1.5.9) from Johnson and Leone (1964)

The University of Adelaide for the quotation from Fisher (1922)

Taylor and Francis for the alpha particle data (Problem 1.5.1) from Rutherford and Geiger (1910)

The American Society of Echocardiography and S. Feinstein for the microbubble data (Problem 1.5.2) from Feinstein et al. (1989)

Springer-Verlag and the authors for the word count data (Problem 3.2.2) from Mosteller and Wallace (1984)

R. Koenker for Figure 5.1

The Press Syndicate of the University of Cambridge for the optical isomer data (Problem 5.1.5) from Cushny and Peebles (1905)

D. Basu for the quotation from Basu (1971).

CHAPTER 1

Statistical Models

Statistical inference is concerned with using *data* to answer *substantive questions.* In the kind of problems to which statistical inference can usefully be applied, the data are *variable* in the sense that, if the data could be collected more than once, we would not obtain identical numerical results each time. It is convenient to illustrate the features of such problems through selected examples.

1.1 SUBSTANTIVE PROBLEMS

1.1.1 Severe Visual Loss Due to Diabetic Retinopathy

Diabetic retinopathy is a complication of diabetes mellitus which has become a leading cause of blindness and visual disability in the United States of America. According to *Duane's Ophthalmology* on CD-ROM, 71% of people who have had diabetes for longer than 10 years have diabetic retinopathy, 90% to 95% who have had diabetes for longer than 30 years have diabetic retinopathy and, of these, about 30% have proliferative diabetic retinopathy.

The Diabetic Retinopathy Study was a large multi-center clinical trial begun in 1971 by The National Eye Institute to evaluate the effectiveness of photocoagulation in delaying the onset of blindness in eyes with diabetic retinopathy. One eye from each of 1742 patients with proliferative or severe nonproliferative diabetic retinopathy and a visual acuity of at least 20/100 in both eyes was randomly assigned to either argon laser or xenon arc photocoagulation and the other eye to an untreated control group. Best corrected visual acuity was measured regularly at 4 month intervals after treatment. Preliminary results based on two-year follow up data which were published by the Diabetic Retinopathy Study Group (1976) led to changes to the research protocol of the trial. One set of preliminary results given in Table 1.1 shows the incidence of severe visual loss (a visual actuity of 5/200 or less in two consecutive visits

Table 1.1. Incidence of Severe Visual Loss After 2 Years of Treatment

	Argon		Xenon	
	Untreated	Treated	Untreated	Treated
Number of eyes showing persistent visual acuity loss	26	10	31	8
Numbers of eyes	175	175	179	179

Reprinted with permission from *The American Journal of Ophthalmology* **81** (1976), 383–96.

Table 1.2. Failure Times for Fiber/Epoxy Pressure Vessels (hours)

274	28.5	1.7	20.8	871	363	1311	1661	236	828
458	290	54.9	175	1787	970	0.75	1278	776	126

4 months apart) for subjects who had completed 2 years of treatment by 1976. The problem is to compare the two-year incidence of visual loss under the different treatments.

Problems involving *incidence data* arise naturally in all fields of application. They are simple in their objectives but touch on a number of fundamental issues in statistics, so their solution is both practically and theoretically important.

1.1.2 Pressure Vessel Failure

Keating et al. (1990) reported data on the *failure times* (i.e. the time to failure or lifetime) in hours of 20 pressure vessels constructed of fiber/epoxy composite materials wrapped around metal liners subjected to an unspecified constant pressure. The data are reproduced in Table 1.2. We may be interested in modeling the failure time distribution or we may be interested in parameters of this distribution such as the typical or usual failure time represented by the median failure time.

The important characteristic of the pressure vessel data is that it involves the time to which an event (which in general need not be a failure) occurs. Data of this kind arises frequently in industrial and medical studies though in the latter case the analysis of such data is more optimistically referred to as *survival analysis*.

1.1.3 The Effects of Caffeine

Bellet et al. (1969) reported the results of an experiment to investigate the effect of caffeine ingestion on the release of hormones associated with stress. Eighteen young males were given 5 g of coffee dissolved in 500 ml of water and 500 ml of plain water in random order at least 5 days apart. For details, see Section 6.1. The volume of urine voided in the 3 hours after ingestion of the fluid was recorded and the collected urine was analyzed for catecholamine content (epinephrine and norepinephrine). The data are presented in Table 1.3. The problem is to determine the effect of caffeine ingestion on excreted epinephrine, norepinephrine, and total catecholamine. We can also use the data to explore the effect of caffeine ingestion on the volume of voided urine.

If we find changes in the total catecholamine in this experiment, we would like to conclude that they are caused by the ingestion of caffeine. However, it is possible that any changes are caused by a change in the volume of voided urine and hence possibly only indirectly by the caffeine. In their paper, Bellet et al. (1969) dealt with this issue by quoting a number of previous studies which established that "the amount of catecholamines excreted per unit time is practically independent of urinary flow rate." If this is the case, a change in total

Table 1.3. Voided Urine (ml/3 hr) and Catecholamine (μg/3 hr)

Total Catecholamine		Epinephrine		Norepinephrine		Urine	
Coffee	Control	Coffee	Control	Coffee	Control	Coffee	Control
7.74	4.22	1.14	0.47	6.60	3.75	585	350
12.68	11.15	1.96	0.99	10.72	10.16	594	684
9.53	7.18	2.00	0.43	7.53	6.75	840	430
7.67	4.26	1.36	1.57	6.31	2.69	592	475
10.22	4.47	3.12	0.77	7.10	3.70	520	292
7.25	3.21	2.54	0.41	4.71	2.80	405	151
10.62	3.75	2.88	2.02	7.74	1.73	835	200
7.92	5.96	1.75	1.19	6.17	4.77	995	670
3.18	3.14	1.24	0.49	1.94	2.65	310	315
7.68	6.70	0.88	0.55	6.80	6.15	390	160
3.67	3.21	0.83	0.39	2.84	2.82	295	248
6.39	4.83	2.94	2.07	3.45	2.76	202	115
4.69	4.57	1.42	0.83	3.27	3.74	185	110
6.50	3.56	0.64	0.58	5.86	3.06	155	200
6.23	1.61	1.30	0.24	0.93	1.37	670	115
4.73	4.85	1.21	1.86	3.52	3.69	480	500
3.67	4.21	0.39	0.68	3.28	3.61	505	510
10.69	9.16	0.41	0.23	10.28	8.93	380	340

catecholamine should not be related to the change in the volume of urine produced. A question we can explore with this data is whether there is a relationship between the change in total catecholamine and the change in the volume of urine produced.

Problems involving the determination of the nature and magnitudes of *effects* and problems involving the exploration of *relationships* between variables occur in many fields of application, and methods for solving these problems are of very wide applicability.

1.1.4 Stellar Kinematics

One theory about the formation of our Galaxy is that it started as a large, near-spherical blob of almost pure hydrogen and helium gas which collapsed, forming stars in the process. Some (rare) stars were formed early on, while the gas cloud was still spherical and almost pure, but as the collapse proceeded, various physical processes caused the gas cloud to start to rotate and assume a flattened shape. Stars were formed more often in the later stages of evolution, and these stars follow more disciplined, near-circular paths around the center of the Galaxy. Also, as stars evolve and die (as they do many times in the history of our Galaxy) they form heavier elements such as iron in the nuclear reactions in their centers, and these heavy elements are ejected into the gas which has not yet formed stars so the stars which form later have more heavy elements (higher [Fe/H]) than the older stars. The younger stars are called *disk stars* because of the flattened, rotating shape they form, whereas the old, rare stars which form a spherical shape and have more random motions are called *halo stars*.

In a study of stellar kinematics, Morrison et al. (1990) measured V_z, the component of velocity which is perpendicular to the Galactic Plane in km/s, for a sample of 72 RR Lyrae variables. These are stars which are often used as distance gauges in the Milky Way because their brightness fluctuates rapidly and their true brightness (and hence distance from earth) can be worked out from their fluctuation period. They are also most easily made as a late stage in the evolution of an old star and hence can be used as tracers of the oldest stars in the Milky Way. The stars were chosen so that 17 had [Fe/H] values typical of disk stars, 19 had [Fe/H] values typical of halo stars and 36 had intermediate [Fe/H] values. The data are given in Table 1.4. The problem is to describe and *compare the distribution* of velocities of these three types of stars in order to clarify the status of the intermediate stars.

1.1.5 Corrosion Resistance of Steel Plates

Johnson and Leone (1964, p. 439) described an experiment to test the corrosion resistance of enamel covered steel plates which are used in the manufacture of hot water heating equipment. The experiment involved allowing 10% hydrochloric acid to run over the plates at four different temperatures (140, 160,

Table 1.4. The Component of Stellar Velocity Orthogonal to the Galactic Plane (km/s) for Three Types of RR Lyrae Variable Stars

Disk stars	4, 44, −23, −32, 26, 13, 34, −24, −10, −34, 72, −26, −32, −144, 3, 0, −43
Intermediate stars	0, 45, 6, −48, 65, 55, −69, −77, 117, 33, −17, 64, 5, −286, −175, −29, −63, 23, 58, 69, 7, −1, 25, −268, −44, 87, −102, −42, 25, 16, 62, 31, 5, 21, −77, −63
Halo stars	214, 129, 34, −31, 155, 76, −18, −96, 33, −81, −6, 20, 95, −72, −110, −90, −118, −61, −4

Previously unpublished. Reprinted with the permission of H.L. Morrison. Private Communication.

180 and 200°F) for four different exposure times (4, 6, 8, 10, and 12 hours) and measuring the weight loss for each plate. Hydrochloric acid was used instead of water because its corrosive effect on enamel covered plates is similar but occurs more quickly than with water. The data for one type of plate (for which only three levels of temperature were used) are given in Table 1.5. The problem is to describe the relationship between the response weight loss and the two explanatory variables time and temperature. Another interesting problem with this kind of accelerated test (which we will not consider here) is to relate the results obtained with the hydrochloric acid to those obtained with water.

The problem of *modeling the relationship* between a *response* and several *explanatory variables* so that we can determine the effect of the explanatory variables, predict the response at different values of the explanatory variables, and so on, arises in many fields of application and is very important in statistics.

1.2 INITIALLY PLAUSIBLE MODELS

Fundamental to statistical inference is the recognition that the data appropriate to answering the substantive question are typically variable and that this variability can be represented by a *probability distribution* F_0. This means that the data $\mathbf{z}$ can be regarded as a *realization of a random variable* $\mathbf{Z}$ which takes values in a set $\mathscr{Z}$ called the *sample space*, and which has a probability distribution F_0 that represents the uncertainty in the value $\mathbf{Z}$ will realize in any particular realization. We identify the substantive questions with questions about F_0 so that the objective of inference is to use the data $\mathbf{z}$ to answer questions about F_0.

The interpretation of $\mathbf{z}$ as a realization of a random variable $\mathbf{Z}$ can be useful even in circumstances when it is not literally true. For example, computer generated random numbers satisfy a simple deterministic relationship, but the kind of *chaotic behavior* they exhibit can still for some purposes be usefully treated as of stochastic origin. Similarly, perfect thumb control can produce deterministic coin tossing but variation in the initial conditions leads to

Table 1.5. Weight Loss in Enamel Covered Steel Plates

Weight Loss (10^{-4} g)	Time (hrs)	Temperature (°F)	Weight Loss (10^{-4} g)	Time (hrs)	Temperature (°F)
68	4	160	88	8	180
76	4	160	96	8	180
81	4	160	157	8	180
96	4	180	265	8	200
92	4	180	123	8	200
91	4	180	175	8	200
115	4	200	100	10	160
133	4	200	95	10	160
124	4	200	84	10	160
90	6	160	125	10	180
209	6	160	122	10	180
387	6	160	144	10	180
100	6	180	222	10	200
106	6	180	195	10	200
764	6	180	159	10	200
148	6	200	413	12	160
394	6	200	86	12	160
130	6	200	100	12	160
76	8	160	134	12	180
77	8	160	174	12	180
83	8	160	146	12	180

Reprinted from *Statistics and Experimental Design: In Engineering and the Physical Sciences*, Volume 1 (1964, p. 440). New York: Wiley. With the permission of the authors.

variation in the outcomes and can produce a sequence of tosses which is usefully treated as a realization of a stochastic process. It is neither necessary, nor necessarily useful, to distinguish these cases from others in which a probabilistic approach is useful.

If $\mathscr{Z}$ is countable, the sample space is said to be *discrete* and a probability distribution on $\mathscr{Z}$ is defined by the *probability mass* associated with each element of $\mathscr{Z}$. The probability of an event E is then obtained by summing the probabilities over events in E. If $\mathscr{Z}$ is not discrete, a probability distribution can often be defined on $\mathscr{Z}$ by a *probability density function* (a non-negative function whose integral over $\mathscr{Z}$ is 1) such that the probability of an event E is obtained by integrating the density function over E. We will refer to both types of defining functions as *density functions* and unify their presentation by noting that integrals in general expressions should be replaced by sums in the discrete case.

We do not know the distribution F_0 but we can often suggest a set of candidate distributions $\mathscr{F}$ for F_0 which we call a *statistical model* for $\mathbf{z}$. If we represent the candidate distributions by their density functions and label them by a *parameter* θ (which could be the density functions themselves) taking values in a set Ω called the *parameter space*, we can write our model as

$$\mathscr{F} = \{f(\mathbf{z}; \theta), \mathbf{z} \in \mathscr{Z} : \theta \in \Omega\}.$$

For each possible value of $\theta \in \Omega$, $f(\mathbf{z}; \theta)$ is a density function which specifies a probability distribution for $\mathbf{Z}$ on the sample space $\mathscr{Z}$. We often treat the components $Z_1, \ldots, Z_n$ of $\mathbf{Z}$ as *independent* and, in this case, we can represent the model in terms of the *marginal densities* of the Z_i as

$$\mathscr{F} = \left\{f(\mathbf{z}; \theta) = \prod_{i=1}^{n} f_i(z_i; \theta), \mathbf{z} \in \mathscr{Z} : \theta \in \Omega\right\},$$

where $f_i(\cdot\,; \theta)$ is the density of Z_i. If the Z_i are *identically distributed*, $f_i(\cdot\,; \theta) = f(\cdot\,; \theta)$ and we obtain a further simplification of the model. However, we do need to model data that is both dependent and not identically distributed (some examples are given in Section 1.3), so the general formulation is important. To emphasize that models describe the *joint distribution* of $\mathbf{Z}$, we define models in these terms rather than through equivalent statements involving the marginal distributions of individual observations and the dependence structure.

Notice that we have used lower case $\mathbf{z}$ in two ways: firstly as the realized value of a random variable and secondly as the argument of a probability density function. This convention is potentially confusing but the context should clarify the intended meaning.

The choice of a model depends on the objectives of the analysis (the substantive questions) as well as experience in the analysis of data of similar types, optimistic assumptions about the data generating process to produce a simple model, concerns about the effects of particular inadequacies in standard models, and possibly the results of exploratory data analyses. We can always expand a model (improving the fit to the data) by increasing the number of parameters, but at the cost of an increase in complexity. Model choice inevitably involves a compromise between simplicity or *parsimony* and explanatory power or *quality of fit*.

We always need to question the empirical validity of a model $\mathscr{F}$. We can do this formally by adopting a more complicated model $\mathscr{F}_1$ which includes $\mathscr{F}$ as a special case and then asking whether we can reduce $\mathscr{F}_1$ to $\mathscr{F}$, or we can use informal, graphical methods (Section 1.5). In either case, we need an initially plausible model to start the process. Initially plausible models (and indeed all models) are always subject to modification in the light of additional experience.

1.2.1 Severe Visual Loss Due to Diabetic Retinopathy

Let r_i denote the number of patients receiving treatment i in one eye, where $i = 1$ represents argon treatment and $i = 2$ represents xenon treatment, and let z_{ij} denote the number of untreated ($j = 1$) or treated ($j = 2$) eyes experiencing severe visual loss within 2 years for patients receiving treatment i. In this notation, Table 1.1 is a special case of Table 1.6 in which z_{ij} and r_i are given particular numerical values. Suppose initially that we are interested in the probability of severe visual loss within 2 years in untreated eyes in subjects meeting the eligibility criteria of the Diabetic Retinopathy Study and receiving argon treatment in the other eye. In our data, we observe the number of untreated eyes z_{11} in a sample of size r_1 which experience severe visual loss within 2 years. Each eye either suffers severe visual loss or does not, so we can think of z_{11} as a realization of a random variable Z_{11} which takes on a value in the sample space $\mathscr{Z}_1 = \{0, 1, \ldots, r_1\}$. (Note that r_1 on its own is uninformative about the probability of severe visual loss, so we treat it as fixed and build a *conditional model* for Z_{11} given r_1. The issue of conditioning is discussed further in Sections 3.9 and 7.4.) A model for Z_{11} should depend on the unknown probabilities π_{11k} that the kth eye suffers severe visual loss within 2 years, $k = 1, \ldots, r_1$, and possibly other unknown parameters. If $\pi_{11k} = \pi_{11}$ is constant for each eye (i.e., the patients are homogeneous) and the outcome for each subject is independent of that for all other subjects, simple distribution theory yields the *binomial model* (5b in the Appendix)

$$\mathscr{F} = \left\{ f(z_{11}; \pi_{11}) = \binom{r_1}{z_{11}} \pi_{11}^{z_{11}} (1 - \pi_{11})^{r_1 - z_{11}}, z_{11} = 0, 1, \ldots, r_1 : 0 \le \pi_{11} \le 1 \right\} \tag{1.1}$$

for z_{11}.

The key assumptions for the binomial model (1.1) to hold are that the probability π_{11k} is constant for every subject and the outcomes for each subject are independent of each other. In principle, we can validate the model by passing judgement on the validity of these two assumptions. This is, however, rarely unambiguous because we can often construct scenarios under which the conditions fail; for example, the subjects are not homogeneous if the probability of severe visual loss is affected by age, race, health status, etc. Provided we have

Table 1.6. Incidence of Severe Visual Loss After 2 Years of Treatment

	Argon		Xenon	
	Untreated	Treated	Untreated	Treated
Number of eyes showing persistent visual acuity loss	z_{11}	z_{12}	z_{21}	z_{22}
Numbers of eyes	r_1	r_1	r_2	r_2

measured any variables which might give rise to heterogeneity, we can check the homogeneity assumption. In this case, we could allow π_{11k} to depend on the observed value of these variables for the kth subject and then explore the validity of the homogeneity assumption by exploring whether these variables actually contribute to the model. The independence condition is more difficult to evaluate and so is often simply asserted. This has been described by Box et al. (1978, p. 86) as "the declaration of independence." At the very least, the plausibility of the assumption should be considered by careful thought about the data generation and collection processes.

Now suppose that we want to compare the two groups of control eyes. If the assumptions for the binomial model apply separately to the two groups of eyes, we can model the data for the two groups by

$$\mathscr{F} = \left\{ f(z_{11}, z_{21}; \pi_{11}, \pi_{21}) = \binom{r_1}{z_{11}} \pi_{11}^{z_{11}} (1 - \pi_{11})^{r_1 - z_{11}} \binom{r_2}{z_{21}} \pi_{21}^{z_{21}} (1 - \pi_{21})^{r_2 - z_{21}}, \right.$$
$$\left. z_{11} = 0, 1, \ldots, r_1, z_{21} = 0, 1, \ldots, r_2: 0 \le \pi_{11}, \pi_{21} \le 1 \right\} \quad (1.2)$$

and then compare π_{11} to π_{21}. The parameter of interest is therefore $\pi_{11} - \pi_{21}$. If we can conclude that $\pi_{11} = \pi_{21}$, we can reduce (1.2) to a single binomial model like (1.1) for the number of untreated eyes $z = z_{11} + z_{21}$ in a sample of size $r = r_1 + r_2$ which experience severe visual loss within 2 years.

An important aspect of (1.2) is that the assumptions that (i) the two groups of eyes are independent and (ii) π_{11} is not functionally related to π_{21} mean that we can model each population separately using models like (1.1) and then set up a combined model for comparing the populations by multiplying the separate models together. This is a very useful simplification. When the groups we wish to compare are not independent, we need to incorporate the dependence into the model from the start. For example, while it seems reasonable to treat patients as independent, it is likely that the outcomes for eyes from the same patient are dependent. Thus, to compare the treated eyes to the control eyes requires us to model the paired responses for each subject rather than the aggregated data in Table 1.1 (Problem 1.3.5). However, a crude alternative approach is to compare the argon-treated eyes to the xenon controls and vice versa using models like (1.2).

1.2.2 Pressure Vessel Failure

Let z_i denote the failure time of the ith pressure vessel, $i = 1, \ldots, n$, so the data in Table 1.2 corresponds to a set of $n = 20$ particular z_i values. We regard z_i as a realization of a random variable Z_i which can take on any non-negative real value so the sample space for a sample $\mathbf{Z} = (Z_1, \ldots, Z_n)$ of size n is $\mathscr{Z} = [0, \infty)^n$. It seems reasonable to treat the time to failure of each vessel as independent of the time to failure of any other vessel. In this case, the simplest initial plausible model for the failure times is the *exponential model* (6b in the Appendix)

$$\mathscr{F} = \left\{ f(\mathbf{z}; \lambda) = \prod_{i=1}^{n} \lambda \exp(-\lambda z_i), z_i > 0: \lambda > 0 \right\}. \tag{1.3}$$

Under the exponential model (1.3), the median time between losses is

$$\theta = \frac{\log(2)}{\lambda}. \tag{1.4}$$

We can reparameterize the model in terms of θ as

$$\mathscr{F} = \left\{ f(\mathbf{z}; \theta) = \prod_{i=1}^{n} \frac{\log(2)}{\theta} \exp\left(\frac{-z_i \log(2)}{\theta} \right), z_i > 0: \theta > 0 \right\},$$

but it is simpler to work with (1.3) and apply (1.4) to derive inferences for θ when required. There are of course infinitely many other possible parameterizations for (1.3).

Experience shows that the exponential model (1.3) is often not flexible enough to describe real failure time data. We can explore the adequacy of the model by considering more flexible models which contain the exponential model as a special case. If we retain the independence assumption, we can consider the *gamma model* (6b in the Appendix)

$$\mathscr{F} = \left\{ f(\mathbf{z}; \kappa, \lambda) = \prod_{i=1}^{n} \frac{1}{\Gamma(\kappa)} \lambda^{\kappa} z_i^{\kappa-1} \exp(-\lambda z_i), z_i > 0: \kappa, \lambda > 0 \right\} \tag{1.5}$$

or the *Weibull model* (6g in the Appendix)

$$\mathscr{F} = \left\{ f(\mathbf{z}; \kappa, \lambda) = \prod_{i=1}^{n} \kappa\lambda(\lambda z_i)^{\kappa-1} \exp\{-(\lambda z_i)^{\kappa}\}, z_i > 0: \kappa, \lambda > 0 \right\}. \tag{1.6}$$

Both models reduce to the exponential model (1.3) when $\kappa = 1$ and the extra parameter κ increases the range of shapes allowed under the model. There are a number of other models we can consider but we will not pursue any of them here.

Models for the analysis of independent failure time data like (1.3), (1.5) and (1.6) which are of the form

$$\mathscr{F} = \left\{ f(\mathbf{z}; \theta) = \prod_{i=1}^{n} f(z_i; \theta): \theta \in \Omega \right\}$$

are often usefully classified by their *hazard functions* $h(z; \theta) = f(z; \theta)/\{1 - F(z; \theta)\}$, where $f(z; \theta)$ is the marginal failure time density function and $F(z; \theta)$ is the marginal failure time distribution function. (Here 'marginal' means over units in the sample.) The hazard function gives the instantaneous rate of failure at z given survival to z. The exponential model is characterized by the fact that it has a constant hazard function

$$h(z; \lambda) = \lambda.$$

This implies that the probability of failure within a specified time interval is the same regardless of how long the unit has been on trial, a property that is often referred to as the lack of memory property of the exponential distribution. The hazard functions for the gamma model (1.5) and the Weibull model (1.6) are

$$h(z; \lambda, \kappa) = \frac{z^{\kappa-1} \exp(-\lambda z)}{\int_z^\infty x^{\kappa-1} \exp(-\lambda x)\, dx},$$

and

$$h(z; \lambda, \kappa) = \lambda\kappa(\lambda z)^{\kappa-1},$$

respectively. Both of these functions are monotone decreasing for $\kappa < 1$ and monotone increasing for $\kappa > 1$.

With continuous data (i.e., data for which the model density is absolutely continuous) it is usually not possible to evaluate the model by thinking about primitive assumptions from which the model can be deduced so we consider evaluation (at least of the shape of the model) more as an empirical matter. We will explore the validity of the exponential model (1.3) informally in Section 1.5 and more formally in Sections 4.2–4.3.

Whatever model we finally adopt, the parameter of interest is the median of the underlying distribution and we want to specify a range of plausible values for this parameter.

1.2.3 The Effects of Caffeine

Suppose initially that we are interested in the effect of caffeine on the volume of urine voided. We assume that the subject responses are independent of each other. The standard approach to analyzing data of this type is to assume that the effect of the caffeine is to increase the volume of urine voided by a constant amount μ, say, over the baseline level. We do not expect the increase in urine to be exactly μ in each case because there is variability in the responses. Thus, for each of our $n = 18$ subjects, if subject i has a baseline urine level equal to z_{i0}, the urine level after ingesting caffeine z_{i1} can be written

$$z_{i1} = z_{i0} + \mu + u_i, \qquad 1 \le i \le n,$$

where u_i is the realized value of a random variable U_i. Equivalently, for the pairwise differences $z_i = z_{i1} - z_{i0}$, we have

$$z_i = \mu + u_i, \qquad 1 \leq i \leq n. \tag{1.7}$$

Potentially, the U_i can take on any real value so the sample space $\mathscr{Z} = \mathbb{R}^n$. The distribution of the U_i should be symmetric about the origin (the probability of a positive or negative deviation is the same) and probably the same for each subject. The classical choice is to assume that the U_i all have identical *Gaussian distributions* (6a in the Appendix) with mean 0 and variance σ^2 or, equivalently, that the Z_i all have identical Gaussian distributions with mean μ and variance σ^2. Formally, we write

$$\mathscr{F} = \left\{ f(\mathbf{z}; \mu, \sigma) = \prod_{i=1}^{n} \frac{1}{\sigma} \phi\left(\frac{z_i - \mu}{\sigma}\right), -\infty < z_i < \infty : \mu \in \mathbb{R}, \sigma > 0 \right\}, \tag{1.8}$$

where $\phi(x) = (2\pi)^{-1/2} \exp(-x^2/2)$, μ is the constant amount about which the caffeine effects vary, and σ describes the magnitude of the variability. The parameter μ is of direct interest in answering the substantive question, but σ is not and so is called a *nuisance parameter*.

Now consider the effect of caffeine on the total catecholamine. As noted in Section 1.1.3, we want to explore the question of whether the ingestion of caffeine changes the amount of catecholamine in the voided urine and whether the change in the amount of total catecholamine is related to a change in the volume of voided urine.

Let y_i denote the change in the total catecholamine and z_i the change in the volume of urine produced by the ingestion of caffeine in subject i, $1 \leq i \leq n$. A scatterplot of y_i against z_i shown in Figure 1.1 shows that the change in total catecholamine increases with the change in the volume of urine produced and that there is variability about this relationship which is roughly constant. The scatterplot can be enhanced by scatterplot smoothing (Section 6.9.4) but this is not essential.

We can construct a model for the relationship between y and z by treating (y_i, z_i) as realizations of independent random variables (Y_i, Z_i). If we fix the explanatory variable Z_i and think of modeling the distribution of the response Y_i given $Z_i = z_i$ as a function of the fixed value z_i, it is natural to adopt a model which implies that

$$\mathrm{E}(Y_i \mid Z_i = z_i) = \alpha + \beta z_i \quad \text{and} \quad \mathrm{Var}(Y_i \mid Z_i = z_i) = \sigma^2.$$

It is conventional to assume further that the conditional distributions are Gaussian so that

$$Y_i \mid Z_i = z_i \sim \mathrm{N}(\alpha + \beta z_i, \sigma^2), \qquad 1 \leq i \leq n,$$

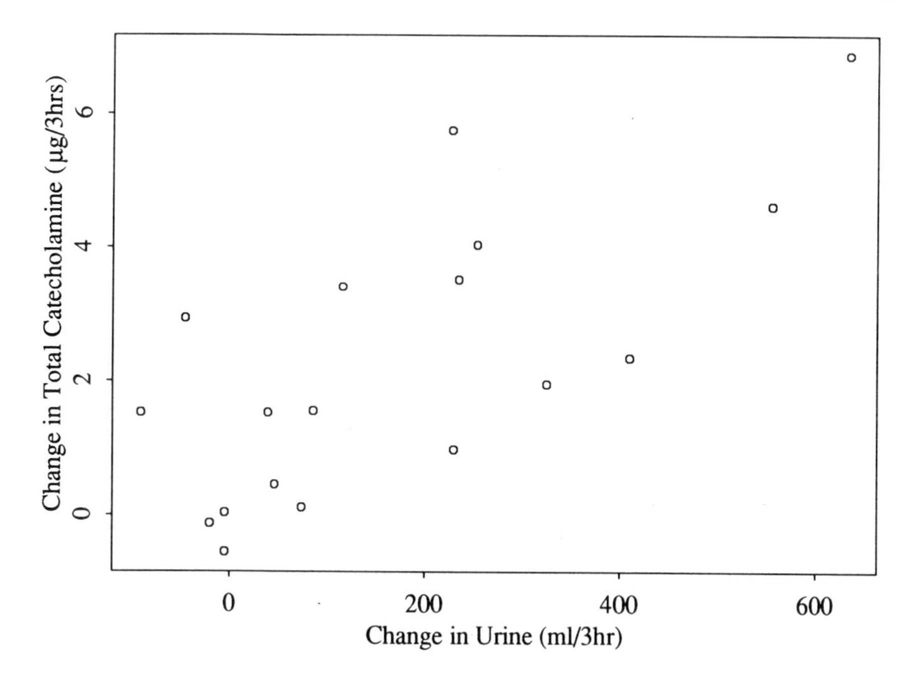

Figure 1.1. A scatterplot of the change in total catecholamine against the change in the volume of urine produced.

or, formally,

$$\mathscr{F} = \left\{ f(\mathbf{y}; \alpha, \beta, \sigma) = \prod_{i=1}^{n} \frac{1}{\sigma} \phi\left(\frac{y_i - \alpha - \beta z_i}{\sigma}\right), -\infty < y_i < \infty : \alpha, \beta \in \mathbb{R}, \sigma > 0 \right\}, \tag{1.9}$$

where $\phi(x) = (2\pi)^{-1/2} \exp(-x^2/2)$. This is called the *simple regression model.* Notice that if the slope $\beta = 0$, there is no relationship between the variables and, in this case, α is the mean of the change in total catecholamine Y_i. On the other hand, if the slope $\beta \neq 0$, there is a relationship between the change in total catecholamine Y_i and the change in the volume of urine produced Z_i and, in this case, the intercept α represents the mean of Y_i after removing the effect of Z_i.

A notable feature of the simple regression model is that it is a model for the conditional distribution of the response given the explanatory variable rather than for the joint distribution of these two variables. The joint distribution can be obtained by multiplying the model distribution by the marginal model for the distribution of the explanatory variable but we do not do this here because the marginal distribution of the explanatory variable contains no additional information about the relationship of interest. Note that this is only appropriate if, as we assume here, the explanatory variable is observed without error. For further discussion, see Sections 3.9 and 7.4.

Finally, the process of modeling the effect of caffeine on epinephrine and norepinephrine is similar to modeling the effect of caffeine on total catecholamine.

1.2.4 The Stellar Velocity Data

It seems reasonable to assume that the velocities of the stars are independent and to treat the three groups of stars as independent. As noted in Section 1.2.1, the assumptions that the three groups of stars are independent and that the parameters of the distributions of the three groups are not functionally related means that we can model each population separately and then set up combined models for comparing the populations by multiplying the separate models together.

Let z_i denote the component of velocity away from the Galactic Plane for the ith star of any one of the three groups of stars. Then we regard z_i as a realization of a random variable Z_i, $1 \leq i \leq n$. The fact that our Galaxy is stable implies that the distribution of Z_i should be symmetric with location zero. This makes sense because, if the location was nonzero, there would be systematic motion away from the Galactic Plane which would cause the Galaxy to tear itself apart. We expect the scales of the distributions to be very different, increasing in the order disk < intermediate < halo because there is least random motion away from the Galactic Plane for the disk stars and most for the halo stars. Thus these populations are distinguished by the spread of their velocity distributions. The traditional model for this situation is that the Z_i have Gaussian distributions with common mean zero and different variances. That is,

$$\mathscr{F} = \left\{ f(\mathbf{z}; \sigma) = \prod_{i=1}^{n} \frac{1}{\sigma} \phi\left(\frac{z_i}{\sigma}\right), \; -\infty < z_i < \infty; \sigma > 0 \right\}, \tag{1.10}$$

where $\phi(x) = (2\pi)^{-1/2} \exp(-x^2/2)$, and we allow different variances for the three groups of stars.

Examination of the data in Table 1.4 (which is confirmed by the graphical analysis presented in Section 5.4) shows that the disk and intermediate groups contain some stars with extremely large negative velocities away from the Galactic Plane. These extreme stars are referred to as *outliers*. Generally, outliers are observations which are extreme relative to the bulk of the data. Chambers (1986) distinguished between two types of outliers. *Representative outliers* are observations which are extreme but which contain information relevant to the substantive question. In contrast, *nonrepresentative outliers* do not contain information about the substantive question. They can be viewed as the result of errors or contamination of the process of interest rather than the process of interest. Representative outliers are interesting in the context of the substantive question and, as in extreme value problems for example, can even be the most important aspect of the data; nonrepresentative outliers are also interesting in that their study may reveal unexpected phenomena which account for the nonrepresentativeness or ways to improve the data collection technique. There is some ambiguity in Chambers' classification of outliers, particularly if both kinds are present in a data set, but the distinction between the two types of outliers is conceptually useful.

The outliers in the stellar velocity data can plausibly arise from errors of measurement, from misclassified stars, or simply from stars which belong to a third population of very rare stars with large velocities. The outliers may be of interest if they are stars from a rare third population, but their very rarity makes their study difficult. On the other hand, if they arise from errors, they are typically of much less interest in terms of the substantive question. In the present context, it seems clear that we want to compare the velocity distributions of the "typical" stars in each group irrespective of the apparently aberrant behavior of a small number of stars. By definition, the typical stars which we are interested in constitute the bulk of the data. Thus we find ourselves in the not uncommon situation where elegance and simplicity lead us to suggest a model $\mathscr{F}$ which is appropriate for the bulk of the data but not for all the data.

To model data containing outliers, we can try to construct a model with an additional parameter or parameters which capture the idea of a contaminated core model. One way to do this when the core distribution is Gaussian and the contamination is symmetric (so the overall distribution remains symmetric) is to consider not the Gaussian model but the *Student t model* (6c in the Appendix)

$$\mathscr{G} = \left\{ g(\mathbf{z}; \sigma, \nu) = \prod_{i=1}^{n} \frac{1}{\sigma} g_\nu\left(\frac{z_i}{\sigma}\right), -\infty \le z_i \le \infty; \sigma > 0, \nu > 0 \right\} \quad (1.11)$$

where $g_\nu(x) = [\Gamma\{(\nu+1)/2\}/(\pi\nu)^{1/2}\Gamma(\nu/2)]\{1 + x^2/\nu\}^{-(\nu+1)/2}$ denotes the density of Student's t-distribution with ν degrees of freedom. Here ν is the "*non-Gaussianity parameter*" which determines the tail behavior. The t-distribution does not in fact have very long tails so this model captures only a restricted range of tail behavior. More severe but less tractable alternatives are provided by the class of stable distributions. There is no reason in general to assume symmetry, so we should consider introducing further parameters to allow for asymmetric tails while keeping the central part of the distribution roughly Gaussian. We could try to allow different transformations in the two tails but there is no entirely satisfactory way to parameterize this kind of asymmetry, and the models quickly become less parsimonious.

A different kind of model to accommodate outliers was proposed by Tukey (1960). In this model we think of the data as actually arising from

$$\mathscr{G} = \left\{ g(\mathbf{z}; \sigma, \varepsilon, c) = \prod_{i=1}^{n} \left[(1-\varepsilon)\frac{1}{\sigma}\phi\left(\frac{z_i}{\sigma}\right) + \varepsilon c(z_i) \right], -\infty \le z_i \le \infty; \sigma > 0, \right.$$
$$\left. 0 < \varepsilon < 1, c \in \mathscr{C} \right\}, \quad (1.12)$$

where $\mathscr{C}$ is some class of long-tailed distributions. This *contamination model* is an attractive model for modeling outliers; we regard most of the data (a fraction $1 - \varepsilon$) as following the *core distribution* in $\mathscr{F}$ but a small portion of the data (a fraction ε) as following a different unknown distribution. If c has long tails, some of the ε fraction of observations with distribution c will be extreme with high probability and hence may be regarded as outliers, whereas if c has similar tails to the distribution in $\mathscr{F}$, the contamination model represents a number of small deviations from $\mathscr{F}$.

Another model for outliers is the so-called *mean-shift model* which assumes that the outliers shift the mean of the core distribution. This is easy to apply when we specify which observations are shifted. Since in practice we may not know which observations are shifted, we have to allow possibly different mean shifts for each observation. That is,

$$\mathscr{M} = \left\{ f(\mathbf{z}; \sigma, \mu_1, \ldots, \mu_n) = \prod_{i=1}^{n} \frac{1}{\sigma} \phi\left(\frac{z_i - \mu_i}{\sigma}\right), -\infty \leq z_i \leq \infty; \right.$$
$$\left. \sigma > 0, -\infty \leq \mu_i \leq \infty \right\}.$$

There are obviously too many parameters in this model. A convenient way to overcome this difficulty is to let the mean shifts be independent random variables with common distribution function H. The model in this case is

$$\mathscr{M} = \left\{ f(\mathbf{z}; \sigma) = \prod_{i=1}^{n} \int_{-\infty}^{\infty} \frac{1}{\sigma} \phi\left(\frac{z_i - w}{\sigma}\right) dH(w), -\infty \leq z_i \leq \infty; \sigma > 0 \right\}.$$

This is not the same as the contamination model (1.12) but it is conceptually similar to it. The contamination model (1.12) is simpler to work with and is more widely used.

If the distributions G in $\mathscr{G}$ are known up to a finite number of parameters, then the contamination model (1.12) is a parametric model (like (1.11)). Although the parametric approach to modeling outliers has its uses, it suffers from the disadvantage that we have to put considerable effort into modeling the outliers which are typically few in number and, when nonrepresentative, a nuisance rather than fundamental to our problem. It seems preferable to consider models in which we are much less specific about the departures from the core model. An advantage of the contamination model is that we can achieve this by letting $\mathscr{G}$ be a large class of distributions, possibly so large that we cannot parameterize it with a finite dimensional parameter. These models are less parsimonious than parametric models but it turns out that we can avoid having to estimate the infinite dimensional nuisance parameters in our inferences. This is discussed in detail in Chapter 5.

The introduction of explicit outlier models enables us to clarify the importance of the distinction between representative and nonrepresentative outliers. Under the Student t model (1.11), the variance of an observation is

$$\sigma^2(G) = \frac{\nu\sigma^2}{(\nu - 2)}, \qquad \nu > 2,$$

while under the contamination model (1.12), it is

$$\sigma^2(G) = (1 - \varepsilon)\sigma^2 + \varepsilon\sigma^2(C) + \varepsilon(1 - \varepsilon)\mu(C)^2,$$

where $\mu(C)$ is the mean and $\sigma^2(G)$ and $\sigma^2(C)$ are the variances of the distributions in $\mathscr{G}$ and $\mathscr{C}$ respectively. If the outliers are representative, then we are interested in $\sigma^2(G)$, the variance of the distribution of the data, whereas if the outliers are nonrepresentative, we are interested in σ^2, the variance of the uncontaminated core distribution. In our analysis, we treat σ^2 as the parameter of interest.

1.2.5 Corrosion Resistance of Steel Plates

The range of values for the response variable (weight loss) is quite large, indicating that it may be preferable to model the data with weight loss on the log scale. Even on the log scale, there appear to be a number of outliers in the data. All the available information about the data is included in Table 1.5 and there is no additional information about the outliers. This makes it difficult to determine whether they are representative or not. If, as is frequently the case, we are interested in the relationship followed by the bulk of the data, we view the outliers as nonrepresentative and proceed on this basis.

If we examine the structure of the data carefully, we see that the experiment involved 14 treatments (the set of all possible combinations of the 3 levels of temperature and 5 of exposure time other than the 12 h, 200°F experiment which was not done) and 3 observations on each treatment, so the treatments provide an (incomplete) two-way classification of the data and we can consider a simple model with 14 different response means. That is, if we let μ_{ij} denote the expected log weight loss at temperature i and time j, then y_{ijk}, the log weight loss of the kth plate at the ith level of time and the jth level of temperature is

$$y_{ijk} = \mu_{ij} + e_{ijk}, \qquad i = 1, \ldots, 5 \quad \text{and} \quad j, k = 1, 2, 3, (i, j) \neq (5, 3), \quad (1.13)$$

where $\{e_{ijk}\}$ are independent and identically distributed random variables with location 0 and unknown spread σ which represents the between plate variation. We need to model the variation to complete the formulation of the model and we will discuss this after exploring the implications of the mean structure we have adopted.

The model (1.13) is adequate for many purposes, but it is often useful to parameterize the means μ_{ij}. A useful structured way of doing this is to represent the 14 means as shown in Table 1.7.

Table 1.7. A Representation of the Mean Response at Different Levels of Temperature and Time in the Enamel Covered Steel Plate Experiment

	Temperature (°F)		
Time (hrs)	160	180	200
4	μ	$\mu + \beta_2$	$\mu + \beta_3$
6	$\mu + \alpha_2$	$\mu + \alpha_2 + \beta_2 + \gamma_{22}$	$\mu + \alpha_2 + \beta_3 + \gamma_{23}$
8	$\mu + \alpha_3$	$\mu + \alpha_3 + \beta_2 + \gamma_{32}$	$\mu + \alpha_3 + \beta_3 + \gamma_{33}$
10	$\mu + \alpha_4$	$\mu + \alpha_4 + \beta_2 + \gamma_{42}$	$\mu + \alpha_4 + \beta_3 + \gamma_{43}$
12	$\mu + \alpha_5$	$\mu + \alpha_5 + \beta_2 + \gamma_{52}$	No data collected

The representation in Table 1.7 can be written as

$$\mu_{ij} = \mu + \alpha_i + \beta_j + \gamma_{ij}, \qquad i = 1, \ldots, 5 \quad \text{and} \quad j = 1, 2, 3, (i, j) \neq (5, 3)$$
$$\text{with } \alpha_1 = \beta_1 = \gamma_{1j} = \gamma_{i1} = 0. \qquad (1.14)$$

Notice that $\mu = \mu_{11}$, $\alpha_i = \mu_{i1} - \mu_{11}$, $\beta_j = \mu_{1j} - \mu_{11}$, and $\gamma_{ij} = \mu_{ij} - \mu_{i1} - \mu_{1j} + \mu_{11}$. The parameters α_i and β_j are called the *main effects* of time and temperature, respectively, and the γ_{ij} are called *interactions*. The interactions γ_{ij} are often suggestively written $(\alpha\beta)_{ij}$ to emphasize that they are interactions and to preserve notation; we use γ_{ij} to emphasize the fact that the interactions are distinct from the main effects α_i and β_j. The interaction terms allow the effect of time to depend on the temperature setting and vice versa. If the interactions $\gamma_{ij} = 0$, the response means are additive in the main effect parameters α_i and β_j which implies that the effect of time is constant across the temperature settings and the effect of temperature is constant across the time settings. This is shown schematically in Figure 1.2. Of course, with data, we also need to take the variability into account and we do this by fitting the model. When we fit the model, this representation of the means makes it convenient for us to compare mean responses, to explore whether the time and temperature effects interact and, if they do not, the magnitudes of the separate time and temperature effects.

There is nothing special about the parameterization we have adopted for the mean structure in (1.14), and other parameterizations can also be used. At least when the two-way classification is complete, a popular alternative is to write

$$\mu_{ij} = \mu + \alpha_i + \beta_j + \gamma_{ij}, \qquad i = 1, \ldots, 5 \quad \text{and} \quad j = 1, 2, 3,$$
$$\text{with } \sum_{i=1}^{5} \alpha_i = \sum_{j=1}^{3} \beta_j = \sum_{j=1}^{3} \gamma_{ij} = \sum_{i=1}^{5} \gamma_{ij} = 0, \qquad (1.15)$$

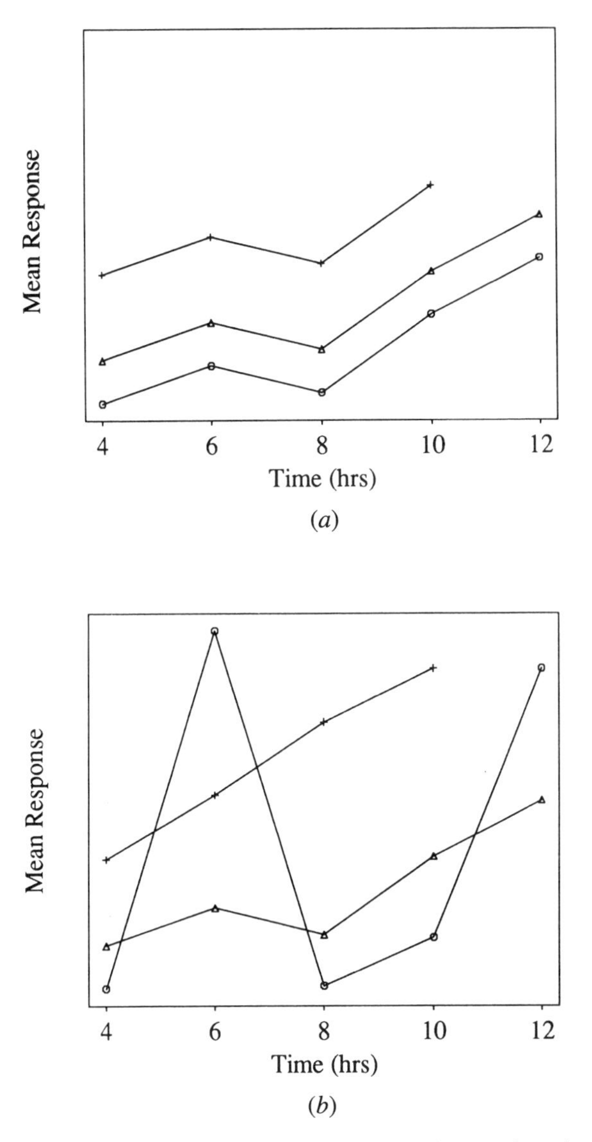

Figure 1.2. Schematic representation of the mean log weight loss against time with symbols representing the three levels of temperature (○ = 160°F, △ = 180°F, and + = 200°F) to illustrate the absence and presence of interactions in the model. (*a*) Additive effects (no interaction) (*b*) Interaction.

which is of basically the same form but with different constraints (and hence different interpretations) on the parameters. The constraints in (1.14) are called *cornerpoint constraints*, while those in (1.15) are called *mean constraints*. The mean constraints are more convenient for hand calculation when the two-way classification is complete, but this is not the case here and the cornerpoint constraints are simpler to use.

The *two-way classification models* we have introduced above ignore the fact that the explanatory variables time and temperature are quantitative variables and treat them as qualitative. This is clearer if we rewrite the *mean structure* using *indicator variables* as

$$\begin{aligned}\mu_{ij} = \mu &+ \alpha_2 I(i=2) + \alpha_3 I(i=3) + \alpha_4 I(i=4) + \alpha_5 I(i=5) \\ &+ \beta_2 I(j=2) + \beta_3 I(j=3) + \gamma_{22} I(i=2, j=2) + \gamma_{23} I(i=2, j=3) \\ &+ \gamma_{32} I(i=3, j=2) + \gamma_{33} I(i=3, j=3) + \gamma_{42} I(i=4, j=2) \\ &+ \gamma_{43} I(i=4, j=3) + \gamma_{52} I(i=5, j=2).\end{aligned}$$

(Here $I(A) = 1$ if A holds and 0 otherwise.) This suggests that we are ignoring information in the nature of the explanatory variables and that we try to model the relationship between the response and the explanatory variables directly.

A *scatterplot matrix* for the data is shown in Figure 1.3. Although the

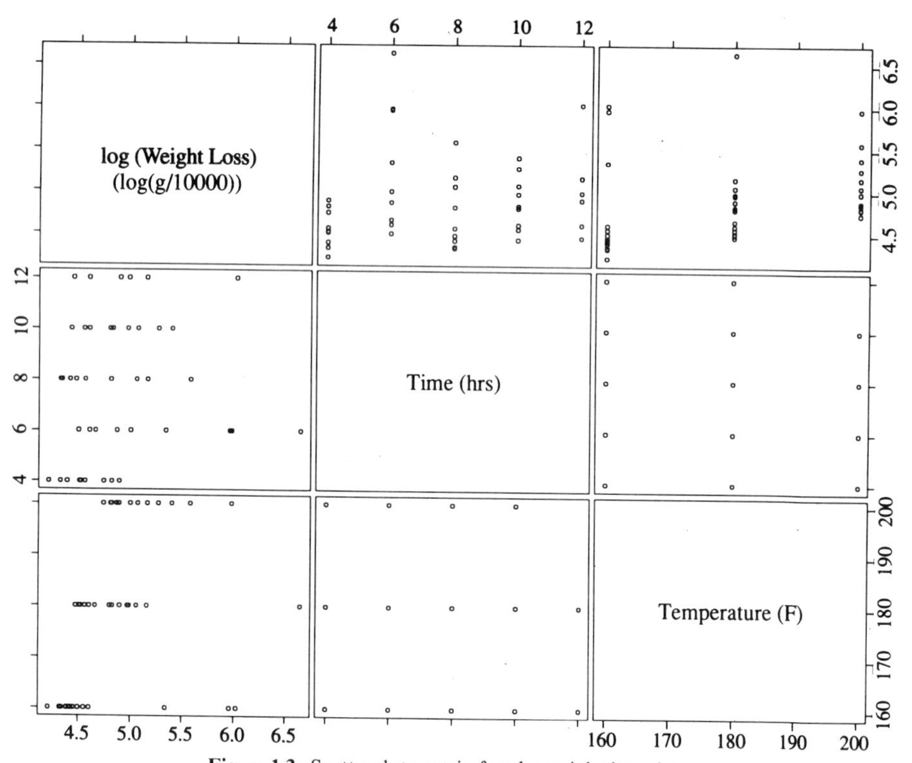

Figure 1.3. Scatterplot matrix for the weight loss data.

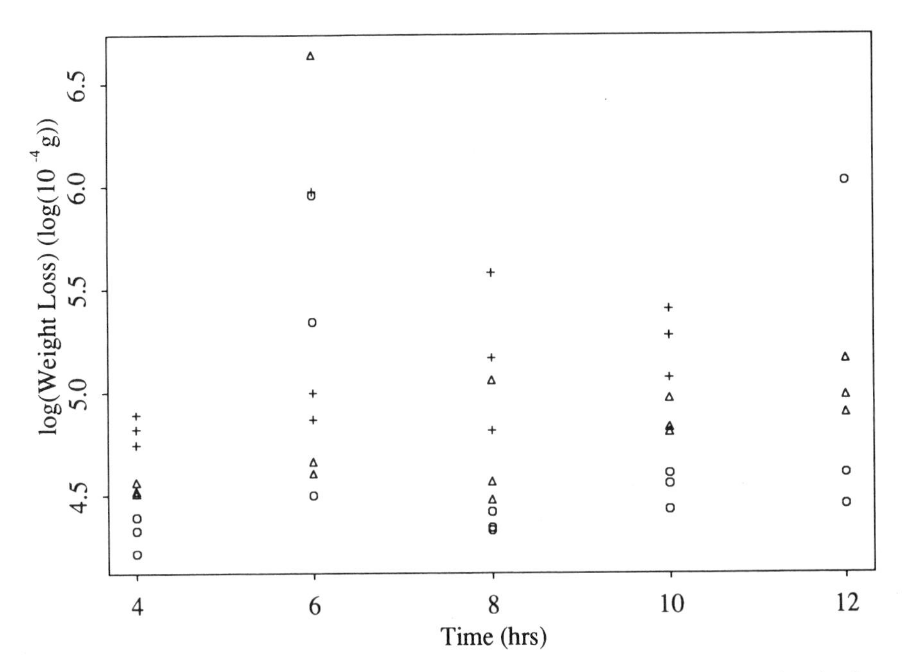

Figure 1.4. Plot of the logarithm of weight loss against time with symbols representing the three levels of temperature (○ = 160°F, △ = 180°F, and + = 200°F).

marginal plots are not informative about the nature of the joint relationship between the log weight loss, time and temperature, they do provide some insights. They show that log weight loss is linearly related to time and temperature separately, but there appear to be some outliers. This is confirmed by Figure 1.4 which shows all three variables in a single plot. Other graphical techniques can usefully be applied here but their discussion would take us too far from our central concerns. The power and elegance of these and other graphical methods for exploring relationships between variables is shown in Cleveland (1993, Chapters 3–5).

Preliminary graphical analysis suggests that we can consider the *regression model*

$$y_i = \alpha + \beta_1 x_{1i} + \beta_2 x_{2i} + e_i, \qquad 1 \le i \le 42, \tag{1.16}$$

where y_i is the log weight loss, x_{1i} is the time, x_{2i} is the temperature, and $\{e_i\}$ are independent and identically distributed random variables with location 0 and unknown spread σ.

Whether we decide to parameterize the mean structure by (1.14), (1.15), or (1.16), we complete the model specification by modeling the random variation about the conditional mean response. In the absence of outliers in the response, it is common to assume that the errors $\{e_i\}$ have a Gaussian distribution.

Formally, in the case of (1.16), we write

$$\mathscr{F} = \left\{ f(\mathbf{y}; \alpha, \beta_1, \beta_2, \sigma) = \prod_{i=1}^{n} \frac{1}{\sigma} \phi\left(\frac{y_i - \alpha - \beta_1 x_{1i} - \beta_2 x_{2i}}{\sigma}\right), -\infty < y_i < \infty: \alpha, \beta_1, \beta_2 \in \mathbb{R}, \sigma > 0 \right\},$$

where $\phi(x) = (2\pi)^{-1/2} \exp(-x^2/2)$. However, in accordance with the discussion in Section 1.2.4, with outliers present in the data, we may assume a contaminated version of this Gaussian regression model, namely

$$\mathscr{G} = \left\{ g(\mathbf{y}; \alpha, \beta_1, \beta_2, \sigma, \varepsilon, c) = \prod_{i=1}^{n} \left[(1-\varepsilon) \frac{1}{\sigma} \phi\left(\frac{y_i - \alpha - \beta_1 x_{1i} - \beta_2 x_{2i}}{\sigma}\right) + \varepsilon c(y_i) \right], -\infty \leq y_i \leq \infty; \alpha, \beta_1, \beta_2 \in \mathbb{R}, \sigma > 0, 0 < \varepsilon < 1, c \in \mathscr{C} \right\}.$$

Obvious modifications to these models complete the specification of the two-way classification models.

Note that in contrast to the situation in Section 1.1.3, the explanatory variables in (1.16) were chosen by the experimenter and so are not stochastic. In this context, the model represents the distribution of the response and the issue of specifying a conditional as opposed to a joint model does not arise. It also means that the possibility of outliers in the explanatory variables does not arise. This affords considerable simplification in both the modeling and the analysis of these data.

Both the regression model (1.16) and the two-way classification models (1.14) and (1.15) are particular cases of the *general linear model* which are distinguished by the different ways they represent the mean structure. The different representations correspond to treating the explanatory variables as qualitative (the two-way classification model) or as quantitative (the regression model). The two-way classification model is less informative (it describes the value of the response surface at the observed levels of the explanatory variables but does not describe the overall surface) but more flexible in that it has 15 parameters in the mean compared to 3 in the regression model. Generally, if the explanatory variables are quantitative and we can find a simple functional form for the conditional mean structure, the regression model is preferable to the two-way classification model.

PROBLEMS

1.2.1. For each of the following problems, discuss briefly whether the conditions for a binomial model are likely to be met or not:

(a) The number of street lights along a road which fail.

(b) The number of times an operator commits an error in 10 trials of a new machine.

(c) The number of people who smoke in a sample of size 30 drawn without replacement from a population of size 300.

(d) The number of people who smoke in a sample of size 30 drawn without replacement from a population of size 30,000.

(e) The number of wallabies trapped on an island.

1.2.2. The conditions for a Poisson model for the number of events which occur in a set time, area or volume are:

(a) The number of events which occur during nonoverlapping intervals are independent random variables.

(b) The distribution of the number of events during any interval depends only on the length of the interval and not on the endpoints of the interval.

(c) For a sufficiently small interval, the probability of obtaining exactly one event during that interval is proportional to the length of the interval.

(d) The probability of obtaining two or more events during a sufficiently small interval is negligible.

For each of the following problems, discuss briefly whether the conditions for a Poisson model are met or not:

(a) The number of telephone calls in set time periods.

(b) The number of car accidents at an intersection during set time periods.

(c) The number of car accidents in a day in Sydney.

(d) The number of stars in a fixed volume of sky.

(e) The number of times a word is used in a text of set length.

1.2.3. The Diabetic Retinopathy Study (1976) reported the data given in Table 1.8 on eyes recovering from severe visual loss. Propose and justify a model

Table 1.8. Incidence of Recovery from Severe Visual Loss

	Untreated	Treated
Number of eyes with evidence of recovering visual acuity	10	12
Number of eyes	82	42

Reprinted with permission from *The American Journal of Ophthalmology* **81** (1976), 383–96.

which can be used to make inferences about the probability of recovery from severe visual loss for untreated eyes. Discuss any difficulties which arise in trying to compare the probability of recovery from severe visual loss for untreated with treated eyes.

1.2.4. Reparameterize the binomial model (1.1) in terms of the parameter $\theta = \log\{\pi_{11}/(1-\pi_{11})\}$. Hence or otherwise, parameterize (1.2) in terms of θ and Δ, where $\theta = \log\{\pi_{11}/(1-\pi_{11})\}$ and $\theta + \Delta = \log\{\pi_{21}/(1-\pi_{21})\}$. Interpret this parameterization. Can you see any advantages for this parameterization?

1.3 CLASSES OF MODELS

The models proposed in Section 1.2 have a number of important features in common. Rather than deal separately with every single model we encounter, it is useful to group models with similar features into classes of models for which we can obtain general results which can then be specialized to particular cases of interest.

1.3.1 The Exponential Family

A useful and important class of models is the *k-parameter exponential family model* which is given by

$$\mathscr{F} = \left\{ f(\mathbf{z}; \theta) = \prod_{i=1}^{n} \exp\left[\psi_1 a_1(z_i) + \cdots + \psi_k a_k(z_i) + \phi + b(z_i)\right], \mathbf{z} \in \mathscr{Z}: \right.$$
$$\left. \theta = (\psi_1, \ldots, \psi_k, \phi) \in \Omega \right\}. \quad (1.17)$$

Here $a_1, \ldots, a_k$ and b are known real valued functions and Ω is chosen so that $f(\cdot; \theta)$ integrates to 1 for each $\theta \in \Omega$. A number of common models including the binomial (1.1), exponential (1.3), gamma (1.5) and Gaussian models (1.8) and (1.10) are in the exponential family. For example, for the binomial model

$$\mathscr{F} = \left\{ f(z; \pi) = \binom{r}{z} \pi^z (1-\pi)^{r-z}, z = 0, 1, \ldots, r: 0 \le \pi \le 1 \right\},$$

we have

$$\log f(z; \pi) = z \log \pi + (r - z) \log(1 - \pi) + \log\binom{r}{z}$$
$$= z \log\left(\frac{\pi}{(1-\pi)}\right) + r \log(1-\pi) + \log\binom{r}{z}, \qquad z = 0, 1, \ldots, r, \quad 0 \le \pi \le 1,$$

so $k = 1$, $a_1(z) = z$, $\psi_1 = \log\{\pi/(1-\pi)\}$, and $\phi = r\log(1-\pi)$. (Notice that $n = 1$ here; this is the reason we use r, not n, to denote the number of subjects.) Similarly, for the Gaussian model

$$\mathscr{F} = \left\{ f(\mathbf{z}; \mu, \sigma) = \prod_{i=1}^{n} \frac{1}{(2\pi\sigma^2)^{1/2}} \exp\left\{ -\frac{(z_i-\mu)^2}{2\sigma^2} \right\}, -\infty < z_i < \infty : \mu \in \mathbb{R}, \sigma > 0 \right\}.$$

we have

$$\log f(z; \mu, \sigma) = -\frac{(z-\mu)^2}{2\sigma^2} - \frac{\log(2\pi\sigma^2)}{2}$$

$$= \left(\frac{-1}{2\sigma^2}\right) z^2 + \left(\frac{\mu}{\sigma^2}\right) z - \frac{\left(\log(2\pi\sigma^2) + \dfrac{\mu^2}{\sigma^2}\right)}{2}, \qquad \theta = (\mu, \sigma^2) \in \mathbb{R} \times \mathbb{R}^+$$

so $k = 2$, $a_1(z) = z^2$, $a_2(z) = z$, $\psi_1 = -1/2\sigma^2$, $\psi_2 = \mu/\sigma^2$ and $\phi = -(\log(2\pi\sigma^2) + \mu^2/\sigma^2)/2$.

The Weibull model (1.6) considered in the pressure vessel example is not in the exponential family. For this model

$$\mathscr{F} = \left\{ f(\mathbf{z}; \kappa, \lambda) = \prod_{i=1}^{n} \kappa\lambda(\lambda z_i)^{\kappa-1} \exp\{-(\lambda z_i)^\kappa\}, z_i > 0 : \lambda > 0 \right\}$$

so

$$\log f(z; \kappa, \lambda) = \kappa \log(z) - \lambda^\kappa z^\kappa - \log(z) + \kappa \log(\lambda) + \log(\kappa), \qquad z > 0; \quad \lambda > 0,$$

but this is not in the exponential family when κ is unknown because then z^κ is not a known function.

1.3.2 The Location–Scale Family

A second useful and important class of models is the *location–scale model* which is given by

$$\mathscr{F} = \left\{ f(\mathbf{z}; \mu, \sigma) = \prod_{i=1}^{n} \frac{1}{\sigma} h\left(\frac{z_i - \mu}{\sigma}\right), -\infty < z_i < \infty, \mu \in \mathbb{R}, \sigma > 0 \right\}. \quad (1.18)$$

where h is a fixed known density function. The effect of changing μ is simply to shift the distribution, so we call μ a *location parameter*. The effect of changing

σ is to make the distribution more or less concentrated, so we call σ a *scale parameter*. If the scale σ is known we call $\mathscr{F}$ a *location model* and if the location μ is known we call $\mathscr{F}$ a *scale model*. The Gaussian model (1.8) is a special case of a location–scale model with $h(z) = \phi(z) = (2\pi)^{-1/2} \exp(-z^2/2)$; (1.10) is a scale model. The exponential model (1.3) is a scale model with scale parameter $\sigma = \lambda^{-1}$ and $h(z) = \exp(-z)$. The gamma (1.5) and Weibull (1.6) models are not location–scale families because they include an unknown shape parameter κ.

1.3.3 Classification of Models

In most of the models we have considered so far, the parameter space Ω has been a subset of the finite dimensional Euclidean space $\mathbb{R}^k$ with k small and finite. Such models are said to be parametric, and the problem of making inferences about the parameters in such models can be described as a *parametric problem*. However, as we saw in Sections 1.2.4 and 1.2.5, not all statistical problems are of this type. By way of another example, suppose that we consider a location–scale model (1.18) but that we treat the density function h as an unknown parameter. Formally, let $\mathscr{H}$ be a nonempty class of density functions and set

$$\mathscr{F} = \left\{ f(\mathbf{z}; \mu, \sigma, h) = \prod_{i=1}^{n} \frac{1}{\sigma} h\left(\frac{z_i - \mu}{\sigma}\right), -\infty < z_i < \infty : \mu \in \mathbb{R}, \sigma > 0, h \in \mathscr{H} \right\}.$$

If $\mathscr{H}$ is the class of all distributions which are symmetric about the origin, then the parameter space is the product of a two dimensional Euclidean space and an infinite dimensional space of density functions. The problem of making inference about the infinite dimensional parameter h is a *nonparametric problem*. The situation in which h is a nuisance parameter and we want to make inference about the finite dimensional parameter (μ, σ) is a *semiparametric problem*, although in the literature this is often also referred to as a nonparametric problem (see for example Sections 5.1.1, 6.9.1–6.9.3). Throughout this book, with the exception of Section 1.5 and Chapters 5 and 6, we will focus on parametric problems, though much of our discussion is also relevant to the intrinsically far more complicated nonparametric and semiparametric problems.

1.3.4 Models Incorporating Explanatory Variables

The classes of models considered above all treat the observations as realizations of independent and identically distributed random variables. Models for data with neither of these attributes are important and widely used in practice. For example, if in addition to our observations Z_i we have covariate information x_i on the ith unit, $i = 1, \ldots, n$, which is either observed without error or known (i.e., nonstochastic), we can allow the parameters in the model to be functions of x. Thus for the one parameter exponential family model which, from (1.17), is

$$\mathscr{F} = \left\{ f(\mathbf{z}; \theta) = \prod_{i=1}^{n} \exp\left[\psi a(z_i) + \phi + b(z_i)\right], z \in \mathscr{F} : \theta = (\psi, \phi) \in \Omega \right\},$$

we can allow ψ and/or ϕ to depend on x. If we let $g(\mathrm{E}Z) = g^*(\psi) = x^{\mathrm{T}}\beta$, for a known link function g, we obtain the important class of *generalized linear models* (McCullagh and Nelder, 1989). The regression models (1.9) and (1.16) as well as the two-way classification models (1.14) and (1.15) are in this class.

Similarly, for the location–scale family model (1.18), with $\mu = \mu(x)$, we obtain the regression model

$$\mathscr{F} = \left\{ f(\mathbf{z}; \mu, \sigma) = \prod_{i=1}^{n} \frac{1}{\sigma} h\left(\frac{z_i - \mu(x_i)}{\sigma}\right), -\infty < z_i < \infty : \mu \in \mathscr{M}, \sigma > 0 \right\}.$$

We can take $\mu(x) = x^{\mathrm{T}}\gamma$ to obtain a linear regression model like (1.9) and (1.16) or a two-way classification model like (1.14) or (1.15). We can also take $\mu(x)$ to be a nonlinear function known up to a finite number of unknown parameters to obtain a *nonlinear regression model* or $\mu(x)$ an unknown parameter in a space $\mathscr{M}$ of smooth functions to obtain a *nonparametric regression model.* More generally, we can let σ be a function of the covariates too (usually not linear) so that if h is treated as a parameter we can have up to three distinct infinite dimensional parameters in the model.

1.3.5 Temporal Dependence

Dependence arises commonly in models for data which is developing in time or has a spatial arrangement. For example, a useful model for temporal dependence is provided by the *autoregressive process* $\{Z_i\}$ which has the property that the ith observation depends on the $(i - 1)$th observation and an independent innovation. Explicitly, for $|\rho| < 1$,

$$Z_1 \sim \mathrm{N}\left(\mu, \frac{\sigma^2}{1 - \rho^2}\right)$$

and

$$Z_i - \mu = \rho(Z_{i-1} - \mu) + \sigma U_i, \qquad i = 2, \ldots, n,$$

with $\{U_i\}$ a sequence of independent Gaussian random variables with mean 0 and variance 1. The observations are clearly dependent and in fact the *autocovariance function*

$$\gamma(h) = \mathrm{Cov}\,(Z_i, Z_{i+h}) = \frac{\sigma^2 \rho^{|h|}}{1 - \rho^2}.$$

The model can be written as

$$\mathscr{F} = \left\{ f(\mathbf{z}; \mu, \rho, \sigma) = \frac{(1-\rho^2)^{1/2}}{(2\pi\sigma^2)^{1/2}} \exp\left\{ -\frac{(1-\rho^2)(z_1-\mu)^2}{2\sigma^2} \right\} \right.$$
$$\left. \times \prod_{i=2}^{n} \frac{1}{(2\pi\sigma^2)^{1/2}} \exp\left[-\frac{\{z_i - \mu - \rho(z_{i-1}-\mu)\}^2}{2\sigma^2} \right], -\infty < z_i < \infty : \mu \in \mathbb{R}, \sigma > 0 \right\}. \tag{1.19}$$

1.3.6 Cluster Dependence

A useful model which allows for independent clusters in the data with constant positive correlation within clusters can be written as

$$Z_{ij} = \mu + A_i + U_{ij}, \qquad j = 1, \ldots, m, \quad i = 1, \ldots, g,$$

where the A_i are independent with identical Gaussian distributions with mean 0 and variance σ_a^2 and are independent of the U_{ij} which are themselves independent with identical Gaussian distributions with mean 0 and variance σ_u^2. The A_i are called *random effects* and are sometimes of interest in their own right; in other problems, the so-called *variance components* σ_a^2 and σ_u^2 are of interest. This model entails

$$\operatorname{Cov}(Z_{ij}, Z_{kl}) = \begin{cases} \sigma_a^2 + \sigma_u^2 & \text{if } i = k, j = 1 \\ \sigma_a^2 & \text{if } i = k, j \neq 1 \\ 0 & \text{if } i \neq k. \end{cases}$$

Moreover, if we let Σ denote the block diagonal matrix with blocks $\sigma_a^2 J + \sigma_u^2 I$, where J is the $m \times m$ matrix with all elements equal to 1 and I is the $m \times m$ identity matrix, we can write the model formally as

$$\mathscr{F} = \left\{ f(\mathbf{z}; \mu, \sigma_a, \sigma_u) = \frac{1}{(2\pi)^{mg/2}|\Sigma|^{1/2}} \exp\{-(\mathbf{z}-\mu)^{\mathrm{T}}\Sigma^{-1}(\mathbf{z}-\mu)/2\}, \right.$$
$$\left. -\infty < z_{ij} < \infty : \mu \in \mathbb{R}, \sigma_a \geq 0, \sigma_u > 0 \right\}, \tag{1.20}$$

where $\mathbf{z}^T = (z_{11}, z_{12}, \ldots, z_{im}, \ldots, z_{gm})$. Obviously, any multivariate Gaussian distribution with a nondiagonal covariance matrix is a possible model for dependent data. The technique of introducing a random variable in common to a group of random variables is a very useful way of modeling clustering.

1.3.7 Paired Dependence

The Gaussian models (1.8) and (1.9) for the effect of caffeine derived in Section 1.2.3 can be derived from models like (1.20). Suppose that the effect on the ith subject of treatment t is

$$Z_{it} = \alpha + \mu t + A_i + U_{it}, \qquad t = 0, 1, \quad i = 1, \ldots, n, \tag{1.21}$$

where the A_i are independent with identical Gaussian distributions with mean 0 and variance σ_a^2 and are independent of the U_{ij} which are themselves independent with identical Gaussian distributions with mean 0 and variance σ_u^2. In this model, the random variable A_i represents the effect of the ith subject, so σ_a^2 is the between subject variance, and the random variable U_{it} represents the within subject error so σ_u^2 is the within subject variance. Differencing the data leads to

$$Z_i = Z_{i1} - Z_{i0} = \mu + U_{i1} - U_{i0} = \mu + U_i, \qquad i = 1, \ldots, n, \tag{1.22}$$

where U_i are independent Gaussian random variables with mean 0 and variance $\sigma^2 = 2\sigma_u^2$. The model (1.22) is identical to (1.8). In the differencing approach, we see from (1.22) that we eliminate and then ignore A_i so the distribution of A_i can be allowed to be arbitrary. (See also Section 6.1.1.)

The model (1.9) can similarly be derived from

$$Y_{it} = \alpha_0 + \alpha t + \beta Z_{it} + A_i + V_{it}, \qquad t = 0, 1, \quad i = 1, \ldots, n,$$

where the A_i are independent with identical Gaussian distributions with mean 0 and variance σ_a^2 and are independent of the V_{it} which are themselves independent with identical Gaussian distributions with mean 0 and variance σ_v^2.

PROBLEMS

1.3.1. Show that the gamma model (1.5) is in the exponential family.

1.3.2. Establish whether the Poisson model (5d in the Appendix)

$$\mathscr{F} = \left\{ f(\mathbf{y}; \lambda) = \prod_{i=1}^{n} \frac{\lambda^{y_i} \exp(-\lambda)}{y_i!}, y_i = 0, 1, 2, \ldots : \lambda > 0 \right\}$$

and/or the *negative binomial model* (5e in the Appendix)

$$\mathscr{F} = \left\{ f(z; \pi) = \binom{z-1}{r-1} \pi^r (1-\pi)^{z-r}, z = r, r+1, \ldots : 0 \le \pi \le 1 \right\}$$

are in the exponential family. What about the other forms of the negative binomial model given in (5e) in the Appendix?

1.3.3. Determine which of the following density functions give rise to models in the exponential family and/or the location-scale family and which do not.

1. $f(x; \theta) = \exp\{-2\log(\theta) + \log(2x)\} I(0 \leq x \leq \theta); \qquad \theta > 0.$
2. $f(x; \theta) = \dfrac{2(x + \theta)}{1 + 2\theta}, \qquad 0 < x < 1; \quad \theta > 0.$
3. $$f(x; \sigma, \nu) = \frac{\Gamma\left(\frac{\nu + 1}{2}\right)}{(\pi\nu\sigma^2)^{1/2}\Gamma\left(\frac{\nu}{2}\right)} \frac{1}{\left\{1 + \dfrac{x^2}{\sigma^2\nu}\right\}^{(\nu+1)/2}},$$ $$-\infty \leq x \leq \infty; \quad \sigma > 0, \quad \nu > 0.$$
4. $f(x; \alpha, \beta) = \dfrac{\Gamma(\alpha + \beta)}{\Gamma(\alpha)\Gamma(\beta)} x^{\alpha - 1}(1 - x)^{\beta - 1}, \qquad 0 < x < 1; \quad \alpha > 0, \quad \beta > 0.$

1.3.4. Consider the general regression model

$$Y_j = g(x_j) + e_j, \qquad 1 \leq j \leq n,$$

where $\{x_j\}$ is a known sequence of p-vectors and $\{e_j\}$ is a sequence of independent and identically distributed random variables with distribution function F. For each of the following models, specify the dimensions of any unknown parameters and then classify the possible inference problems associated with the parameters as parametric, semiparametric, or nonparametric.

1. $g(x) = x'\beta, \quad F = \mathrm{N}(0, 1)$
2. $g(x) = x'1, \quad F = \mathrm{N}(0, \sigma^2)$
3. $g(x) = x'\beta, \quad F =$ a symmetric distribution
4. $g(x) =$ a smooth function, $\quad F = \mathrm{N}(0, 1)$
5. $g(x) =$ a smooth function, $\quad F = \mathrm{N}(0, \sigma^2)$
6. $g(x) =$ a smooth function, $\quad F =$ a symmetric distribution.

1.3.5. Suppose that for the ith subject in the Diabetic Retinopathy Study receiving argon treatment in one eye (see Section 1.1), we observe (y_i, x_i), where $y_i = 1$ if the treated eye suffers severe visual loss in 2 years and 0 otherwise, and $x_i = 1$ if the untreated eye suffers severe visual loss within 2 years and 0 otherwise. Changing notation from that used in Section 1.2.1 for simplicity, we can regard the data for n subjects as n independent realizations of (Y, X), where Y has a binomial$(1, \theta)$ distribution and

X has a binomial$(1, \eta)$ distribution. One way to take the possible dependence of eyes from the same subject into account is to represent the distribution of (Y, X) as a multinomial distribution with four outcomes $(1, 1)$, $(1, 0)$, $(0, 1)$ and $(0, 0)$ with associated probablities π_{11}, π_{10}, π_{01} and π_{00}. Express θ and η as functions of π_{11}, π_{10}, π_{01} and π_{00}. Show that

$$\mathrm{Cov}\,(Y, X) = \pi_{11} - \eta\theta.$$

and hence that

$$\pi_{11} = \lambda + \theta\eta, \qquad \pi_{10} = \theta(1 - \eta) - \lambda$$

$$\pi_{01} = \eta(1 - \theta) - \lambda, \qquad \pi_{00} = (1 - \eta)(1 - \theta) + \lambda,$$

where $\lambda = \mathrm{Cov}\,(Y, X)$. Write down the joint distribution of (Y_i, X_i), $1 \le i \le n$, in the $(\pi_{11}, \pi_{10}, \pi_{01}, \pi_{00})$ parameterization and then in the (θ, η, λ) parameterization. Interpret the parameters (θ, η, λ).

1.3.6. Consider a two-state Markov process $\{Z_i\}$ which takes on the values 0 or 1 only. Suppose that $Z_0 = 1$ and the probability of transition from state j to k is given by

$$\mathrm{P}\{Z_{i+1} = k \mid Z_i = j\} = \theta_{kj},$$

where $\theta_{0j} + \theta_{1j} = 1$ for $j = 0$ and $j = 1$. Suppose that we observe realizations of $Z_1, Z_2, \ldots, Z_n$. Use the fact that we can always decompose a joint distribution into a product of conditional distributions as

$$f(\mathbf{z}; \theta) = f(z_0; \theta) f(z_1 \mid z_0; \theta) f(z_2 \mid z_1, z_0; \theta) \cdots f(z_n \mid z_{n-1}, \ldots, z_0; \theta)$$

and a Markov process satisfies $f(z_k \mid z_{k-1}, \ldots, z_0; \theta) = f(z_k \mid z_{k-1}; \theta)$ for every $k > 1$, to write down a model for the joint distribution of $\{M_{00}, M_{10}, M_{01}, M_{11}\}$, where M_{kj} is the number of transitions from j to k.

1.4 STATISTICAL INFERENCE

The components of the inference problem are

- a substantive question
- data $\mathbf{z}$ which we interpret as a realization of a random variable $\mathbf{Z}$ with a distribution F_0
- a model $\mathscr{F}$ for F_0.

The objective of inference is to answer the substantive question by reformulating it as a question about the underlying distribution F_0 and then using the data $\mathbf{z}$, the model $\mathcal{F}$ and any other information we have to answer the question about F_0. The kinds of questions we ask about F_0 are typically of one or both of two types:

- Is $F_0 \in \mathcal{F}$? i.e. can the model be viewed as a reasonably close approximation to the data generating process?

or

- Can we determine a set of plausible values for a parameter $\theta(F_0)$ or can we determine whether a particular value of a given parameter $\theta(F_0)$ is plausible?

The answers to these questions are derived from the data through the calculation and interpretation of the realized values $t(\mathbf{z})$ of statistics $t(\mathbf{Z})$, which are functions of the data which do not depend on any unknown parameters.

1.4.1 The Abstract Framework

Most theoretical discussion of inference is concerned with the problem of providing inferences to address questions about the model $\mathcal{F}$ which is assumed to hold exactly so that $F_0 \in \mathcal{F}$. (This assumption is useful both in developing inference procedures and for evaluating their theoretical properties under ideal conditions but, in practice, it is best to think of a model $\mathcal{F}$ as a useful approximation rather than as an exact description of the data generating process.) The point of departure is this abstract framework but without reference to the substantive question or other contextual information. Attention is focused on how we choose and then use statistics to explore the model questions, how we assess the uncertainty in our conclusions, and how we interpret the results. This abstract framework emphasizes the wide applicability of statistical inference and is convenient for an idealized mathematical discussion of the properties of inference procedures but avoids many important issues including how and why models are formulated, how models are interpreted, how model questions are formulated, data collection, and the existence of unquantifiable uncertainties in the final inferences. An advantage of introducing data into theoretical discussion is that it forces us to confront these issues.

1.4.2 The Scope of Inference

A characteristic of statistical inference is the fact that we usually want any conclusions we reach about a substantive question to apply more broadly than to the present data $\mathbf{z}$. For this to be possible, we need to be able to relate the present data $\mathbf{z}$ to at least similar sets of observations, similar experiments etc. We achieve this by treating $\mathbf{z}$ as a realization of a random variable $\mathbf{Z}$ whose underlying distribution F_0 describes the mechanism generating $\mathbf{Z}$ and thereby the scope of the inference.

In assessing the scope of an inference, it can be useful to think of F_0 as describing a population of possible observations (or units on which observations can be made) from which the observations in our data set are selected. In this framework, inferences about F_0 are conclusions about the *underlying population*. We usually distinguish between two types of population of interest:

(1) *Finite existent populations*, such as people meeting the eligibility criteria for the Diabetic Retinopathy Study at a specified time and place, pressure vessels produced at one plant on a given day, or healthy 18–22 year old male students attending a specified college on a specified date.

(2) *Hypothetical populations*, such as past and future people meeting the eligibility criteria for the Diabetic Retinopathy Study, past and future pressure vessels produced by the same process, or healthy 18–22 year old male students.

It is precisely the vagueness and flexibility inherent in the notion of a hypothetical population that makes this interpretation useful.

Deming (1950, 1953) classified studies of underlying populations as either *enumerative* or *analytic* according to whether the objectives of the study can, at least in principle, be achieved without error or not. Studies of hypothetical populations are always analytic but studies of finite existent populations can be either analytic or enumerative. For a finite existent population, we can in principle create a list (called a *frame*) of every unit in the population. If we observe every unit in the frame, we can make precise statements about the population. That is, there is no uncertainty about the conclusions. In this case, the inferences are enumerative. If however, we are interested in reaching conclusions which apply not only to the present finite existent population but also to other similar (possibly hypothetical) populations in a different time or place, then, by definition, no frame can be constructed and we cannot eliminate uncertainty by observing the entire population. In this case, the inferences are analytic.

Even in a enumerative study of a finite existent population, it is generally impractical to make observations on the entire population, so we observe a subset or *sample* of units from the population. Since we require conclusions reached on the basis of the observations in the sample to apply to the entire population, we need to know the relationship of the sample to the population. This may be achieved if the sample is a *random sample* from the population in the sense that every possible sample has a calculable (but not necessarily equal) chance of selection (see Section 6.5.3) and/or the model incorporates this information (see Section 6.5.7). In the absence of an explicit model relating the sample to the population, a nonrandom *purposive* or *convenience sample* is only useful if it is "like" a random sample. This is a difficult nonstatistical judgement to make, and there are notorious examples of such judgements that have subsequently proved to be unfounded (see for example Bryson, 1976) so it is generally worth the effort to obtain random samples.

The examples we considered in Section 1.1, and indeed most scientific problems, are analytic studies because implicitly the population includes observations on presently unavailable and unspecified units. We can specify the hypothetical population by asking the question "Of what population can this sample be regarded as a random sample?" and then deeming this to be the population to which our inferences apply. Fisher (1922, p. 313) justified this process:

> It should be noted that there is no falsehood in interpreting any set of independent measurements as a random sample from an infinite population; for any such set of numbers are a random sample from the totality of numbers produced by the same matrix of causal connections: the hypothetical population which we are studying is an aspect of the totality of the effects of these conditions, of whatever nature they may be. (Reprinted with permission. Copyright © (1964) by the University of Adelaide.)

To use an inference which has been justified in this way, we need to ensure that the data generating process either has not changed or has changed in a known way so that the current sample is relevant to future samples of interest. In other words, a random sample of current units is a convenience sample (in the sense that it is readily available) from the entire hypothetical population of interest and so may be very different (in unknown ways) from a sample drawn at a different time and place. These judgements are often not made explicitly though they are clearly important to any use of the inference.

1.4.3 Predictive Inference

A different formulation of the inference problem has been advocated by Geisser (1993, pp. 1–4) and others who argue that inferences about the unobservable parameters in a model should be replaced by inferences about observable future observations. For example, instead of making inferences about the typical time to failure of a pressure vessel or the typical change in the volume of urine produced by the ingestion of caffeine, we should formulate the problem as one of predicting time to failure of a pressure vessel or the change in the volume of urine produced by the ingestion of caffeine. An inference framework based on parameters does not preclude making predictions when it is natural (for example, in problems in which we can interpret the problem of making inferences about finite population parameters as a prediction problem (Section 6.5), in time series problems, in regression problems such as if we want to predict the weight loss of enamel covered steel plates subjected to 10% hydrochloric acid for a specified time at a specified temperature and so on, but the predictivist approach asserts that we should only make predictions.

The predictivist view is motivated by a desire to accommodate the approximate nature of models and to allow the possibility of validating the inferences. Although in a strict predictivist framework there is no need to formulate models involving unknown parameters, it is usually simpler and more convenient to do so. In this case, inference about parameters can often be viewed as an intermediate step to prediction. In fact, inferences about parameters can also be viewed as a limiting case of predictive inference (because parameters are observable given all possible observations) and can therefore be a legitimate endpoint in predictive inference. This undermines the claims that predictions naturally overcome the approximate nature of models and can be readily validated.

PROBLEMS

1.4.1. Suppose we have a finite population of size $N = 50$ students in a statistics class and we want to estimate the proportion π who are doing science degrees. Note that $N\pi$ is an integer. Suppose we draw a sample of size $n = 10$ without replacement and we observe that z/n of these students are doing science degrees. Then the distribution of Z, the number of students in the sample doing science degrees is the *hypergeometric distribution* for which

$$\mathrm{P}\{Z = z\} = \frac{\binom{N\pi}{z}\binom{N(1-\pi)}{n-z}}{\binom{N}{n}}, \qquad \max(0, n - N(1-\pi)) \leq z \leq \min(n, N\pi),$$

$$0 \leq \pi \leq 1.$$

On the other hand, if we sample independently with replacement, the binomial model is applicable. Find the expectation and variance of the statistic Z/n for the two situations. Use the problem of making inference about π in these two models to illustrate and explain the difference between an enumerative and an analytic study.

1.4.2. Suppose that we are interested in comparing the median income of employed Australians in 1995 and 1990. Is this an analytic or an enumerative study? Justify your answer. Now suppose that we are interested in predicting the median income in 2000. Discuss briefly whether this is an analytic or an enumerative study.

1.5 INFORMAL INFERENCES

It is useful to distinguish between *primary questions* about $\mathscr{F}$ which relate directly to the substantive question of interest and *secondary questions* which relate to the *nuisance aspects* of a model (such as its shape) which support the analysis but are not directly concerned with the question of interest. Secondary questions are of indirect but not necessarily lesser importance because analyses which focus purely on the primary aspects of a model may be invalidated by the failure of secondary assumptions.

Answers to primary questions about $\mathscr{F}$ are usually provided through *formal inferences* which include an attempt to assess the uncertainty in the conclusion. Secondary questions can also be addressed formally but are often addressed through the use of *graphical methods* which provide simpler interpretations of deviations from an assumed model than formal inferences and often suggest changes to the model which result in better fit to the data. The analysis is *informal* in the sense that at best a vague attempt is made to quantify the uncertainty associated with the conclusions. This means that, in interpreting plots, we need to keep in mind that, even if the model is valid, sampling variability can generate apparent departures from the model. The problem is further complicated by the fact that models are approximations and therefore we should expect to see departures from candidate models. We need to judge the adequacy of models by interpreting and assessing the importance (in the context of the substantive problem) of these departures. The effect of informal inference is that any subsequent formal inferences use over optimistic assessments of uncertainty.

The main concern of informal inference is with the shape of the model distribution. For independent and identically distributed random variables $Z_1, \ldots, Z_n$, each Z_i has the same underlying distribution. We can obtain information about the underlying distribution from the distribution of the data and, in particular, evaluate the appropriateness of the shape of the model by comparing the distribution of the data to the model distribution. These comparisons can reasonably be based on density, distribution, or quantile functions. We discuss generic methods based on these functions for a generic sample $z_1, \ldots, z_n$ and illustrate them using the pressure vessel failure data (Section 1.2), before considering the remaining data sets from Section 1.1.

For the pressure vessel failure times we considered the exponential model (1.3). Under this model, the density, distribution and quantile functions are

$$f(x; \lambda) = \lambda \exp(-\lambda x), \qquad x > 0, \tag{1.23}$$

$$F(x; \lambda) = \int_0^x f(t; \lambda)\, dt = 1 - \exp(-\lambda x), \qquad x > 0, \tag{1.24}$$

and

$$Q(u; \lambda) = F^{-1}(u; \lambda) = -\frac{1}{\lambda} \log(1 - u), \qquad 0 \leq u \leq 1, \tag{1.25}$$

respectively. An obvious difficulty in comparing these to other density, distribution, and quantile functions is that we may need to specify the value of λ. It is often sensible to use a range of reasonable values but when we need a specific value, we set λ equal to $\hat{\lambda} = 1/\bar{z}$, where $\bar{z} = n^{-1} \sum_{j=1}^{n} z_i = 575.53$. (The reasons for this choice of λ are discussed in Section 4.1.)

1.5.1 Histograms and Densities

The density of the data is readily obtained by calculating the relative frequencies with which the distinct z_i in the data set are observed. That is, by calculating

$$f_n(z_i) = n^{-1} \sum_{j=1}^{n} I(z_j = z_i), \qquad i = 1, \ldots, n,$$

where $I(\cdot)$ is the indicator function, $I(A) = 1$ if A holds and 0 otherwise. This function is shown for the pressure vessel data in Figure 1.5. It consists of n spikes of equal height so is uninformative about the underlying density.

We can achieve a more meaningful representation of the density by calculating the relative frequencies after grouping observations which are close

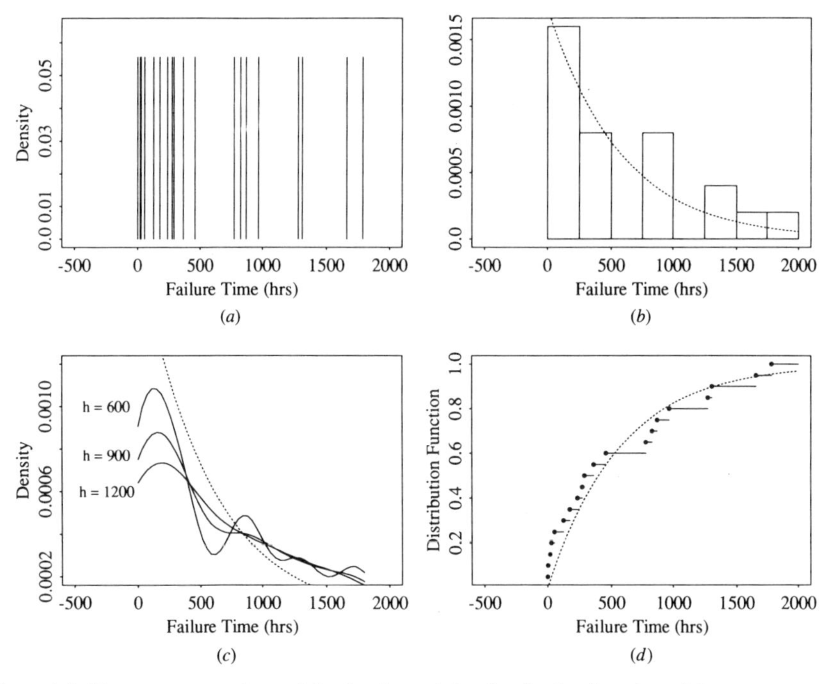

Figure 1.5. Three representations of the density and the distribution function of the pressure vessel data: (*a*) density function; (*b*) density-scale histogram; (*c*) kernel density estimate; (*d*) distribution function. The superimposed curve is an exponential density in (*b*) and (*c*) and an exponential distribution function in (*d*).

to each other. The result is a *density-scale histogram* which, in its simplest form is constructed by choosing a regular grid of points over the range of the data such that $x_k = x_{k-1} + 2h$ for some $h > 0$, constructing a set of bins $B_h(x_k) = [x_k - h, x_k + h]$ which are of width h and centered at x_k, calculating the number of observations in $B_h(x_k)$ and plotting

$$f_{nh}(x_k) = \frac{1}{2nh} \sum_{i=1}^{n} I\left(\left|\frac{z_i - x_k}{h}\right| \leq 1\right),$$

the number of observations in $B_h(x_k)$ divided by $2nh$ against $x \in B_h(x_k)$. The histogram is a piecewise constant function which is often represented as a set of vertical bars as shown for the pressure vessel data in Figure 1.5. Its shape is affected by the choice of both the grid $\{x_k\}$ and the bin width or *window width* h. The exponential density (1.23) with λ set to $\hat{\lambda}$ is superimposed on the histogram. It is not easy to evaluate the approximation because the shape of the histogram depends on the choice of grid and/or window width and the histogram is not smooth like the exponential density.

We can reduce the difficulty of having to choose the grid for a histogram and obtain a smoother representation of the density by letting the bins overlap and plotting the density $f_{nh}(x)$ for each x in the range of the data rather than for $x = x_k$. The resulting curve $f_{nh}(x) = (2nh)^{-1} \sum_{i=1}^{n} I\{|(z_i - x)/h| \leq 1\}$ can be made smoother by replacing the box kernel $I(|u| \leq 1)/2$ by a continuous density function or kernel. For a general function K, we obtain the *kernel density estimate*

$$f_{nh}(x) = \frac{1}{nh} \sum_{i=1}^{n} K\left(\frac{z_i - x}{h}\right), \qquad h > 0.$$

The choice of K in a kernel density estimate is not critical and we can use the standard Gaussian density $\phi(u) = (2\pi)^{-1/2} \exp(-u^2/2)$ for convenience. In contrast, the choice of h is very important. Plots of the kernel density estimate for the pressure vessel data for three choices of h are shown in Figure 1.5. The choice of h reflects a compromise between closeness to the density of the data represented by the relative frequencies $f_n(z_i)$ (corresponding to $h \to 0$) and smoothness represented by K (corresponding to $h \to \infty$). For many purposes, it is adequate to plot the kernel density estimate for different values of h to explore the effect of changing h and, if a single density is required, to choose one of these. A number of methods of choosing h are discussed by Silverman (1986, pp. 43–60).

Both the histogram and the kernel density estimates in Figure 1.5 suggest that the density of the pressure vessel failure data is approximately exponential. However, we need moderately large samples to obtain good estimates of densities and this conclusion depends on the window width. Also, the density scale is not the best scale for comparing distributions visually.

1.5.2 The Empirical Distribution Functions

As an alternative to comparing density functions, we can consider comparing distribution functions. We can sum or integrate any of the three densities presented in Section 1.5.1 to produce a distribution function. The easiest to work with and the most important is the sample or empirical distribution function obtained by summing the density of the data. The *empirical distribution function* is

$$F_n(x) = n^{-1} \sum_{i=1}^{n} I(z_j \leq x), \qquad -\infty < x < \infty,$$

a step function with jumps at the distinct observed z_i which exhibits the cumulative relative frequency of occurrence of each distinct observed z_i.

An important property of the empirical distribution which we will exploit subsequently (and one of the reasons it is so important) is that integrals with respect to the empirical distribution function are averages over the sample. In particular, for any function g we have that

$$\int g(x)\, dF_n(x) = n^{-1} \sum_{i=1}^{n} g(z_i),$$

so if we take $g(x) = x$, we see that the sample mean is the mean of the empirical distribution etc.

The empirical distribution function for the pressure vessel data is shown in Figure 1.5 with the exponential distribution function (1.24) with λ set to $\hat{\lambda}$ superimposed. Working with the empirical distribution function enables us to avoid smoothing but we still need to specify λ and it is still not an ideal scale for visual comparison.

1.5.3 Quantile–Quantile Plots

The *empirical quantile function* is the inverse of the empirical distribution function F_n which is defined to be

$$F_n^{-1}(u) = \inf\{x: F_n(x) \geq u\}, \qquad 0 < u < 1,$$

to allow for the fact that F_n is a step function. It follows that

$$F_n^{-1}(u) = \begin{cases} z_{n,nu} & \text{if } nu \text{ is an integer} \\ z_{n,[nu]+1} & \text{otherwise,} \end{cases}$$

where $z_{n1} \leq z_{n2} \leq \cdots \leq z_{nn}$ are the ordered observations, otherwise called the *order statistics*. The quantile function is therefore equivalent to the order statistics.

For a model in the location–scale family (1.18), the distribution function is of the form $H\{(x - \mu)/\sigma\}$ with H known, and the quantile function is

$$Q(u; \mu, \sigma) = \mu + \sigma H^{-1}(u). \tag{1.26}$$

If we were to plot $Q(u; \mu, \sigma)$ against $Q(u; 0, 1) = H^{-1}(u)$, we would obtain a straight line with intercept μ and slope σ. If we were to plot any other quantile function against $H^{-1}(u)$, we would obtain a nonlinear curve. This is the basis for the *quantile–quantile* or *qq-plot* which is a plot of $F_n^{-1}(u)$ against $H^{-1}(u)$ and which enables us to assess the quality of location–scale family models without having to specify values for μ and σ. Essentially, a linear relationship in the plot shows that the empirical distribution approximately satisfies the relationship (1.26) and deviations from linearity indicate departures from this form. There is some arbitrariness in the choice of the argument u to associate with z_{ni} but the common choices are $(i - 0.5)/n$ or $i/(n + 1)$. We will use the first choice and plot z_{ni} against $Q((i - 0.5)/n, 1)$.

The exponential qq-plot for the pressure vessel data with $H^{-1}(u) = -\log(1 - u)$ from (1.24) is shown in Figure 1.6. The plot is quite linear though the curve flattens out slightly in the top right hand corner. This means that the empirical distribution of the failure times can plausibly be approximated by an exponential distribution or perhaps by a distribution with a slightly shorter tail

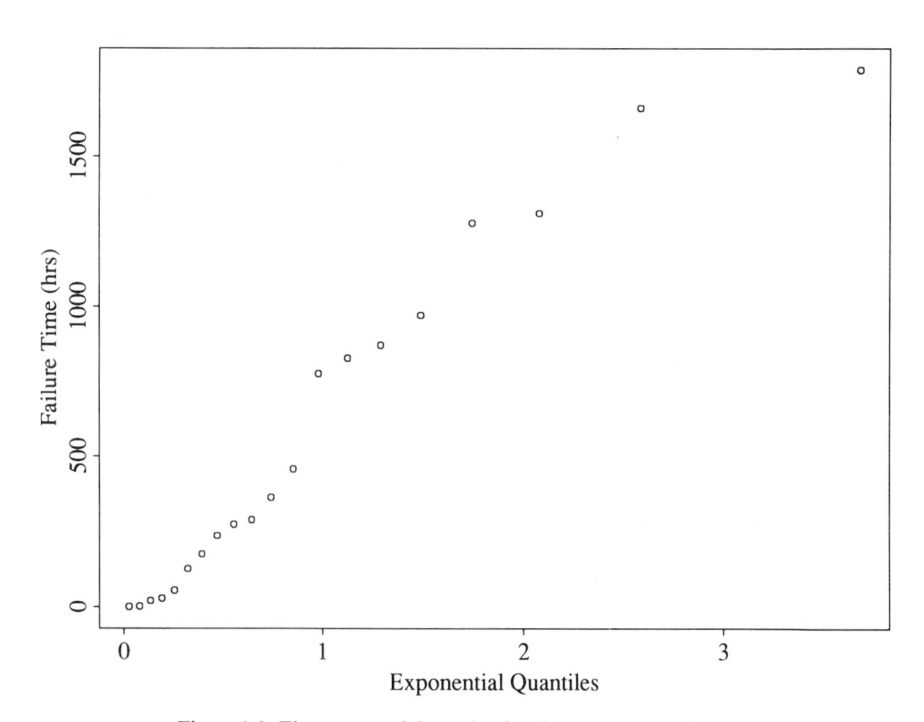

Figure 1.6. The exponential qq-plot for the pressure vessel data.

than the exponential distribution. We will consider modeling the failure times with either the gamma or the Weibull model to explore this issue further in Sections 4.2–4.3. However, models with slightly longer tails than the true distribution often lead to conservative inferences (see Chapter 5) so, for parsimony and simplicity, we may tentatively adopt the exponential model.

We can also explore the appropriateness of the gamma model (1.4) for the pressure vessel failure time data. Under (1.4), the quantile function is nonlinear in the shape parameter κ so we cannot just use arbitrary values for κ. We can choose a value for κ graphically, by adjusting the value of κ until a good fit is obtained or, possibly, by robust estimation (see Chapter 5). In any case, it is sensible to construct qq-plots for a range of values of κ. An interesting feature of the gamma qq-plot is that the asymmetry of the distribution tends to force the lower quantiles to bunch up in the plot. A better plot is obtained by applying the *Wilson–Hilferty symmetrizing transformation* (see Section 4.1.10) to both axes, i.e. plot $(Q^*\{(i-0.5)/n\}^{1/3}, z_{ni}^{1/3})$.

Deviations from linearity in qq-plots indicate differences between the proposed model distribution and the empirical distribution. Selected possibilities are illustrated in Figure 1.7. If a small number of points in the top right and/or

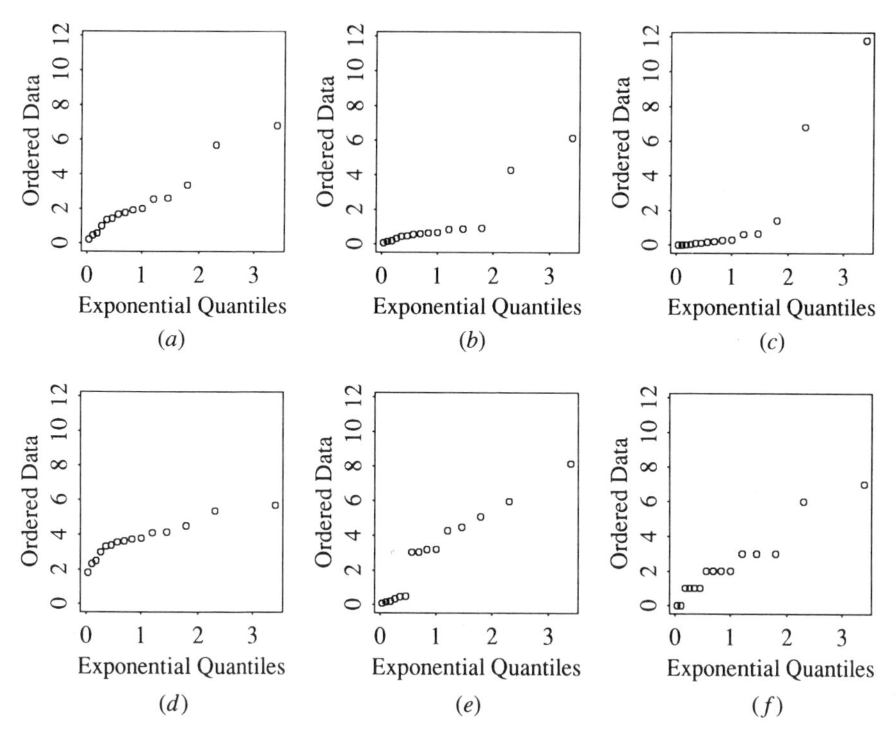

Figure 1.7. Examples of exponential qq-plots showing deviations from exponentiality: (*a*) good fit; (*b*) outliers; (*c*) long upper tail; (*d*) short upper tail; (*e*) clustering; (*f*) rounding.

bottom left corners of the plot depart from a straight line, we may conclude that the approximating distribution approximates the distribution of the bulk of the data but that there are outliers which it does not fit. Actually, qq-plots are least reliable in the tails so we should not be too quick to declare points to be outliers. If there is curvature at the end(s) of the plot, the empirical distribution has shorter or longer tails than the approximating distribution. An S curve indicates *shorter tails* while a chair shape indicates *longer tails.* (Outliers occur in the tails of the distribution, so tend to create a similar affect. The distinction between the two cases is based on the number of points causing the departure and the extent of their separateness from the bulk of the data.) If the approximating distribution is symmetric and one tail of the empirical distribution is rather different from the other, this will be reflected in the curve of the qq-plot. Finally, qq-plots can reveal *rounding* (horizontal steps) and *clusters* (gaps) in the data.

1.5.4 The Effect of Caffeine

The Gaussian model (1.8) for the effect of caffeine on the volume of urine voided assumes that the effect is additive. An informal check of this assumption can be made by examining a scatterplot of the paired differences against either the control or treatment value. An additive effect should result in a random cloud of points with no particular structure and exhibiting no relationship between the variables. For the urine data, Figure 1.8 supports the assumption that the coffee effect is additive in this data set.

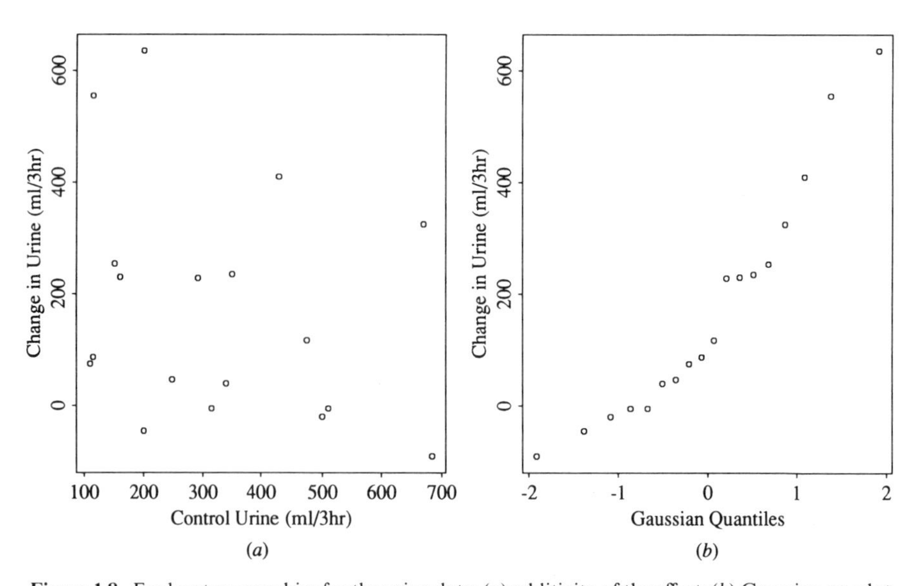

Figure 1.8. Exploratory graphics for the urine data: (*a*) additivity of the effect; (*b*) Gaussian qq-plot.

The model (1.8) also assumes that the pairwise differences are Gaussianly distributed. The qq-plot based on the Gaussian distribution is slightly curved in the left tail but otherwise roughly linear so the differences seem to come from a distribution which has a short left tail but is otherwise approximately Gaussian. If we adopt the conservative strategy of ignoring the left tail, we may proceed under the assumption that the differences can be modelled as independent observations from a Gaussian distribution with unknown mean μ and variance σ^2.

We cannot check the distributional assumptions of the simple regression model (1.9) relating the change in total catecholamine to the change in the volume of urine voided by examining a qq-plot of the response because we would be examining the marginal rather than the conditional distribution of the response. (The distinction between marginal and conditional distributions is not important if the explanatory variables are in fact nonstochastic, but then departures from Gaussianity cannot be separated from the effect of the changing mean structure so it is still not a suitable plot.)

If we write the model (1.9) in the equivalent form

$$y_i = \alpha + \beta z_i + e_i, \qquad 1 \leq i \leq 18,$$

where $\{e_i\}$ are independent and identically distributed Gaussian random variables with mean 0 and unknown variance σ^2, we see that, aside from the linearity assumption, the distributional assumptions are invested in the unobserved errors $\{e_i\}$. Although we do not observe the errors, we can obtain information about them from the residuals obtained by computing estimates $\hat{\alpha}$ and $\hat{\beta}$ of α and β in the model, computing the *fitted values*

$$\hat{y}_i = \hat{\alpha} + \hat{\beta} z_i, \qquad 1 \leq i \leq 18,$$

and then the *residuals*

$$r_i = y_i - \hat{y}_i = y_i - \hat{\alpha} - \hat{\beta} z_i, \qquad 1 \leq i \leq 18.$$

If the model describes the structural aspects of the relationship between the response y_i and the explanatory variable z_i, then the plot of the residuals against the fitted values (or equivalently the explanatory variable) should show no structure in the sense that the points should be randomly scattered within a horizontal band about the x-axis. Deviations from this pattern in the *residual plot* indicate that the relationship is nonlinear, that the variation about the relationship is nonconstant and/or that outliers are present in the data; see for example Weisberg (1985, Chapter 6, especially p. 132). Once we have established that a simple linear relationship with constant variation about the linear relationship provides a satisfactory approximation to the structure of the data, we can use a qq-plot to examine the shape of the error distribution. We can also use the residuals to explore temporal and/or spatial dependence in the errors.

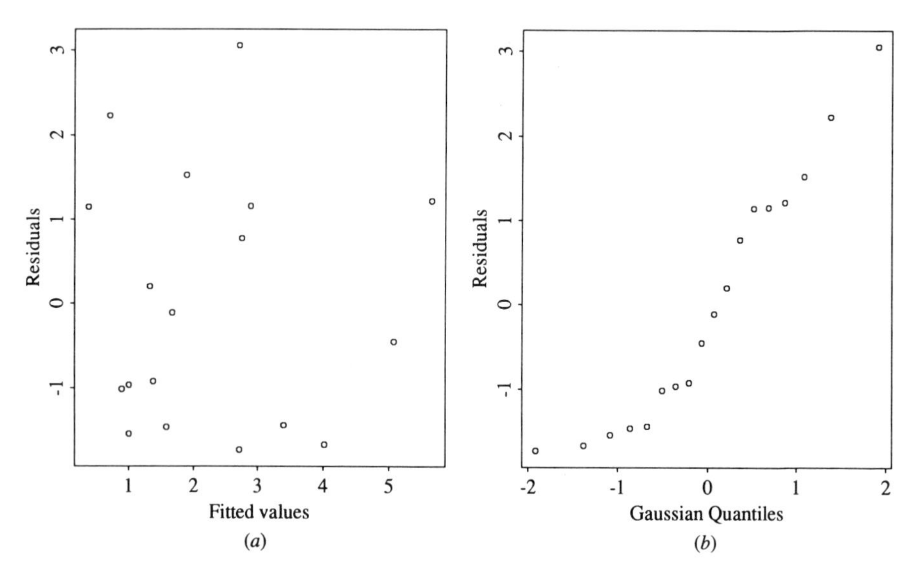

Figure 1.9. Diagnostics from the least squares fit of the simple linear regression model relating the change in total catecholamine to the change in the volume of urine produced: (*a*) residual plot; (*b*) Gaussian qq-plot.

The remaining problem is to construct $\hat{\alpha}$ and $\hat{\beta}$. We will deal with this question in more detail later. Nonetheless, the *least squares estimates* which are obtained by minimizing the sum of the squared residuals

$$\sum_{i=1}^{n} (Y_i - \alpha - \beta z_i)^2$$

often provide reasonable estimates of α and β. The residual plot and the qq-plot of the least squares residuals in Figure 1.9 show that the simple linear relationship with constant variability about the relationship is a reasonable approximation for these data and that the residual distribution is asymmetric with a shorter lower tail than the Gaussian distribution.

The residual plot and the qq-plot of the residuals are often referred to as *diagnostic plots*. A number of other diagnostic plots have been proposed (see for example Atkinson, 1985, or Cook and Weisberg, 1994) but the residual and qq-plots are adequate for our present purposes.

1.5.5 The Model for the Stellar Velocity Data

We can examine the validity of the Gaussian model (1.10) for the stellar velocity data by constructing qq-plots for the three data sets as shown in Figure 1.10. We see that none of the distributions is exactly Gaussian. The disk and

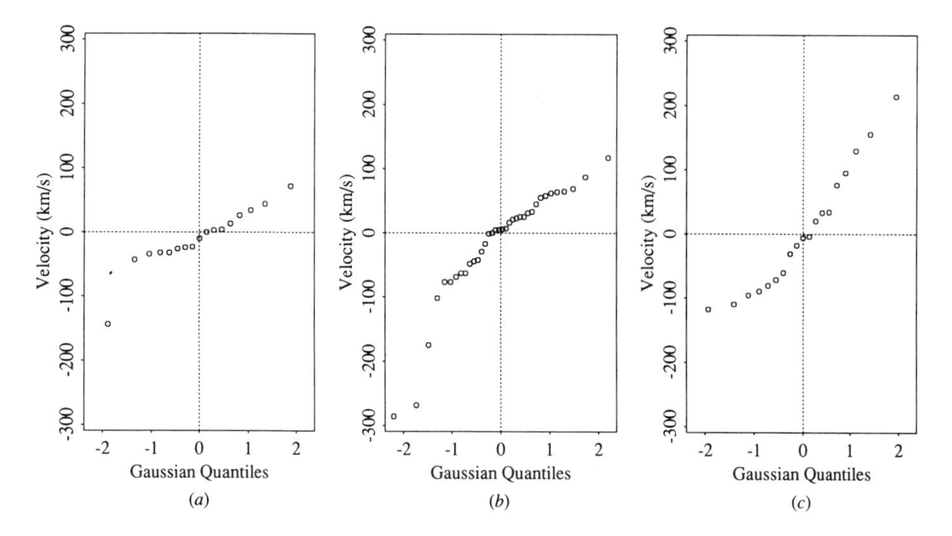

Figure 1.10. Gaussian qq-plots for the stellar velocity data: (*a*) disk stars; (*b*) intermediate stars; (*c*) halo stars.

intermediate stars have some extreme stars in the left tail but are otherwise Gaussian, whereas the halo stars have a slightly short left tail. The extreme stars mentioned in Section 1.2.4 are clearly different (by virtue of their extremeness) from the rest of the stars. With the exception of the outliers, the qq-plots are roughly linear. The lines pass through or close to the origin and have increasing slopes as we move from the disk through the intermediate to the halo stars. This confirms the assumption that the velocity distributions have mean zero and that the spreads increase across the groups of stars.

1.5.6 The Model for Corrosion Resistance

The linear model (1.16) proposed for the steel plate data relates the corrosion resistance (on the log scale) to exposure time and temperature. We need to make sure that the response (for the bulk of the data) has constant variance, or consider either alternative transformations or a different model with non-constant variance. We need to consider the possibility that the conditional mean structure assumed in the model for the bulk of the data may be too simple (i.e. we may need to include interaction terms $x_{1i}x_{2i}$, quadratic terms x_{1i}^2 and/or x_{2i}^2, or further nonlinear terms) or the model may be more complicated than we require (i.e., one of the explanatory variables does not have an effect of the response).

The presence of outliers in the data can complicate the use of diagnostic methods based on the fitted values and residuals because the outliers can affect the parameter estimates (and hence the fitted values and residuals) in such a way that they mask themselves. We therefore need to fit the model using methods which deal sensibly with outliers before we examine the diagnostics. This is the general topic of Chapter 5 and we will defer further discussion of (1.16) until Section 5.5.

PROBLEMS

1.5.1. Rutherford and Geiger (1910) observed the number of scintillations due to alpha particles from the radioactive decay of 2608 polonium samples in a fixed time interval of 7.5 seconds. Since the data is discrete, it is convenient to report the number of observed time intervals containing each number of particles. The observed data are given in Table 1.9.

Physicists argue that the conditions for a Poisson model are met by radioactive decay so Z_i, the number of scintillations from the ith sample, is a random variable with a Poisson distribution. That is, the data can be modeled by

$$\mathscr{F} = \left\{ f(\mathbf{y}; \lambda) = \prod_{i=1}^{n} \frac{\lambda^{y_i} \exp(-\lambda)}{y_i!}, y_i = 0, 1, 2, \ldots : \lambda > 0 \right\},$$

where λ is the mean number of particles emitted in 7.5 seconds. A natural substantive question to ask is whether the data support the Poisson model. Hoaglin (1980) suggested plotting $\log f_k + \log(k!)$ against k for all k such that $f_k > 0$ to obtain a plot in which linearity corresponds to Poissonness. Explore the validity of the Poisson model.

1.5.2. Feinstein et al. (1989) reported data on counts of the number of micro-bubbles of various diameters created by ultrasonic sonication of physiologic solutions. The diameters (in microns) are discretized to integer values. The counts were obtained by a scanning laser particle counter which produced rounded percentages of bubbles in each size class and the total number of particles/ml^3. In one trial, 1792 particles/ml^3 were observed with the distribution shown in Table 1.10.

Table 1.9. Number of Scintillations in 7.5 Seconds and the Frequency of Occurrence

k	0	1	2	3	4	5	6	7	8	9	10	11	12	13	14
f_k	57	203	383	525	532	408	273	139	45	27	10	4	0	1	1

Table 1.10. The Distribution of Microbubble Size

Size (μm)	1	2	3	4	5	6	7	8	9	10
f_k (rounded %)	13	16	20	18	14	11	5	1	0	1

It may be helpful to add 0.1% to each frequency to ensure that the frequencies sum to 100%. Note that no bubbles of diameter zero can be observed. How would you modify a model which includes the zero class to obtain one which does not? Does a truncated Poisson model seem a priori reasonable for this kind of data? Use Hoaglin's plot (Problem 1.5.1) to explore its validity.

1.5.3. Suppose that the 20 pressure vessel measurements reported in Section 1.1.2 were obtained in the order shown in Table 1.2. Plot the failure times against the order in which the experiments were carried out. What patterns in the plot would indicate dependence in the data? Do the data provide any evidence against the assumption of independence?

1.5.4. Schmoyer (1991) gave the bursting times in hours of Kevlar 49/epoxy vessels subjected to a stress of 80% of an original estimated bursting strength. The data which were originally analysed by Barlow et al. (1984) are given in Table 1.11. Propose and explore the validity of simple models for these data.

1.5.5. Use the data presented in Table 1.3 to develop separate models for the change in epinephrine and the change in norepinephrine due to the ingestion of caffeine. Interpret the parameters in your models.

1.5.6. Use the data in Table 1.3 to explore the relationship between epinephrine excretion and the volume of urine produced. What do you conclude? Suppose there is a relationship between these two variables. Explain the difference between carrying out an analysis on the effect of caffeine on epinephrine excretion (i) ignoring the effect of volume changes and

Table 1.11. Bursting Times of Kevlar 49/Epoxy Vessels

19.1	24.3	69.8	71.2	136.0	199.1	403.7	432.2
453.4	514.1	514.2	541.6	544.9	554.2	664.5	694.1
876.7	930.4	1254.9	1275.6	1536.8	1755.5	2046.2	6177.5

(ii) adjusting for the effect of volume changes. Formulate a model for the relationship which would enable you to adjust for the effect of volume changes in the analysis and interpret the parameters in your model.

1.5.7. Use the data in Table 1.3 to explore the relationship between norepinephrine excretion and the volume of urine produced. Formulate a model for the relationship which would enable you to adjust for the effect of volume changes in the analysis and interpret the parameters in your model.

1.5.8. For the enamel covered steel plate data presented in Table 1.5, calculate and plot the treatment means and medians against time using different symbols to represent temperature. Is there evidence of an interaction between the effects of time and temperature in the plots? Explore the distribution of the residuals from the means and the medians.

1.5.9. Data from a second experiment of the type described in Section 1.1.5 but on a different type of enamel covered steel plate are given in Table 1.12. Repeat Problem 1.5.8 for these data.

Table 1.12. Weight Loss in Enamel Covered Steel Plates

Weight Loss (10^{-4} g)	Time (hrs)	Temp. (°F)	Weight Loss (10^{-4} g)	Time (hrs)	Temp. (°F)
143	4	160	147	8	160
184	4	160	207	8	160
203	4	160	171	8	160
191	4	180	253	8	180
216	4	180	250	8	180
169	4	180	217	8	180
212	4	200	272	8	200
228	4	200	265	8	200
245	4	200	290	8	200
122	6	140	194	10	160
153	6	140	199	10	160
148	6	140	181	10	160
146	6	160	240	10	180
140	6	160	253	10	180
197	6	160	237	10	180
229	6	180	217	10	200
209	6	180	197	10	200
219	6	180	196	10	200
243	6	200	292	12	200
230	6	200	197	12	200
219	6	200	311	12	200

Reprinted from *Statistics and Experimental Design*: In *Engineering and the Physical Sciences*, Volume 1 (1964), p. 440. Wiley, New York. With the permission of the authors.

1.5.10. For the data of Problem 1.5.9, explore the relationship between weight loss, time and temperature, and formulate a model for the relationship.

1.5.11. Proschan (1963) gave data on the times (in operating hours) between successive failures of air conditioning equipment in Boeing 720 aircraft. For one particular aircraft, the data were (to be read across):

50	44	102	72	22	39	3	15	197	188	79	88
46	5	5	36	22	139	210	97	30	23	13	14

Noting that the data is essentially a *time series*, what plots would you use to examine the possibilities of autocorrelation and trends? Formulate and check the appropriateness of a model based on the exponential distribution.

1.5.12. Computers generate random numbers which purport to be uniformly distributed on (0, 1) (see 6f in the Appendix). For each initial number (called a *seed*), computer generated random numbers actually follow a deterministic sequence of finite length. In principle, for any seed, we could run the generator and determine the set of numbers exactly. Nonetheless, the sequence appears remarkably stochastic. (This is what is meant by chaotic behaviour.) Set up a stochastic model for computer generated random numbers and explore its validity using generated samples of size 10 and 100. What do you conclude?

1.6 INDUCTIVE ARGUMENTS

The inherent difficulties in developing appropriate methods for inductive arguments can be daunting. *Induction*, by its very nature, is uncertain. Statistical inference uses probability (expressed in and derived from the statistical model $\mathscr{F}$) in an attempt to quantify this uncertainty but, as we have seen, there are also unquantifiable uncertainties which cannot be ignored. In a sense, statistical inference shows the limits of what can be quantified about uncertainty and thereby emphasizes the need to assess nonstatistical features such as the reasonableness of a theory or the quality of the experimental technique, as well as the unquantifiable uncertainties which arise in the data collection process and the statistical analysis. No single inference should be viewed as definitive, but a sequence of results based on genuinely independent replication (ideally involving investigating the phenomenon by different means, from different viewpoints) can provide confirmatory evidence, the weight of which may ultimately be convincing.

FURTHER READING

Cox and Snell (1981, pp. 33–50) is an excellent practically oriented introduction to statistics, which is well worth reading. More discussion of the role and interpretation of statistical models is provided by Cox (1990), Lehmann (1990) and Bernado and Smith (1994, pp. 237–238). The predictivist viewpoint is presented and developed by Geisser (1993). Hahn and Meeker (1993) discuss the practical requirements for statistical inference for both enumerative and analytic studies in some detail. The recent books by Cleveland (1993) and Cook and Weisberg (1994) discuss and illustrate the use of graphical methods in general and, in particular, in the context of exploring the relationship between a response and explanatory variables.

CHAPTER 2

Bayesian, Fiducial, and Likelihood Inference

Consider the problem of making inferences about the probability of severe visual loss within 2 years in untreated eyes in subjects meeting the eligibility criteria of the Diabetic Retinopathy Study and receiving argon treatment in the other eye. In the notation of Section 1.2.1, the data consist of the number of untreated eyes $z_{11} = 26$ in a sample of size $r_1 = 175$ which experience severe visual loss within 2 years. We decided to model the data with the binomial model (1.1), which we restate for convenience as

$$\mathscr{F} = \left\{ f(z_{11}; \pi_{11}) = \binom{r_1}{z_{11}} \pi_{11}^{z_{11}} (1 - \pi_{11})^{r_1 - z_{11}}, z_{11} = 0, 1, \ldots, r_1 \colon 0 \leq \pi_{11} \leq 1 \right\}. \tag{2.1}$$

In the context of (2.1), the problem is to make inferences about π_{11}. We begin by considering the approach to statistical inference presented by Bayes (1763) and developed further by Laplace (1774; 1812).

2.1 THE BAYESIAN PARADIGM

The Bayesian approach is based on the idea that we should treat π_{11} as a realization of an unobserved random variable which has a distribution with density g_0. This stochastic interpretation of the parameter necessitates a reinterpretation of the model (2.1) as the conditional distribution of Z_{11} given the parameter value π_{11}. That is, we regard the data z_{11} as a realization of the unobserved random variable Z_{11} given π_{11} and $f(\cdot; \pi_{11})$ as the conditional density of Z_{11} given π_{11}. As is natural in this framework, inference is based on the conditional distribution of π_{11} given $Z_{11} = z_{11}$ which is found from Bayes' theorem to have density function

$$g(\pi_{11} \mid Z_{11} = z_{11}) \propto f(z_{11}; \pi_{11}) g_0(\pi_{11})$$
$$\propto \pi_{11}^{z_{11}} (1 - \pi_{11})^{r_1 - z_{11}} g_0(\pi_{11}), \qquad (2.2)$$

where the proportionality constant is chosen so the density integrates (or sums) to 1. We refer to the conditional distribution of π_{11} given z_{11} as the *posterior distribution* of π_{11}. The posterior distribution can be used to derive specific probability statements about π_{11} or about functions of π_{11}. In fact, in the Bayesian paradigm, all inferential statements about π_{11} are derived from the posterior distribution so it represents a complete solution to the inference problem.

2.1.1 The Likelihood Function

Notice that the data z_{11} appears in the posterior density only through the function

$$f(z_{11}; \pi_{11}) \propto \pi_{11}^{z_{11}} (1 - \pi_{11})^{r_1 - z_{11}} = \pi_{11}^{26} (1 - \pi_{11})^{149}, \qquad 0 \leq \pi_{11} \leq 1, \quad (2.3)$$

which is the model density evaluated at the observed data $z_{11} = 26$ and viewed as a function of the parameters, in this case π_{11}. This function is called the *likelihood* (Fisher, 1922) and plays a central role in statistical inference. The density of the posterior distribution is proportional to the product of the likelihood function and the density g_0.

2.1.2 The Prior Density

The calculations to obtain the posterior distribution require knowledge of g_0 which is usually unknown. Thus in practice, it is necessary to assign a density g to be used in place of g_0 in these calculations. Since this distribution has to be specified before we can proceed and is usually specified even before we collect the data, it is called a *prior distribution* for π. The interpretation placed on the resulting posterior distribution (and hence on the inferential statements derived from it) depends on the way we arrive at and the interpretation we place on the prior distribution g. (This issue is discussed in detail in Sections 2.2 and 2.4.)

For the analysis of severe visual loss within 2 years for the control eye of argon-treated subjects, lack of knowledge about the rate of severe visual loss makes it difficult to specify a prior. One possibility favored by Laplace for the binomial model in precisely this kind of situation (see Section 2.2.10) is to take the *uniform prior* $g(\pi) = I(0 \leq \pi \leq 1)$ to reflect our prior beliefs.

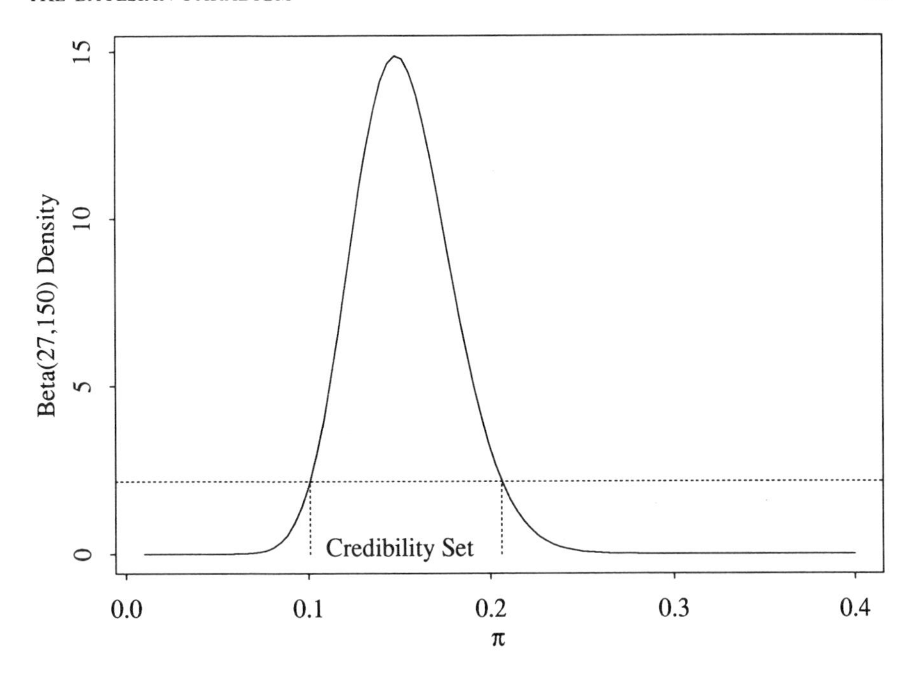

Figure 2.1. The posterior density of π from a uniform prior. The 95% Bayesian credibility set for π lies between the dotted vertical lines.

2.1.3 Analysis for the Control Eye of Argon-Treated Subjects

Combining the likelihood (2.3) with the uniform prior $g(\pi) = I(0 \leq \pi \leq 1)$ as per (2.2), we obtain the posterior density

$$g(\pi_{11} \mid Z_{11} = 26) \propto \pi_{11}^{26}(1 - \pi_{11})^{149}, \qquad 0 \leq \pi_{11} \leq 1. \tag{2.4}$$

The posterior density (2.4) is of the same form as the likelihood (2.3) and is recognisably the density of the *beta(27, 150) distribution* (6e in the Appendix). The posterior density (2.4) is plotted in Figure 2.1. Given the data, our beliefs about π_{11} are represented by a beta(27, 150) distribution and any specific statement we wish to make about π_{11} is made by referring to this distribution.

2.1.4 Parameter Estimation

If we need to estimate a single plausible value for π_{11}, we can calculate the posterior mode of (2.4) which is $26/175 = 0.149$ or the posterior mean of (2.4) which is $27/177 = 0.153$. The *posterior mode* is the most likely value of π_{11} given the data, whereas the *posterior mean* (also called the *Bayes estimate*) is the estimate $\hat{\pi}_{11}$ which minimizes the posterior mean squared error $\mathrm{E}\{(\hat{\pi}_{11} - \pi_{11})^2 \mid z_{11}\}$ when π_{11} has the prior distribution g.

2.1.5 Credibility Sets

The contours of the posterior density (2.4) define nested subsets of the parameter space called *credibility sets* which can be used to summarize the posterior distribution. The integral of the posterior density over any set in the parameter space gives the posterior probability that π_{11} is in the set so we can choose the sets to achieve desired posterior probabilities. If the endpoints of the set correspond to contours of the posterior density, the points in the set will have higher posterior density than any point outside the set and the set can be called a *highest posterior density* credibility set.

The beta(27, 150) distribution is unimodal so the $100(1-\alpha)\%$ credibility set is the interval $[a, b]$, where

$$a^{26}(1-a)^{149} = b^{26}(1-b)^{149} \tag{2.5}$$

and

$$1-\alpha = \mathrm{P}\{a \le \pi_{11} \le b \mid Z_{11} = 26\} = \mathrm{B}(b; 27, 150) - B(a; 27, 150), \tag{2.6}$$

where $\mathrm{B}(x; r, s)$ is the distribution function of the beta(r, s) distribution. (Clearly a and b are functions of both $z_{11} = 26$ and α but we suppress this dependence in our notation.) Equation (2.5) ensures that a and b correspond to contours of the posterior density (i.e., the density at a equals the density at b) and (2.6) ensures that the credibility set has the specified posterior probability content. Equations (2.5)–(2.6) do not have an explicit solution but we can easily find a numerical solution. We find that the 95% credibility set for π_{11} is

$$[0.10, 0.21].$$

Given the data, we believe with posterior probability 0.95 that π_{11} is in the set $[0.10, 0.21]$.

Single sets are generally of less interest than the family of sets indexed by α because the set of contours describes the surface. This relationship to the surface defined by the posterior density means that if we require a credibility set for a function of π_{11}, we need to transform the posterior and calculate a new credibility set from the new posterior distribution; we cannot just transform the set.

2.1.6 Comparing Independent Groups of Eyes

Now suppose that we want to compare the two groups of control eyes. We noted in Section 1.2.1 that we can model the data for the two groups using separate binomial models which can be combined as in (1.2) to provide a single model for both groups. If the parameters π_{11} and π_{21} for the two groups are *a priori* independent, they are also *a posteriori* independent and the joint posterior distribution of (π_{11}, π_{21}) is the product of the two posterior distributions. Thus if we use uniform priors for both π_{11} and π_{21}, we can apply the result of Section 2.1.3 and obtain the *joint posterior density* of (π_{11}, π_{21}) as

$$g(\pi_{11}, \pi_{21} \mid z_{11}, z_{21}) \propto \pi_{11}^{z_{11}}(1 - \pi_{11})^{r_1 - z_{11}}\pi_{21}^{z_{21}}(1 - \pi_{21})^{r_2 - z_{21}}, \quad 0 \leq \pi_{11}, \pi_{21} \leq 1. \tag{2.7}$$

The *marginal posterior density* of the parameter of interest $\eta = \pi_{11} - \pi_{21}$ is obtained by making the transformation $\pi_{11} = \eta + \theta, \pi_{21} = \theta$ which has Jacobian 1 in (2.7) (see 8 in the Appendix), integrating over θ to produce

$$h(\eta \mid z_{11}, z_{21}) = \begin{cases} \displaystyle\int_{-\eta}^{1} (\eta + \theta)^{z_{11}}(1 - \eta - \theta)^{r_1 - z_{11}}\theta^{z_{21}}(1 - \theta)^{r_2 - z_{21}}\, d\theta & \text{if } -1 \leq \eta \leq 0 \\ \displaystyle\int_{0}^{1-\eta} (\eta + \theta)^{z_{11}}(1 - \eta - \theta)^{r_1 - z_{11}}\theta^{z_{21}}(1 - \theta)^{r_2 - z_{21}}\, d\theta & \text{if } 0 \leq \eta \leq 1 \end{cases}$$

and then normalizing so that the integral of $h(\eta \mid z_{11}, z_{21})$ is 1. The marginal posterior density of η is then

$$\frac{h(\eta \mid z_{11}, z_{21})}{\int_{-1}^{1} h(\eta \mid z_{11}, z_{21})\, d\eta}. \tag{2.8}$$

We can actually find the denominator in (2.8) explicitly because it follows from (2.7) that

$$\int_{-1}^{1} h(\eta \mid z_{11}, z_{21})\, d\eta = \frac{\Gamma(z_{11} + 1)\Gamma(r_1 - z_{11} + 1)\Gamma(z_{21} + 1)\Gamma(r_2 - z_{21} + 1)}{\Gamma(r_1 + 2)\Gamma(r_2 + 2)},$$

where Γ denotes the gamma function.

Note that we obtain the same marginal posterior density for η if we first reparameterize (1.2) in terms of η and θ so the likelihood is

$$f(z_{11}, z_{21}; \eta, \theta) \propto (\eta + \theta)^{z_{11}}(1 - \eta - \theta)^{r_1 - z_{11}}\theta^{z_{21}}(1 - \theta)^{r_2 - z_{21}}, \tag{2.9}$$

transform the prior distribution to

$$g(\eta, \theta) = I(0 \leq \eta + \theta \leq 1)I(0 \leq \theta \leq 1), \tag{2.10}$$

integrate the product of (2.9) and (2.10) over θ, and then normalize as in (2.8).

We can use the marginal posterior density (2.8) to construct estimates of η or credibility sets for η once we have evaluated the integrals required for (2.8). In fact, the operation of integration is fundamental to Bayesian inference: We integrate to evaluate the normalizing constant in the posterior density, the Bayes estimate and other posterior moments, the probability content of credibility sets, and marginal posterior distributions. Explicit integration is possible only in simple problems so in general we need to resort to numerical methods. We can consider using analytic approximations (which are discussed in Section 4.6), Monte Carlo methods (which are discussed in Section 6.8.5) or numerical integration.

2.1.7 Numerical Integration

Numerical integration methods for univariate integrals are based on approximating an integral by a weighted sum of the form

$$\int_c^d f(x)\,dx \approx \sum_{i=0}^{k} w_i f(x_i).$$

We can either subdivide $[c, d]$ into k equal subintervals of length Δx with endpoints $x_0, x_1, \ldots, x_k$ and then choose the weights $w_1, \ldots, w_k$ or choose both the endpoints $x_0, x_1, \ldots, x_k$ and the weights. In the first case, we obtain methods such as *Simpson's rule*:

$$\int_c^d f(x)\,dx \approx \tfrac{1}{3}(f(x_0) + 4f(x_1) + 2f(x_2) + 4f(x_3) + 2f(x_4) + \cdots + 4f(x_{k-1}) + f(x_k))\Delta x, \tag{2.11}$$

(for k even) which is exact when f is a quadratic or cubic polynomial, while in the second, we obtain methods such as the *Gauss–Legendre formula*

$$\int_c^d f(x)\,dx \approx \tfrac{1}{2}(d - c) \sum_{i=0}^{k} w_i f\{\tfrac{1}{2}(c + d) + \tfrac{1}{2}(d - c)x_i\}, \tag{2.12}$$

where x_i is the ith zero of the *Legendre polynomial* $P_k(x)$ and $w_i = 2/(1 - x_i^2) \times [P_k'(x_i)]^2$, which is exact for polynomials of degree $2k + 1$ and less. The zeros and weights are tabulated in Table 25.4 of Abramowitz and Stegun (1970). Other choices of orthogonal polynomials lead to other approximations; see Abramowitz and Stegun (1970, Chapter 25).

The simplest though not necessarily the most economical way to extend these methods to higher dimensional integrals is as

$$\int_{c_2}^{d_2} \int_{c_1}^{d_1} f(x_1, x_2)\,dx_1\,dx_2 \approx \sum_{j=0}^{k_1} \sum_{i=0}^{k_1} w_i^{(1)} w_j^{(2)} f(x_i^{(1)}, x_j^{(2)}), \tag{2.13}$$

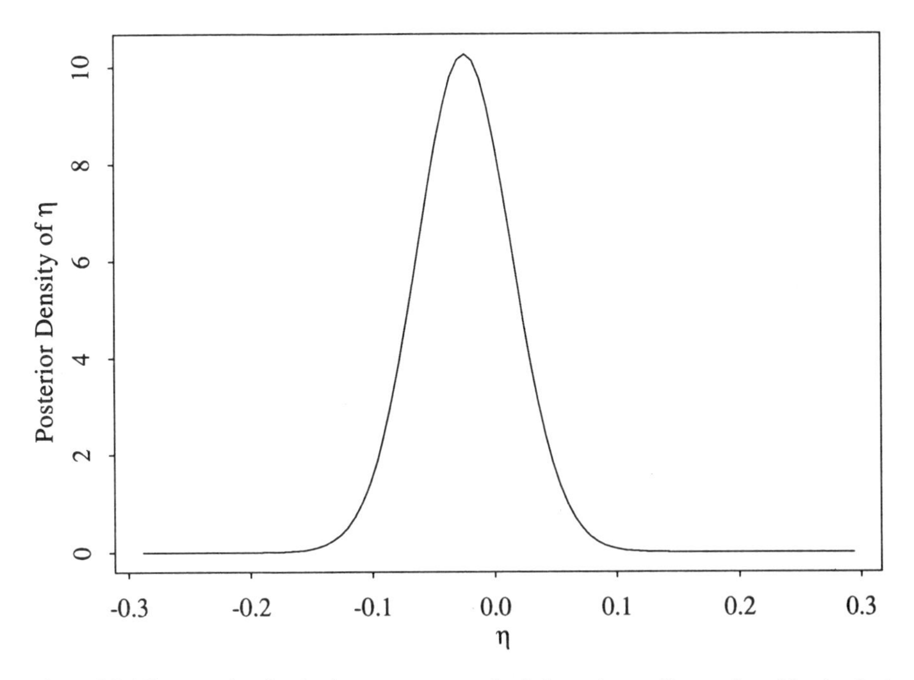

Figure 2.2. The posterior density for $\eta = \pi_{11} - \pi_{21}$ for independent uniform priors. The density is obtained by numerical integration using Simpson's rule.

A detailed discussion of the use of numerical integration methods would take us far beyond our present scope but further discussion on the use of numerical integration (using the *Gauss–Hermite approximation*) in multiparameter Bayesian analysis can be found in Smith et al. (1985, 1987).

In the problem of comparing the two groups of control eyes, we approximated the joint density of η given in (2.8) by applying Simpson's rule (2.11) with $k = 96$ over a fine grid of η values. The marginal posterior density of η is plotted in Figure 2.2.

PROBLEMS

2.1.1. For the data from the Diabetic Retinopathy Study in Table 1.1, use uniform prior distributions to construct posterior distributions for the probability of severe visual loss within 2 years for the argon- and xenon-treated eyes separately. Plot the posterior distributions and construct separate 95% credibility intervals for the probability of severe visual loss within 2 years.

2.1.2. Use the results from Problem 2.1.1 to compare the probability of severe visual loss within 2 years for xenon- and argon-treated eyes.

2.1.3. The Poisson model

$$\mathscr{F} = \left\{ f(\mathbf{y}; \lambda) = \prod_{i=1}^{n} \frac{\lambda^{y_i} \exp(-\lambda)}{y_i!}, \qquad y_i = 0, 1, 2, \ldots : \lambda > 0 \right\}$$

provides a reasonable model for Rutherford and Geiger's (1910) alpha particle data presented in Problem 1.5.1. Using a $\Gamma(\theta, \kappa)$ prior, obtain the posterior distribution and explore the form of the Bayes estimate and the posterior variance. Suppose from previous experiments, we known that λ has a distribution with mean approximately equal to 3.5 and variance 0.2. Find the gamma prior with these moments. Plot your prior and the resulting posterior distribution for λ. What is the effect of the data on the prior information?

2.1.4. Consider the binomial model

$$\mathscr{F} = \left\{ f(z; \pi) = \binom{r}{z} \pi^z (1 - \pi)^{r-z} z = 0, 1, \ldots, r : 0 \leq \pi \leq 1 \right\}.$$

Suppose we adopt the beta(a, b) prior for π and then observe z 1s in r trials. Let Y denote the outcome of one additional trial, independent of the trials we have already observed. Find $\mathrm{P}\{Y = 1 \mid z\}$. For the case $a = b = 1$, $\mathrm{P}\{Y = 1 \mid z\}$ is called *Laplace's rule of succession.* Interpret the rule. What does the rule say about the untreated eye of a patient meeting the criteria for inclusion in the Diabetic Retinopathy study and receiving argon treatment in the other eye? Is Laplace's rule of succession sensible when $r = z = 1$?

2.1.5. In the context of the binomial model of Problem 2.1.4, suppose that after observing z out of r 1s, we make m further independent trials. What is the probability that we observe y out of m 1s?

2.2 PRIOR DISTRIBUTIONS

The fundamental idea behind the Bayesian paradigm is that the unknown parameters should be treated as random variables. One argument for this idea is that all uncertainties (including those concerning the parameters) should be represented by probability distributions and this requires the parameters to be treated as random variables. A deeper argument given by de Finetti (1937) shows that stochastic parameters are in a sense intrinsic to model formulation. The argument is based on treating the data as a realization of exchangeable rather than independent random variables.

2.2.1 Exchangeable Random Variables

One way to obtain the binomial model (2.1) is to consider an infinite sequence of binary random variables $\{X_i\}$ taking only the values 0 and 1. If these random variables are independent with $\mathrm{P}(X_i = 1) = \pi$, then the probability that in any finite subsequence $\{X_1, \ldots, X_r\}$ of length r there are exactly z 1s is

$$\mathrm{P}\left\{\sum_{i=1}^{r} X_i = z\right\} = \binom{r}{z}\pi^z(1-\pi)^{r-z}, z = 0, 1, \ldots, r, \qquad 0 \leq \pi \leq 1.$$

Aside from the reference to an infinite sequence of random variables, this is the standard derivation of the binomial model.

Now suppose that we assume that the random variables $\{X_i\}$ are exchangeable rather than independent. This means that the random variables are identically distributed but dependent. Formally, an infinite sequence of random variables $\{X_i\}$ is *exchangeable* if and only if the joint distribution of any r variables in the sequence is invariant to permutation of the indices i. Sequences of independent and identically distributed random variables are trivially exchangeable but so are sequences like $\{X_i = Y_i B\}$, where $\{Y_i\}$ is a sequence of independent and identically distributed binary random variables and B is an independent binary random variable. To see that $\{X_i\}$ is exchangeable, notice that the X_i are dependent but are *conditionally independent* and identically distributed given B. It follows that the conditional distribution of any set of r of the X_i given B is invariant to permutation and hence that the unconditional distribution of any set of r of the X_i is invariant to permutation.

The conditioning argument used above to establish exchangeability is quite general because conditionally independent and identically distributed sequences are equivalent to exchangeable sequences (see for example Taylor et al., 1985, Chapter 1).

2.2.2 de Finetti's Theorem

A consequence of the equivalence of exchangeability and conditional independence is that the probability of observing a specified sequence of exchangeable random variables can always be written as a mixture of the probabilities from the independent case. Specifically, de Finetti (1937) showed that, for any infinite exchangeable sequence of the binary random variables there exists a probability measure G_0 on the interval [0, 1] such that

$$\mathrm{P}\left\{\sum_{i=1}^{r} X_i = z\right\} = \int_0^1 \binom{r}{z}\pi^z(1-\pi)^{r-z}\, dG_0(\pi).$$

An elementary proof of this result is given by Heath and Sudderth (1976) and a general presentation is given by Taylor et al. (1985, Chapter 1). That is, for any infinite exchangeable sequence of binary random variables $\{X_i\}$, there exists a random variable π with distribution G_0 such that $\sum_{i=1}^{r} X_i$ given π has a binomial(r, π) distribution so the concept of a stochastic parameter is intrinsic to the model. We call G_0 either the *mixing distribution* or the *true prior distribution*.

The case in which the sequence of random variables is independent is degenerate in the sense that the distribution G_0 is a degenerate distribution putting all its mass on a single value of π. Thus in this case, we cannot produce a prior distribution. de Finetti argued that unconditional independence excludes the possibility of learning through experience because, by definition, independence requires the future to be unaffected by the past. He argued further that we need to assume some form of dependence and that exchangeability is often a reasonable assumption to make (de Finetti, 1970, Chapter 11).

2.2.3 Interpreting a prior

de Finetti's theorem provides a justification for the Bayesian formulation of the model but it does not tell us how to find g_0 or how to interpret what happens when we use a different g in lieu of g_0.

If we have previous observations of the same type as the present data and the data generating process is either stable or changing in a known way over time, then we can estimate g_0 and adopt a frequency interpretion for the prior and hence posterior distribution. In its simplest form, this means that we can think of a sequence of independent replications of the data generating process yielding realizations of π and then interpret the prior distribution as representing the limiting frequency of particular events involving π. (Frequency interpretations of statistical inference are discussed further in Chapter 3.)

In the absence of previous observations, we regard a prior distribution as a mathematical expression of our degree of belief in propositions about π before we observe the data. There are two different approaches to selecting a prior to represent our beliefs:

(1) Following Ramsey (1931), de Finetti (1937), Good (1950), Savage (1954; 1961; 1962a), and others, we can choose a prior distribution to reflect the strength of our subjective beliefs about the parameters before we collect the data. A formal process of real or hypothetical betting games (which may be quite difficult to carry out in practice) can be used to quantify beliefs. Each individual is entitled to construct their own prior provided they follow the rules for self-consistent or coherent betting behavior in deriving the prior.

(2) Following Jeffreys (1946; 1939/1961), we determine a prior distribution to reflect beliefs which, given the same information, any reasonable person ought to hold. In principle, this approach can produce a distribution in the absence of any information so that the prior describes a state of ignorance about the parameters.

These two approaches are sometimes described as the "subjective" and "rational" approaches but, if we use this terminology, we should avoid any pejorative implication.

2.2.4 Prior Elicitation

The process of determining an appropriate prior distribution to reflect subjective beliefs about unknown parameters is known as *prior elicitation*.

A simple method is to construct a histogram by breaking the parameter space into intervals and assigning to each interval the probability with which we believe the parameter to be in that interval. In trying to assess the probability of severe visual loss within 2 years, we can break the interval (0, 1) into 10 subintervals of width 0.1 and then assign probabilities to each of these. It is unlikely that π_{11} will be large, so these intervals get low probability. We might assign highest probability between 0 and 0.3% though we might have anecdotal evidence (such as would be used in designing the study) which enables us to be even more specific. The precise probabilities to be assigned to the intervals can be determined from the odds we would be willing to bet (for stakes that are small but worth considering) that π_{11} is in the interval. Another useful method is to try to specify selected quantiles of the prior distribution and then use as the prior a distribution which fits the quantiles. These and other approaches to prior elicitation are discussed in Berger (1985, Chapter 3).

Whether we use the histogram, quantile, or any other method to elicit a prior distribution, we can consider introducing a convenient approximation to the prior distribution. The approximating distribution is usually chosen for its mathematical tractability but of course should still represent our actual prior beliefs.

2.2.5 Conjugate Priors

A useful class of priors for approximating prior beliefs about a binomial model is the beta(a, b) family of distributions which have density

$$g(\pi) \propto \pi^{a-1}(1-\pi)^{b-1}, \qquad 0 \le \pi \le 1. \tag{2.14}$$

To distinguish the known, fixed parameters a and b of the prior distribution from the unknown parameters of the model, the parameters of the prior distribution are sometimes called *hyperparameters*. The posterior distribution for this class of priors is

$$g(\pi \mid Z = z) \propto \pi^{z+a-1}(1-\pi)^{r-z+b-1}, \qquad 0 \le \pi \le 1, \tag{2.15}$$

which is the beta($z + a$, $r - z + b$) distribution. The prior and the posterior are in the same class of distributions so the prior is said to be *conjugate* for the model. For this to occur, the prior density must be of the same

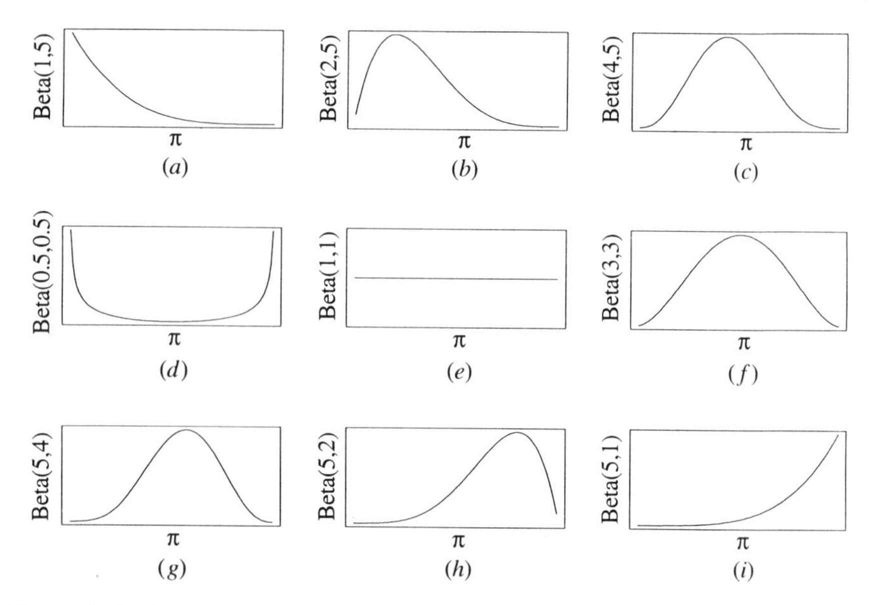

Figure 2.3. The beta(a, b) family of prior distributions: (a) $a = 1$, $b = 5$; (b) $a = 2$, $b = 5$; (c) $a = 4$, $b = 5$; (d) $a = 0.5$, $b = 0.5$; (e) $a = 1$, $b = 1$; (f) $a = 3$, $b = 3$; (g) $a = 5$, $b = 4$; (h) $a = 5$, $b = 2$; (i) $a = 5$, $b = 1$.

functional form as the likelihood function but with the data replaced by hyerparameters.

The class of beta priors is quite flexible and can represent a wide range of prior beliefs. The uniform prior corresponds to choosing $a = b = 1$, large values of $a = b$ correspond to a prior belief that π is near $\frac{1}{2}$, larger a than b corresponds to a prior belief that π is more likely to be large than small, and so on. Some of the possibilities are illustrated in Figure 2.3.

Conjugate prior distributions are not always tractable but when, as in this example, they are, they have some nice features. One of these is that continued updating is straightforward. For example, if we continue to study the incidence of severe visual loss within 2 years, the posterior from the study we have already done can be used as the prior for the next study. By starting with the conjugate prior, we ensure that the posterior distribution is always a beta distribution so is reasonably tractable. Thus, if after the first study, our posterior distribution is beta$(z_1 + a, r_1 - z_1 + b)$ and we observe z_2 out of r_2 conversions in the second, the posterior becomes beta$(z_1 + z_2 + a, r_1 + r_2 - z_1 - z_2 + b)$. This updating process provides a nice model for learning from the accumulation of data. A second nice feature of conjugate priors (at least for exponential family models (1.17)) is that the Bayes estimate (Section 2.1.4)

$$\mathrm{E}(\pi \mid z) = \frac{z + a}{r + a + b} = \frac{r}{r + a + b}\left(\frac{z}{r}\right) + \frac{a + b}{r + a + b}\left(\frac{a}{a + b}\right)$$

is a linear function of the sample estimate z/r of the rate of severe visual loss and the prior estimate $a/(a + b)$ with weights determined by the values of the hyperparameters and the sample size. The posterior variance or *precision* also has a simple, intuitive structure. Thus we can see how the data combines with the prior information to modify our beliefs about π.

2.2.6 Hierarchical Priors

Although in theoretical discussions we avoid assigning specific values to the hyperparameters in (2.14), it is important to keep in mind that in any actual application we have to specify numerical values for the hyperparameters before we can use (2.15). If we do not want to commit ourselves to particular values, we can incorporate uncertainty about the value of the hyperparameters by treating the prior as the conditional distribution of the parameters given the hyperparameters and assigning a distribution (a *hyperprior*) to the hyperparameters (Good, 1975; Lindley and Smith, 1972). For example, we could treat the hyerparameters a and b in our beta prior as independent and uniformly distributed on $[0, 2]$. Then we have that

$$h(a, b) = \tfrac{1}{4}I(0 \le a \le 2)I(0 \le b \le 2)$$

so the prior distribution for π is

$$g(\pi) = \frac{1}{4}\int_0^2 \int_0^2 \frac{\Gamma(a + b)}{\Gamma(a)\Gamma(b)} \pi^{a-1}(1 - \pi)^{b-1}\, da\, db. \tag{2.16}$$

We applied (2.13) with Simpson's rule (2.11) to approximate (2.16). The result presented in Figure 2.4 is stable over $k = 6$, 12, and 24. The density resembles the beta prior with $a = b < 1$ but is flatter in the middle.

We can sometimes ensure that the prior distribution in a hierarchical specification has a convenient form by using a distribution for the hyperparameters which is conjugate to the prior. Thus for example, we might be prepared to assume that given the hyperparameter μ, $\log\{\pi/(1 - \pi)\}$ is a priori $N(\mu, 100)$ and that μ is $N(0, 100)$. The marginal distribution of $\log\{\pi/(1 - \pi)\}$ is then $N(0, 200)$ which incorporates the uncertainty about μ. In this circumstance, it is feasible to extend the number of layers in the hierarchical model structure and increase the uncertainty in the prior by pushing the hyperparameter specification back to deeper levels in the hierarchical structure.

2.2.7 Posterior Robustness

If our beliefs about the parameters are sufficiently imprecise, it may be easier to determine a set of plausible prior distributions which are consistent with our

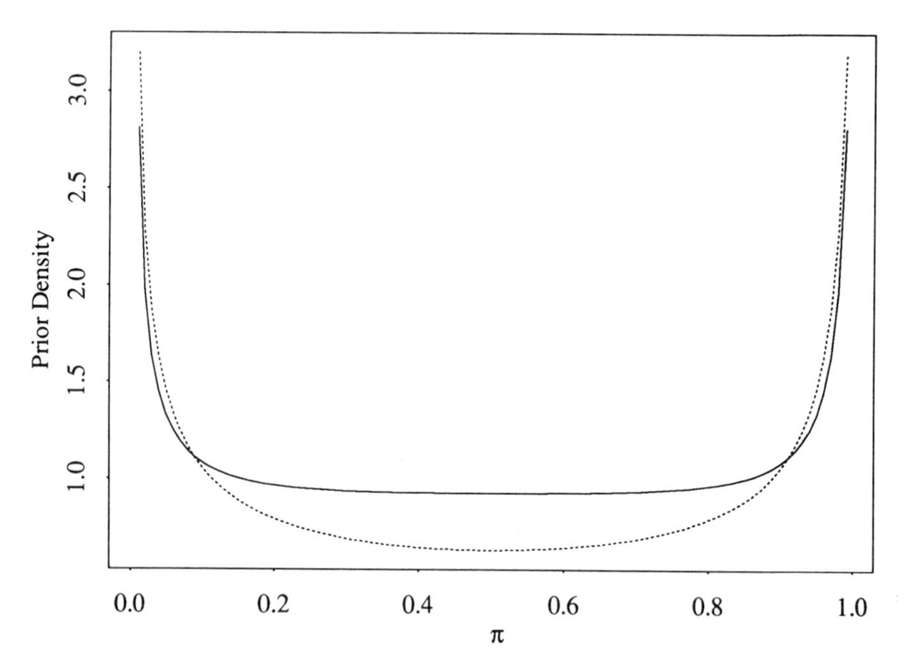

Figure 2.4. The prior density for the hierarchical model in which π given a and b has a beta(a, b) distribution and the hyperparameters a and b have independent U(0, 2) distributions. The density is obtained by numerical integration using Simpson's rule. The dashed curve represents Jeffreys' prior.

knowledge and beliefs than a single prior distribution. For example, in studying the incidence of severe visual loss within 2 years, we can consider the class of beta priors with $a, b \in [0, 2]$. An alternative to constructing a hierarchical prior is to calculate a posterior distribution for each prior. If the posterior distribution based on each of the prior distributions in the plausible class is similar, then the analysis is said to be *posterior robust* and the precise choice of prior in the plausible class is not critical. For the class of beta priors with $a, b \in [0, 2]$, the posterior distribution varies over the class of beta$(z + a, r - z + b)$ distributions with $a, b \in [0, 2]$. For large r and z, changing a and b has little effect on the posterior so the analysis is posterior robust. This is no longer true if r and z are small so we see that posterior robustness depends on both the observed data and the class of priors. Generally, if the class of priors is small, posterior robustness will usually obtain whereas if it is too large, posterior robustness will rarely obtain. Nonetheless, it is important that the class of priors covers a range of tail behavior consistent with the prior beliefs. This generally means that the class of conjugate priors is too restrictive and we need to extend this class by allowing mixtures of distributions. A detailed discussion of posterior robustness is given by Berger (1984; 1985, pp. 195–252). See also Section 5.8.4.

2.2.8 Vague Priors

Prior distributions with support over the parameter space which have densities which are fairly flat over the support and dominated by the likelihood are variously called *vague*, *dominated*, or *noninformative priors*. For such prior distributions, the posterior distribution mainly reflects the effect of the data through the likelihood and is relatively insensitive to small changes in the prior and hence to the precise form of the chosen prior. Consequently, inferences based on similar vague priors are posterior robust in the sense of Section 2.2.7. The use of vague priors is often sensible in inference problems because the beliefs to be represented by the prior are usually fairly imprecise and we want the final conclusions to reflect the data rather more than the prior.

2.2.9 The Principle of Precise Measurement

The *Principle of precise measurement* (Savage 1962a, pp. 20–25) or the *principle of stable measurement* (Edwards et al., 1963, pp. 101–111) states that vague priors (in the sense of Section 2.2.8) which are strongly dominated by the likelihood can be approximated by locally uniform priors provided the resulting posterior density is integrable. Specifically, if

(a) there is a region D such that the likelihood function is negligible outside D

(b) within D the prior density g changes little

(c) the prior density g is not substantially larger outside D than it is in D,

then the prior distribution can be approximated by a uniform distribution. A schematic diagram of the likelihood and prior density in a situation in which the principle of precise measurement applies is shown in Figure 2.5.

The generality of the principle is increased by noting that there may be a transformation of the parameters to a scale on which it applies. In this case, it is usually worth working with the parameters on the transformed scale.

Typical problems in which the principle of precise measurement does not apply include those in which low prior probability is assigned to regions for which the likelihood $f(z; \pi)$ is relatively large and those in which high prior probability is assigned to regions over which the likelihood $f(z; \pi)$ is small. They also include problems in which both the prior and the data are diffuse so there is no region D clearly favored by the data. This occurs when the sample size is very small so the likelihood is not sufficiently dominant, or when there is no data as in the design stage of a study.

2.2.10 The Principle of Insufficient Reason

The discussion in Sections 2.2.4–2.2.9 on the problem of representing our prior knowledge and beliefs in a prior distribution assumed that we do have beliefs

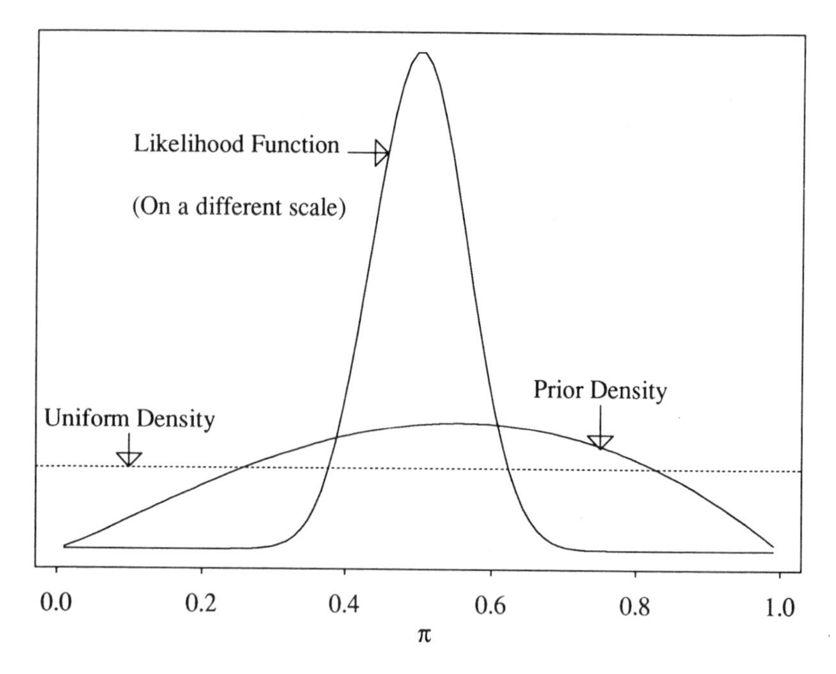

Figure 2.5. A situation in which the principle of precise measurement is applicable. The scale on the y axis is omitted because the likelihood function is not on the same scale as the prior densities.

about the parameters. What happens if we feel we have no beliefs or we are reluctant to use our beliefs? The problem then is of trying to represent a state of ignorance about the parameters. This problem is as old as Bayesian analysis itself. Bayes (1763) gave an explicit argument for adopting a uniform prior distribution (sometimes referred to as *Bayes' postulate*) but a number of interpretations of this work are possible; see Stigler (1986, Chapter 3) and Geisser (1993, pp. 46–9). Laplace (1774) dealt with the problem by means of what later came to be known as the *principle of insufficient reason* or the *principle of indifference* (Keynes, 1921, p. 41). If we have a discrete parameter which can take on one of k distinct values and there is no reason to believe one is more likely than any other, then we should treat the outcome as equally likely. In other words, we assign equal probability or a discrete uniform distribution to the parameter space. In the continuous case, we assign a uniform distribution over the space of possible outcomes. This is the principle of insufficient reason: we should use uniform distributions unless we have good reason not to. As a consequence, we should use uniform distributions to represent ignorance. Note that this is a very different argument for the use of uniform priors from that presented by the principle of precise measurement in Section 2.2.9.

2.2.11 Jeffreys' Priors

The greatest practical difficulty in the use of uniform priors is to decide on what scale a parameter can be taken to be uniform. One method for choosing the scale on which to assign the uniform distribution was developed by Jeffreys (1946).

For a single observation Z from a model $\mathscr{F} = \{f(z;\theta), z \in \mathscr{Z}; \theta \in \Omega\}$ for which $\mathscr{Z}$ does not depend on θ and $\partial \log f(z;\theta)/\partial\theta$ exists and is finite for all $z \in \mathscr{Z}$ and $\theta \in \Omega$, let

$$I(\theta) = \mathrm{E}\left\{\frac{\partial \log f(Z;\theta)}{\partial\theta}\right\}^2 = \int_{-\infty}^{\infty} \left\{\frac{\partial \log f(z;\theta)}{\partial\theta}\right\}^2 f(z;\theta)\,dz. \qquad (2.17)$$

We call $I(\theta)$ the *Fisher information* for θ. Jeffreys recommended that when we have no prior information about θ we should choose our prior to be

$$g(\theta) \propto I(\theta)^{1/2}. \qquad (2.18)$$

That is, the prior density is proportional to the square root of the Fisher information for θ in a single observation. (The motivation for this recommendation is presented in Section 2.4.2.)

For the binomial model (2.1), we have

$$\log f(z;\pi) \propto z \log(\pi) + (r - z)\log(1 - \pi),$$

or

$$\frac{\partial \log f(z;\pi)}{\partial\pi} = \frac{z}{\pi} - \frac{n - z}{1 - \pi} = \frac{z - n\pi}{\pi(1 - \pi)}$$

and hence by (2.17)

$$\begin{aligned} I(\pi) &= \sum_{z=0}^{n} \left(\frac{z - n\pi}{\pi(1 - \pi)}\right)^2 \binom{n}{z} \pi^z (1 - \pi)^{n-z} \\ &= \left(\frac{1}{\pi(1 - \pi)}\right)^2 \mathrm{Var}(Z) \\ &= \frac{n}{\pi(1 - \pi)}. \end{aligned}$$

That is, from (2.18), the Jeffreys prior is $g(\pi) \propto \pi^{-1/2}(1 - \pi)^{-1/2}$ which is a beta(1/2, 1/2) distribution. This can be interpreted as a uniform prior for $\arcsin(\pi^{1/2})$.

For large z and r, the beta$(z + 1/2, r - z + 1/2)$ posterior distribution from the Jeffreys prior is very similar to the beta$(z, r - z)$ posterior from the uniform prior so the inferences are not greatly affected by whether we choose the uniform or the Jeffreys prior.

PROBLEMS

2.2.1. Suppose that we have a model of the form

$$\mathscr{F} = \{f(z; \pi): 0 \leq \pi \leq 1\}$$

and we adopt the prior $g(\pi) \propto I(\pi = \pi_0)$. Obtain and then interpret the posterior distribution. Discuss whether or not this outcome is desirable in an inferential problem.

2.2.2. One interpretation of the negative binomial model is that it represents the number of independent binomial$(1, \pi)$ trials required to observe r 1s. The model can be written

$$\mathscr{F} = \left\{f(z; \pi) = \binom{z-1}{r-1}\pi^r(1-\pi)^{z-r}, z = r, r+1, \ldots: 0 \leq \pi \leq 1\right\}.$$

Find the conjugate family of prior distributions for this model. Obtain the posterior distribution and explore the form of the Bayes estimate and the precision. Choose a conjugate prior to represent your beliefs about the probability that a tossed coin will come up heads. Toss a coin until you observe six heads. Plot your prior and the posterior distribution for π. What is the Bayes estimate of π in this case? Repeat the experiment and explore how it modifies your conclusions.

2.2.3. One criticism of the use of beta priors with binomial models is that they are unimodal. This criticism can be met by using a prior distribution which is a mixture of beta distributions, i.e., the prior density is of the form

$$g(\pi) = \varepsilon g_{\mathrm{B}}(\pi; a, b) + (1 - \varepsilon) g_{\mathrm{B}}(\pi; c, d).$$

where $g_{\mathrm{B}}(\cdot; a, b)$ is the density of the beta(a, b) distribution. Show that this prior is conjugate for the binomial model. Find the Bayes estimate and the precision. Suppose that our prior beliefs about the rate of severe visual loss in untreated eyes in subjects in which the other eye is receiving argon treatment are represented by the above prior with $\varepsilon = 0.7$, $a = 10$, $b = 20$, $c = 20$, $d = 10$. Plot and interpret this prior. Find and plot the posterior distribution. Does the principle of precise measurement apply in this case?

2.2.4. The truncated Poisson model

$$\mathscr{F} = \left\{f(\mathbf{y}; \lambda) = \prod_{i=1}^{n} \frac{\lambda^{y_i}\exp(-\lambda)}{y_i!\{1 - \exp(-\lambda)\}}, y_i = 1, 2, \ldots: \lambda > 0\right\}$$

provides a reasonable model for the Feinstein et al. (1989) microbubble data presented in Problem 1.5.2. Find the conjugate family of distributions for this model. Is this a convenient family of distributions? Suppose from previous experiments, we know that λ has a distribution with mean approximately equal to 3 and variance 0.1. Find the gamma prior with these moments and then obtain the posterior. Use numerical integration to normalize the posterior density. Then plot your prior and the resulting posterior distribution for λ. What is the effect of the data on the prior information?

2.2.5. The Poisson model

$$\mathscr{F} = \left\{ f(\mathbf{y}; \lambda) = \prod_{i=1}^{n} \frac{\lambda^{y_i} \exp(-\lambda)}{y_i!}, y_i = 0, 1, 2, \ldots, : \lambda > 0 \right\}$$

provides a reasonable model for Rutherford and Geiger's (1910) alpha particle data presented in Problem 1.5.1. Find the Jeffreys prior for λ. Find and then plot the posterior distribution of $\lambda \mid \mathbf{Z} = \mathbf{z}$. Show how to find a credibility interval for λ.

2.2.6. Suppose we adopt the exponential model

$$\mathscr{F} = \left\{ f(\mathbf{y}; \lambda) = \prod_{i=1}^{n} \lambda \exp(-\lambda y_i), 0 \leq y_i \leq \infty: \lambda > 0 \right\}$$

for Proschan's (1963) air conditioning data presented in Problem 1.5.11. Find the Jeffreys prior for λ and then the corresponding posterior distribution of $\lambda \mid \mathbf{Z} = \mathbf{z}$ and hence of $\theta = (1/\lambda) \log(2) \mid \mathbf{Z} = \mathbf{z}$. Plot the posterior distribution for θ. Show how to find an exact 95% credibility interval for θ and compare the posterior probabilities of $H_0: \theta \leq \theta_0$ and $H_1: \theta > \theta_0$.

2.2.7. Twelve random numbers generated on a random number generator are given below:

0.86972163 0.83693480 0.79465413 0.32566555 0.01770435 0.97561784

0.20400030 0.47204266 0.51425063 0.29038457 0.58089520 0.68833185

The generated numbers are purportedly uniformly distributed on (0, 1) but may actually be distributed on a proper subset of (0, 1). That is, the data may be modeled by

$$\mathscr{F} = \left\{ f(\mathbf{y}; \theta) = \prod_{i=1}^{n} \frac{1}{\theta} I(0 \leq y_i \leq \theta); 0 \leq \theta \right\},$$

where $I(\cdot)$ is the indicator function, and we want to make inferences about θ. Show that θ is a scale parameter. Use the prior $g(\theta) \propto \theta^{-1} I(0 \leq \theta)$ to obtain the posterior distribution of $\theta \mid \mathbf{Z} = \mathbf{z}$. Plot the posterior density and use it to derive a $100(1 - \alpha)\%$ credibility set for θ. Compare the posterior probabilities of H_0: $\theta \geq 1$ and H_1: $\theta < 1$.

2.2.8. Suppose we are concerned with the lower endpoint of the distribution of computer generated random numbers which purport to be uniformly distributed on (0, 1). Suppose that the observations can be modeled as realizations of independent random variables with density

$$\mathscr{F} = \left\{ f(\mathbf{y}; \theta) = \prod_{i=1}^{n} \frac{1}{1 - \xi} I(\xi \leq y_i \leq 1); 0 \leq \xi < 1 \right\},$$

where $I(\cdot)$ is the indicator function. The data are the same 12 observations as in Problem 2.2.7. Suppose we adopt the prior distribution with density

$$g(\xi) = (a + 1)(1 - \xi)^a, \qquad 0 \leq \xi \leq 1, \quad a > -1.$$

What kind of prior beliefs does this density represent? Find and plot the posterior distribution of $\xi \mid \mathbf{Z} = \mathbf{z}$. Derive a $100(1 - \alpha)\%$ credibility set for ξ.

2.2.9. Show that the *predictive distribution* of a single additional observation independent of the observed data under a model for which a single observation has density $f(y \mid \lambda)$ satisfies

$$p(y \mid \mathbf{z}) = \int f(y \mid \lambda) g(\lambda \mid \mathbf{z}) \, d\lambda,$$

where $g(\lambda \mid \mathbf{z})$ is the posterior density of λ. In the context of Problem 2.2.5, using the Jeffreys prior for λ, find the predictive distribution of the number of alpha particles emitted in a fixed time interval of 7.5 seconds, given the observed data. Plot the predictive distribution.

2.2.10. Suppose we adopt the exponential model of Problem 2.2.6 for Barlow et al.'s (1984) pressure vessel failure data presented in Problem 1.5.4. Use the Jeffreys prior to find the predictive distribution of the time to failure of an independent pressure vessel. Plot the predictive distribution and show how to find an exact 95% Bayesian prediction interval for a single observation.

2.3 THE EFFECT OF CAFFEINE ON THE VOLUME OF URINE

To develop the Bayesian paradigm further, consider the data presented in Table 1.3 on the effect of caffeine on the volume of urine voided. We decided in Section 1.2.3 to model the change in the volume of urine voided by the ingestion of caffeine using the Gaussian model

$$\mathscr{F} = \left\{ f(\mathbf{z}; \mu, \sigma) = \prod_{i=1}^{n} \frac{1}{(2\pi\sigma^2)^{1/2}} \exp\left\{ -\frac{(z_i - \mu)^2}{2\sigma^2} \right\}, -\infty < z_i < \infty : \mu \in \mathbb{R}, \sigma > 0 \right\}. \tag{2.19}$$

The problem in the context of (2.19) is to make inferences about μ in the presence of the nuisance parameter σ.

2.3.1 The Likelihood Function

Multiplying through and gathering terms in (2.19), we can rewrite the model density as

$$\begin{aligned} f(\mathbf{z}; \mu, \sigma) &= \frac{1}{(2\pi\sigma^2)^{n/2}} \exp\left\{ -\sum_{i=1}^{n} \frac{(z_i - \mu)^2}{2\sigma^2} \right\} \\ &= \frac{1}{(2\pi\sigma^2)^{n/2}} \exp\left\{ -\frac{\nu s^2}{2\sigma^2} - \frac{n(\bar{z} - \mu)^2}{2\sigma^2} \right\}, \end{aligned} \tag{2.20}$$

where

$$\bar{z} = n^{-1} \sum_{i=1}^{n} z_i, \; \nu = n - 1 \quad \text{and} \quad s^2 = \nu^{-1} \sum_{i=1}^{n} (z_i - \bar{z})^2.$$

This is a convenient form for the likelihood function for (μ, σ) which is obtained by replacing the argument $\mathbf{z}$ by the observed differences $\mathbf{z} = (235, -90, 410, \ldots, 40)$ or equivalently by replacing $\bar{z}$ and s^2 by the mean $\bar{z} = 170.72$ and variance $s^2 = 42719.86$ of the observed differences.

2.3.2 The Prior Distribution

For the situation in which we lack strong prior knowledge about the parameters (μ, σ), we can consider using the Jeffreys prior (2.18). It is convenient to derive the Jeffreys prior for a general location-scale family (1.18) which includes the Gaussian model as a particular case. The density of a single observation from a location–scale family is of the form

$$f(x; \mu, \sigma) = \frac{1}{\sigma} h\left(\frac{x - \mu}{\sigma}\right),$$

where h is a known density function.

If σ is known,

$$\frac{\partial \log f(x; \mu)}{\partial \mu} = -\frac{1}{\sigma} h'\left(\frac{x-\mu}{\sigma}\right) \Big/ h\left(\frac{x-\mu}{\sigma}\right)$$

so by (2.17)

$$I(\mu) = \frac{1}{\sigma^2} \mathrm{E}\left\{\mathrm{h}'\left(\frac{Z-\mu}{\sigma}\right) \Big/ h\left(\frac{Z-\mu}{\sigma}\right)\right\}^2 = \frac{1}{\sigma^2} \int_{-\infty}^{\infty} \left\{\frac{h'(x)}{h(x)}\right\}^2 h(x)\, dx$$

(on changing variables) which is a constant. Hence the Jeffreys prior for a location parameter is, by (2.18),

$$g(\mu) \propto 1, \qquad \mu \in \mathbb{R}. \tag{2.21}$$

The prior (2.21) is not a proper density in that its integral is infinite but it can be interpreted as a *locally uniform prior*. (We discuss some of the issues involved in using improper prior distributions in Section 2.4.)

If μ is known,

$$\frac{\partial \log f(x; \mu, \sigma)}{\partial \sigma} = -\frac{1}{\sigma}\left\{1 + \left(\frac{x-\mu}{\sigma}\right) h'\left(\frac{x-\mu}{\sigma}\right) \Big/ h\left(\frac{x-\mu}{\sigma}\right)\right\}$$

so that by (2.17)

$$I(\sigma) = \frac{1}{\sigma^2} \int_{-\infty}^{\infty} \left\{1 + x \frac{h'(x)}{h(x)}\right\}^2 h(x)\, dx \propto \frac{1}{\sigma^2}.$$

Hence the Jeffreys prior for a scale parameter is by (2.18)

$$g(\sigma) \propto \frac{1}{\sigma}, \qquad \sigma > 0, \tag{2.22}$$

which is again improper. We can interpret (2.22) as a locally uniform prior for $\log(\sigma)$.

If μ and σ are unknown but independent, we can take the product of (2.21) and (2.22) to obtain

$$g(\mu, \sigma) \propto \frac{1}{\sigma}, \qquad \mu \in \mathbb{R}, \quad \sigma > 0. \tag{2.23}$$

Generally, in multiparameter problems, the Fisher information $I(\theta)$ is a matrix

$$I(\theta) = \int_{-\infty}^{\infty} \frac{\partial \log f(z; \theta)}{\partial \theta} \left\{ \frac{\partial \log f(z; \theta)}{\partial \theta} \right\}^{\mathrm{T}} f(z; \theta)\, dz$$

and the *Jeffreys rule* is to choose as the prior the square root of the determinant of the Fisher information matrix. For the location–scale model, the off diagonal element of the Fisher information matrix is

$$I_{\mu,\sigma} = \frac{1}{\sigma^2} \int_{-\infty}^{\infty} \left\{ \frac{h'(x)}{h(x)} \right\} \left\{ 1 + x \frac{h'(x)}{h(x)} \right\} h(x)\, dx,$$

so the Fisher information is

$$I(\mu, \sigma) = \frac{1}{\sigma^2} \begin{pmatrix} \int_{-\infty}^{\infty} \left\{ \frac{h'(x)}{h(x)} \right\}^2 h(x)\, dx & \int_{-\infty}^{\infty} \left\{ \frac{h'(x)}{h(x)} \right\} \left\{ 1 + x \frac{h'(x)}{h(x)} \right\} h(x) \\ \int_{-\infty}^{\infty} \left\{ \frac{h'(x)}{h(x)} \right\} \left\{ 1 + x \frac{h'(x)}{h(x)} \right\} h(x)\, dx & \int_{-\infty}^{\infty} \left\{ 1 + x \frac{h'(x)}{h(x)} \right\}^2 h(x)\, dx \end{pmatrix}$$

which has determinant proportional to $1/\sigma^4$. Hence, application of Jeffreys' rule leads to the prior

$$g(\mu, \sigma) \propto \frac{1}{\sigma^2}, \qquad \mu \in \mathbb{R}' \sigma > 0,$$

which can be interpreted as a locally uniform prior for σ^{-1}. However, Jeffreys argued that for this problem the parameters should be treated as a priori independent. Thus the prior (2.23) is treated as the Jeffreys prior for location–scale models.

2.3.3 The Marginal Posterior Distribution

The joint posterior density is the product of the likelihood (2.20) and the Jeffreys prior (2.23), namely

$$\begin{aligned} g(\mu, \sigma \mid \mathbf{Z} = \mathbf{z}) &\propto \frac{1}{\sigma^{n+1}} \exp\left\{ -\frac{\nu s^2}{2\sigma^2} - \frac{n(\bar{z} - \mu)^2}{2\sigma^2} \right\} \\ &= \frac{1}{\sigma^{n+1}} \exp\left[-\frac{\nu s^2\{1 + n(\bar{z} - \mu)^2/\nu s^2\}}{2\sigma^2} \right], \end{aligned} \qquad (2.24)$$

which is illustrated in Figure 2.6. The marginal posterior density for μ is

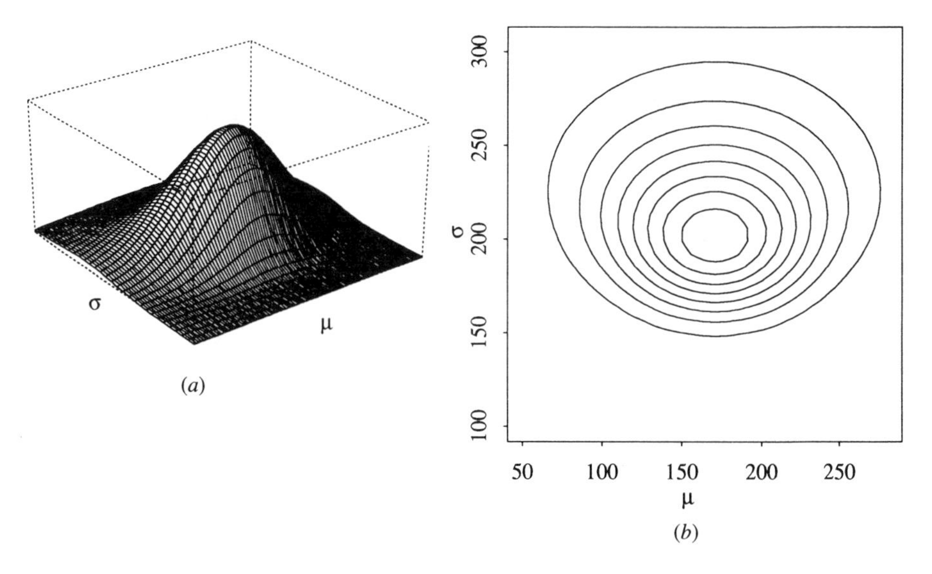

Figure 2.6. The joint posterior density of (μ, σ) from the Jeffreys prior: (*a*) perspective plot; (*b*) contour plot.

obtained by integrating (2.24) over σ as

$$g(\mu \mid \mathbf{Z} = \mathbf{z}) \propto \int_0^\infty \frac{1}{\sigma^{n+1}} \exp\left[-\frac{\{vs^2 + n(\bar{z} - \mu)^2\}}{2\sigma^2}\right] d\sigma \propto \frac{1}{\left\{1 + \dfrac{n(\bar{z} - \mu)^2}{vs^2}\right\}^{n/2}}. \tag{2.25}$$

(See 4c in the Appendix). That is $n^{1/2}(\mu - \bar{z})/s \mid \mathbf{Z} = \mathbf{z} \sim t_v$, where t_v denotes the Student t distribution with $v = n - 1$ degrees of freedom (6c in the Appendix). This is a proper distribution and so can be used like any proper posterior distribution to make inferences about the parameters. In the caffeine example, $(\mu - 170.72)/48.72 \mid \mathbf{Z} = \mathbf{z} \sim t_{17}$. This marginal posterior distribution is plotted in Figure 2.7. (The marginal posterior distribution for σ is discussed in Section 5.8.3.)

The Jeffreys prior corresponds to taking $\log(\sigma)$ to be uniformly distributed. However, if we took σ^2, σ, σ^{-1} or σ^{-2} rather than $\log(\sigma)$ to be uniformly distributed, the above analysis shows that the marginal posterior distribution of $n^{1/2}(\mu - \bar{z})/s \mid \mathbf{Z} = \mathbf{z}$ would be Student t with $n - 3$, $n - 2$, n, or $n + 1$ degrees of freedom respectively. Thus the choice of scale affects the inference.

2.3.4 Credibility Sets

The most plausible value of the parameter μ is that which maximizes the posterior density (2.25), namely $\hat{\mu} = \bar{z} = 170.72$. It is straightforward to obtain

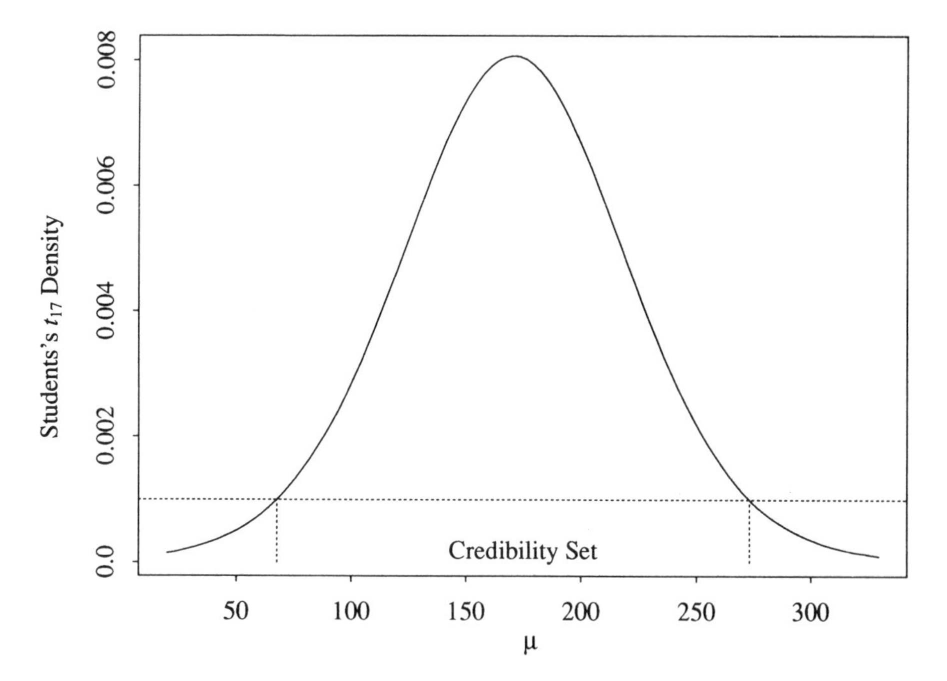

Figure 2.7. The marginal posterior density of μ from the Jeffreys prior. The 95% Bayesian credibility set for μ lies between the dotted vertical lines.

credibility sets for μ because the posterior density (2.25) is unimodal and symmetric about $\bar{z}$ so the contours of (2.25) define intervals of the form $[\bar{z} - a, \bar{z} + a]$. To ensure that μ is in this set with probability $1 - \alpha$, we require

$$\begin{aligned} 1 - \alpha &= \mathrm{P}\{\bar{z} - a \leq \mu \leq \bar{z} + a \mid \mathbf{Z} = \mathbf{z}\} \\ &= G_v(n^{1/2}a/s) - G_v(-n^{1/2}a/s) \\ &= 1 - 2G_v(-n^{1/2}a/s) \end{aligned}$$

or

$$a = -n^{-1/2}sG_v^{-1}(\alpha/2) = n^{-1/2}sG_v^{-1}(1 - \alpha/2),$$

where G_v is the distribution function of the Student t distribution with v degrees of freedom. Thus the $100(1 - \alpha)\%$ credibility interval for μ is

$$[\bar{z} - n^{-1/2}sG_v^{-1}(1 - \alpha/2), \bar{z} + n^{-1/2}sG_v^{-1}(1 - \alpha/2)]. \tag{2.26}$$

In particular, the 95% credibility interval for μ, the mean change in the volume of urine voided after ingesting caffeine, is

$$[67, 274].$$

Given the data, we believe μ to be in the interval $[67, 274]$ with probability 0.95.

PROBLEMS

2.3.1. Suppose that we observe $\{Z_1, \ldots, Z_n\}$ on the Gaussian model (2.19). Supposed initially that the variance σ^2 is known and that we adopt the conjugate prior $\mu \sim \mathrm{N}(\lambda, \tau^2)$. Find the posterior distribution of μ given the data. Explore the nature of the Bayes estimate of μ and the precision.

2.3.2. Suppose that the Gaussian model (2.19) holds and that we adopt the hierarchical model in which $\mu \mid \lambda \sim \mathrm{N}(\lambda, \tau^2)$ with τ known and $\lambda \sim \mathrm{N}(\eta, \omega^2)$ with η and ω known. Find the prior distribution (the marginal distribution of μ) and the posterior distribution.

2.3.3. Now suppose that the model variance σ in the Gaussian model of Problem 2.3.1 is unknown. Show that the "*Gaussian–inverse gamma*" distribution

$$g(\mu, \sigma) \propto \frac{1}{\sigma\tau^{k+1}} \exp\left\{-\frac{\kappa\tau^2}{2\sigma^2} - \frac{\rho(\mu - \lambda)^2}{2\sigma^2}\right\},$$

with known hyperparameters κ, τ, ρ, and λ is the conjugate family of distributions. Find the marginal posterior distribution of μ.

2.3.4. Find the conjugate family of distributions for the exponential family model which is given by

$$\mathscr{F} = \left\{f(\mathbf{z}; \theta) = \prod_{i=1}^{n} \exp\left[\psi_1 a_1(z_i) + \cdots + \psi_k a_k(z_i) + \phi + b(z_i)\right],\right.$$
$$\left. z_i \in A \colon \theta = (\psi_1, \ldots, \psi_k, \phi) \in \Omega\right\}.$$

Specialise the result to the Gaussian and gamma models.

2.3.5. Suppose that we observe $\{Z_1, \ldots, Z_n\}$ on the Gaussian model (2.19). Suppose initially that the variance σ^2 is known and we adopt the Jeffreys prior. Show that the predictive distribution for a new independent observation is $\mathrm{N}(\bar{z}, \sigma^2(n + 1)/n)$. Modify this result to allow for the case of unknown σ.

2.3.6. Suppose that we observe $\{(Y_1, x_1), \ldots, (Y_n, x_n)\}$ on the Gaussian regression model

$$\mathscr{F} = \left\{ f(y; \alpha, \beta, \sigma) = \prod_{i=1}^{n} \frac{1}{(2\pi\sigma^2)^{1/2}} \exp\left\{ -\frac{(y_i - \alpha - x_i\beta)^2}{2\sigma^2} \right\}, \right.$$

$$\left. y_i \in \mathbb{R}; \alpha, \beta \in \mathbb{R}, \sigma > 0 \right\}.$$

Show that

$$\sum_{i=1}^{n} (y_i - \alpha - x_i\beta)^2 = \sum_{i=1}^{n} (y_i - \hat{\alpha} - x_i\hat{\beta})^2 + n(\hat{\gamma} - \gamma)^{\mathrm{T}} C(\hat{\gamma} - \gamma)$$

where

$$\hat{\gamma} = (\hat{\alpha}, \hat{\beta})^{\mathrm{T}} = \left(\bar{y} - \bar{x}\hat{\beta}, \sum_{i=1}^{n} (x_i - \bar{x})(y_i - \bar{y}) \Big/ \sum_{i=1}^{n} (x_i - \bar{x})^2 \right)^{\mathrm{T}},$$

and

$$C = \begin{pmatrix} 1 & n^{-1} \sum_{i=1}^{n} x_i \\ n^{-1} \sum_{i=1}^{n} x_i & n^{-1} \sum_{i=1}^{n} x_i^2 \end{pmatrix}$$

Suppose initially that σ is known. Show that if $g(\alpha, \beta) \propto 1$, the posterior distribution of β given the data is $\mathrm{N}(\hat{\beta}, (\sum_{i=1}^{n} (x_i - \bar{x})^2)^{-1})$.

2.3.7. In the context of Problem 2.3.6, show that if $g(\alpha, \beta, \sigma) \propto \sigma^{-1}$, the posterior distribution of $\gamma = (\alpha, \beta)^{\mathrm{T}}$ given the data has density

$$g(\gamma \mid \mathbf{z}) \propto \frac{1}{[(n-2)s^2 + (\gamma - \hat{\gamma})^{\mathrm{T}} C(\gamma - \hat{\gamma})]^{n/2}},$$

where $s^2 = (n-2)^{-1} \sum_{i=1}^{n} (y_i - \hat{\alpha} - x_i\hat{\beta})^2$. This is the bivariate Student t distribution with $n - 2$ degrees of freedom. Obtain the marginal posterior distribution for $(\beta - \bar{\beta})\{\sum_{i=1}^{n} (x_i - \hat{x})^2\}^{1/2}/s$ and show how to set a $100(1 - \alpha)\%$ credibility interval for β.

2.3.8. For the Gaussian regression model of Problem 2.3.6, show that the predictive distribution for an additional observation at $x_0^{\mathrm{T}} = (1, x_0)$ has density

$$p(y_0 \mid \mathbf{z}, x_0) \propto \frac{1}{\left[(n-2)s^2 + \dfrac{(y_0 - x_0^{\mathrm{T}}\hat{\gamma})^2}{1 + n^{-1} x_0^{\mathrm{T}} C^{-1} x_0} \right]^{n/2}}, \qquad -\infty < y_0 < \infty.$$

Show how to set a $100(1 - \alpha)\%$ prediction interval for y_0.

2.3.9. Suppose that we observe $\{Z_1, \ldots, Z_n\}$ on the n-dimensional multivariate Student t model.

$$\mathscr{F} = \left\{ f_{St}(\mathbf{y}; \mu, \sigma, \nu) = \int_0^\infty f_N\left(\mathbf{y}; \mu, \frac{\sigma^2}{h}\right) f_G\left(h; \frac{\nu}{2}, \frac{\nu}{2}\right) dh, \right.$$
$$\left. -\infty < y_i < \infty: \mu \in \mathbb{R}, \sigma, \nu > 0 \right\},$$

where

$$f_N(\mathbf{y}; \mu, \tau) = \prod_{i=1}^n \frac{1}{(2\pi\tau)^{1/2}} \exp\left(-\frac{y_i^2}{2\tau}\right)$$

and

$$f_G(h; a, b) = \frac{1}{\Gamma(a)} b^a h^{a-1} \exp(-bh), \qquad h > 0,$$

is the density of the gamma distribution with parameters $a, b > 0$, holds. Show that the $\{Z_i\}$ are exchangeable but not independent. Show that, if the degrees of freedom ν is known, under the usual Jeffreys prior, the marginal posterior density of μ is

$$g(\mu \mid \mathbf{z}) \propto \int_0^\infty \frac{1}{\theta} f_N(\mathbf{z}; \mu, \theta)\, d\theta.$$

That is, the marginal posterior for μ under the multivariate Student t model is the same as under the Gaussian model. (Hint: Use the Jeffreys prior for σ^2 and make the transformation $(\sigma^2, h) \rightarrow (\theta = \sigma^2/h, z = h)$.)

2.4 IMPROPER PRIORS

We noted in Section 2.3.2 that the Jeffreys prior is not always a density function because its integral over the parameter space can be infinite. We call such functions *improper densities* and describe the corresponding distributions as improper prior distributions.

2.4.1 Technical Issues

When improper priors lead to proper posterior distributions as in (2.24), they can be used to make Bayesian inferences. Nonetheless, the use of improper distributions requires us to extend our view of probability theory to admit improper distributions and there are sometimes technical problems which arise from the use of improper distributions. In particular, integrals, sequences and limiting operations need to be treated carefully because the properties of proper distributions cannot simply be assumed to hold for improper distributions. Simple examples are easy to find.

(a) We can multiply marginal densities to obtain a joint density but we cannot integrate an improper joint density to recover the marginal densities.

(b) For any finite $k_1 < k_2$, the probability of $\{k_1 < \mu \leq k_2\}$ relative to the probability of $\{\mu \leq k_1\} \cup \{\mu > k_2\}$ is always 0 when μ has an improper distribution but can be arbitrarily large when μ has a proper distribution.

(c) Suppose $Z \mid \mu \sim \mathrm{N}(\mu, 1)$ and $g(\mu) \propto 1$ so that $\mu \mid Z \sim \mathrm{N}(Z, 1)$. The statement $(Z - \mu) \mid \mu \sim \mathrm{N}(0, 1)$ can be interpreted as meaning that $Z - \mu$ is independent of μ while $(\mu - Z) \mid Z \sim N(0, 1)$ can be interpreted as meaning that $Z - \mu$ is independent of Z!

(d) The usual transformation formula need not apply for improper distributions Recall that if $g(\theta)$ is a proper distribution for Θ and we make the one-to-one transformation $\theta \to \lambda(\theta)$, the density of Λ is given by

$$\begin{aligned} h(\lambda) &= \frac{\partial \mathrm{P}\{\Lambda \leq \lambda\}}{\partial \lambda} \\ &= \frac{\partial \mathrm{P}\{\theta(\Lambda) \leq \theta(\lambda)\}}{\partial \lambda} \\ &= \frac{\partial \mathrm{P}\{\Theta \leq \theta(\lambda)\}}{\partial \theta(\lambda)} \left| \frac{\partial \theta(\lambda)}{\partial \lambda} \right| \\ &= g(\theta(\lambda)) \left| \frac{\partial \theta(\lambda)}{\partial \lambda} \right|. \end{aligned} \tag{2.27}$$

This simple relationship does not in general hold for improper priors because the implicit probability integrals need not exist.

2.4.2 Jeffreys' Priors and Their Limitations

Jeffreys (1946) argued that when we use improper priors, we should require them to satisfy the transformation formula because then the procedure for producing the prior does not depend on the parameterization. Since

$$I(\theta) = \int_{-\infty}^{\infty} \left\{ \frac{\partial \log f(z; \lambda(\theta))}{\partial \lambda(\theta)} \frac{\partial \lambda(\theta)}{\partial \theta} \right\}^2 f(z; \lambda(\theta))\, dz = I(\lambda(\theta)) \left| \frac{\partial \lambda(\theta)}{\partial \theta} \right|^2,$$

Jeffreys' priors satisfy the transformation formula (2.27) whether they are proper or not.

Bayesian analysis is generally difficult in problems in which the parameter space is high dimensional because it becomes very difficult to specify meaningful prior distributions. In this situation, the use of Jeffreys' priors seems particularly attractive. Unfortunately, while Jeffreys' improper priors generally produce acceptable results in low dimensional problems, they can lead to unexpected results in high dimensional problems. These results include marginalization paradoxes (Stone and Dawid, 1972) in which the marginal posterior distribution of a function of the parameters $g(\theta \mid \mathbf{Z})$ depends on the data only through a simple statistic, say $t(Z)$, and $g(\theta \mid \mathbf{Z}) = g(\theta \mid t(\mathbf{Z}))$ differs from the posterior distribution obtained starting from the model $h(t(\mathbf{Z}) \mid \theta)$, and strong inconsistency (Stein, 1959; Stone, 1976) in which credibility sets have poor repeated sampling properties (see Section 3.8, Problems 2.4.5 and 3.8.2–3.8.3).

The work of Bernado (1979) is directed towards the construction of priors which avoid the paradoxes of Jeffreys' priors by ordering the parameters in order of interest, writing the prior density as the product of conditional priors and defining the prior as the limit of a sequence of proper distributions. Some difficulties with this approach are discussed in Berger and Bernado (1989). A number of other approaches to choosing priors to use in the absence of strong prior beliefs are available. See for example Jaynes (1968), Novick (1969), Box and Tiao (1973, p. 34 ff.), Zellner (1977), and Akaike (1978).

It is possible in practice to avoid the use of improper prior distributions altogether but they can provide formal connections between different approaches to inference so they do at the very least have an important theoretical role in discussions of statistical inference.

2.4.3 The Representation of Ignorance

The claim of Laplace, Jeffreys, and others noted in Sections 2.2.10–2.2.11 that uniform priors can be used to quantify a state of ignorance about the parameters and so may be called ignorance priors has been contested by Boole (1854, p. 370 ff.), Fisher (1922; 1935a, pp. 6–7; 1956a, Chapter 2) and others who argued that knowing nothing about two events is not the same as knowing that they are equally likely and that is not possible to quantify ignorance because uniformity on one scale implies knowledge on another. For example, if we assume that a parameter π is uniformly distributed so the density of π is

$$g(\pi) = I(0 \leq \pi \leq 1),$$

then the logit $\tau = \log\{\pi/(1-\pi)\}$ has a logistic distribution with density

$$h(\tau) = \frac{e^{\tau}}{(1+e^{\tau})^2}, \qquad -\infty \leq \tau \leq \infty.$$

This is a symmetric density with mode at the origin (corresponding to $\pi = 1/2$) which suggests that we believe π is more likely to be near 1/2 than 0 or 1. That is, the values of π are not equally likely on the logit scale and we seem to have prior knowledge about π.

More strongly, if we follow Jeffreys' rule (2.18) for choosing a prior, the scale on which the prior distribution is uniform depends on the model $\mathscr{F}$. (An example is given in Section 7.2.3.) This means that the same ignorance is quantified differently depending on the nature of the data (see for example Lindley, 1972, p. 71). This is clearly objectionable if one holds that ignorance is an absolute as opposed to a relative state.

The modern approach is to try to avoid the problem of representing ignorance by arguing that "ignorance" priors should be viewed as reference priors which provide a standard or reference analysis in situations in which we want the contribution of the likelihood to dominate that of the prior to the posterior distribution. That is, they are simply standard vague priors. See for example, Bernado and Smith (1994, p. 298).

The arguments are important because a number of authors (including LeCam, 1977; Kiefer, 1977a; Efron, 1986) have argued that the Bayesian approach (given a model and a prior, we multiply these together according to (2.2) and integrate to compute the posterior distribution) is too simple and inflexible to handle the range of problems to which we would like to apply statistical methods. Many of the arguments can be expressed in terms of the practical difficulties of obtaining prior distributions in difficult problems, including high dimensional problems and particularly problems in which we are either reluctant to use our prior beliefs or have none.

PROBLEMS

2.4.1. Consider the negative binomial model

$$\mathscr{F} = \left\{ f(z; \pi) = \binom{z-1}{r-1} \pi^r (1-\pi)^{z-r}, z = r, r+1, \ldots : 0 \leq \pi \leq 1 \right\}.$$

Show that the Jeffreys prior is $\pi^{-1}(1-\pi)^{-1/2}$. How does the use of this prior change the conclusions reached in Problem 2.2.1?

2.4.2. Suppose that we observe $\{Z_1, \ldots, Z_n\}$ on the Gaussian model (2.19). Suppose initially that μ is known. If we adopt the Jeffreys prior for σ, show that the posterior distribution of σ^2 is $\sum_{i=1}^n z_i^2 / \chi_n^2$. Now suppose that the mean μ is also unknown. Show that if we adopt the prior $g(\mu, \sigma) \propto 1/\sigma^2$ suggested by Jeffreys' rule, the marginal posterior distribution of σ^2 is $\sum_{i=1}^n (z_i - \bar{z})^2/\chi_n^2$ so no degree of freedom is lost when we do not know μ. What happens if we follow Jeffreys' recommendation and use $g(\mu, \sigma) \propto 1/\sigma$?

2.4.3. Suppose that Z has a gamma distribution with density

$$f(z; \kappa, \lambda) = \frac{1}{\Gamma(\kappa)} \lambda^\kappa z^{\kappa-1} \exp(-\lambda z), \qquad z > 0, \quad \kappa, \lambda > 0.$$

Show that the following relationships hold:

$$\begin{aligned} \mathrm{E} \log(\lambda Z) &= \psi(\kappa) \\ \mathrm{E}\{\log(\lambda Z)\}^2 &= \psi'(\kappa) + \psi(\kappa)^2 \\ \mathrm{E}\{\lambda Z \log(\lambda Z)\} &= \kappa\psi(\kappa + 1), \end{aligned}$$

where $\psi(\kappa) = \partial \log \Gamma(\kappa)/\partial\kappa$ is the digamma function. Hence use the fact that $\psi(\kappa + 1) = \psi(\kappa) + \kappa^{-1}$ to show that the prior obtained from Jeffreys' rule for the gamma model is

$$g(\kappa, \lambda) \propto \frac{\{\kappa\psi'(\kappa) - 1\}^{1/2}}{\lambda}, \qquad \kappa, \lambda > 0.$$

2.4.4. (Kahn, 1987). Let $Z \sim \text{binomial}(n, p)$. It is not a common problem, but there are situations in which we need to make inference about n rather than p. Suppose that we adopt the prior distribution with density

$$g(n, p) \propto \frac{p^{\alpha-1}(1-p)^{\beta-1}}{n^{\gamma}}, \qquad 0 \le p \le 1, \quad n = 1, 2, \ldots$$

where α, β, and γ are fixed. Find the marginal posterior density of $N \mid Z = z$. Use the fact that for any fixed r,

$$\lim_{m \to \infty} \frac{m^r \Gamma(m)}{\Gamma(m + r)} = 1$$

to show that

$$\lim_{n \to \infty} n^{\alpha+\gamma} g(n \mid Z = z) = \text{const.}$$

Interpret this result in terms of the tails of the posterior density of $N \mid Z = z$. Show that for the uniform improper prior g with $\alpha = \beta = 1$ and $\gamma = 0$, the posterior density of $N \mid Z = z$ is improper. For what values of α, β, and γ is the posterior density proper? Is it sensible to use improper posteriors to calculate credibility sets and posterior odds ratios?

2.4.5. (Stein, 1959) Suppose that we observe $\{Z_1, \ldots, Z_n\}$ on the model

$$\mathscr{F} = \left\{ f(\mathbf{z}; \mu_1, \ldots, \mu_n) = \prod_{i=1}^{n} \frac{1}{(2\pi)^{1/2}} \exp\left\{ -\frac{(z_i - \mu_i)^2}{2} \right\}, \ -\infty < z_i < \infty : \mu \in \mathbb{R} \right\}.$$

(i.e., $Z_1, \ldots, Z_n$ are independent Gaussian random variables with means $\mu_1, \ldots, \mu_n$ and common known variance 1.) Suppose that we are interested in obtaining a $100(1-\alpha)\%$ credibility set of the form $[\theta_\alpha, \infty)$ for $\theta = \sum_{i=1}^n \mu_i^2$. Show that if $\mu_1, \ldots, \mu_n$ are a priori independent and we assign the Jeffreys prior to each μ_i, we obtain the approximate $100(1-\alpha)\%$ credibility interval

$$[\hat{\theta} + n + \Phi^{-1}(\alpha)(2n + 4\hat{\theta})^{1/2}, \infty),$$

where $\hat{\theta} = \sum_{i=1}^n Z_i^2$. (Hint: Show that $\theta \mid \mathbf{Z} = \mathbf{z}$ has a noncentral chi-squared distribution and use the approximation suggested in (5bii) of the Appendix.)

2.5 BAYESIAN HYPOTHESIS TESTING

A natural way to compare hypotheses H_0 and H_1 expressed in terms of the parameters in the model is to compare the posterior probabilities of the hypotheses. We often do this by examining the *posterior odds ratio* $P(H_0 \mid \mathbf{z})/P(H_1 \mid \mathbf{z})$ which is the ratio of the posterior probability of H_0 to that of H_1. Large values of the posterior odds ratio provide evidence in favor of H_0 over H_1 while small values provide evidence in favor of H_1 over H_0.

We can compare the posterior odds ratio to the prior odds ratio to see how the data have changed our beliefs about H_0 and H_1. The amount by which the data (under our prior specification) has changed our beliefs is reflected in the *Bayes factor*

$$\frac{P(H_0 \mid \mathbf{z})/P(H_1 \mid \mathbf{z})}{P(H_0)/P(H_1)}$$

which is the ratio of the posterior odds ratio to the prior odds ratio. The Bayes factor generally depends on the prior so cannot be interpreted in terms of the data alone.

2.5.1 Testing a Sharp Null Hypothesis

Bayesian hypothesis testing is straightforward whenever the hypotheses H_0 and H_1 specify sets of values or both specify single values. However, difficulties arise when we compare a null hypothesis which specifies a single value for a parameter (a *sharp null hypothesis*) to an alternative which specifies a set of values for the parameter.

To bring out the issues in the simplest case, consider first the admittedly artificial problem of testing the hypothesis H_0: $\pi_{11} = \pi_0$ against H_1: $\pi_{11} \neq \pi_0$ under the binomial model (2.1). If we calculate the posterior odds ratio

$$\frac{P\{H_0 \mid Z_{11} = z_{11}\}}{P\{H_1 \mid Z_{11} = z_{11}\}},$$

we find that it equals 0 because the fact that the posterior distribution of π_{11} is absolutely continuous implies that $P\{\pi_{11} = \pi_0 \mid Z_{11} = z_{11}\} = 0$ for any $0 \le \pi_0 \le 1$. This result seems to provide strong evidence in favour of H_1. However, the posterior distribution (2.4) is absolutely continuous whenever the prior distribution is absolutely continuous so this result reflects the nature of the null hypothesis and the choice of prior rather than anything fundamental about the data.

Two approaches have been proposed to circumvent this difficulty.

1. *Change the hypothesis.* Change the problem to one of testing H_0: $|\pi_{11} - \pi_0| \le \delta$ against H_1: $|\pi_{11} - \pi_0| > \delta$ for some small $\delta > 0$ so that we have the regular problem of comparing two set hypothesis. Then, under a uniform prior for π_{11}, we have

$$\begin{aligned} P\{H_0 \mid Z_{11} = z_{11}\} &= P\{-\delta \le \pi_{11} - \pi_0 \le \delta \mid Z_{11} = z_{11}\} \\ &= B(\pi_0 + \delta; 27, 150) - B(\pi_0 - \delta; 27, 150) \end{aligned}$$

so that the posterior odds ratio is

$$\frac{B(\pi_0 + \delta; 27, 150) - B(\pi_0 - \delta; 27, 150)}{1 - B(\pi_0 + \delta; 27, 150) + B(\pi_0 - \delta; 27, 150)}. \tag{2.28}$$

The prior odds ratio is $2\delta/(1 - 2\delta)$ so the Bayes factor is just (2.28) divided by $2\delta/(1 - 2\delta)$.

The choice of δ is problematical but in practical problems can often be chosen to reflect the smallest detectable or meaningful difference from 0. Even in analysing experiments to demonstrate psychokinesis in which a sharp null hypothesis is meaningful because psychokinesis either exists or does not and a "little" psychokinesis is not a sensible null hypothesis to entertain, the experiment is likely to be affected by small biases from imperfections in technique, equipment, etc. and in this case, δ reflects our beliefs in an upper bound for this bias. (An experiment to detect psychokinesis is discussed in Sections 3.4.5 and 3.5.3.)

2. *Change the prior.* Jeffreys (1939/1961) suggested that we assign a prior probability p_0 to H_0 and a prior distribution $(1 - p_0)h(\pi_{11})$ to $\pi_{11} \ne \pi_0$. This prior assigns π_{11} the value π_0 with probability p_0 and asserts that π_{11} has density h given that $\pi_{11} \ne \pi_0$ with probability $1 - p_0$. The posterior distribution of π_{11} is then obtained from (2.2) as

$$k(\pi_{11} \mid Z_{11} = z_{11}) = \frac{p_0 f(z_{11}; \pi_0) I(\pi_{11} = \pi_0) + (1 - p_0) f(z_{11}; \pi_{11}) h(\pi_{11}) I(\pi_{11} \ne \pi_0)}{c(z_{11})},$$

where

$$c(z_{11}) = p_0 f(z_{11}; \pi_0) I(\pi_{11} = \pi_0) + (1 - p_0) \times \int_0^1 f(z_{11}; \pi_{11}) h(\pi_{11}) I(\pi_{11} \neq \pi_0)\, d\pi_{11},$$

so the posterior odds ratio is

$$\frac{p_0}{1 - p_0} \frac{f(z_{11}; \pi_0)}{\int_{t \neq \pi_0} f(z_{11}; t) h(t)\, dt}. \tag{2.29}$$

Notice that the normalizing constant $c(z_{11})$ cancels in the posterior odds ratio. The prior odds ratio is $p_0/(1 - p_0)$ so the Bayes factor is just (2.29) divided by $p_0/(1 - p_0)$. If we believe that the two hypotheses are a priori equally likely, we can take $p_0 = 1/2$ so that $p_0/(1 - p_0) = 1$ and the posterior odds ratio (2.29) is then just the Bayes factor. With a uniform prior under H_1, the Bayes factor is just the posterior density evaluated at π_0.

2.5.2 Comparing Independent Sets of Eyes

For a more realistic (and hence more complicated) application, consider comparing the hypothesis $H_0: \eta = 0$ to the general alternative hypothesis $H_1: \eta \neq 0$ in the context of Section 2.1.6. The simplest way to proceed is to take the prior

$$g(\eta, \theta) = I(0 \leq \eta + \theta \leq 1, 0 \leq \theta \leq 1)$$

from the formulation represented by (2.9) and (2.10) and modify it as suggested by Jeffreys to

$$\begin{aligned} h(\eta, \theta) &= p_0 g(0, \theta) I(\eta = 0) + (1 - p_0) g(\eta, \theta) I(\eta \neq 0) \\ &= p_0 I(\eta = 0, 0 \leq \theta \leq 1) + (1 - p_0) I(\eta \neq 0, 0 \leq \eta + \theta \leq 1, 0 \leq \theta \leq 1). \end{aligned}$$

Under H_0 which we believe holds with probability p_0, θ is uniformly distributed on (0, 1) while under H_1 which we believe holds with probability $1 - p_0$, η and θ have density $g(\eta, \theta)$, $\eta \neq 0$. (In terms of the original (π_{11}, π_{21}) parameterization, under H_0 which we believe holds with probability $1 - p_0$, $\pi_{11} = \pi_{21}$ is uniformly distributed on (0, 1) while under H_1 which we believe holds with probability p_0, π_{11} and π_{12} are independent and uniformly distributed on (0, 1) given that $\pi_{11} \neq \pi_{12}$.)

The joint posterior distribution of (η, θ) obtained from (2.2) is proportional to

$$\begin{aligned} k(\eta, \theta \mid Z_{11} = z_{11}, Z_{21} = z_{21}) &= p_0 f(z_{11}, z_{21}; 0, \theta) I(\eta = 0, 0 \leq \theta \leq 1) \\ &\quad + (1 - p_0) f(z_{11}, z_{21}; \eta, \theta) I(\eta \neq 0, 0 \leq \eta + \theta \leq 1, 0 \leq \theta \leq 1), \end{aligned}$$

so the marginal posterior distribution of η is

$$k(\eta \mid Z_{11} = z_{11}, Z_{21} = z_{21})$$

$$= \frac{\int_0^1 k(\eta, \theta \mid Z_{11} = z_{11}, Z_{21} = z_{21})\, \mathrm{d}\theta}{c(z_{11}, z_{21})}$$

$$= \frac{p_0 h(0 \mid z_{11}, z_{21}) I(\eta = 0) + (1 - p_0) h(\eta \mid z_{11}, z_{21}) I(\eta \neq 0)}{c(z_{11}, z_{21})},$$

where h is given by (2.8) and

$$c(z_{11}, z_{21}) = p_0 h(0 \mid z_{11}, z_{21}) + (1 - p_0) \int_{-1}^{1} h(\eta \mid z_{11}, z_{21}) I(\eta \neq 0)\, d\eta.$$

The posterior odds ratio is therefore

$$\frac{p_0}{1 - p_0} \frac{h(0 \mid z_{11}, z_{21})}{\int_{-1}^{1} h(\eta \mid z_{11}, z_{21}) I(\eta \neq 0)\, d\eta}$$

$$= \frac{p_0}{1 - p_0} \frac{\Gamma(z_{11}+z_{21}+1)\Gamma(r_1+r_2-z_{11}-z_{21}+1)\Gamma(r_1+2)\Gamma(r_2+2)}{\Gamma(r_1+r+2)\Gamma(z_{11}+1)\Gamma(r_1-z_{11}+1)\Gamma(z_{21}+1)\Gamma(r_2-z_{21}+1)}. \tag{2.30}$$

Note that the prior odds ratio is $p_0/(1 - p_0)$ and the Bayes factor is just the posterior density evaluated at $\eta = 0$.

For the two groups of control eyes, $r_1 = 175$, $z_{11} = 26$, $r_2 = 179$, and $z_{21} = 31$ so with $p_0 = 1/2$, the posterior odds ratio (2.30) is 8.43 which provides strong evidence in favour of the null hypothesis.

2.5.3 Difficulties With Improper Priors

The first difficulty in using an improper prior in a Bayesian test is that, as noted in (2) in Section 2.4.1, we usually cannot write down the prior odds ratio. Nonetheless, if we are prepared to overlook this difficulty, the improper prior we use results in a proper posterior distribution and we are comparing two set hypotheses, we can compute a meaningful posterior odds ratio. Unfortunately, the situation is much more complicated when we test a sharp hypothesis.

Suppose we observe $\mathbf{Z}$ on the Gaussian model (2.19) but for simplicity we assume that the variance $\sigma^2 = 1$ is known and we want to test the sharp hypothesis H_0: $\mu = 0$ against the alternative H_1: $\mu \neq 0$. If, we assign prior probability p_0 to H_0 and suppose μ is uniformly distributed over $[-c, c]$ when

H_0 does not hold so that

$$g(\mu) = p_0 I(\mu = 0) + (1 - p_0)\frac{1}{2c} I(|\mu| \leq c, \mu \neq 0),$$

the posterior density is

$$g(\mu \mid z) = \frac{p_0 \exp\left(-\frac{n\bar{z}^2}{2}\right) I(\mu = 0) + (1 - p_0) \exp\left\{-\frac{n(\bar{z} - \mu)^2}{2}\right\} \frac{1}{2c} I(|\mu| \leq c, \mu \neq 0)}{k(z)},$$

where

$$\begin{aligned} k(z) &= p_0 \exp\left(-\frac{n\bar{z}^2}{2}\right) + (1 - p_0)\frac{1}{2c}\int_{-c}^{c} \exp\left\{-\frac{n(\bar{z} - \mu)^2}{2}\right\} d\mu \\ &= p_0 \exp\left(-\frac{n\bar{z}^2}{2}\right) + (1 - p_0)\frac{1}{2c}\left(\frac{2\pi}{n}\right)^{1/2} [\Phi\{n^{1/2}(c - \bar{z})\} - \Phi\{-n^{1/2}(c + \bar{z}\}], \end{aligned}$$

and $\Phi(x)$ is the distribution function of the standard Gaussian distribution. The posterior odds ratio is by (2.29)

$$\begin{aligned} P(H_0 \mid z)/P(H_1 \mid z) &= \frac{p_0}{1 - p_0} \frac{2c \exp(-n\bar{z}^2/2)}{\int_{-c}^{c} \exp\{-n(\bar{z} - t)^2/2\}\, dt} \\ &= \frac{p_0}{1 - p_0} \frac{2c(n/2\pi)^{1/2} \exp(-n\bar{z}^2/2)}{\Phi\{n^{1/2}(c - \bar{z})\} - \Phi\{-n^{1/2}(c + \bar{z})\}}. \end{aligned} \tag{2.31}$$

Bartlett (1957) pointed out that if we let $c \to \infty$, the denominator in (2.31) tends to $1 - p_0$ and the numerator tends to infinity so the posterior odds ratio tends to infinity. That is, as c increases (making our beliefs about μ under the alternative hypothesis vaguer), the evidence in favour of H_0 increases. This is known as *Bartlett's paradox*. (The Jeffreys/Lindley paradox which prompted Bartlett's discovery is presented in Section 3.5.3.)

If we use Jeffreys' improper prior for μ (so the prior density of μ equals a constant k_1 say) under the alternative hypothesis, the prior density of μ is

$$g(\mu) = p_0 I(\mu = 0) + (1 - p_0)k_1 I(\mu \neq 0),$$

and the posterior density is

$$g(\mu \mid z) = \frac{p_0 \exp\left(-\frac{n\bar{z}^2}{2}\right) I(\mu = 0) + (1 - p_0) \exp\left\{-\frac{n(\bar{z} - \mu)^2}{2}\right\} k_1 I(\mu \neq 0)}{k(z)},$$

where

$$k(z) = p_0 \exp\left(-\frac{n\bar{z}^2}{2}\right) + (1 - p_0)k_1 \int_{-\infty}^{\infty} \exp\left\{-\frac{n(\bar{z} - \mu)^2}{2}\right\} d\mu$$
$$= p_0 \exp\left(-\frac{n\bar{z}^2}{2}\right) + (1 - p_0)k_1\left(\frac{2\pi}{n}\right)^{1/2}.$$

The posterior odds ratio is by (2.29)

$$\frac{p_0}{1 - p_0}\left(\frac{n}{2\pi}\right)^{1/2} \exp\left(-\frac{n\bar{z}^2}{2}\right)\frac{1}{k_1} = \frac{p_0}{1 - p_0}\frac{n^{1/2}\phi(n^{1/2}\bar{z})}{k_1}, \tag{2.32}$$

where ϕ is the standard Gaussian density. The evidence concerning H_0 depends on the value of $\bar{z}$ and the arbitrary constant k_1 (which plays roughly the role of $1/2c$ in (2.31)). It seems natural to set $k_1 = 1$ but the arbitrariness of this choice needs to be acknowledged.

The arbitrariness of the normalizing constant k_1 in any improper prior makes the use of improper priors when testing sharp hypotheses problematic. We can avoid the difficulty by refusing to test sharp null hypotheses because when we test set hypotheses the constant k_1 cancels out in constructing the posterior density. If we are committed to testing a sharp hypothesis, we then have to use an arbitrary constant, a proper prior or else find a different way of defining a "Bayes factor" so that the constant k_1 is eliminated (see for example Aitken, 1991 and O'Hagan, 1995).

Jeffreys (1939/1961, p. 274) recommended using a vague but proper prior. For the present Gaussian problem, he suggested using the *Cauchy prior* for the value of μ under H_1, so the prior is

$$g(\mu) = p_0 I(\mu = 0) + (1 - p_0)\frac{1}{\pi(1 + \mu^2)} I(\mu \neq 0).$$

The posterior density is

$$g(\mu \mid z) = \frac{p_0 \exp\left(-\frac{n\bar{z}^2}{2}\right) I(\mu = 0) + (1 - p_0) \exp\left\{-\frac{n(\bar{z} - \mu)^2}{2}\right\}\frac{1}{\pi(1 + \mu^2)} I(\mu \neq 0)}{k(z)},$$

where

$$k(z) = p_0 \exp\left(-\frac{n\bar{z}^2}{2}\right) + (1 - p_0) \int_{-\infty}^{\infty} \exp\left\{-\frac{n(\bar{z} - \mu)^2}{2}\right\}\frac{1}{\pi(1 + \mu^2)} d\mu,$$

and the posterior odds ratio is by (2.29)

$$\frac{\dfrac{p_0}{1-p_0}\exp\left(-\dfrac{n\bar{z}^2}{2}\right)}{\displaystyle\int_{-\infty}^{\infty}\exp\left\{-\frac{n(\bar{z}-\mu)^2}{2}\right\}\frac{1}{\pi(1+\mu^2)}\,d\mu}.$$

For n large, Jeffreys used Laplace's method (see Section 4.6.2 and Problem 4.6.5) to show that the Bayes factor is approximately

$$\left(\frac{\pi n}{2}\right)^{1/2}(1+\bar{z}^2)\exp\left(-\frac{n\bar{z}^2}{2}\right). \tag{2.33}$$

Further discussion, particularly concerning the relationship of Bayesian tests to significance tests (Section 3.2) and the effect of sample size on posterior odds ratios (the Jeffreys/Lindley paradox) is given in Section 3.5.3.

2.5.4 Testing for a Change in the Volume of Urine Voided

For the change in the volume of urine voided due to the ingestion of caffeine, the hypothesis H_0: $\mu = 0$ corresponds to the assertion that the ingestion of caffeine has no systematic effect on the volume of urine voided. To test this assertion under the Gaussian model (2.19), we compare the hypothesis H_0: $\mu = 0$ to the general alternative hypothesis H: $\mu \neq 0$.

If we assign prior probability $p_0 g(\sigma)$ to H_0 and a prior distribution $(1-p_0)h(\mu \mid \sigma)g(\sigma)$, where $\int_{-\infty}^{\infty} h(\mu \mid \sigma)\,d\mu < \infty$, to H_1, the posterior distribution of (μ, σ) is proportional to

$$k(\mu, \sigma \mid \mathbf{Z}=\mathbf{z}) = p_0 f(\mathbf{z}; 0, \sigma)g(\sigma)I(\mu=0) + (1-p_0)f(\mathbf{z}; \mu, \sigma)h(\mu \mid \sigma)g(\sigma)I(\mu\neq 0).$$

The marginal posterior for μ is then

$$k(\mu \mid \mathbf{Z}=\mathbf{z}) = \frac{\displaystyle\int_0^{\infty} k(\mu, \sigma \mid \mathbf{Z}=z)/\mathrm{d}\sigma}{\mathrm{k}(\mathbf{z})},$$

where

$$k(\mathbf{z}) = p_0\int_0^{\infty} f(\mathbf{z}; 0, \sigma)g(\sigma)\,d\sigma + (1-p_0)\int_{-\infty}^{\infty}\int_0^{\infty} f(\mathbf{z}; \mu, \sigma)h(\mu \mid \sigma)g(\sigma)\,d\sigma I(\mu\neq 0)\,d\mu.$$

It follows that the posterior odds ratio is

$$\frac{p_0}{1-p_0} \frac{\int_0^\infty f(\mathbf{z}; 0, \sigma) g(\sigma)\, d\sigma}{\int_{-\infty}^\infty \int_0^\infty f(\mathbf{z}; \mu, \sigma) h(\mu \mid \sigma) g(\sigma)\, d\sigma I(\mu \neq 0)\, d\mu}. \tag{2.34}$$

Jeffreys (1939/1961, p. 268) suggested using the priors

$$g(\sigma) \propto \sigma^{-1}$$

$$h(\mu \mid \sigma) = \frac{1}{\sigma\pi(1 + \mu^2/\sigma^2)}.$$

In this case, the posterior odds ratio (2.34) involves the ratio of

$$\int_0^\infty \frac{1}{\sigma^{n+1}} \exp\left\{-\frac{\nu s^2}{2\sigma^2} - \frac{n\bar{z}^2}{2\sigma^2}\right\} d\sigma \propto \frac{1}{\left(1 + \dfrac{n\bar{z}^2}{\nu s^2}\right)^{n/2}} \frac{1}{\left(\dfrac{\nu s^2}{n}\right)^{n/2}}$$

as in (2.25) and

$$\int_{-\infty}^\infty \int_0^\infty \frac{1}{\sigma^n} \exp\left\{-\frac{\nu s^2}{2\sigma^2} - \frac{n(\bar{z} - \mu)^2}{2\sigma^2}\right\} \frac{1}{\sigma^2\pi(1 + \mu^2/\sigma^2)}\, d\sigma I(\mu \neq 0)\, d\mu.$$

This second integral cannot be evaluated explicitly but Jeffreys used Laplace's method (see Section 4.6.2 and Problem 4.6.5) to show that for n large the denominator integral is proportional to

$$\left(\frac{2}{\pi n}\right)^{1/2} \frac{1}{\left(1 + \dfrac{n\bar{z}^2}{\nu s^2}\right)} \frac{1}{\left(\dfrac{\nu s^2}{n}\right)^{n/2}}$$

and hence that the Bayes factor is approximately

$$\left(\frac{\pi n}{2}\right)^{1/2} \frac{1}{\left(1 + \dfrac{n\bar{z}^2}{\nu s^2}\right)^{n/2-1}}.$$

For the change in the volume of urine voided after the ingestion of caffeine, the Bayes factor in is 0.069 which provides strong evidence against the null hypothesis H_0: $\mu = 0$. That is, there is clearly an effect on the volume of urine voided.

PROBLEMS

2.5.1. In the context of Problem 2.2.7, obtain the posterior odds ratio for testing H_0: $\theta = 1$ against the general alternative. Compare the conclusions to that obtained from a one-sided test of H_0: $\theta \geq 1$ against H_1: $\theta < 1$.

2.5.2. In the context of Problem 2.2.8, test the null hypothesis H_0: $\xi = 0$ against H_1: $\xi > 0$.

2.5.3. Suppose that we observe **Z** on the Gaussian model (2.19) but for simplicity we assume that the variance $\sigma^2 = 1$ is known. Under the Jeffreys prior $g(\mu) = k_1$, construct the posterior odds ratio for testing H_0: $\mu \leq \mu_0$ against H_0: $\mu > \mu_0$. Explain why k_1 does not appear in the posterior odds ratio.

2.5.4. Suppose that we observe **Z** on the Gaussian model (2.19) and want to test H_0: $\mu = 0$ against H_1: $\mu \neq 0$. Suppose that under H_0 we adopt the prior $p_0 k_1/\sigma$ and that under H_1 we adopt the prior $(1 - p_0)(k_1/\sigma)(1/2c)I(-c \leq \mu \leq c)$. Obtain the posterior odds ratio and discuss the effect on it of different choices of k_1 and c.

2.5.5. Suppose that in the context of Problem 2.3.6 we want to test H_0: $\beta = \beta_0$ against the general alternative H_1: $\beta \neq \beta_0$. Suppose that under H_0 we adopt the prior $p_0 k_1$ and that under H_1 we adopt the prior $(1 - p_0)k_1 k_2$. Obtain the posterior odds ratio and discuss the effect on it of different choices of k_1 and k_2 and of $\{x_i\}$.

2.6 FIDUCIAL THEORY

Fisher's (1922; 1935a, pp. 6–6; 1956a, Chapter 2) objections to the Bayesian paradigm led him to try to develop alternative approaches to inference. While he objected strongly to the fact that the Bayesian approach requires the specification of a prior distribution, he recognized the value of the posterior distribution in generating sets of plausible parameter values which admit a simple interpretation. He made a considerable effort to overcome the need to specify the prior in the Bayesian approach, particularly for situations in which we have no prior information about the parameters, by developing an approach called *fiducial inference* (Fisher, 1930) which produces a distribution (called the *fiducial distribution*) for the parameters without the need to specify a prior distribution. Savage (1962b) colourfully described this approach as "an attempt to make the Bayesian omlette without breaking the Bayesian eggs."

2.6.1 The Fiducial Argument

Before we collect any data, the parameters of the model are regarded as fixed unknown constants. For the Gaussian model (2.19), the likelihood function

$$f(\mathbf{z}; \mu, \sigma) \propto \frac{1}{\sigma^n} \exp\left\{-\frac{vs^2}{2\sigma^2} - \frac{n(\bar{z} - \mu)^2}{2\sigma^2}\right\}, \qquad (\mu, \sigma) \in \mathbb{R} \times [0, \infty),$$

depends on the data only through $(\bar{z}, s^2)$, a statistic of the same dimension as the parameter space. Just as we interpret the data $\mathbf{z}$ as a realization of the random variable $\mathbf{Z}$, we can interpret $(\bar{z}, s^2)$ as a realization of $(\bar{Z}, S^2)$ which has a distribution with density function

$$s(x, y; \mu, \sigma) \propto y^{(n-3)/2} \frac{1}{\sigma^n} \exp\left\{-\frac{vy}{2\sigma^2} - \frac{n(x - \mu)^2}{2\sigma^2}\right\}, \qquad (x, y) \in \mathbb{R} \times [0, \infty)$$

(see 9 in the Appendix). In the absence of any prior information about (μ, σ), Fisher asserted that we have the fiducial distribution

$$f_f(\mu, \sigma) \propto \frac{1}{\sigma^{n+1}} \exp\left\{-\frac{vs^2}{2\sigma^2} - \frac{n(\bar{z} - \mu)^2}{2\sigma^2}\right\}, \qquad (\mu, \sigma) \in \mathbb{R} \times [0, \infty),$$

for (μ, σ). Fisher argued that the fiducial distribution should be treated as an ordinary distribution which obeys the laws of probability theory. The marginal fiducial distribution of μ is therefore

$$f_f(\mu) \propto \int \frac{1}{\sigma^{n+1}} \exp\left\{-\frac{vs^2}{2\sigma^2} - \frac{n(\bar{z} - \mu)^2}{2\sigma^2}\right\} d\sigma$$

or $n^{1/2}(\mu - \bar{z})/s \mid \mathbf{Z} = \mathbf{z} \sim t_{n-1}$.

The fiducial distribution is supposed to describe our beliefs about the unknown parameter (μ, σ). Its main use is to generate *fiducial sets* which are derived from the contours of the fiducial distribution in the same way as Bayesian credibility sets are derived from the contours of the posterior distribution. Since the fiducial Gaussian distribution for this problem is identical to the posterior distribution obtained from the Jeffreys prior, the $100(1 - \alpha)\%$ fiducial interval for μ is identical to the Bayesian credibility interval with the Jeffreys prior (2.26).

Three important questions arise. First, why do we base the analysis on $(\bar{Z}, S^2)$? Second, how do we interpret the logical change of status of (μ, σ) from fixed unknown constants to random variables with a fiducial distribution? And third, what is the basis for multiplying the likelihood by σ^{-1} to obtain the fiducial density?

2.6.2 The Role of Sufficiency

One of Fisher's important contributions to statistics was to highlight the role of the likelihood function. He argued (in Fisher, 1922) that the likelihood function contains all the information in the data about the model. He formalized this by arguing that the likelihood is *sufficient* in the sense that the conditional distribution of the data $\mathbf{Z}$ given the likelihood is the same for all the distributions in the model $\mathscr{F}$ and hence the data contains nothing further about the model. Since the Gaussian likelihood is a function of the data only through $(\bar{z}, s^2)$, it follows that all the information about the model is contained in $(\bar{z}, s^2)$. We say that $(\bar{Z}, S^2)$ is sufficient for the model and refer to it as a *sufficient statistic*. From this perspective, it is natural to base inference on $(\bar{Z}, S^2)$. (The concept of sufficiency is discussed in more detail in Section 7.3.)

2.6.3 The Status of the Parameters

The fiducial approach to inference involves a change in the status of the parameters which is difficult to understand, and to date no widely accepted formal justification has been given. Consequently, the interpretation of the fiducial argument is problematical and no general theory is available. Fisher regarded the fiducial argument as obviously reasonable and so never explained it in much detail. The best explanation he gave involved the following argument quoted by Basu (1973):

- Suppose that under the model we have $\mathrm{P}\{Y > \psi\} = 0.5$.
- Suppose that we draw a single observation from the model and that the observation is $Y = 5$.
- In the absence of any other knowledge about ψ, it is reasonable to assert that $\mathrm{P}_{\mathrm{f}}\{\psi < 5\} = 0.5$.

That is, if Y is equally likely to be greater than or less than a location parameter ψ whatever the value of ψ may be, then ψ is equally likely to be greater than or less than any particular realized value of Y. Basu (1973) described this argument as an example of a fallacy of five terms, i.e., like the logically flawed argument that team A beat team B and team B beat team C so team A must have beaten team C.

2.6.4 Multiplication of the Likelihood by σ^{-1}

The reason for multiplying the likelihood by σ^{-1} is unclear. It is apparently based on combining two separate single-parameter arguments.

1. For the Gaussian model (2.19) with σ known, the sample mean $\bar{Z}$ is

sufficient for μ and has distribution function

$$F(u; \mu) = \mathrm{P}(\bar{Z} \leq u; \mu) = \int_{-\infty}^{u} \frac{n^{1/2}}{\sigma} \phi\left(\frac{n^{1/2}(x-\mu)}{\sigma}\right) dx$$
$$= \Phi\left(\frac{n^{1/2}(u-\mu)}{\sigma}\right), \qquad u \in \mathbb{R},$$

where Φ and ϕ are the standard Gaussian distribution and density functions respectively. Following the argument of Section 2.6.3, after we observe $\bar{z}$, the fiducial distribution for μ has distribution function $1 - \Phi(n^{1/2}(\bar{z} - \mu)/\sigma)$ and hence density function

$$f_{\mathrm{f}}(\mu) = -\frac{\partial F(\bar{z}; \mu)}{\partial \mu} = \frac{n^{1/2}}{\sigma} \phi\left(\frac{n^{1/2}(\bar{z}-\mu)}{\sigma}\right).$$

That is, the fiducial argument transforms the density $(n^{1/2}\sigma)\phi(n^{1/2}(\bar{z} - \mu)/\sigma)$ of $\bar{Z}$ to the fiducial density $(n^{1/2}/\sigma)\phi(n^{1/2}(\bar{z} - \mu)/\sigma)$ of μ by switching the arguments.

2. For the Gaussian model (2.19) with μ known, $S^2(\mu) = \sum_{i=1}^{n} (Z_i - \mu)^2$ is sufficient for the model and has distribution function

$$F(v, \sigma) = \mathrm{P}(S^2(\mu) \leq v; \sigma) = \int_{-\infty}^{v} \frac{1}{\sigma} k_n\left(\frac{x}{\sigma}\right) dx = K_n\left(\frac{v}{\sigma}\right), \qquad v \geq 0,$$

where K_n is the distribution function of the χ_n^2 distribution, and after we observe $s(\mu)^2$, the fiducial distribution for σ has density function

$$f_{\mathrm{f}}(\sigma) = -\frac{\partial F(s(\mu)^2; \sigma)}{\partial \sigma} = \frac{s^2(\mu)}{\sigma^2} k_n\left(\frac{s^2(\mu)}{\sigma}\right).$$

That is, the fiducial argument transforms the density $\sigma^{-1}k_n(s^2(\mu)/\sigma)$ of $S^2(\mu)$ to the fiducial density $(s^2(\mu)/\sigma^2)k_n(s^2(\mu)/\sigma)$ of σ by switching the arguments and multiplying by $s^2(\mu)/\sigma$.

Fisher's result for the two parameter case is obtained by putting these two separate results together; that is, switching arguments and multiplying by σ^{-1}. Hence the fiducial distribution is the same as the posterior distribution with Jeffreys' prior in this case. (That this is not generally true was shown by Lindley, 1958). However, the distribution function of $(\bar{Z}, S^2)$ is

$$F(u, v; \mu, \sigma) = \mathrm{P}(\bar{Z} \leq u, S^2 \leq v; \mu, \sigma) = \Phi\left(\frac{n^{1/2}(u-\mu)}{\sigma}\right) K_{n-1}\left(\frac{v}{\sigma}\right)$$

and differentiating the right-hand side with respect to μ and σ does not lead to the same result because σ appears in both factors on this side. There seems no obvious basis for treating the two parameters separately.

2.6.5 Further Limitations of Fiducial Inference

Even if we are prepared to ignore the conceptual difficulties underlying fiducial inference, there are still a number of operational difficulties. First, the method does not apply to discrete data because of the difficulty of interchanging the role of discrete data and continuous parameters. Second, Fisher confined the use of the fiducial argument to models for which there is a sufficient statistic of the same dimension as the parameter space. This can be overcome by working directly from the likelihood. In this case, the fiducial density is just a normalized likelihood function (after multiplication by σ^{-1} if σ is a scale parameter) but of course it has a very different interpretation. Finally, Fisher gave no general explanation of how to handle the multiparameter case. Tukey (1957a) and Brillinger (1962) showed that fiducial distributions for multiparameter problems are not necessarily unique and it is often unclear how to proceed. Nonetheless, the appeal of the fiducial approach is that, if we could interpret the fiducial distribution, it would yield intervals which have a very direct interpretation in terms of that distribution. For a positive modern viewpoint, see Wilkinson (1977). Some further examples of fiducial intervals are given in Section 3.8.

2.6.6 The Impact of Fiducial Inference

Whatever the present view of fiducial inference, there is no doubt that it had a major impact on statistics through the work it stimulated. This includes Neyman's (1934, 1937) work on confidence intervals (see Section 3.6) which in turn motivated Pitman's (1938) work on conditional inference (see Section 3.9). Ironically perhaps, conditional inference provides a partial justification for fiducial inference. A related approach called *structural inference* (Fraser, 1968) arose from an attempt to make sense of fiducial theory. A structural model is a model of the form $X = \theta E$ where E has a known distribution and $\theta \in \mathcal{G}$ for some transformation group $\mathcal{G}$. Inference is based on the fact that $\theta^{-1}X$ is a *pivotal quantity* (an invertible function of θ which has a known distribution, see Section 3.6.1) with the same distribution as E which is called the *structural distribution*. The importance of pivotal quantities was further emphasized by Barnard (1973) who introduced a theory of inference based on pivotal quantities.

2.7 LIKELIHOOD THEORY

Fisher's (1922) argument that the likelihood function contains all the information in the data about the model (Section 2.6.2) suggests that we can base inference on the likelihood alone. This idea was adopted by Barnard (1949),

Barnard et al. (1962), Birnbaum (1962), Edwards (1972) and others as the basis for likelihood theory.

2.7.1 Relative "Plausibility"

If we follow Fisher (1922) and interpret the likelihood function as expressing the relative "plausibility" of different parameters in the absence of any other information about them, then, for the effect of caffeine on the volume of urine voided (Sections 1.1.3 and 1.2.3), the Gaussian likelihood (2.20)

$$f(\mathbf{z}; \mu, \sigma) = \frac{1}{(2\pi\sigma^2)^{n/2}} \exp\left\{-\frac{vs^2}{2\sigma^2} - \frac{n(\bar{z}-\mu)^2}{2\sigma^2}\right\}$$

$$= \frac{1}{(2\pi\sigma^2)^9} \exp\left\{-\frac{384478.7}{\sigma^2} - \frac{9(170.72-\mu)^2}{\sigma^2}\right\}, \qquad (\mu, \sigma) \in \mathbb{R} \times [0, \infty)$$

expresses the *relative "plausibility"* of values of μ and σ given the data.

The relative "plausibility" concept suggests that the shape of the likelihood is important but the scaling constant is not so the obvious way to compare the likelihood of pairs of parameters is by examining the ratio of their likelihoods. Even though the scaling constant is unimportant, in plotting likelihoods it is useful to scale them so that, for example, the maximum value equals 1. Differentiating the log of the likelihood (2.20) with respect to μ and σ^2 and equating the result to 0, we find that the likelihood has an extremum at $(\hat{\mu}, \hat{\sigma})$ satisfying

$$\frac{n(\bar{z}-\hat{\mu})}{\hat{\sigma}^2} = 0$$

$$-\frac{n}{2\hat{\sigma}^2} + \frac{vs^2}{2\hat{\sigma}^4} + \frac{n(\bar{z}-\hat{\mu})^2}{2\hat{\sigma}^4} = 0,$$

namely, $(\hat{\mu}, \hat{\sigma}) = (\bar{z} = 170.72, \{v/n\}^{1/2}s = 200.86)$. The matrix of second partial derivatives of the log-likelihood (which is called the *Hessian*) evaluated at $(\hat{\mu}, \hat{\sigma})$, is

$$\begin{pmatrix} -n/\hat{\sigma}^2 & -2n(\bar{z}-\hat{\mu})/2\hat{\sigma}^4 \\ -2n(\bar{z}-\hat{\mu})/2\hat{\sigma}^4 & n/2\hat{\sigma}^4 - vs^2/2\hat{\sigma}^6 - n(\bar{z}-\hat{\mu})^2/2\hat{\sigma}^6 \end{pmatrix} = \begin{pmatrix} -0.0896 & 0 \\ 0 & -6.7e-10 \end{pmatrix}$$

which is negative definite so $(\hat{\mu}, \hat{\sigma})$ maximizes the likelihood.

Incidentally, if we reparameterize the likelihood in terms of $\lambda = \lambda(\mu, \sigma)$ and $\tau = \tau(\mu, \sigma)$, then $\hat{\lambda} = \lambda(\hat{\mu}, \hat{\sigma})$ and $\hat{\tau} = \tau(\hat{\mu}, \hat{\sigma})$ maximize the likelihood of (λ, τ). This means that we can pass readily between parameterizations and we can maximize the likelihood in the most convenient one.

We can scale the likelihood (2.20) by its maximum value $f(\mathbf{z}; \hat{\mu}, \hat{\sigma}) = (2\pi)^{-n/2}\{vs^2/n\}^{-n}\exp(-n/2)$ to obtain the *likelihood ratio*

$$\frac{f(\mathbf{z}; \mu, \sigma)}{f(\mathbf{z}; \hat{\mu}, \hat{\sigma})} = \frac{(206.6878)^{18}}{\sigma^{18}} \exp\left\{-384478.7\sigma^2 - \frac{9(170.72-\mu)^2}{\sigma^2}\right\} + 9,$$

$$(\mu, \sigma) \in \mathbb{R} \times [0, \infty). \quad (2.35)$$

The shape of this function is essentially the same as that of the joint posterior density plotted in Figure 2.7.

Since the Gaussian likelihood is a function of the data only through $\bar{z}$ and s^2, the information in the data contained in the likelihood is mathematically equivalent to the information contained in $\bar{z}$ and s^2. However, in general, the likelihood does not provide such a spectacular simplification of the data and, in likelihood theory, use is made of the form of the likelihood function so we do not simply extract $\bar{z}$ and s^2 from the likelihood function.

2.7.2 Profile Likelihood

In the caffeine–effect problem, we are interested in μ and σ is a nuisance parameter which we would like to eliminate. One possibility is to maximize the likelihood (2.20) over σ for fixed μ and then replace σ by this maximizing function to obtain the *concentrated* or *profile likelihood* which is a function of μ alone. We find that

$$\tilde{\sigma}(\mu)^2 = n^{-1}\sum_{i=1}^{n}(z_i - \mu)^2 = \frac{vs^2}{n} + (\bar{z} - \mu)^2$$

so the profile likelihood is

$$f(\mathbf{z}; \mu, \tilde{\sigma}(\mu)) = \frac{1}{(2\pi)^{n/2}\left\{\dfrac{vs^2}{n} + (\bar{z}-\mu)^2\right\}^{n/2}} e^{-n/2}, \qquad \mu \in \mathbb{R}. \quad (2.36)$$

From (2.36), the rescaled likelihood or likelihood ratio equals

$$\frac{f(\mathbf{z}; \mu, \tilde{\sigma}(\mu))}{f(\mathbf{z}; \hat{\mu}, \hat{\sigma})} = \left\{1 + \frac{n(\bar{z}-\mu)^2}{vs^2}\right\}^{-n/2} = \left\{1 + \frac{(170.72-\mu)^2}{(200.86)^2}\right\}^{-9}, \qquad \mu \in \mathbb{R}, \quad (2.37)$$

using the fact that $\tilde{\sigma}(\bar{z})^2 = s^2 = \hat{\sigma}^2$, and this function is plotted in Figure 2.8.

2.7.3 Likelihood Sets

The contours of a likelihood surface are nested sets of plausible parameter values which may be called *likelihood sets* and have the property that each set

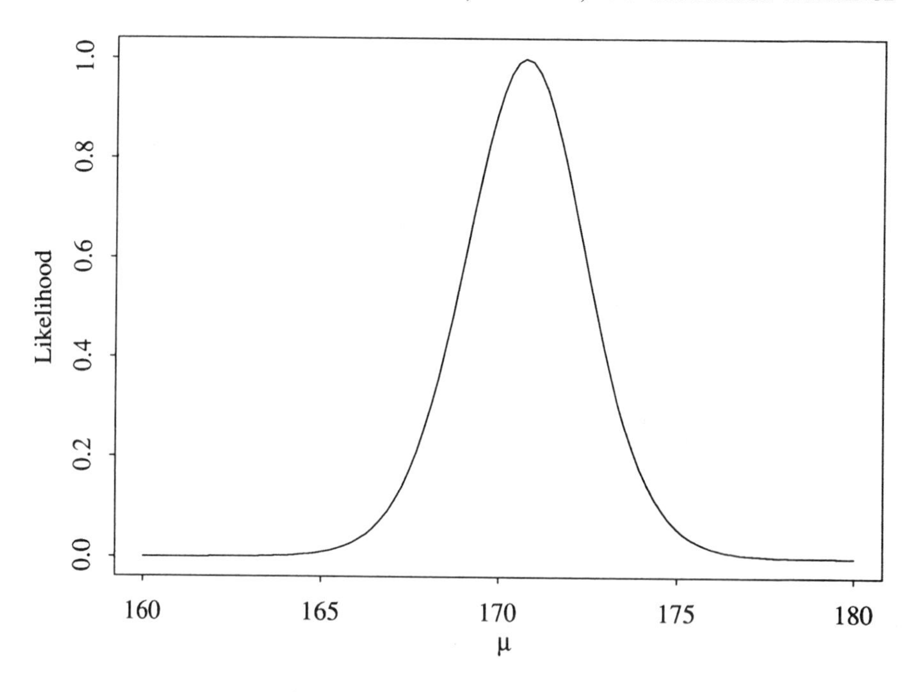

Figure 2.8. The profile likelihood for μ.

is the set of parameter values which are relatively more "plausible" than the excluded values.

Likelihood sets for μ can be read off the plot of the profile likelihood (Figure 2.8) or, in this case, can be obtained analytically. Formally, for some $k > 0$, we have from (2.37) the likelihood set

$$\begin{aligned} L_k(\mu) &= \left\{ \mu: \frac{1}{\left\{1 + \dfrac{n(\bar{z} - \mu)^2}{vs^2}\right\}^{n/2}} \geq k \right\} \\ &= \{\mu: \bar{z} - n^{-1/2}v^{1/2}s(k^{-2/n} - 1)^{1/2} \leq \mu \leq \bar{z} + n^{-1/2}v^{1/2}s(k^{-2/n} - 1)^{1/2}\} \\ &= \{\mu: 170.72 - 200.86(k^{-1/9} - 1)^{1/2} \leq \mu \leq 170.72 + 200.86(k^{-1/9} - 1)^{1/2}\}. \end{aligned}$$

The endpoints of the set are k^{-1} times less "plausible" than the most "plausible" parameter values. Suggested values of k are $k = 1/20$ and $k = 1/100$ (corresponding to the endpoints being 20 and 100 times less "plausible" than the most "plausible" parameter values) although, as with other contour sets, the family of sets indexed by k is of more interest than any single set. The likelihood sets are intervals in this problem but, in general, need not be.

2.7.4 Likelihood Hypothesis Tests

We can compare the relative "plausibility" of two values μ_0 and μ_1 of μ by comparing the likelihood at these values by computing the ratio

$$\frac{f(\mathbf{z}; \mu_0, \tilde{\sigma}(\mu_0))}{f(\mathbf{z}; \mu_1, \hat{\sigma}(\mu_1))} = \frac{\left\{1 - \dfrac{n(\bar{z} - \mu_1)^2}{vs^2}\right\}^{n/2}}{\left\{1 + \dfrac{n(\bar{z} - \mu_0)^2}{vs^2}\right\}^{n/2}}.$$

If we represent the likelihood of a set by the maximum value of the likelihood over that set, we can easily compare the likelihoods of sets. In particular, to compare the relative "plausibility" of H_0: $\mu = \mu_0$ to H_1: $\mu \neq \mu_0$, we examine $f(\mathbf{z}; \mu, \tilde{\sigma}(\mu))/f(\mathbf{z}; \hat{\mu}, \hat{\sigma})$. In interpreting these ratios, we take the view that the likelihood ratio defines a scale for measuring relative "plausibility" which, as with any other arbitrary scale, we need to calibrate against experience.

2.7.5 Elimination of Nuisance Parameters

In many inference problems the likelihood function involves nuisance parameters which we need to eliminate before we can make inferences about the parameters of interest. One approach is to maximize the likelihood over the nuisance parameters as we did in Section 2.7.2 to obtain the profile likelihood (2.36). The use of maximization makes the profile likelihood relatively easy to obtain and it turns out to be fundamental to frequentist (see Sections 3.3, 3.5 and 4.5) as well as likelihood inference. However, the profile likelihood cannot always be treated as a "likelihood" (see Section 4.5.15), a fact which may in some applications imply that the profile likelihood is not a satisfactory method of eliminating nuisance parameters.

The failings of the profile likelihood can often be overcome (at least approximately) by modifications which produce modified profile likelihoods (see Section 4.5.15) but other "likelihoods" of admittedly more limited availability may also be worth considering. While these other "likelihoods" are primarily motivated from frequentist concerns, it is convenient to mention them at this stage.

A different way to eliminate nuisance parameters is to use the marginal distribution of an appropriately chosen subset of the data rather than the model distribution (the joint distribution of the data) as the basis for the likelihood. The subset of the data is chosen so that the resulting *marginal likelihood* involves only the parameters of interest. The most famous marginal likelihood proposed by Patterson and Thompson (1971) (which is also a modified profile likelihood) is obtained for the variance parameters by eliminating the mean parameters from a Gaussian model. See Section 4.5.15 and Problems 4.5.5–4.5.6. As an alternative to marginal likelihood, we can sometimes condition on an appropriate function of the data and nuisance parameters to eliminate the nuisance parameters. The "likelihood" resulting from the conditional distribution of the data given the appropriate quantity is called a *conditional likelihood.* McCullagh and Nelder (1989, Chapter 7) provide a useful introduction to marginal and conditional likelihoods. Finally, Cox (1972; 1975) introduced the *partial likelihood* which also has the effect of eliminating nuisance parameters.

2.7.6 Relationship of Likelihood to Bayesian Inference

The likelihood approach is similar to the Bayesian approach except that the likelihood approach does not involve the use of a prior distribution. This has at least two important consequences:

1. Likelihood inference depends on the nonprobablistic concept of relative "plausibility" which is determined from the likelihood ratio whereas Bayesian inference has a simple probability interpretation.
2. Likelihood inference uses the maximum value of the likelihood over a set to represent the likelihood of the set whereas Bayesian inference uses the posterior probability of a set obtained by integrating the posterior density over that set.

The use of maximization instead of integration is rather arbitrary and we could consider integrating likelihoods with respect to weight functions. However, the weight functions have to be specified in the same way as prior distributions so this approach offers no advantages over the Bayesian approach.

The profile likelihood function (2.36) obtained by maximizing the joint likelihood (2.20) is of the same form as the Student t density (2.25) obtained by integrating the joint posterior distribution from the Jeffreys prior for the Gaussian model. There is a close connection between integration and maximization which is exploited in Laplace's method for approximating an integral by the maximum of the integrand. Laplace's method for approximating integrals is presented in Sections 4.6.4–4.6.5 but for the present we note that it works well when the integrand is unimodal and peaked so that the main contribution of the integrand to the integral occurs in the neighbourhood of its maximum value. This is the case for the Gaussian model (2.19) so the results of maximizing and integrating the likelihood with respect to the Jeffreys prior are the same. However, in other circumstances, the results may be less comparable.

2.7.7 Numerical Maximization of the Likelihood

Just as the integration required for Bayesian inference cannot always be done explicitly, the maximization in likelihood inference also cannot always be done explicitly. In this circumstance, we usually resort to numerical maximization methods.

One of the simplest numerical maximization methods is based on constructing a quadratic approximation to the likelihood surface and then maximizing the quadratic approximation. Let $\theta_{(m)}$ be a tentative value for the value of θ at which the likelihood is maximized and let $\eta(x, \theta) = \partial \log \{f(x; \theta)\}/\partial\theta$ be the first derivative of the log likelihood with respect to θ in a model for independent and identically distributed Z_i. Then for θ in a neighbourhood of $\theta_{(m)}$, we have the quadratic approximation (see 2 in the Appendix)

$$\log\{f(\mathbf{z}; \theta)\} \approx \log\{f(\mathbf{z}; \theta_{(m)})\} + (\theta - \theta_{(m)})^{\mathrm{T}} \sum_{i=1}^{n} \eta(\mathbf{z}_i, \theta_{(m)})$$
$$+ \frac{1}{2}(\theta - \theta_{(m)})^{\mathrm{T}} \sum_{i=1}^{n} \eta'(\mathbf{z}_i; \theta_{(m)})(\theta - \theta_{(m)}).$$

The matrix $n^{-1}\sum_{i=1}^{n} \eta'(\mathbf{z}_i; \theta)$ is the Hessian of the log-likelihood function. Provided the Hessian is nonsingular, the quadratic approximation to the log-likelihood is maximized at $\theta_{(m+1)}$, where

$$\theta_{(m+1)} = \theta_{(m)} - \left\{n^{-1} \sum_{i=1}^{n} \eta'(\mathbf{z}_i, \theta_{(m)})\right\}^{-1} n^{-1} \sum_{i=1}^{n} \eta(\mathbf{z}_i, \theta_{(m)}).$$

Iteration of this procedure until $n^{-1}\sum_{i=1}^{n} \eta(\mathbf{z}_i, \theta_{(m)}) \approx 0$ defines a sequence $\{\theta_{(m)}\}$ converging to an extremum of the log-likelihood function which is a maximum if the Hessian evaluated at the extremum is negative definite. This is called the *Newton–Raphson algorithm*. The convergence properties of the algorithm depend both on the initial value (so it is often valuable to try a range of different starting values) and the quality of the quadratic approximation to the log-likelihood function. There are a number of different approaches to numerical maximization and in particular cases, the form of the log-likelihood function may suggest alternative, preferable algorithms.

PROBLEMS

2.7.1. Suppose that we observe $\{(Y_1, x_1), \ldots, (Y_n, x_n)\}$ on the model

$$\mathscr{F} = \left\{f(y; \beta, \sigma) = \prod_{i=1}^{n} \frac{1}{(2\pi\sigma^2 v(x_i))^{1/2}} \exp\left\{-\frac{(y_i - x_i\beta)^2}{2\sigma^2 v(x_i)}\right\}, \right.$$
$$\left. y_i \in \mathbb{R}; \beta \in \mathbb{R}, \sigma > 0\right\}$$

which is used in finite population problems. The function $v(\cdot)$ allows the variance of the response to depend on the explanatory variable and is often taken as known from previous surveys. Find the values of β and σ at which the likelihood is maximized. Find the profile likelihood for β.

2.7.2. Show that

$$\sum_{i=1}^{g}\sum_{j=1}^{m}(z_{ij}-\bar{z})^2=\sum_{i=1}^{g}\sum_{j=1}^{m}(z_{ij}-\bar{z}_i)^2+m\sum_{i=1}^{g}(\bar{z}_i-\bar{z})^2,$$

where $\bar{z}_i = m^{-1}\sum_{j=1}^{m} z_{ij}$ and $\bar{z} = g^{-1}\sum_{i=1}^{g}\bar{z}_i$. Consider the variance component model presented in Section 1.5:

$$\mathscr{F}=\left\{f(\mathbf{z};\mu,\sigma_a,\sigma_u)=\frac{1}{(2\pi)^{mg/2}|\Sigma|^{1/2}}\exp\{-(\mathbf{z}-\mu)^{\mathrm{T}}\Sigma^{-1}(\mathbf{z}-\mu)/2\},\right.$$

$$\left.-\infty<z_{ij}<\infty:\mu\in\mathbb{R},\sigma_a>0,\sigma_u>0\right\},$$

where Σ is the block diagonal matrix with blocks $\sigma_a^2 J + \sigma_u^2 I$, J is the $m \times m$ matrix with all elements equal to 1 and I is the $m \times m$ identity matrix. Show that the likelihood can be written as

$$f(\mathbf{z};\mu,\tau_a,\tau_u)=\frac{1}{(2\pi)^{mg/2}\tau_u^{g(m-1)/2}\tau_a^{g/2}}$$

$$\times\exp\left\{-\sum_{i=1}^{g}\sum_{j=1}^{m}\frac{(z_{ij}-\bar{z}_i)^2}{2\tau_u}-\frac{m}{2\tau_a}\left\{\sum_{i=1}^{g}(\bar{z}_i-\bar{z})^2+g(\bar{z}-\mu)^2\right\}\right\},$$

where $\tau_u = \sigma_u^2$ and $\tau_a = m\sigma_a^2 + \sigma_u^2$. Hence show that the likelihood is maximized at

$$(\hat{\mu},\hat{\tau}_a,\hat{\tau}_u)=\left(\bar{z},g^{-1}m\sum_{i=1}^{g}(\bar{z}_i-\bar{z})^2,\{g(m-1)\}^{-1}\sum_{i=1}^{g}\sum_{j=1}^{m}(z_{ij}-\bar{z}_i)^2\right)$$

if $\hat{\tau}_a \geq \hat{\tau}_u$

and

$$(\hat{\mu},\hat{\tau}_u)=\left(\bar{z},(mg)^{-1}\sum_{i=1}^{g}\sum_{j=1}^{m}(z_{ij}-\bar{z})^2\right)$$

otherwise.

2.7.3. In the context of Problem 2.7.2, find the likelihood ratio for testing the hypothesis H_0: $\mu = 0$ against H_1: $\mu \neq 0$. Also find the likelihood ratio for testing the hypothesis that H_0: $\sigma_a = 0$ against H_1: $\sigma_a > 0$ by testing H_0: $\tau_a = \tau_u$ against H_0: $\tau_a > \tau_u$.

2.7.4. For the model presented in Problem 2.7.2, write the likelihood as a function of $\eta = \tau_u$ and $\theta = \tau_a/\tau_u$. Then find the profile likelihood of θ.

Now suppose we adopt the prior distribution $g(\mu, \tau_u, \tau_a) \propto 1/\tau_u\tau_a$, $\tau_a \geq \tau_u$. Find the marginal posterior distribution of θ. How does this compare to the profile likelihood? (For a discussion of some criticism of this Bayesian analysis, see Box and Tiao, 1973, pp. 303–15.)

2.7.5. Consider the complete two-way classification model in which we observe

$$y_{ijk} = \mu + \alpha_i + \beta_j + \gamma_{ij} + e_{ijk}, \qquad i = 1, \ldots, a, \quad j = 1, \ldots, b, \quad k = 1, \ldots, n,$$

with

$$\sum_{i=1}^{a} \alpha_i = \sum_{j=1}^{b} \beta_j = \sum_{i=1}^{a} \gamma_{ij} = \sum_{j=1}^{b} \gamma_{ij} = 0,$$

and $\{e_i\}$ are independent $N(0, \sigma^2)$ random variables. Show that the analysis of variance decomposition

$$\sum_i \sum_j \sum_k (y_{ijk} - \bar{y}_{...}) = n \sum_i \sum_j (\bar{y}_{ij.} - \bar{y}_{...})^2 + \sum_i \sum_j \sum_k (y_{ijk} - \bar{y}_{ij.})^2$$

and

$$n \sum_i \sum_j (\bar{y}_{ij.} - \bar{y}_{...})^2 = nb \sum_i (\bar{y}_{i..} - \bar{y}_{...})^2 + na \sum_j (\bar{y}_{.j.} - \bar{y}_{...})^2$$

$$+ n \sum_i \sum_j (\bar{y}_{ij.} - \bar{y}_{i..} - \bar{y}_{.j.} + \bar{y}_{...})^2,$$

where a dot subscript indicates averaging over that subscript, holds. Obtain the values of μ, α_i, β_j, and γ_{ij} which maximize the likelihood. Interpret the terms in the analysis of variance decomposition by expressing them in terms of these values.

FURTHER READING

The Bayesian paradigm is well presented by a number of authors including Savage (1962a), Cornfield (1969), and Lindley (1972). Box and Tiao (1973) provide useful discussion and develop a number of "standard" analyses. Bernado and Smith (1994, Chapter 5) examine the topics we consider in this book in more detail and provide extensive references. Mosteller and Wallace (1964/1984) present an elegant example of a serious Bayesian analysis in practice. Jeffreys (1939/1961) is a classic book which is well worth reading. Fisher (1956a, Chapter 2) contains a clear statement of Fisher's criticism of the Bayesian paradigm, and Zabell (1992) gives an interesting perspective on

Fisher's thought. Further useful discussion of fiducial theory may be found in Dempster (1964) and Kendall and Stuart (1979, Chapter 21). Likelihood theory is expounded by Edwards (1972), whereas Basu (1973) presents a critical discussion of both likelihood and fiducial theory. Cox and Hinkley (1974) and Barnett (1982) contain much useful material presenting and comparing the different approaches to inference.

CHAPTER 3

Frequentist Inference

To analyze the effect of caffeine on the volume of urine voided (Section 1.1.3), we decided in Section 1.2.3 to model the change in the volume of urine produced by the ingestion of caffeine using the Gaussian model

$$\mathscr{F} = \left\{ f(\mathbf{z}; \mu, \sigma) = \prod_{i=1}^{n} \frac{1}{(2\pi\sigma^2)^{1/2}} \exp\left\{ -\frac{(z_i - \mu)^2}{2\sigma^2} \right\}, \right.$$
$$\left. -\infty < z_i < \infty : \mu \in \mathbb{R}, \sigma > 0 \right\}. \quad (3.1)$$

The Bayesian, fiducial and likelihood approaches to making inferences about μ in the presence of the nuisance parameter σ were discussed in Sections 2.3, 2.6 and 2.7 respectively, and we now consider the frequentist approach.

The frequentist approach is based on the computation of statistics. A *statistic* $t = t(\mathbf{z})$ is a function of the data $\mathbf{z}$ which involves no unknown parameters. Since the data $\mathbf{z}$ is a realization of the random variable $\mathbf{Z}$, the value of a statistic $t = t(\mathbf{z})$ is a realization of $T = t(\mathbf{Z})$ which is a random variable with a distribution, called the *sampling distribution* of T, determined by the distribution F_0 of $\mathbf{Z}$. We do not know F_0 but we can use the model $\mathscr{F}$ to derive a sampling distribution which will be valid at least whenever $F_0 \in \mathscr{F}$. The sampling distribution of T under $\mathscr{F}$ can be interpreted as describing the behavior we can expect of T under $\mathscr{F}$. More precisely, the sampling distribution describes the outcome of the thought experiment in which we construct the distribution of T from the realizations of an infinite sequence of hypothetical samples generated under $\mathscr{F}$.

Frequentist methods have a *sampling* or *frequency basis* in the sense that they use the sampling distribution to evaluate and interpret inferences. According to Neyman (1977), the frequentist approach involves the use of hypothesis tests (Section 3.3) and confidence intervals (Sections 3.6–3.7) to make inferences or, as Neyman preferred, to guide inductive behavior. We adopt a broader view by considering all approaches with a frequency basis to be frequentist. This definition allows us to treat significance tests (Section 3.2) and conditional inferences (Section 3.9) as frequentist methods.

Within frequentist inference, it is useful to make a rough distinction between *testing* and *interval estimation.* The primary interest in most statistical problems is in making informative statements about the magnitude of effects like the magnitude of the change in the volume of urine voided following the ingestion of caffeine (i.e., interval estimation) but we are sometimes interested in addressing simpler questions like "Does the ingestion of caffeine have an effect on the volume of urine produced?" (i.e., testing). Although the latter question can be subsumed within the former, efforts to answer this kind of question have been important to the historical, theoretical, and practical development of statistics. This is not to suggest that tests are more important than confidence intervals, because in fact tests often provide a less satisfactory means of presenting evidence.

Frequentist tests and confidence intervals for a parameter θ are often conveniently derived from statistics which have been constructed with the specific objective of being close to θ. We call such statistics $T = t(\mathbf{Z})$ *estimators* and their realized values $t = t(\mathbf{z})$ *estimates.* The construction of estimators is referred to a *point estimation.* The properties of inferential statements derived from point estimates often depend on the properties of the point estimate so it is useful to discuss point estimation before we discuss testing and and interval estimation.

3.1 POINT ESTIMATION

A characteristic of the frequentist approach is that, in contrast to the Bayesian, fiducial and likelihood methods described in Chapter 2, it does not specify a single method for constructing procedures. Instead, we specify the desirable properties of a procedure and then construct procedures which are in some sense optimal. Nonetheless, it is useful to have general methods for constructing procedures which can at least provide a point of departure. Two general methods of constructing estimators are the *method of moments* and the *method of maximum likelihood.*

3.1.1 Method of Moments Estimation

If we compute the sample moments $m_k = n^{-1} \sum_{j=1}^{n} Z_j^k, k = 1, 2,$ their expectations under the Gaussian model (3.1) which are $\mathrm{E}m_1 = \mu$ and $\mathrm{E}m_2 = \sigma^2 + \mu^2$, and then solve the system of equations

$$m_1 - \hat{\mu}_{\mathrm{m}} = 0$$

$$m_2 - \hat{\sigma}_{\mathrm{m}}^2 - (\hat{\mu}_{\mathrm{m}})^2 = 0,$$

we obtain the *method of moments estimators* (Pearson, 1894)

$$\hat{\mu}_{\mathrm{m}} = m_1 = \bar{Z}$$

$$\hat{\sigma}^2_{\mathrm{m}} = m_2 - m_1^2 = n^{-1} \sum_{i=1}^{n} (Z_i - \bar{Z})^2$$

of μ and σ^2. The method of moments is discussed further in Section 4.3.

3.1.2 Maximum Likelihood Estimation

It follows from (2.20) that the likelihood under the Gaussian model (3.1) can be written as

$$f(\mathbf{z}; \mu, \sigma) = \frac{1}{(2\pi\sigma^2)^{1/2}} \exp\left\{-\frac{n\hat{\sigma}^2}{2\sigma^2} - \frac{n(\bar{z} - \mu)^2}{2\sigma^2}\right\}, \qquad \mu \in \mathbb{R}, \quad \sigma > 0, \tag{3.2}$$

where $\hat{\sigma}^2 = n^{-1} \sum_{i=1}^{n} (z_i - \bar{z})^2$. We saw in Section 2.7.1 that (3.2) is maximized at $\hat{\mu} = \bar{z}$ and $\hat{\sigma}^2 = n^{-1} \sum_{i=1}^{n} (z_i - \bar{z})^2$. The most "plausible" estimators of μ and σ are the *maximum likelihood estimators* (Bernoulli, 1777; Gauss, 1809; Edgeworth, 1908–9; Fisher, 1922) which are therefore $\hat{\mu} = \bar{Z}$ and $\hat{\sigma}^2 = n^{-1} \sum_{i=1}^{n} (Z_i - \bar{Z})^2$. Maximum likelihood estimation is discussed further in Sections 4.2–4.3.

3.1.3 Optimal Point Estimation

The method of moments and maximum likelihood estimators of μ and σ^2 in (3.1) are the same, but in general they need not be. We can also consider other estimators such as the sample median for μ and the median absolute deviation from the median divided by 0.6745 for σ, and the question of which estimators we should use arises.

The best possible estimator T of θ equals θ for all θ so that T contains no error. Since T is a random variable, this means that T equals θ with probability 1 whatever the value of θ, a requirement which is generally impossible to achieve in finite samples. More realistically, we would like an estimator to be as close to θ as possible on average. That is, if we use squared error to measure closeness, we want T to minimize the mean squared error

$$\mathrm{E}(T - \theta)^2 = \mathrm{Var}\,(T) + (\mathrm{E}T - \theta)^2 \tag{3.3}$$

for all θ. This is unfortunately also not a realistic goal because an estimator like $T = 6$ minimizes (3.3) when $\theta = 6$ so minimizing (3.3) for all θ is equivalent to finding an estimator to estimate θ without error. The mathematical solution to these difficulties is to introduce further criteria to restrict the class of estimators under consideration so that estimators like $T = 6$ are precluded.

3.1.4 Unbiased Estimation

A simple restriction of the class of estimators is to the class of *unbiased estimators* of θ which are "on average" close to θ. An estimator T of θ is unbiased for θ if

$$\mathrm{E}T = \theta \qquad \text{for all } \theta. \tag{3.4}$$

If we restrict attention to the class of unbiased estimators, the mean squared error (3.3) reduces to the variance and the member of the class with minimum variance for all θ is called a *uniformly minimum variance unbiased (UMVU) estimator.*

For the Gaussian model (3.1), $\bar{Z}$ is the UMVU estimator of μ (Lehmann, 1983, p. 84). However,

$$\mathrm{E}\hat{\sigma}^2 = \frac{n-1}{n}\sigma^2$$

so the maximum likelihood (and method of moments) estimator of σ^2 is biased. The sample variance

$$S^2 = n(n-1)^{-1}\hat{\sigma}^2 = (n-1)^{-1}\sum_{i=1}^{n}(Z_i - \bar{Z})^2$$

is unbiased and is the UMVU estimator of σ^2 (Lehmann, 1983, p. 84). The UMVU estimator of σ is

$$\frac{\Gamma\left(\dfrac{n-1}{2}\right)}{2^{1/2}\Gamma\left(\dfrac{n}{2}\right)}\left\{\sum_{i=1}^{n}(Z_i - \bar{Z})^2\right\}^{1/2}$$

(Lehmann, 1983, p. 84).

Unbiased estimators have the undesirable property that they can take values outside the parameter space and the class of unbiased estimators excludes otherwise "natural" estimators such as the maximum likelihood estimator of σ^2 in (3.1). In addition UMVU estimators only exist in a limited class of problems.

3.1.5 Equivariant Estimation

An alternative to unbiasedness is to restrict the class of estimators to *equivariant estimators.* For example, to estimate location and scale parameters like μ and σ in (3.1), we can consider the class of *location and scale equivariant estimators* which satisfy

$$\tilde{\mu}(a\mathbf{z} + b) = a\tilde{\mu}(\mathbf{z}) + b \quad \text{and} \quad \tilde{\sigma}(a\mathbf{z} + b) = |a|\tilde{\sigma}(\mathbf{z}), \tag{3.5}$$

for any $a, b \in \mathbb{R}$. In this class of estimators, we try to find the estimators which minimize the mean squared error (3.3) or some other appropriate function.

For location and scale parameters in a location–scale model (1.18), the equivariant estimators with minimum mean squared error are the Pitman (1938) estimators. For any location and scale equivariant estimators $\tilde{\mu}$ and $\tilde{\sigma}$ satisfying (3.4), let $\{C_i = (z_i - \tilde{\mu}(\mathbf{Z}))/\tilde{\sigma}(\mathbf{Z}), i = 1, \ldots, n\}$. We call $\{C_i\}$ a *configuration.* Then we can write

$$\mathrm{E}\frac{(\tilde{\mu}(\mathbf{Z}) - \mu)^2}{\sigma^2} = \mathrm{E}\left(\mathrm{E}\left[\frac{(\tilde{\mu}(\mathbf{Z}) - \mu)^2}{\sigma^2} \mid \{C_j\}\right]\right).$$

We will show in Section 3.9.3 that

$$\mathrm{E}\left[\frac{(\tilde{\mu} - \mu)^2}{\sigma^2} \mid \{C_j\}\right] = \int_0^\infty \int_{-\infty}^\infty \frac{(\tilde{\mu} - t)^2}{s^2} g(t, s)\, ds\, dt, \tag{3.6}$$

where

$$g(t, s) = \frac{\dfrac{1}{s^{n+1}} \displaystyle\prod_{j=1}^n h\left(\frac{z_j - t}{s}\right)}{\displaystyle\int\int \frac{1}{s^{n+1}} \prod_{j=1}^n h\left(\frac{z_j - t}{s}\right) ds\, dt}, \qquad t \in \mathbb{R}, \quad s \geq 0,$$

is the likelihood of (t, s) multiplied by s^{-1} and normalized to integrate to 1. The conditional expectation (3.6) is constant over all equivariant estimators $\tilde{\mu}(\mathbf{Z})$ and does not depend on μ and σ so we can minimize $\mathrm{E}\sigma^{-2}(\tilde{\mu}(\mathbf{Z}) - \mu)^2$ by minimizing (3.6). The minimum is

$$\tilde{\mu}(\mathbf{Z}) = \frac{\displaystyle\int_0^\infty \int_{-\infty}^\infty \frac{1}{s^2} t g(t, s)\, dt\, ds}{\displaystyle\int_0^\infty \int_{-\infty}^\infty \frac{1}{s^2} g(t, s)\, dt\, ds}$$

which is called the *Pitman estimator* of μ. Similarly, the equivariant estimator which minimizes $\mathrm{E}\sigma^{-2r}(\tilde{\sigma}(\mathbf{Z})^r - \sigma^r)^2$ is

$$\tilde{\sigma}(\mathbf{Z})^r = \frac{\displaystyle\int_0^\infty \int_{-\infty}^\infty \frac{1}{s^r} g(t, s)\, dt\, ds}{\displaystyle\int_0^\infty \int_{-\infty}^\infty \frac{1}{s^{2r}} g(t, s)\, dt\, ds}$$

which is called the Pitman estimator of σ^r. Lehmann (1983, Chapter 3) gives an alternative exposition of this approach.

Under the Gaussian model (3.1), the Pitman estimator of μ is $\bar{Z}$ and the Pitman estimator of σ^2 is $\tilde{\sigma}^2 = n(n+1)^{-1}\hat{\sigma}^2 = (n+1)^{-1}\sum_{i=1}^{n}(Z_i - \bar{Z})^2$. The Pitman estimator of σ is

$$\frac{\Gamma\left(\frac{n}{2}\right)}{2^{1/2}(n-1)^{1/2}\Gamma\left(\frac{n+1}{2}\right)}\left\{\sum_{i=1}^{n}(Z_i - \bar{Z})^2\right\}^{1/2}$$

See Problem 3.1.7.

3.1.6 The Class in Which a Procedure is Optimal

The search for optimal estimators can provide a challenging set of mathematical problems and it is easy to lose sight of the statistical issues. Ultimately, we need to assess how compelling we find particular optimality results by thinking about the optimality criterion and the class of estimators to which it is applied.

Generally, the smaller the class of estimators, the less interesting the result. The classic example of this is the *Gauss–Markov theorem* which asserts that, if we have independent and identically distributed observations $Z_1, \ldots, Z_n$ with mean μ and variance σ^2, irrespective of the actual distribution of $Z_1, \ldots, Z_n$, the sample mean is the best linear, unbiased estimator of μ. The class of linear estimators of μ is the class of weighted sums of the observations, where the weights do not depend on the observations. The unbiasedness requirement implies that the weights must be normalized to sum to 1. Thus, the Gauss–Markov theorem applies to the class of estimators $T_w = \sum_{i=1}^{n} w_i Z_i$, where $\sum_{i=1}^{n} w_i = 1$, and states that the variance $\sigma^2 \sum_{i=1}^{n} w_i^2$ of T_w is minimized when $w_i \equiv n^{-1}$. Given that we have to specify the weights in advance of examining the data, without any information about μ, it is not surprising that it is optimal to assign equal weights. If we are allowed to make the weights depend on the data, we can obtain "better" nonlinear, biased estimators, a point that is often overlooked. As Lindley (1972) pointed out, optimality criteria which restrict the class of estimators to (say unbiased estimators) often restrict the class to only one "interesting" estimator which is then, not surprisingly, optimal.

The optimality criteria considered in this section require the exact sampling distributions of estimators so they are often difficult to implement or are only applicable in rather restrictive situations. We extend the concepts of optimal estimation to allow large sample approximations to sampling distributions in Section 4.3.

PROBLEMS

3.1.1. Consider the negative binomial model (5e in the Appendix) parameterized as

$$\mathscr{F} = \left\{ f(\mathbf{z}; \lambda, \delta) = \prod_{i=1}^{n} \left\{ z_i! \Gamma\left(\frac{\lambda}{\delta}\right) \right\}^{-1} \Gamma\left(z_i + \frac{\lambda}{\delta}\right) \delta^{z_i} (1 + \delta)^{-(z_i + \lambda/\delta)}, \right.$$

$$\left. z_i = 0, 1, 2, \ldots, \lambda > 0, \delta > 0 \right\}.$$

Here $\mathrm{E}Z_i = \lambda$ is the mean of the distribution, $\mathrm{Var}\,(Z_i) = \lambda(1 + \delta)$ is the variance of the distribution and $\delta = 0$ corresponds to the Poisson model so λ is the mean occurrence rate and δ is the non-Poissonness rate. Obtain method of moments estimators for λ and δ. How would you obtain the maximum likelihood estimator in this case?

3.1.2. Graphical methods indicate that the truncated Poisson model

$$\mathscr{F} = \left\{ f(\mathbf{y}; \lambda) = \prod_{i=1}^{n} \frac{\lambda^{y_i} \exp(-\lambda)}{y_i! \{1 - \exp(-\lambda)\}}, \; y_i = 1, 2, \ldots : \lambda > 0 \right\}$$

provides a reasonable model for the Feinstein et al. (1989) microbubble particle data presented in Problem 1.5.2. Obtain the method of moments estimator $\hat{\lambda}_{\mathrm{m}}$ for λ. How would you obtain the maximum likelihood estimator?

3.1.3. Suppose we have n observations $\mathbf{Z}$ on the Gaussian model (3.1) and we are interested in the variance parameter σ^2. Suppose we consider only estimators of σ^2 of the form cS^2, where $S^2 = \sum_{j=1}^{n} (Z_j - \bar{Z})^2$. Show that the mean squared error of cS^2 is

$$\mathrm{E}\{cS^2 - \sigma^2\}^2 = \sigma^4\{c^2(n^2 - 1) - 2c(n - 1) + 1\}$$

and that this is minimized at $c = (n + 1)^{-1}$, the Pitman estimator of σ^2 for this model. Show that the unbiased estimator of σ^2 has $c = (n - 1)^{-1}$ and compare its mean squared error to that of the Pitman estimator.

3.1.4. Suppose that we have observations $\mathbf{Z}$ on the uniform (location) model

$$\mathscr{F} = \left\{ f(\mathbf{y}; \theta) = \prod_{i=1}^{n} I(\theta - \tfrac{1}{2} \le y_i \le \theta + \tfrac{1}{2}) : \theta \in \mathbb{R} \right\}.$$

Compare the estimators $(Z_{n1} + Z_{nn})/2$, where $Z_{n1} \le \cdots \le Z_{nn}$, and $\bar{Z}$.

3.1.5. Suppose that we have observations $\mathbf{Z}$ on the exponential model

$$\mathscr{F} = \left\{ f(\mathbf{y}; \lambda) = \prod_{i=1}^{n} \lambda \exp(-\lambda y_i), \; y_i > 0 : \lambda > 0 \right\}.$$

Find the Pitman estimator for $1/\lambda$ and compare its mean squared error to that of the corresponding maximum likelihood estimator.

3.1.6. Suppose that we have observations $\mathbf{Z}$ on the uniform model

$$\mathscr{F} = \left\{ f(\mathbf{y}; \theta) = \prod_{i=1}^{n} \frac{1}{\theta} I(0 \leq y_i \leq \theta) : \theta > 0 \right\}.$$

Obtain the Pitman estimator of θ and compare it to the maximum likelihood estimator and the unbiased estimator based on Z_{nn}, where $Z_{n1} \leq \cdots \leq Z_{nn}$.

3.1.7. Suppose that we have observations $\mathbf{Z}$ on the Gaussian model (3.1). Obtain the Pitman estimators of μ, σ^2, and σ.

3.2 SIGNIFICANCE TESTS

The idea behind a significance test is to formulate a null hypothesis and then examine the extent to which the data provide evidence to refute this hypothesis. In this form, significance tests have been in use at least since Arbuthnott (1710). They were given their modern justification and then popularized by Fisher (1925a).

The key steps in carrying out a significance tests are

1. formulating the null hypothesis H_0
2. choosing a statistic which has a known sampling distribution under H_0
3. using the sampling distribution to compute the significance probability or p-value of the test.

3.2.1 The Null Hypothesis

The first step in a significance test is to formulate the null hypothesis. Fisher felt strongly that we collect data in order to find effects so that initially we should take the view that there are no effects and change our view only if the data provide evidence against it. Thus the *null hypothesis* is a precise statement of what "no effect" in the substantive problem means in terms of the model. For the volume of urine problem, the null hypothesis is that the ingestion of caffeine has no effect on the volume of urine produced, which can be written H_0: $\mu = 0$ under (3.1).

3.2.2 The Test Statistic

If the null hypothesis H_0: $\mu = 0$ is true and the model (3.1) holds, we expect (amongst other things) the mean of the data (which we saw in Section 3.1 is estimating $\mu = 0$) to be close to zero. The traditional recommendation for this problem is to compute the so-called *studentized mean*

$$t = \frac{n^{1/2}\bar{z}}{s} = \frac{170.72}{48.72} = 3.50$$

which is the sample mean $\bar{z} = 170.72$ divided by the standard error of the mean $s/n^{1/2} = 48.72$ and which should also be close to 0 under H_0. The studentized mean in this case seems far from 0, but we need some way to calibrate this value.

3.2.3 The Reference Distribution

Suppose we are told that the rental cost of an apartment in Sydney is \$1130/week and we want to know whether this is expensive. A sensible way to decide is to build up a *reference distribution* for the rental price of similar apartments in Sydney and then see where the particular apartment fits into the reference distribution. If the rent falls in the upper tail of the reference distribution, the rent is expensive; otherwise it is not. The reference distribution quantifies what we should expect similar apartments to cost and enables us to quantify the extent to which the rent for a particular apartment is atypical. The sampling distribution of T under $\mathscr{F}$ and H_0 can be interpreted as describing the behavior we can expect of T under $\mathscr{F}$ and H_0 and so is a plausible reference distribution.

The distribution of T under (3.1) and the null hypothesis $H_0: \mu = 0$, was treated as approximately normal until "Student" (1908) derived the exact distribution which is called the Student t distribution with $\nu = n - 1$ degrees of freedom (see 9 in the Appendix) and Fisher (1925b) tabulated the distribution function. The density of T with $n = 18$ is plotted in Figure 3.1. We observe that $t = 3.50$ is in the tail of the distribution.

3.2.4 The p-value

To quantify the extremeness or unusualness of the observed value $t = 3.50$, we compute the *significance probability* or *p-value*. The p-value is the limiting proportion of realizations of T as or more extreme than the observed value (3.50) in an infinite sequence of hypothetical repeated trials under $\mathscr{F}$ and H_0 and is given by

$$\begin{aligned} p &= P\{|T_\nu| \geq |t|; H_0\} \\ &= 2\{1 - G_\nu(|t|)\} \\ &= 2\{1 - G_{17}(3.50)\} \\ &= 0.0027. \end{aligned} \tag{3.7}$$

If the p-value is small, Fisher argued that either a rare chance event (leading to

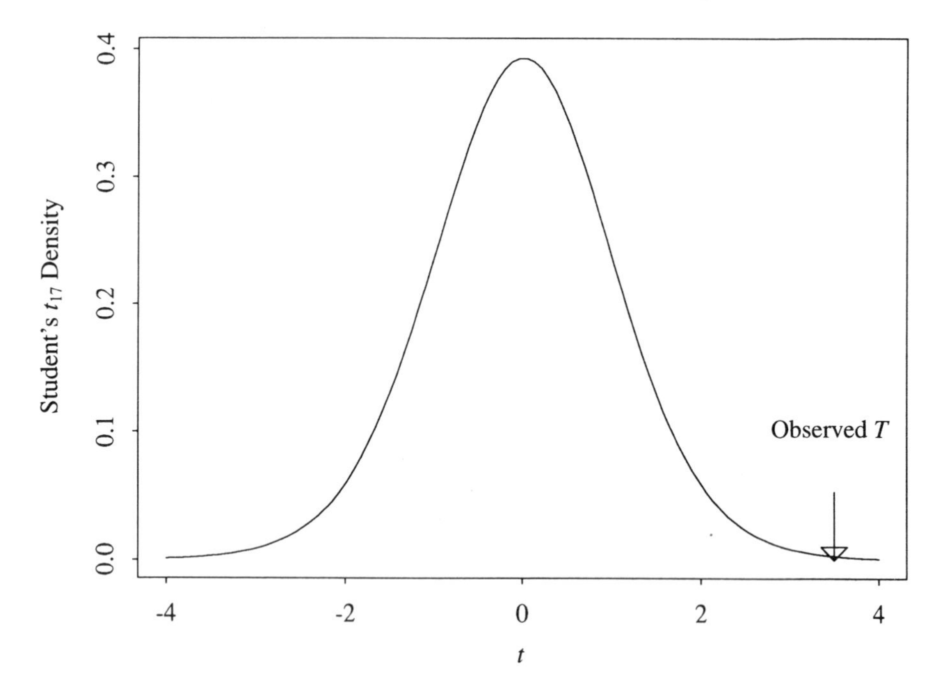

Figure 3.1. The sampling distribution of T under the Gaussian model and H_0: $\mu = 0$. The arrow shows the observed value of T.

the observed t) has occurred or that the assumed distribution (and hence H_0) is incorrect. (Strictly, the assumed distribution depends on $\mathscr{F}$ and H_0 so the problem may be with $\mathscr{F}$ rather than H_0.) Thus small p-values constitute evidence against H_0.

We have transformed the problem from the studentized mean scale to the more convenient p-value scale but we still have a calibration problem. What constitutes a small p-value depends on the experimeter and the context, but Fisher's general advice (which stemmed from copyright and other limitations on the publication of tables in the 1920s – see Barnard, 1990) that p-values of 0.01 or 0.05 are small has become standard in many fields. Our p-value of 0.0027 is clearly small by Fisher's standard, so we say that the test is significant and conclude that the data provide evidence that the ingestion of caffeine has an effect on the volume of urine produced.

If a p-value is very large, Fisher suggested that we should also consider rejecting the null hypothesis as being "too good". This is the basis for his famous suggestion that Mendel's data is too good to have arisen naturally (Fisher, 1936).

The p-value is not interpretable as a probability reflecting the falsity or belief in the falsity of H_0. In our formulation, H_0 is either true or false so has probability 0 or 1. The probability underlying the p-value calculation is derived from the reference distribution which reflects the stochastic nature of the observations and has the limiting frequency or frequentist interpretation. This makes it difficult to compare p-values obtained from different experiments or using different statistics.

3.2.5 Learning from Significance Tests

Fisher argued strongly that a nonsignificant test does not establish the null hypothesis; it simply means that the present experiment does not provide evidence to refute it. He argued further that a significant result does not on its own establish the falsity of the null hypothesis. Since this can only be done by repeating the experiment at other places and under other conditions, both significant and nonsignificant results should be published to let the literature reflect the empirical results correctly. This is in accord with Fisher's belief that although we learn from experience our knowledge is always provisional and subject to possible revision.

Fisher's ideas are obviously closely related to the Popperian concept of *falsifiability*. Popper (1935, pp. 40–42, Chapter 4) argued that scientific theories can be refuted but not verified by empirical experience and that we advance science by formulating null hypotheses which we then attempt to refute. This notion that a hypothesis is set up only in the hope that it will be refuted prompted Jeffreys (1939/1961, p. 377) to describe a null hypothesis as "merely something set up like a coconut to stand until it is hit".

3.2.6 The Choice of Test Statistic

Fisher was never explicit about how we should choose a statistic on which to base the test. Although his concept of sufficiency (Sections 2.6.2 and Section 7.3) and the discussion of point estimation in Section 3.1 provide some justification for using the sample mean and standard deviation in the volume of urine problem, they had been used in this kind of problem by Laplace (Stigler, 1986, p. 151) and their use was regarded as standard well before modern attempts at formal justification. In any case, to Fisher, the general process of choosing test statistics depended on constructive imagination, knowledge, and experience and it could never be reduced to a mathematical problem.

PROBLEMS

3.2.1. We used a graphical method to examine the fit of the Poisson model to Rutherford and Geiger's (1910) alpha particle data presented in Problem 1.5.1. Alternatively, a formal goodness of fit test of H_0: the data are realizations of Poisson random variables can be based on the famous χ^2

statistic (Pearson, 1900) of the form

$$X^2 = \sum_{k=1}^{K} \frac{(f_k - \mathrm{E}_k)^2}{\mathrm{E}_k},$$

where K is the number of distinct frequencies, $\mathrm{E}_k = 2608\hat{\lambda}^k \exp(-\hat{\lambda})/k!$, $k = 1, \ldots, K$ and $\hat{\lambda}$ is an estimator of λ. Here E_k, the expected number of observations equal to k under the null hypothesis, is estimated by the number of samples times an estimate of the probability that the number of scintillations in a single sample under the model, namely $\mathrm{P}\{Z = k\} = \lambda^k \exp(-k)/k!$. It can be shown that (in large samples) under the null hypothesis, X^2 has a chi-squared distribution with $K - 1 - p$ degrees of freedom, where p is the number of estimated parameters in E_k. Standard advice for implementing the test is to pool categories so that $\mathrm{E}_k \geq 5$ for all k.

Show that $\hat{\lambda} = \bar{Z}$ is the maximum likelihood estimator of λ under the Poisson model. Compute and interpret the p-value of the χ^2 goodness of fit test without pooling any categories. How does the result change if we pool the last four categories? Which analysis is preferable? Explain.

3.2.2. Mosteller and Wallace (1964/1984, p. 33) presented data on the number of occurrences of the word "may" in 262 blocks of text, each of approximately 200 words, written by James Madison. The data are given in Table 3.1.

A possible model (though not the only one and arguably not the best one) for the number of occurrences of distinct words in a document is the Poisson distribution with parameter equal to the mean number of occurrences. Use the χ^2 test of Problem 3.2.1 to test the hypothesis that the data follow the Poisson model.

3.2.3. An alternative model for the word counts presented in Problem 3.2.2 is the negative binomial model introduced in Problem 3.1.1. Given estimators $\hat{\lambda}$ of λ and $\hat{\delta}$ of δ, show how to modify the chi-squared statistic of Problem 3.2.1 to test the hypothesis that the data follow a negative binomial model. Use the method of moments estimators for λ and δ from Problem 3.1.1 to test the goodness of fit of the negative binomial model.

Table 3.1. Number of Occurrences of "May" and the Frequency of Occurrence

k	0	1	2	3	4	5	6
f_k	156	63	29	8	4	1	1

Reprinted with permission from Mosteller, F. and Wallace, D.L. *Inference and Disputed Authorship: The Federalist* (1964/1984, p. 33).

3.2.4. Graphical methods indicate that the truncated Poisson model discussed in Problem 3.1.2 provides a reasonable model for the Feinstein et al. (1989) microbubble particle data presented in Problem 1.5.2. Given an estimator $\hat{\lambda}$ of λ, show how to modify the chi-squared statistic of Problem 3.1.1 to test the hypothesis that the data follow a truncated Poisson model. Use the method of moments estimator of λ obtained in Problem 3.1.2 to test the null hypothesis that the data are realizations of truncated Poisson random variables. Interpret the result.

3.2.5. Suppose we observe the number of occurrences of distinct words in an unattributed document of length n words to be Y and the number of occurrences of the same word in a similar document of length m words known to be written by the suspected author of the first document to be X. If we assume $Y \sim \text{Poisson}(n\lambda)$ and $X \sim \text{Poisson}(m\mu)$, we are interested in testing the hypothesis that the two documents are written by the same author or $\text{H}_0\colon \lambda = \mu$. We can show that

$$\text{P}\{Y = y \mid Y + X = t\} = \binom{t}{y}\pi^y(1-\pi)^{t-y}, \qquad y = 0, 1, \ldots, t,$$

where $\pi = n\lambda/(n\lambda + m\mu)$ so a conditional test of H_0 is obtained by testing $\text{H}_0\colon \pi = n/(n+m)$. Thisted and Efron (1987) report that in the "Shakespearian" poem discovered in 1985, there are $y = 258$ distinct words out of $n = 429$. In "The Phoenix and Turtle" which is attributed to Shakespeare, we observe $z = 216$ distinct words out of $m = 352$. Write down an expression for the exact conditional p-value for testing H_0. Use the fact that for large t, we can approximate the binomial(t, π) distribution by the Gaussian distribution with mean $t\pi$ and variance $t\pi(1-\pi)$ to obtain the p-value. Interpret your result.

3.3 HYPOTHESIS TESTING

Neyman and Pearson (1928; 1933) set out to provide significance tests with the logical justification they felt was lacking. They introduced the concept of an *alternative hypothesis* H_1, and reformulated the testing problem as one of choosing between H_0 and H_1. One of the hypotheses is assumed true and the result of the test is a statement of which is true. The statement is, of course, not necessarily correct but, Neyman and Pearson argued, we should behave as if it is and proceed on this basis.

A Neyman–Pearson test to choose between H_0 and H_1 is constructed by considering the sample space $\mathscr{X}$ of all possible samples of size n and partitioning the space into two regions, one of which corresponds to the samples which will result in a decision to reject H_0 (and accept H_1) and its complement which corresponds to the samples which will result in a decision to accept H_0. The region $C(\mathbf{z})$ for which we reject H_0 in favor of H_1 is called the *critical* or *rejection region* and its complement is called the *acceptance region*.

3.3.1 Types of Hypotheses

To carry out a Neyman–Pearson test for a caffeine effect in the volume of urine problem, we need to formulate both a null hypothesis and an alternative hypothesis for μ in the Gaussian model (3.1). We will consider the Fisherian null hypothesis of no difference, namely $H_0: \mu = 0$, but it is worth noting that the Neyman–Pearson approach permits us to consider more general null hypotheses. For example, as in Section 2.5.1, we could consider the null hypothesis that the effects are only nearly the same to allow for small differences which are deemed practically unimportant. This hypothesis can be formulated as $H_0: |\mu| \leq \delta$ for small $\delta > 0$. Alternatively, we could consider the null hypothesis that caffeine lowers the volume of urine produced which is expressed as $H_0: \mu \leq 0$. The choice of a null hypothesis is dependent on both the substantive question and its context. Similar issues arise in the choice of an alternative hypothesis; suppose that we consider $H_1: \mu \neq 0$.

The simplest possible hypothesis testing problem occurs when both hypotheses are *simple hypotheses* in the sense that they specify the model completely. In the context of the Gaussian model (3.1), hypotheses that specify unique values μ_0 and σ_0 for μ and σ respectively, are simple; otherwise they are *composite hypotheses*. The Fisherian null hypothesis $H_0: \mu = 0$ is a composite hypothesis because the value of σ is unspecified but it is a sharp hypothesis about μ because μ takes only a single value. The alternative hypothesis $H_1: \mu \neq 0$ is also a composite hypothesis.

When both hypotheses are simple, the choice of tests is straightforward and we will consider this situation in Sections 3.3.2–3.3.3 before considering the more general problem in Sections 3.3.4–3.3.7.

3.3.2 Most Powerful Tests

Suppose for simplicity that the general model

$$\mathscr{F} = \{f(\mathbf{z}; \theta): \theta \in \Omega\} \tag{3.8}$$

holds and consider testing the simple null hypothesis $H_0: \theta = \theta_0$ against the simple alternative hypothesis $H_1: \theta = \theta_1$.

A test of H_0 against H_1 has four possible outcomes, two of which are correct and two of which are incorrect. Schematically

	Reality	
Decision	H_0 true	H_0 false
Accept H_0	Correct	Type II
Reject H_0	Type I	Correct

Neyman and Pearson called type I an *error of the first kind* (we reject H_0 when it is true) and type II an *error of the second kind* (we accept H_0 when it is false). They argued that for a good test, the probabilities of type I and II errors which are

$$P\{C(\mathbf{Z}); H_0\} = \int_{C(\mathbf{z})} f(\mathbf{z}; \theta_0)\, d\mathbf{z} \tag{3.9}$$

and

$$P\{C(\mathbf{Z})^c; H_1\} = \int_{C(\mathbf{z})^c} f(\mathbf{z}; \theta_1)\, d\mathbf{z} \tag{3.10}$$

respectively, should be small. (The integrals in (3.9) and (3.10) and the expressions which follow should be interpreted as sums in the discrete case.) We can write the probability of a type II error as

$$P\{C(\mathbf{Z})^c; H_1\} = 1 - P\{C(\mathbf{Z}); H_1\}$$

so minimizing the probability of a type II error is equivalent to maximizing

$$P\{C(\mathbf{Z}); H_1\} = \int_{C(\mathbf{z})} f(\mathbf{z}; \theta_1)\, d\mathbf{z} \tag{3.11}$$

which is called the *power* of the test.

It turns out that we cannot, for fixed n, minimize the probability (3.9) of a type I error and maximize the power (3.11) simultaneously, so Neyman and Pearson suggested that we should fix the probability of a type I error at a level α called the *size* or *significance level* of the test and then maximize the power of the test. Thus in mathematical terms, out of the set of all critical regions $C(\mathbf{z})$ satisfying

$$P\{C(\mathbf{Z}); H_0\} = \int_{C(\mathbf{z})} f(\mathbf{z}; \theta_0)\, d\mathbf{z} = \alpha, \tag{3.12}$$

we should base the test on the region which maximizes (3.11). The test with maximum power subject to (3.12) is called the *most powerful α level* test.

The Neyman–Pearson theory does not discuss the possibility that neither hypothesis is in reality correct. This is an example of what Kimball (1957) called an *error of the third kind*: giving the “right” answer to the wrong question. Notwithstanding their neglect in the theory, errors of the third kind are very important in practice. See also Wolfowitz (1967).

3.3.3 The Neyman–Pearson Lemma

To find a most powerful α level test for testing $H_0: \theta = \theta_0$ against $H_1: \theta = \theta_1$ under the model (3.8), we need to find the region $C(\mathbf{Z})$ in the sample space $\mathscr{Z}$ which maximizes (3.11) subject to (3.12). That is, we need to find a region $C(\mathbf{Z})$ such that for any other region $D(\mathbf{Z})$ in the sample space $\mathscr{Z}$, we have

$$\int_{C(\mathbf{z})} f(\mathbf{z}; \theta_1)\, d\mathbf{z} - \int_{D(\mathbf{z})} f(\mathbf{z}; \theta_1)\, d\mathbf{z} \geq 0. \tag{3.13}$$

Since the only difference between the integrals on the left-hand side of (3.13) is in their regions of integration, we can rewrite the left-hand side of (3.13) as

$$\int_{C(\mathbf{z}) \cap D(\mathbf{z})^c} f(\mathbf{z}; \theta_1)\, d\mathbf{z} - \int_{C(\mathbf{z})^c \cap D(\mathbf{z})} f(\mathbf{z}; \theta_1)\, d\mathbf{z}. \tag{3.14}$$

To make further progress, we need to relate the two integrals in (3.14) to similar integrals under the null hypothesis (or effectively $f(\mathbf{z}; \theta_1)$ to $f(\mathbf{z}; \theta_0)$) so that we can use (3.12). Since we are trying to establish an inequality in (3.14), it is enough to enforce an inequality between $f(\mathbf{z}; \theta_1)$ and $f(\mathbf{z}; \theta_0)$. Upon reflection, if under $C(\mathbf{z}) \cap D(\mathbf{z})^c \subset C(\mathbf{z})$ we have $\mathbf{f}(\mathbf{z}; \theta_1) \geq k f(\mathbf{z}; \theta_0)$ for some $k > 0$, and if under $C(\mathbf{z})^c \cap D(\mathbf{z}) \subset C(\mathbf{z})^c$ we have the opposite inequality, namely $\mathbf{f}(\mathbf{z}; \theta_1) < k f(\mathbf{z}; \theta_0)$, we see that (3.14) satisfies

$$\int_{C(\mathbf{z}) \cap D(\mathbf{z})^c} f(\mathbf{z}; \theta_1)\, d\mathbf{z} - \int_{C(\mathbf{z})^c \cap D(\mathbf{z})} f(\mathbf{z}; \theta_1)\, d\mathbf{z}$$
$$\geq k \int_{C(\mathbf{z}) \cap D(\mathbf{z})^c} f(\mathbf{z}; \theta_0)\, d\mathbf{z} - k \int_{C(\mathbf{z})^c \cap D(\mathbf{z})} f(\mathbf{z}; \theta_0)\, d\mathbf{z}.$$

Reversing the passage from (3.13) to (3.14), we have

$$\int_{C(\mathbf{z}) \cap D(\mathbf{z})^c} f(\mathbf{z}; \theta_1)\, d\mathbf{z} - \int_{C(\mathbf{z})^c \cap D(\mathbf{z})} f(\mathbf{z}; \theta_1)\, d\mathbf{z}$$
$$\geq k \int_{C(\mathbf{z})} f(\mathbf{z}; \theta_0)\, d\mathbf{z} - k \int_{D(\mathbf{z})} f(\mathbf{z}; \theta_0)\, d\mathbf{z} \tag{3.15}$$

to which we can apply (3.12). Since both tests $C(\mathbf{z})$ and $D(\mathbf{z})$ are level α tests satisfying (3.12), the integrals on the right-hand side of (3.15) equal α and the right-hand side of (3.15) is 0. This establishes that (3.12) holds for $C(\mathbf{z}) = \{f(\mathbf{z}; \theta_1) \geq k f(\mathbf{z}; \theta_0)\}$ where k is chosen to satisfy (3.12) and this is therefore the most powerful level α test. This test is called the *likelihood ratio test* and the result that it is the most powerful α level test is called the *Neymann–Pearson lemma*.

Lemma 3.1 (Neyman and Pearson, 1933) *Suppose that the model* $\mathscr{F} = \{f(\mathbf{z}; \theta): \theta \in \Omega\}$ *holds. If there exists a* $0 < k < \infty$ *such that* $\mathrm{P}\{f(\mathbf{Z}; \theta_1)/f(\mathbf{Z}; \theta_0) \geq k; \theta = \theta_0\} = \alpha$, $0 < \alpha < 1$, *then the most powerful* α *level test for testing* $\mathrm{H}_0: \theta = \theta_0$ *against* $\mathrm{H}_1: \theta = \theta_1$ *is given by*

$$C(\mathbf{z}) = \left\{\frac{f(\mathbf{z}; \theta_1)}{f(\mathbf{z}; \theta_0)} \geq k\right\}.$$

We have assumed in our derivation of the Neyman–Pearson lemma that a likelihood ratio test of exact size α exists. With discrete data, the sampling distribution of the likelihood ratio statistic does not take on all possible values between 0 and 1 so a likelihood ratio test with exact size α may not exist. The simplest practical alternative in this case is to replace (3.12) by the inequality constraint

$$\mathrm{P}\{C(\mathbf{Z}); \mathrm{H}_0\} = \int_{C(\mathbf{z})} f(\mathbf{z}; \theta_0)\, d\mathbf{z} \leq \alpha. \tag{3.16}$$

3.3.4 Uniformly Most Powerful Tests

When the alternative hypothesis is composite, the best test, called the *uniformly most powerful test*, is the test that is more powerful than any other test at every alternative parameter value. Thus the uniformly most powerful level α test for testing $\mathrm{H}_0: \theta = \theta_0$ against $\mathrm{H}_1: \theta \neq \theta_0$ under the model (3.8) is the test $C(\mathbf{z})$ which maximizes

$$\mathrm{P}\{C(\mathbf{Z}); \theta\} \qquad \text{for all } \theta \neq \theta_0 \tag{3.17}$$

subject to $\mathrm{P}\{C(\mathbf{Z}); \theta_0\} \leq \alpha$.

In our problem of testing $\mathrm{H}_0: \mu = 0$ against $\mathrm{H}_1: \mu \neq 0$ under the Gaussian model (3.1), the null hypothesis is also composite so (3.17) needs to be generalized slightly. The uniformly most powerful level α for testing $\mathrm{H}_0: \mu = 0$ against $\mathrm{H}_1: \mu \neq 0$ under the model (3.8) is the test $C(\mathbf{z})$ which maximizes

$$\mathrm{P}\{C(\mathbf{Z}); \mu, \sigma\} \qquad \text{for all } \mu \neq 0, \quad \sigma > 0$$

subject to $\mathrm{P}\{C(\mathbf{Z}); \mu = 0, \sigma\} \leq \alpha$ for all $\sigma > 0$.

Uniformly most powerful tests exist only in limited cases and, unfortunately, the problem of testing $\mathrm{H}_0: \mu = 0$ against $\mathrm{H}_1: \mu \neq 0$ under the Gaussian model (3.1) is not such a case; see Lehmann (1959/1991, p. 108 ff). Intuitively, the difficulty is that the parameter space specified by the alternative hypothesis can be partitioned into subsets and we can construct tests which are most powerful for parameter values in each subset but have very low power for parameter values anywhere else. At least over the regions where they are most powerful, these tests dominate tests which are trying to be powerful over the entire parameter space specified by the alternative hypothesis. In particular, we have

$$\{\mu \neq 0, \sigma\} = \{\mu > 0, \sigma\} \cup \{\mu < 0, \sigma\}$$

and the α level tests $C_1(\mathbf{z}) = \{\mathbf{z}: n^{1/2}\bar{z}/s \geq k_1\}$, where $\mathrm{P}\{n^{1/2}\bar{z}/s \geq k_1; H_0\} = \alpha$, and $C_2(\mathbf{z}) = \{\mathbf{z}: n^{1/2}\bar{z}/s \leq k_2\}$, where $\mathrm{P}\{n^{1/2}\bar{z}/s \leq k_2; H_0\} = \alpha$, are more powerful than the intuitively reasonable α level test $C(\mathbf{z}) = \{\mathbf{z}: n^{1/2}|\bar{z}/s| \geq k\}$, where $\mathrm{P}\{n^{1/2}|\bar{Z}/S| \geq k; \mathrm{H}_0\} = \alpha$, over $\{\mu > 0, \sigma\}$ and $\{\mu < 0, \sigma\}$ respectively, but have very low power over $\{\mu < 0, \sigma\}$ and $\{\mu > 0, \sigma\}$ respectively. The power functions of these tests are computed in Section 3.3.6 and Problem 3.3.2, and then plotted in Figure 3.3.

3.3.5 Uniformly Most Powerful Unbiased Tests

The mathematical solution to the fact that uniformly most powerful tests exist only rarely is to introduce additional criteria to restrict the class of tests under consideration and then to try to find the most powerful test in that restricted class. For example, we can exclude tests which are powerful over a region of the parameter space at the cost of having very low power elsewhere. In particular, for testing $\mathrm{H}_0: \mu = 0$ against $\mathrm{H}_1: \mu \neq 0$ under (3.1), we may restrict attention to the class of *unbiased tests* which satisfy

$$\mathrm{P}\{C(\mathbf{Z}); \mu = 0, \sigma\} \leq \alpha \qquad \text{for all } \sigma \geq 0$$

and

$$\mathrm{P}\{C(\mathbf{Z}); \mu, \sigma\} \geq \alpha \qquad \text{for all } \mu \neq 0, \quad \sigma > 0. \tag{3.18}$$

That is, we exclude tests which are very powerful in some region of the space of alternatives but very poor (with power less than the size!) elsewhere. (This is analogous at least in its goals to unbiasedness for estimators in Section 3.1.4.) The tests $C_1(\mathbf{z}) = \{\mathbf{z}: n^{1/2}\bar{z}/s \geq k_1\}$ and $C_2(\mathbf{z}) = \{\mathbf{z}: n^{1/2}\bar{z}/s \leq k_2\}$ do not satisfy (3.18) so they are not unbiased and are excluded from consideration. In the class of unbiased tests, the test $C(\mathbf{z}) = \{\mathbf{z}: n^{1/2}|\bar{z}/s| \geq k\}$, where $\mathrm{P}\{n^{1/2}|\bar{Z}/S| \geq k; \mathrm{H}_0\} = \alpha$ is the most powerful test for testing $\mathrm{H}_0: \mu = 0$ against $\mathrm{H}_1: \mu \neq 0$; see Lehmann (1959/1991, p. 192 ff). We say that it is the *uniformly most powerful unbiased test.*

3.3.6 Testing the Effect of Caffeine on the Volume of Urine Produced

The uniformly most powerful unbiased test for testing $\mathrm{H}_0: \mu = 0$ against $\mathrm{H}_1: \mu \neq 0$ under the Gaussian model (3.1) is given by $C(\mathbf{z}) = \{\mathbf{z}: n^{1/2}|\bar{z}/s| \geq c\}$, where c satisfies

$$\alpha = \mathrm{P}\{|T_\nu| \geq c; \mathrm{H}_0\} = 2\{1 - G_\nu(c)\}.$$

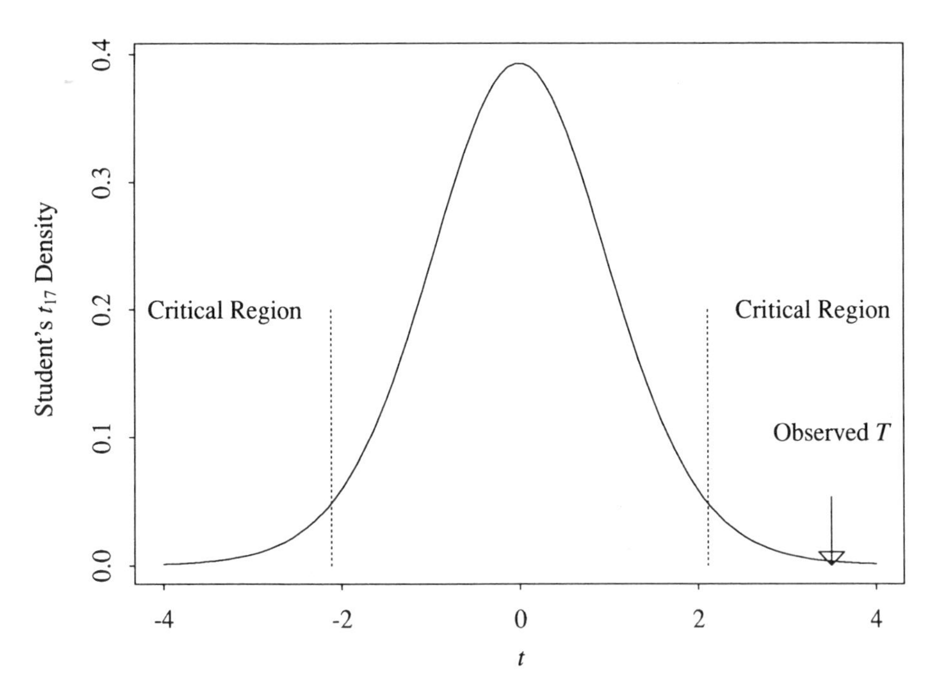

Figure 3.2. The critical region of the 5% level t test for $n = 18$. The density in the plot is that of the sampling distribution of T under the Gaussian model and H_0: $\mu = 0$ and the arrow shows the observed value t of T.

With $n = 18$ and $\alpha = 0.05$, we find $c = 2.11$. That is, we reject H_0 if $|t| \geq 2.11$ and accept H_0 otherwise. The critical region is shown in relation to the t_{17} density in Figure 3.2. The realized value t of T_v for our data is 3.50 so we accept H_1.

The power of this test against the alternative H_1: $\mu = \mu_1$ depends on the distribution of $T_v = n^{1/2}\bar{Z}/S$ when $\mu = \mu_1$. This distribution is called the *noncentral t distribution* with $v = 17$ degrees of freedom and *noncentrality parameter* $t_1 = n^{1/2}\mu_1/\sigma$, and we denote its distribution function by G_{v,t_1}. The power is then

$$P\{|T_v| \geq c; \mu_1, \sigma\} = 1 - G_{v,t_1}(c) + G_{v,t_1}(-c).$$

The noncentral t distribution is not easy to work with but we can approximate it by the *central* (*Student*) *t distribution* to obtain the convenient approximation

$$P\{|T_v| \geq c; \mu_1, \sigma\} \approx 1 - G_v(-t_1 + c) + G_v(-t_1 - c). \tag{3.19}$$

As we vary μ_1 and hence t_1, we obtain a function over the set of possible alternatives called the *power function*. For the volume of urine data, with $n = 18$

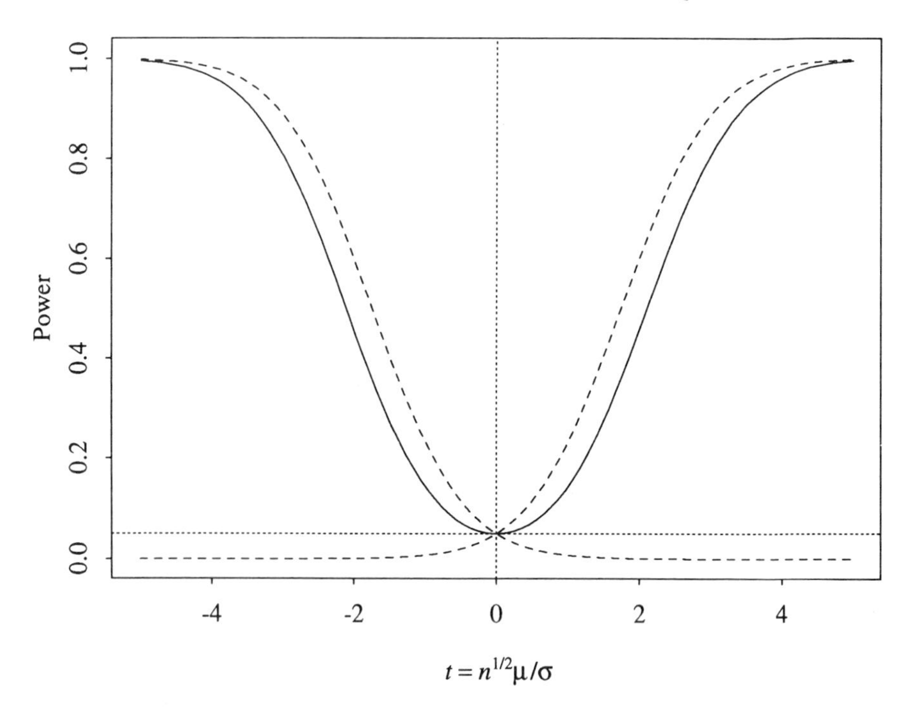

Figure 3.3. The power function of the 5% level t test of $H_0: \mu = 0$ against $H_1: \mu \neq 0$ for $n = 18$. The dashed functions are the power functions of the 5% level t tests of H_0 against $\{\mu > 0, \sigma\}$ and $\{\mu < 0, \sigma\}$ respectively. The horizontal line represents power equal to 0.05, the level of the tests.

and $\alpha = 0.05$, (3.19) reduces to

$$1 - G_{17}(-t_1 + 2.11) + G_{17}(-t_1 - 2.11)$$

which is plotted in Figure 3.3.

3.3.7 One- and Two-Sided Tests

The test $C(\mathbf{z}) = \{\mathbf{z}: n^{1/2}|\bar{z}/s| \geq k\}$ is sometimes called a *two-sided test* to reflect the nature of the two hypotheses. *One-sided tests* such as $C_1(\mathbf{z}) = \{\mathbf{z}: n^{1/2}\bar{z}/s \geq k_1\}$ and $C_2(\mathbf{z}) = \{\mathbf{z}: n^{1/2}\bar{z}/s \leq k_2\}$ are appropriate when the alternative hypothesis is that the ingestion of caffeine increases/reduces the volume of urine produced. So for example, if we test $H_0: \mu = 0$ against $H_1: \mu < 0$ (that caffeine reduces the volume of urine produced) with $\alpha = 0.05$, the critical region is $\{T_\nu \leq c\}$, where

$$0.05 = P\{T_\nu \leq c; H_0\} = G_\nu(c).$$

We find that $c = -1.74$ so we accept H_1. Notice that here we have not explicitly considered the possibility that $\mu > 0$ (i.e., that coffee increases the volume of urine produced); the test has simply accepted the least discordant of the two hypotheses. Thus we can actually interpret the test as one of $H_0: \mu \geq 0$ against $H_1: \mu < 0$.

3.3.8 The Interpretation of Level and Power

The level of a test is the proportion of times, in a large number of tests at the same level, we would reject H_0 when it is in fact true and the power is the proportion of times we would accept the alternative when it is in fact true. We cannot say anything about the performance of a test in any particular implementation so we rely on the fact that it has good properties when we average over the sample space.

3.3.9 The Choice of the Null Hypothesis

The calculations for actually implementing the test as opposed to evaluating its properties depend only on H_0. Thus there is an asymmetry in the treatment of the two hypotheses which imposes a different logical status on them. One view is that the null hypothesis should be the more important of the two hypotheses, the one for which we want precise control of the probability of error. However, this view is only relevant once the data have been collected because, by varying n, we can control both probabilities of error. This suggests correctly that considerations of power may be useful in determining sample sizes (see Section 3.4.5) though, in the absence of extensive knowledge of the phenomena under consideration, such determinations usually depend on strong assumptions.

3.3.10 Achievements and Limitations of Hypothesis Tests

Neyman and Pearson felt that their theory provided a logical basis for testing. First, the explicit introduction of an alternative hypothesis leads to a framework for planning an experiment and for selecting a test. Second, the rejection region concept is closely related to and justifies the use of tail probabilities and significance levels and provides a direct frequency interpretation for them. Fisher rejected these claims throughout his life. He argued against the decision theoretic formulation, the utility considerations, and what he viewed as the attempt to suppress the need for personal judgement. In his view inference should be distinguished from decision making as being provisional and subject to possible revision rather than final. As Pitman (1979, p. 1) explained, inference is persuasive rather than coercive. See also Cox (1958). Neyman argued strongly that it is possible to derive useful inferential procedures using decision theoretic ideas, that the usual end result of inference is the reaching of a decision and that by defining "decision" broadly enough we can regard inference as a part of decision theory. Interestingly, Kiefer (1977a) argued that decision theory is a part of inference.

The utility considerations introduced by Neyman and Pearson constitute a minimal basis for a mathematical treatment. Decision making should depend also on past evidence, past experience, and the possibly unquantifiable costs and benefits of each decision. Although Wald (1950) developed a more complete *decision theory* to overcome these limitations, its complexity and the difficulty of formalizing the more subjective elements of decision making have limited its use in inferential problems. It seems neither possible nor desirable to completely remove the subjective aspects of data analysis from statistical inference.

3.3.11 Mathematical Statistics

The probability calculations for a Neymann–Pearson test do not in any way depend on the data actually observed. Thus while a p-value depends on the observed data, the level is a property of a test without regard to data. Thus the Neyman–Pearson formulation made possible the mathematical comparison of statistical procedures without regard to particular sets of data or contexts and so laid the foundations for a purely mathematical theory of statistics.

PROBLEMS

3.3.1. Suppose that we have observations $\mathbf{Z}$ on the model

$$\mathscr{F} = \left\{ f(\mathbf{y}; \mu, \sigma) = \prod_{i=1}^{n} \frac{1}{(2\pi\sigma^2)^{1/2}} \exp\left\{ -\frac{(y_i - \mu)^2}{2\sigma^2} \right\}, \right.$$
$$\left. -\infty \leq y_i \leq \infty : \mu \in \mathbb{R}, \sigma > 0 \right\}.$$

Consider testing the null hypothesis $H_0: \mu = 0,\ \sigma = \sigma_0$ against the alternative hypothesis $H_1: \mu = \mu_0 > 0,\ \sigma = \sigma_0$. Show that the test $C(\mathbf{z}) = \{\mathbf{z}: n^{1/2}\bar{z}/\sigma_0 \geq c\}$, where c satisfies

$$\alpha = P\left\{ \frac{n^{1/2}\bar{Z}}{\sigma_0} \geq c; 0, \sigma_0 \right\}$$

is the most powerful α level test for this problem. Find an expression for c in terms of α. Find the power of the test and plot it as a function of $\mu_0 > 0$ for fixed σ_0. Explore what happens to the power as α and n change.

3.3.2. Consider the problem of testing $H_0: \mu = 0$ against $H_1: \mu \neq 0$ under the Gaussian model (3.1). Compute the power of the α level tests

$C_1(\mathbf{z}) = \{\mathbf{z}: n^{1/2}\bar{z}/s \geq k_1\}$, where $P\{n^{1/2}\bar{z}/s \geq k_1; H_0\} = \alpha$, and $C_2(\mathbf{z}) = \{\mathbf{z}: n^{1/2}\bar{z}/s \leq k_2\}$, where $P\{n^{1/2}\bar{z}/s \leq k_2; H_0\} = \alpha$.

3.3.3. In Problem 2.2.7 we adopted the uniform model

$$\mathscr{F} = \left\{ f(\mathbf{z}; \theta) = \prod_{i=1}^{n} \frac{1}{\theta} I(0 \leq z_i \leq \theta) : \theta > 0 \right\},$$

where $I(\cdot)$ is the indicator function, for 12 random numbers generated on a random number generator. Derive the likelihood ratio for testing $H_0: \theta = \theta_0$ against $H_1: \theta \neq \theta_0$. Find the critical region and the p-value for the likelihood ratio test. Obtain the power function for the 5% level test with $\theta_0 = 1$ and use it to determine the sample size required to obtain a power of 0.8 against the alternative $\theta = 0.98$.

3.4 IMPLEMENTING TESTS

Various textbooks and applications journals have institutionalized a paradigm for statistical testing which is a hybrid of the conceptually distinct Fisherian and Neyman–Pearson approaches. This hybrid is motivated by the fact that the two theories often lead to the same statistic being used and hence to identical numerical results. However, this formulation has blurred the subtle conceptual differences between the theories.

The connection which makes hybridization possible is that the p-value can be interpreted as the smallest significance level for which the observed data would result in the rejection of the null hypothesis. Thus, an alternative way to carry out a Neyman–Pearson test is to compute the p-value; if it is smaller than a predetermined significance level α, we reject H_0, whereas if it is greater than α we accept H_0.

The hybrid approach adopts the Neyman–Pearson concepts for the theoretical evaluation and comparison of the properties of tests. Tests are actually carried out by computing p-values and referring them to a predetermined significance level. As noted by Gigerenzer et al. (1989, p. 107), it is common to accept the Fisherian view that the null hypothesis cannot be accepted and often also the Fisherian restriction to sharp null hypotheses of no effect. If the p-value is larger than α, the test is said to be "*not significant*"; if the p-value is less than α, the test is "*statistically significant*" and the null hypothesis is said to be "*rejected at the α level*". The Neyman–Pearson interpretation of testing in behavioral terms is often ignored and the p-value is given a Fisherian interpretation which can lead to confusion about the meaning of a significance level.

3.4.1 Naming Tests

Many statistical tests are loosely referred to by the name of the reference distribution used in the test. Thus all the tests constructed in this chapter may be referred to as "*t tests*" or "*paired t tests*". However, there are in fact many t tests corresponding to different hypotheses, different test statistics, and so on, so in the report of an analysis, it needs to be made clear which tests have been used.

3.4.2 The Choice of Significance Level

The widespread convention of choosing levels of 0.05 or 0.01 irrespective of the context of the analysis has neither a scientific nor a logical basis. The choice of level is a question of personal judgement in the Fisherian approach and one of considering type I and II errors in the Neyman–Pearson approach. Since for a given sample size decreasing one error probability increases the other (Figure 3.4), it is possible to argue for a relative balance. In particular, if at $\alpha = 0.05$ the power is very low, one might seriously consider increasing α and so increasing the power.

A simplified example may help illustrate the issues. As part of a claim for

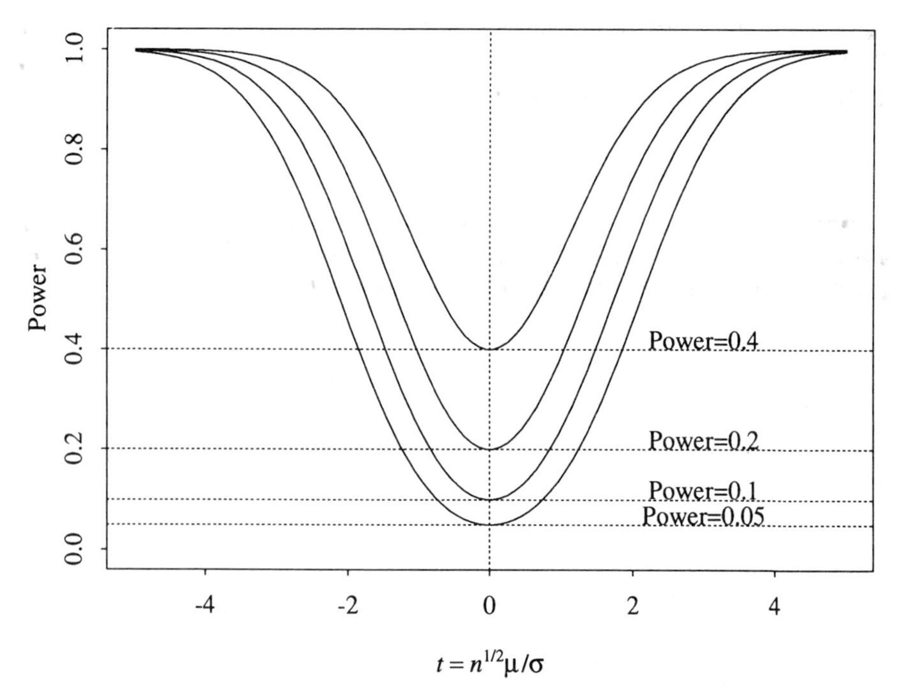

Figure 3.4. The effect on the power function of changing the level of a test while holding the sample size fixed ($n = 18$).

compensation from the government, Vietnam veterans tried to show that they had higher incidences of medical problems arising from their exposure to hazardous chemicals in Vietnam than a control group of contemporaries who did not go to Vietnam. We can think of the outcome as either the government pays compensation or does not. This is a decision theoretic problem in which the null hypothesis of no effect is associated with the action of not paying compensation and the alternative of a positive effect is associated with paying compensation. The two types of errors are

- *Type 1:* Find an effect when there is none or the government pays when it should not.
- *Type 2:* Find no effect when there is one or the veterans should be compensated but are not.

Any effect is expected to be small relative to natural variability and so difficult to detect. A small α yields a test with low power and so places the burden of proof heavily on the veterans. Increasing α increases the power and decreases the burden on the veterans. We have to think about the relative consequences of the two types of errors for the two parties and even entertain discussion about the morality of the burden. By trading off the relative burdens, we should arrive at a reasonable value of α which one could expect to be considerably larger than 0.05. In this type of problem, an explicit context-dependent argument should be given in advance of data collection to justify the choice of α.

3.4.3 Multiple Tests

In the formal presentation of testing, it is usual to assume that everything about the model other than the hypothesis under test is specified before we obtain the data on which the analysis will be based and that the analysis will consist of a single test. In such circumstances, the significance level is well defined and easily interpretable. However, it is also important to allow for testing several hypotheses about different aspects of the underlying population. Clearly, carrying out several formal Neyman–Pearson tests increases the number of possible outcomes and introduces further possible errors. For example, if we have tests $C_1(\mathbf{z})$ and $C_2(\mathbf{z})$ of the hypotheses H_1 and H_2 such that

$$\mathrm{P}\{C_1(\mathbf{Z}); \mathrm{H}_1\} = \mathrm{P}\{C_2(\mathbf{Z}); \mathrm{H}_2\} = \alpha,$$

then the probability of a type I error in at least one of the tests is generally greater than α because

$$\begin{aligned}\mathrm{P}\{C_1(\mathbf{Z}) \cup C_2(\mathbf{Z}); \mathrm{H}_1, \mathrm{H}_2\} &= 1 - \mathrm{P}\{C_1(\mathbf{Z})^c \cap C_2(\mathbf{Z})^c; \mathrm{H}_1, \mathrm{H}_2\} \\ &\geq 1 - \mathrm{P}\{C_1(\mathbf{Z})^c; \mathrm{H}_1\} \\ &= \alpha\end{aligned}$$

and satisfies

$$\begin{aligned} \mathrm{P}\{C_1(\mathbf{Z}) \cup C_2(\mathbf{Z}); \mathrm{H}_1, \mathrm{H}_2\} &\le \mathrm{P}\{C_1(\mathbf{Z}); \mathrm{H}_1\} + \mathrm{P}\{C_2(\mathbf{Z}); \mathrm{H}_2\} \\ &= 2\alpha. \end{aligned}$$

That is, the more tests we make, the more likely we are to make at least one type I error.

The strategy of specifying all aspects of the model other than the one of interest in advance of collecting the data would often lead to absurd results in practice. It is therefore important to allow a more flexible data-analytic approach in which the underlying assumptions are subjected to scrutiny and in which some aspects of the model are specified or modified after examining the data. When we carry out informal tests as in diagnostic checking, the probabilities of various types of error are not well defined. However, the effect of data analysis and multiple tests (formal or informal) is to increase the overall probability of falsely rejecting at least one null hypothesis and hence increases the overall significance level. The practical consequence is that the more testing (formal or informal) that we do, the smaller the p-value has to be before we deem a test to be significant. This means that we need to know exactly what was done in the analysis in order to interpret the p-values.

3.4.4 One- and Two-Sided Tests

It is quite appropriate to consider one- and two-sided tests (referring to the nature of the alternative hypotheses) in the Neyman–Pearson framework. However, many statisticians are sceptical of the widespread use of one-sided tests because they do not fit into the Fisherian framework and because they are open to abuse. This possibility occurs because the critical point for a two-sided test is always more extreme than that for a one-sided test (half the size when the sampling distribution is symmetric) so a significant result is more easily obtained with a one-sided than with a two-sided test.

3.4.5 Sample Size

For reasonable tests with a fixed level of significance, the power increases as the sample size increases. We illustrate this graphically in Figure 3.5 by plotting the approximate power function

$$\mathrm{P}_\mu\{|T_{n-1}| \ge c\} \approx 1 - G_\nu(-n^{1/2}t + c) + G_\nu(-n^{1/2}t - c),$$

where c is chosen so that $\alpha = 2\{1 - G_\nu(c)\}$, as a function of $t = \mu/\sigma 18^{1/2}$ for different sample sizes n. It is clear that when n is small, the power to detect even quite large alternatives to H_0 is small but when n is large, the power to detect alternatives close to H_0 is high.

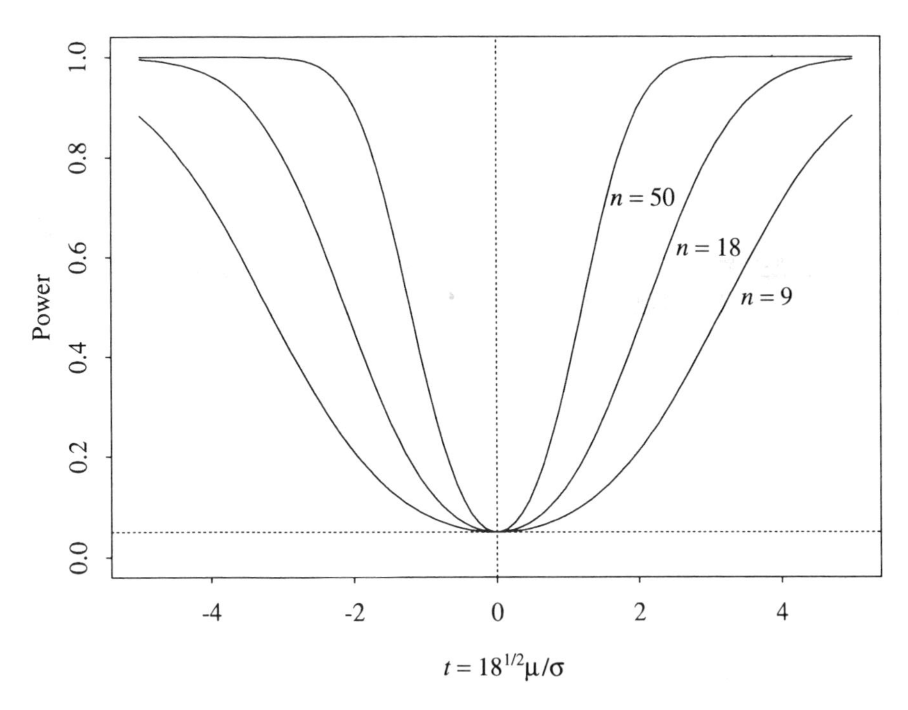

Figure 3.5. The effect on the power function of changing the sample size while holding the level of the test fixed ($\alpha = 0.05$). The horizontal line represents power equal to 0.05, the level of the tests.

The dependence of the test result on sample size is the basis for sample size determination: specify H_1, specify the desired power, and then choose n to achieve this power. But its other implications are perhaps even more important: for small samples, tests tend to be not significant while for large samples they tend to reject H_0. Thus the outcome is largely determined by the sample size irrespective of the actual data! As a simple numerical example, suppose that we have independent observations from a normal distribution with mean μ and variance 100. Then we would accept H_0: $\mu = 0$ at the 5% level if we observed a sample mean of 9.5 when $n = 4$ but we would reject H_0 at the same level if we observed a sample mean of 0.02 when $n = 10^6$. This property is desirable in the sense that it reflects the fact that as n increases we know more about the underlying population but if we conceive the hypothesis that "$\mu = 0$" as an approximation to "μ is small", we can (for large samples) end up rejecting our (untrue) approximation when the actual hypothesis is in fact correct. Consequently some statisticians argue against the use of sharp hypotheses as approximations to hypotheses of more value.

The effect of sample size on a test is not just a hypothetical issue. It arises in Problem 3.2.4 for which $n = 1792$ and even more dramatically in the work of Jahn et al. (1987) which involved $n = 104\,490\,000$ Bernoulli trials to detect psychokinesis. The null hypothesis of no psychokinesis is expressed as $H_0: \pi = 1/2$ in a binomial(n, π) model. The p-value is approximately 0.0003, which appears to provide decisive evidence in favor of psychokinesis. However, as noted by Jefferys (1990; 1992), there were only 0.018% more "successes" than the $n/2$ we would expect to observe under the null hypothesis, so the p-value overstates the evidence against H_0. Jefferys suggested a Bayesian analysis for this problem which is discussed in Section 3.5.3.

An alternative possible resolution is to allow the significance level to depend on n so that it decreases as n increases. This suggestion does make clear that the interpretation of a test depends on the sample size. To make this dependence explicit, Good (1988) has suggested that we report standardized p-values

$$\min\left\{0.5, p\sqrt{\frac{n}{100}}\right\}$$

which are adjusted to a standard sample size of 100. This suggestion has not been widely adopted yet.

For recent discussion of the use of p-values see Royall (1986), Good (1986), Berger and Selke (1987), Casella and Berger (1987), Berger and Delampady (1987), and Barnard (1990).

3.4.6 Substantive Importance

In interpreting the result of a statistical test, it should be kept in mind that statistical significance is not the same as *substantive importance* or, in other words, statistical significance is not the goal of experimentation. As an example, suppose that we regard a caffeine effect of 200 ml/3 h (i.e., slightly less than a cup) as important. The test of $H_0: \mu = 0$ is significant with our data but the effect is not of substantive importance. On the other hand, nonsignificant results can also be important. A judgement of substantive importance is in fact nonstatistical, based on context, knowledge, and experience and the criterion should ideally be specified in advance of data collection.

3.5 LIKELIHOOD RATIO, LIKELIHOOD, AND BAYESIAN TESTS

An important feature of the Neyman–Pearson approach to testing (and to frequentist inference generally) is that it is *nonconstructivist*: it tells us which tests are valid and shows us how to compare tests but it does not tell us how to find optimal tests. While there are, not surprisingly, a number of ways of constructing tests, one of the most general and important approaches uses likelihood ratios to construct tests.

Although use of the likelihood ratio is optional in frequentist theory, there are a number of arguments which suggest that frequentists should base inference on the likelihood ratio. First, as discussed in Sections 2.6.2 and 7.3, the likelihood ratio contains all the information in the data about the model. Second, tests based on likelihood ratios are often optimal as shown by the Neyman–Pearson lemma (Section 3.3.3) and various extensions of it.

3.5.1 The Likelihood Ratio Test for the Effect of Caffeine

From the calculations in Section 2.7.1, the ratio of the likelihood under the Gaussian model (3.1) and $H_1: \mu \neq 0$ to the likelihood under $H_0: \mu = 0$ is given by

$$\frac{f(\mathbf{z}; \hat{\mu}, \hat{\sigma})}{f(\mathbf{z}; 0, \hat{\sigma}(0))} = \left(1 + \frac{n\bar{z}^2}{\nu s^2}\right)^{n/2}. \qquad (3.20)$$

The Neyman–Pearson likelihood ratio test of H_0 against H_1 is given by

$$\left\{\mathbf{z}: \left(1 + \frac{n\bar{z}^2}{\nu s^2}\right)^{n/2} \geq k^*\right\}, \qquad (3.21)$$

where k^* is chosen so that $P\{(1 + n\bar{Z}^2/\nu S^2)^{n/2} \geq k^*; H_0\} = \alpha$.

Since the likelihood ratio $(1 + n\bar{z}^2/\nu s^2)^{n/2}$ is a monotone function of $|n^{1/2}\bar{z}/s|$, the test (3.21) is equivalent to $\{\mathbf{z}: |n^{1/2}\bar{z}/s| \geq k\}$, where $P(|n^{1/2}\bar{Z}/S| \geq k; H_0) = \alpha$, the test we considered in Section 3.3.6. If we decide to implement the likelihood ratio test (3.21) by computing the p-value, we find that the observed likelihood ratio is $(1 + n\bar{z}^2/\nu s^2)^{n/2} = 133.40$, so

$$\begin{aligned} p &= P\left\{\left(1 + \frac{n\bar{Z}^2}{\nu S^2}\right)^{n/2} \geq 133.40\right\} \\ &= P\{|T_{17}| \geq 3.50; H_0\}. \end{aligned}$$

Thus the p-value for the likelihood ratio test of $H_0: \mu = 0$ against $H_1: \mu \neq 0$ is identical to that based on the t-ratio $|n^{1/2}\bar{z}/s|$ in this case.

3.5.2 Relationship to Likelihood Inference

Recall that in the likelihood approach to inference presented in Section 2.7, we compare hypotheses by comparing the ratio of the likelihoods of those hypotheses. In particular, the relative "plausibility" of $H_0: \mu = 0$ to $H_1: \mu \neq 0$ is given by the reciprocal of the likelihood ratio (3.20) which is

$$\frac{f(\mathbf{z}; 0, \hat{\sigma}(0))}{f(\mathbf{z}; \hat{\mu}, \hat{\sigma})} = \left\{1 + \frac{n\bar{z}^2}{\nu s^2}\right\}^{-n/2} = 0.0075,$$

so H_0 is $1/0.0075 \approx 133$ times less "plausible" than H_1.

When we use the likelihood ratio (3.20) as a frequentist test statistic in (3.21), we also use the additional fact that the likelihood ratio is a realization of the random function

$$\frac{f(\mathbf{Z}; 0, \hat{\sigma}(0))}{f(\mathbf{Z}; \hat{\mu}, \hat{\sigma})} = \left\{1 + \frac{n\bar{Z}^2}{\nu S^2}\right\}^{-n/2}$$

to compute its sampling distribution under the model and H_0 and then use this sampling distribution to construct the test. Thus there are two important differences between likelihood and frequentist theory:

1. In likelihood theory, inference must be based on the likelihood whereas in frequentist theory a wide choice of test statistics is allowed.
2. Likelihood theory involves direct examination of the likelihood ratio whereas frequentist theory based on the likelihood ratio requires us to use the sampling distribution to calibrate the magnitude of the likelihood ratio. Paraphrasing a comment of Jeffreys (1939/1961, p. 385), the inference depends on the observed data and also on observable results that have not occurred.

Additionally, a small but practically important difference arises because it is conventional in frequentist discussions to arrange tests so that large values of the test statistic constitute evidence against H_0. This means that frequentists usually base the p-value on the reciprocal of the ratio used in likelihood theory.

The dependence on sampling distributions makes frequentist theory more complex than likelihood theory and is the basis for the very different interpretations of often numerically similar inference statements. However, the inferences can also be quite different as was noted by Dempster (1973). Let Z be a random variable taking values in (0, 1) and consider testing

$$H_0\colon f_0(y) = 1$$

against

$$H_1\colon f_1(y) = \begin{cases} \dfrac{0.001y}{0.95c} & 0 < y \le 0.95 \\[2ex] \dfrac{19.98y}{c} - \dfrac{18.98}{c} & 0.95 < y < 1, \end{cases}$$

where $c \approx 0.0255$ is chosen to ensure that the density integrates to 1. Suppose that we make a single observation and observe $z = 0.95$. Then the likelihood ratio is just

$$\frac{f_0(0.95)}{f_1(0.95)} = \frac{1}{f_1(0.95)} \approx 25.51$$

which provides no evidence against H_0. However, the frequentist calculates the p-value

$$\begin{aligned} p\text{-value} &= P\left\{\frac{f_1(Z)}{f_0(Z)} \geq \frac{f_1(0.95)}{f_0(0.95)}; H_0\right\} \\ &= P\{f_1(Z) \geq f_1(0.95); H_0\} \\ &= P\{Z \geq 0.95; H_0\} \\ &= 0.05 \end{aligned}$$

which provides evidence against H_0. Thus the two approaches are in conflict in this problem.

3.5.3 Relationship to Bayesian Hypothesis Tests

The relationship between significance tests and Bayesian hypothesis tests (Section 2.5) is quite complex. For definiteness, we assume for simplicity throughout this section that the Gaussian model (3.1) with $\sigma = 1$ known holds.

Consider first the problem of testing $H_0: \mu = 0$ against the one-sided alternative hypothesis $H_1: \mu > 0$. If we observe $\bar{z}$, the p-value is

$$p = P(\bar{Z} \geq \bar{z}; H_0) = 1 - \Phi(n^{1/2}\bar{z}),$$

where Φ is the standard Gaussian distribution function. If we adopt the Jeffreys prior $g(\mu) \propto 1$ for μ, the posterior distribution of $\mu \mid \mathbf{Z} = \mathbf{z}$ is $N(\bar{z}, n^{-1})$. Hence the posterior probability of $\{\mu < 0\}$ is

$$P(\mu < 0 \mid \mathbf{Z} = \mathbf{z}) = \Phi(-n^{1/2}\bar{z}) = p$$

by the symmetry of the Gaussian distribution. Thus the one-sided p-value for testing $H_0: \mu = 0$ against $H_1: \mu > 0$ equals the posterior probability (from the Jeffreys prior) of $\{\mu < 0\}$. This relationship is illustrated graphically in Figure 3.6.

Now consider testing $H_0: \mu = 0$ against the general alternative $H_1: \mu \neq 0$. The p-value is now

$$p = 2\{1 - \Phi(n^{1/2}|\bar{z}|)\}. \tag{3.22}$$

Reversing our perspective, we obtain a given p-value p if $|\bar{z}| = n^{-1/2}\Phi^{-1}(1 - p/2)$. The Bayes factor from the Jeffreys Cauchy prior

$$g(\mu) = p_0 I(\mu = 0) + (1 - p_0)\{\pi(1 + \mu^2)\}^{-1} I(\mu \neq 0)$$

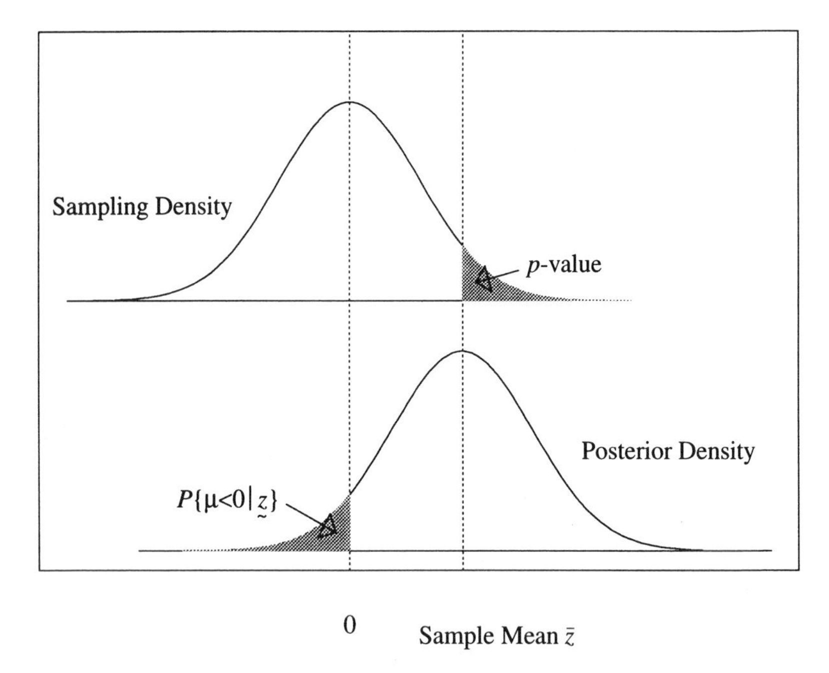

Figure 3.6. Graphical representation of the relationship between the one-sided p-value for testing H_0: $\mu = 0$ against H_1: $\mu > 0$ and the posterior probability that $\mu < 0$ in a Gaussian model.

is from (2.33) approximately

$$\left(\frac{\pi n}{2}\right)^{1/2}(1 + \bar{z}^2)\exp(-n\bar{z}^2/2). \tag{3.23}$$

Two important differences between the p-value (3.22) and the Bayes factor (3.23) are first that the sampling distribution is centered at $\mu = 0$ while the posterior is centered at $\bar{z}$ and second that the p-value is the area under the curve defined by the sampling distribution whereas the Bayes factor is the ordinate of the curve defined by the Bayes factor (3.23).

The most important difference between the two approaches can be seen by expressing the posterior odds ratio (3.23) as a function of the p-value (3.22) as

$$\left(\frac{n\pi}{2}\right)^{1/2}\{1 + n^{-1}\Phi^{-1}(1 - p/2)^2\}\exp\{-\Phi^{-1}(1 - p/2)^2/2\}.$$

This expression shows that for large n (with fixed p), the posterior odds ratio (3.23) can be arbitrarily large (evidence in support of H_0) while p is arbitrarily small (evidence against H_0). This is known as the *Jeffreys/Lindley paradox* (Jeffreys, 1939/1961, p. 248; Lindley, 1957).

The Jeffreys/Lindley paradox is the reason why Jefferys (1990; 1992) reached the opposite conclusion to that of Jahn et al. (1987) on the basis of their psychokinesis experiments (see Section 3.4.5). Jefferys showed that if we adopt a uniform prior under the alternative hypothesis, the Bayes factor equals 12 (so we would have to have very strong prior odds against the null hypothesis of no psychokinesis to obtain evidence against the null hypothesis) despite the strong evidence against the null hypothesis provided by the significance test of Jahn et al. (1987).

Combining Bartlett's paradox (Section 2.5.3) with the Jeffreys/Lindley paradox, we see that Bayesian tests of sharp hypotheses with vague priors on the alternative hypothesis tend to favor the null hypothesis (regardless of sample size) whereas frequentist tests tend to favor the alternative hypothesis in large samples.

3.5.4 Limitations of Inferences Using the Likelihood

All approaches based on the likelihood run into difficulties when applied to models for which no density exists or for which there is a large (increasing) number of parameters. These constraints, mild as they seem, do limit the applicability of these approaches. One advantage of the nonconstructivist frequentist approach is that it does not enforce any single method but simply states the properties that acceptable methods must have. Thus if a method fails, the frequentist task is to try to find another one which does not.

PROBLEMS

3.5.1. Suppose that we have observations $\mathbf{Z}$ on the model

$$\mathscr{F} = \left\{ f(\mathbf{y}; \mu_1, \mu_2, \sigma_1, \sigma_2) = \prod_{j=1}^{2} \prod_{i=1}^{n_j} \frac{1}{(2\pi\sigma_j^2)^{1/2}} \exp\left\{ -\frac{(y_{ji} - \mu_j)^2}{2\sigma_j^2} \right\}, \right.$$

$$\left. -\infty \leq y_{ji} \leq \infty \colon \mu_1, \mu_2 \in \mathbb{R}, \sigma_1, \sigma_2 > 0 \right\}.$$

(That is, we observe realizations of $n_1 + n_2$ independent Gaussian random variables, n_1 of which have mean μ_1 and variance σ_1^2 and n_2 of which have mean μ_2 and variance σ_2^2.) Consider the problem of making inferences about $\delta = \mu_1 - \mu_2$. Let $\bar{z}_1$, s_1, $\bar{z}_2$ and s_2 denote the sample mean and standard deviation for the two sets of observations. Show that the likelihood ratio test for testing $\mathrm{H}_0\colon \delta = \delta_0, \sigma_1 = \sigma_2$ against the general alternative $\mathrm{H}_1\colon \delta \neq \delta_0, \sigma_1 = \sigma_2$ is a monotone function of $T = |\bar{z}_1 - \bar{z}_2 - \delta_0| / \{s_p^2(n_1^{-1} + n_2^{-1})\}^{1/2}$, where

$$s_p^2 = \{(n_1 - 1)s_1^2 + (n_2 - 1)s_2^2\}/(n_1 + n_2 - 2).$$

Use the fact that under the Gaussian model and the null hypothesis T has the Student t distribution with $n_1 + n_2 - 2$ degrees of freedom to obtain the critical region for an α level test of H_0.

3.5.2. Suppose that we have observations $\mathbf{Z}$ on the model of Problem 3.5.1 for which $\sigma_1 = \sigma_2 = \sigma$. Show that the *analysis of variance decomposition*

$$\sum_{j=1}^{2} \sum_{i=1}^{n_j} (z_{ji} - \bar{z})^2 = \sum_{j=1}^{2} n_j(\bar{z}_j - \bar{z})^2 + \sum_{j=1}^{2} \sum_{i=1}^{n_j} (z_{ji} - \bar{z}_j)^2,$$

where

$$\bar{z}_j = n_j^{-1} \sum_{i=1}^{n_j} z_{ji} \qquad \text{and} \quad \bar{z} = \left(\sum_{j=1}^{2} n_j\right)^{-1} \sum_{j=1}^{2} \sum_{i=1}^{n_j} z_{ji}$$

holds. The analysis of variance test of the null hypothesis $H_0: \mu_1 = \mu_2$ is based on the ratio $\sum_{j=1}^{2} n_j(\bar{z}_j - \bar{z})^2 / \sum_{j=1}^{2} \sum_{i=1}^{n_j} (z_{ji} - \bar{z}_j)^2$ with large values of the ratio providing evidence against H_0. Use the fact that under the model and the null hypothesis $H_0: \mu_1 = \mu_2$, the numerator and denominator in this ratio have independent $\sigma^2\chi_1^2$ and $\sigma^2\chi_{n_1+n_2-2}^2$ distributions to construct a test of H_0. How does this test relate to the likelihood ratio test obtained in Problem 3.5.1?

3.5.3. Suppose that we have observations on the model of Problem 3.5.1. Construct the likelihood ratio test for testing $H_0: \sigma_1^2 = \sigma_2^2$ against $H_1: \sigma_1^2 \neq \sigma_2^2$. How does this test relate to the test based on the ratio $F = \sum_{i=1}^{n_1} (z_{1i} - \bar{z}_1)^2 / \sum_{i=1}^{n_2} (z_{2i} - \bar{z}_2)^2$?

3.5.4. Consider the variance component model (1.20) for the caffeine data presented in Section 1.3.7 and reparameterized in Problem 2.7.2 by $\tau_u = \sigma_u^2$ and $\tau_a = m\sigma_a^2 + \sigma_u^2$. The analysis of the caffeine data presented in this chapter is based on the fact that the pairwise differences

$$Z_i = Y_{i1} - Y_{i0} = \mu + u_{i1} - u_{i0}$$

are independent $N(\mu, 2\tau_u)$ random variables. Explore the relationship between the likelihood ratio tests of $H_0: \mu = 0$ against $H_1: \mu \neq 0$ constructed from the mixed model and from the paired differences.

3.6 CONFIDENCE SETS

In the late 1920s, following the success of his work with Pearson on hypothesis testing, Neyman began to develop a frequentist approach to set estimation. Not surprisingly, in view of his hypothesis testing paradigm, Neyman (1934; 1937) adopted a similar, nonconstructivist approach to interval estimation. He defined a $100(1 - \alpha)\%$ *confidence set* for a parameter θ to be a realization $C_\alpha(\mathbf{z})$ of a random set $C_\alpha(\mathbf{Z})$ for which the *confidence level* or *coverage probability*

$$P\{\theta \in C_\alpha(\mathbf{Z}); \theta\} \geq 1 - \alpha. \tag{3.24}$$

To avoid the situation in which a 95% confidence interval does not contain all the points in, say, an 85% confidence interval, we now also usually require the sets to be nested in the sense that $C_\gamma(\mathbf{Z}) \subset C_\alpha(\mathbf{Z})$ for $\alpha < \gamma$.

Given any set $C_\alpha(\mathbf{z})$, we can in principle compute its coverage probability $P\{\theta \in C_\alpha(\mathbf{Z}); \theta\}$ and check whether it is a $100(1 - \alpha)\%$ confidence set, but Neyman's general approach does not specify how we should obtain such sets.

3.6.1 The Pivotal Method

The pivotal method of constructing confidence intervals is based on the inversion of a pivotal quantity. A *pivotal quantity* is a function of $\mathbf{Z}$ and the parameter of interest which has a distribution which does not depend on any other unknown parameters and is monotone in the parameter of interest.

For the Gaussian model (3.1), the function $n^{1/2}(\bar{Z} - \mu)/S \sim t_\nu$ is a pivotal quantity for μ. A $100(1 - \alpha)\%$ confidence interval for μ can be obtained by noting that for any $a < b$,

$$\begin{aligned} G_\nu(b) - G_\nu(a) &= P\left\{a \leq \frac{n^{1/2}(\bar{Z} - \mu)}{S} \leq b\right\} \\ &= P\{\bar{Z} - n^{-1/2}Sb \leq \mu \leq \bar{Z} + n^{-1/2}Sa\} \end{aligned} \tag{3.25}$$

so a and b should satisfy

$$1 - \alpha = G_\nu(b) - G_\nu(a). \tag{3.26}$$

Notice that we have used the fact that a pivotal quantity has a known distribution to obtain a and b and then the fact that it is monotone in μ to invert the pivotal quantity to obtain the set containing μ.

The equation (3.26) defining a and b is a single equation in two unknowns so does not have a unique solution. One possibility is to construct a *central interval* for which $G_\nu(b) = 1 - \alpha/2$ and $G_\nu(a) = \alpha/2$. This yields $b = -a = G_\nu^{-1}(1 - \alpha/2)$ so substituting into (3.25) we obtain

$$C_\alpha(\mathbf{z}) = \{\mu: \bar{z} - n^{-1/2}sG_\nu^{-1}(1 - \alpha/2) \leq \mu \leq \bar{z} + n^{-1/2}sG_\nu^{-1}(1 - \alpha/2)\}. \tag{3.27}$$

In particular, for the caffeine data, the realized interval

$$C_\alpha(\mathbf{z}) = [170.72 - 48.72G_{17}^{-1}(1 - \alpha/2), 170.72 + 48.72G_{17}^{-1}(1 - \alpha/2)]$$

is a $100(1-\alpha)\%$ confidence interval for μ and, taking $\alpha = 0.05$,

$$C_{0.05}(\mathbf{z}) = [67, 274]$$

is a 95% confidence interval for μ.

A general procedure for finding a pivotal quantity is to choose an estimator of the parameter of interest, find its sampling distribution, and then try to find a function of the statistic and the parameter of interest which is pivotal. This is straightforward for models with a nice group structure (such as location–scale models (1.18); see the discussion in Section 2.6.6) but is often impossible. As we will see in Chapter 4 (in for example Section 4.1.11), approximate pivotal quantities are often easy to construct and we can apply the pivotal method to these to obtain approximate confidence intervals.

3.6.2 Confidence Sets Derived by Inverting Tests

Neyman observed that there is a close connection between confidence sets and tests of sharp hypotheses. For the Gaussian model (3.1), the uniformly most powerful unbiased test (Section 3.3.5) for testing $H_0: \mu = \mu_0$ against $H_1: \mu \neq \mu_0$ is given by $C(\mathbf{z}) = \{\mathbf{z}: n^{1/2}|\bar{z} - \mu_0|/s \geq c\}$, where $P\{C(\mathbf{Z}); \mu_0, \sigma\} = \alpha$. Equivalently, in terms of the acceptance region $C(\mathbf{z})^c = \{\mathbf{z}: n^{1/2}|\bar{z} - \mu_0|/s < c\}$, we have

$$P\{C(\mathbf{Z})^c; \mu_0, \sigma\} = 1 - \alpha.$$

Comparing this statement to the definition of a confidence set, we see that, if we fix the data, the set of parameter values for which we accept the null hypothesis, namely $\{\mu_0: n^{1/2}|\bar{z} - \mu_0|/s \leq c\}$, is a $100(1-\alpha)\%$ confidence set for the parameter. That is, if we invert our view of the acceptance region as a set in the sample space for given parameter values to one in the parameter space for a given sample, we obtain a $100(1-\alpha)\%$ confidence set. Notice that this requires us to consider a whole family of sharp null hypotheses rather than a single sharp null hypothesis. Since $c = G_\nu^{-1}(1-\alpha/2)$, we can invert the acceptance region to obtain (3.27).

The inversion step in constructing the confidence interval can be given a useful graphical interpretation. Suppose first that σ is known. Then for each fixed value of μ, we have that

$$P\{\mu - n^{-1/2}\sigma\Phi^{-1}(1-\alpha/2) \leq \bar{Z} \leq \mu + n^{-1/2}\sigma\Phi^{-1}(1-\alpha/2); \mu, \sigma\} = 1 - \alpha.$$

We plot μ on the x-axis and $\bar{z}$ on the y-axis and then plot the lines

$$\bar{z} = \mu - n^{-1/2}\sigma\Phi^{-1}(1-\alpha/2) \quad \text{and} \quad \bar{z} = \mu + n^{-1/2}\sigma\Phi^{-1}(1-\alpha/2)$$

to create Figure 3.7. For each value of μ we can read off a set which contains

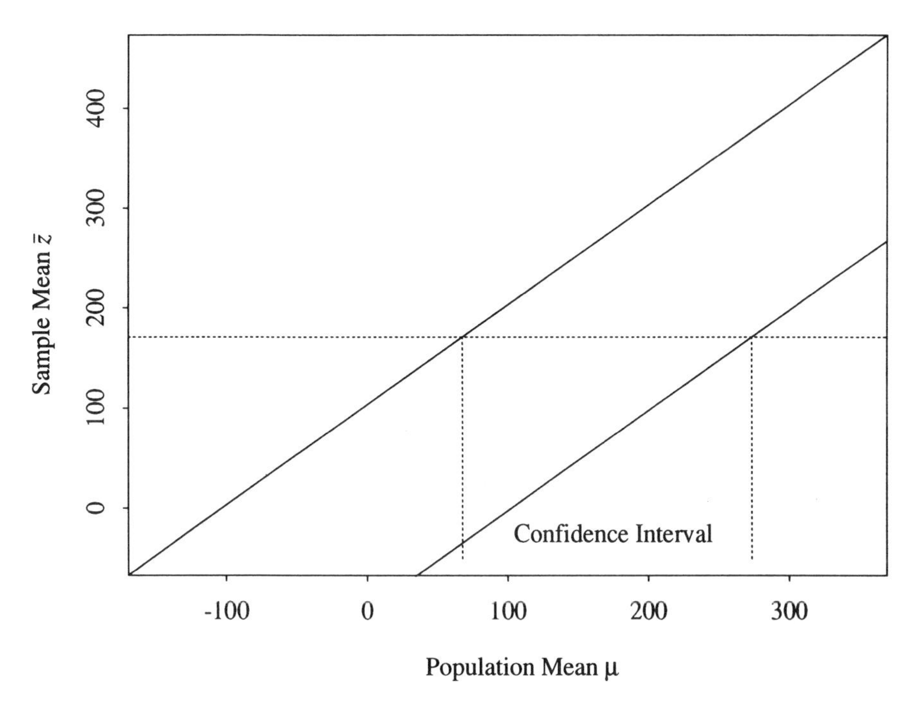

Figure 3.7. A graphical representation of the calculation of confidence limits for the Gaussian mean μ. The 95% confidence interval for μ lies between the dotted vertical lines.

the random variable $\bar{Z}$ with probability $1 - \alpha$. Conversely, for each value of $\bar{z}$, we obtain the confidence interval for μ. Varying n and α yields sets of parallel lines which reflect the effects of changing n and/or α. The probability calculation holds in moving from the x- to the y-axis, the confidence set is obtained in moving from the y- to the x-axis. Whereas in fiducial theory (Section 2.6), Fisher asserted that moving from the y- to the x-axis also has a probability interpretation, Neyman was clear that this is not the case for confidence sets. In the general case that σ is unknown, the confidence lines are again parallel straight lines but a random distance (depending on s) apart. (In fact, in Figure 3.7 we took $\sigma = s$ and G_v instead of Φ_0.) In this case, the length of the interval is a random variable.

In general, test statistics are functions of $\mathbf{Z}$ and the parameter of interest, which have a distribution that does not depend on any other unknown parameters, but they are not necessarily monotone in the parameter of interest and hence not necessarily pivotal quantities. The construction of confidence sets from tests is therefore more general than the construction based on pivotal quantities but the inversion of tests which are not pivotal quantities can be problematical (see Section 3.8.2).

3.6.3 Using Confidence Sets to Carry Out Tests

We have used the Neyman–Pearson test formulation to produce a confidence set but the converse is also possible. In particular an α level test of $H_0: \mu = \mu_0$ can be carried out by computing a $100(1-\alpha)\%$ confidence interval for μ and checking whether the interval contains μ_0 or not. If the interval does not contain μ_0, we can reject H_0. We can even recover the p-value by choosing α so that μ_0 is on the boundary of the confidence interval. For example, setting the boundary of the confidence interval to be zero, we obtain

$$170.72 - 48.72 G_{17}^{-1}(1-\alpha/2) = 0$$

or

$$\alpha = 2\{1 - G_{17}(170.72/48.72)\} = 0.0027$$

which is the p-value for testing $H_0: \mu = 0$. The length of the interval contains information about the precision of the estimate and hence the power of the test so, when possible, it is often preferable to present tests in the form of confidence intervals.

3.6.4 Optimal Confidence Intervals

Neyman took the relationship between tests and confidence sets a step further and defined optimal confidence sets so that those resulting from most powerful tests are optimal. That is, for a fixed coverage probability $1-\alpha$, an *optimal confidence set* is one with the smallest possible probability of containing false parameter values. By this device, he avoided the need to develop a new optimality theory. An alternative approach is to define optimal $100(1-\alpha)\%$ confidence intervals to be those with minimum expected length. This has been explored in large samples by Wilks (1938a). See Section 4.3 for further discussion.

3.6.5 Likelihood-Based Confidence Intervals

Generally, just as it is attractive to base tests on likelihood ratios, it is attractive to ensure that confidence sets are also likelihood sets so that points in the set have higher likelihood than points outside the set. Such a confidence set is said to be *likelihood-based*. Likelihood-based confidence sets are obtained by computing the likelihood set as in Section 2.7.3 but choosing k with reference to the sampling distribution. For example, the likelihood set

$$\left\{\mu: \left\{1 + \frac{n(\bar{Z}-\mu)^2}{\nu S^2}\right\}^{n/2} \geq k\right\}$$

$$= \left\{\mu: \bar{Z} - \nu^{1/2}\frac{S}{n^{1/2}}(k^{-2/n}-1)^{1/2} \leq \mu \leq \bar{Z} + \nu^{1/2}\frac{S}{n^{1/2}}(k^{-2/n}-1)^{1/2}\right\}$$

has coverage probability

$$P\left\{\left\{1+\frac{n(\bar{Z}-\mu)^2}{\nu S^2}\right\}^{n/2} \geq k\right\} = P\left\{\frac{n^{1/2}|\bar{Z}-\mu|}{S} \leq \nu^{1/2}(k^{-2/n}-1)^{1/2}\right\}$$

$$= 2G_\nu\{\nu^{1/2}(k^{-2/n}-1)^{1/2}\} - 1$$

so we can interpret the likelihood set as a $100(2G_\nu\{\nu^{1/2}(k^{-2/n}-1)^{1/2}\} - 1)\%$ confidence interval. Alternatively, the $100(1-\alpha)\%$ likelihood-based confidence set is obtained from

$$G_\nu\{\nu^{1/2}(k^{-2/n}-1)^{1/2}\} = 1 - \alpha/2$$

or

$$k = \left\{1 + \frac{G_\nu^{-1}(1-\alpha/2)^2}{\nu}\right\}^{-n/2}$$

which produces (3.27).

3.6.6 Coverage Properties of Credibility Sets

We can also take a set derived as a *Bayesian credibility set* (i.e., from the posterior distribution of μ given $\mathbf{Z}$) and evaluate its coverage probability (using the distribution of $\mathbf{Z}$ given μ). The frequentist does not care about the way the set was derived but insists that it must satisfy the confidence property. A Bayesian, on the other hand, may be unconcerned about the repeated sampling or frequentist properties of an inference or may regard them as a nonbinding but useful means of calibrating the inference (Rubin, 1984). As we will see in Section 3.8 and Problems 3.8.2–3.8.3, a particular Bayesian credibility set need not be a confidence set at the same level.

3.6.7 The Interpretation of Confidence Sets

A confidence set is a realization of a random set and is not itself random. In particular, the confidence set either does or does not contain μ and the randomness is introduced through the sampling distribution of the quantity from which it is derived. There are two commonly used interpretations of confidence sets:

1. If the data at hand were generated a large number of times by the same mechanism under the same conditions and for each data set we calculated $C_\alpha(\mathbf{z})$, then at most a proportion α of the sets would fail to contain μ.
2. If over many different data sets we calculate $100(1-\alpha)\%$ confidence sets, then at most a proportion α of the sets would fail to contain the parameter of interest.

The difference between these two interpretations is that in the first interpretation the problem is held fixed whereas in the second it is not. Thus the second interpretation involves a "lifetime average" for the statistician. The fact that confidence sets (like Neyman–Pearson tests, see Section 3.3.8) are interpreted in terms of their long run average behavior and not in terms of the properties of individual sets, complicates their use and interpretation. Indeed, in the discussion of the paper in which Neyman introduced confidence sets, Bowley (1934) described them as a "confidence trick".

PROBLEMS

3.6.1. Consider the uniform model of Problem 3.1.3 for 12 random numbers generated on a random number generator. Plot the likelihood ratio for testing $H_0: \theta = \theta_0$ against $H_1: \theta \neq \theta_0$ obtained in Problem 3.3.3 and obtain an explicit expression for the contours of the likelihood ratio. Set a $100(1-\alpha)\%$ likelihood-based confidence interval for θ and compare this interval to the $100(1-\alpha)\%$ Bayesian credibility interval for θ obtained in Problem 2.2.7.

3.6.2. Suppose we are concerned with the lower endpoint of the distribution of computer generated random numbers which purport to be uniformly distributed on (0, 1). Suppose that the observations presented in Problem 2.2.7 can be modeled by

$$\mathscr{F} = \left\{ f(\mathbf{z}; \xi) = \prod_{i=1}^{n} \frac{1}{1-\xi} I(\xi \leq z_i \leq 1) : 0 \leq \xi < 1 \right\},$$

where $I(\cdot)$ is the indicator function. Find a $100(1-\alpha)\%$ likelihood-based confidence interval for ξ. Compare the interval to those obtained in Problem 2.2.8. Explain briefly how to interpret the three intervals. What do you conclude about ξ?

3.6.3. Suppose we adopt the exponential model of Problem 3.1.5 for Proschan's (1963) air conditioning data presented in Problem 1.5.11. Find a $100(1-\alpha)\%$ confidence interval for λ and then for $\theta = (1/\lambda)\log 2$.

3.6.4. Suppose that we have observations $\mathbf{Z}$ on the Gaussian model (3.1). Show that $\sum_{i=1}^{n}(z_i - \bar{z})^2/\sigma^2$ is a pivotal quantity for σ^2 and hence or otherwise find a $100(1-\alpha)\%$ confidence interval for σ^2. How would you make this a likelihood-based confidence interval? How do you find a $100(1-\alpha)\%$ confidence interval for σ?

3.6.5. Suppose that we set $m \geq 1$, $100(1-\alpha)\%$ confidence intervals. Show that the simultaneous coverage probability (the probability that all the intervals contain their target true parameter) is between $1 - m\alpha$ and $1 - \alpha$.

3.6.6. Suppose that we have observations $\mathbf{Z}$ on the multivariate Student t model

$$\mathscr{F} = \left\{ f(\mathbf{y}; \mu, \sigma) = \frac{\Gamma\left(\frac{v+n}{2}\right)}{(v\pi\sigma^2)^{n/2}\Gamma\left(\frac{v}{2}\right)} \frac{1}{\left\{1 + \sum_{i=1}^{n} \frac{(y_i - \mu)^2}{v\sigma^2}\right\}^{(v+n)/2}}, \right.$$

$$\left. -\infty \leq y_i \leq \infty : \mu \in \mathbb{R}, \sigma > 0 \right\},$$

where $v > 0$ is known. Find the maximum likelihood estimators of μ and σ. Use the representation $Z_i = \mu + \sigma Y_i / h^{1/2}$, where Y_i are independent standard Gaussian random variables which are independent of the random variable h to find the exact sampling distribution of $(n-1)^{1/2}(\hat{\mu} - \mu)/\hat{\sigma}$. Obtain a $100(1-\alpha)\%$ confidence interval for μ. What is the relationship between this interval and that obtained under the Gaussian model?

3.6.7. Suppose that we have observations $\{(Y_1, x_1), \ldots, (Y_n, x_n)\}$ on the model

$$\mathscr{F} = \left\{ f(\mathbf{y}; \alpha, \beta, \sigma) = \prod_{i=1}^{n} \frac{1}{(2\pi\sigma^2)^{1/2}} \exp\left\{ -\frac{(y_i - \alpha - x_i\beta)^2}{2\sigma^2}, \right. \right.$$

$$\left. -\infty \leq y_i \leq \infty : \alpha, \beta \in \mathbb{R}, \sigma > 0 \right\}.$$

Show that the maximum likelihood estimator $\hat{\gamma} = (\hat{\alpha}, \hat{\beta})^{\mathrm{T}} = n^{-1}C^{-1}X^{\mathrm{T}}y$, where X is the $n \times 2$ matrix with ith row $(1, x_i)^{\mathrm{T}}$, $y = (Y_1, \ldots, Y_n)^{\mathrm{T}}$, and

$$C = \begin{pmatrix} 1 & n^{-1}\sum_{i=1}^{n} x_i \\ n^{-1}\sum_{i=1}^{n} x_i & n^{-1}\sum_{i=1}^{n} x_i^2 \end{pmatrix}$$

Find the maximum likelihood estimator $\hat{\sigma}$ of σ. What is the sampling distribution of $\hat{\beta}$? Use the fact that $\hat{\sigma}^2 \sim \sigma^2\chi^2_{n-2}$ is independent of $\hat{\beta}$ to construct a pivotal quantity for β and show how to construct a $100(1-\alpha)\%$ confidence interval for β. Apply your result to set a 95%

confidence interval for the slope in the regression model of Section 1.2.3 relating catecholamine excretion to the volume of urine produced during the ingestion of caffeine. Is the relationship statistically significant?

3.6.8. Suppose that we have observations $\mathbf{Z}$ on the Gaussian model (3.1) and we wish to predict the value of a future independent observation Y. Suppose initially that σ is known. Construct a pivotal quantity based on the difference $Y - \bar{Z}$ and use it to construct a $100(1-\alpha)\%$ *prediction interval* for Y. How should we interpret this interval? Now suppose that σ is unknown. Show how to construct a $100(1-\alpha)\%$ prediction interval for Y in this case. Use the caffeine data of Section 1.1.3 to predict the change in the volume of urine produced in an 18 year old male by the ingestion of caffeine under the same circumstances as in the experiment.

3.6.9. Suppose that we have observations $\mathbf{Z}$ on the exponential model of Problem 3.1.5 and we wish to predict the value of a future independent observation Y. Show that the ratio $Y/\bar{Z}$ is a pivotal quantity and use it to construct a $100(1-\alpha)\%$ prediction interval for Y. Use the pressure vessel data of Barlow et al. (1984) presented in Problem 1.5.4 to predict the bursting time of a similar vessel subjected to the same conditions.

3.6.10. Suppose that we have observations $\{(Y_1, x_1), \ldots, (Y_n, x_n)\}$ on the simple regression model of Problem 3.6.7. Suppose we wish to predict the value of a future independent observation Y_0 at x_0. Suppose initially that σ is known. Construct a pivotal quantity based on the difference $Y_0 - \hat{\alpha} - x_0\hat{\beta}$, where $\hat{\alpha}$ and $\hat{\beta}$ are the maximum likelihood estimators of α and β respectively, and use it to construct a $100(1-\alpha)\%$ prediction interval for Y_0. Now use the fact that $s^2 \sim \sigma^2\chi^2_{n-2}$ independent of $\hat{\gamma}$ to construct a $100(1-\alpha)\%$ prediction interval for Y_0 when σ is unknown.

3.7 CONFIDENCE SETS FROM DISCRETE DATA

The construction of confidence sets is complicated when the sample and parameter spaces are different in nature because then the implicit inversion is not straightforward. The simplest example of the kind of difficulties which can arise is provided by considering the construction of confidence intervals for a continuous parameter based on discrete data.

In the Diabetic Retinopathy Study described in Section 1.1.1, we observed that $z = 26$ out of $r = 175$ control eyes of subjects receiving argon treatment in the other eye suffered severe visual loss within 2 years and we adopted the binomial model

$$\mathscr{F} = \left\{ f(z;\pi) = \binom{r}{z}\pi^z(1-\pi)^{r-z}, z = 0, 1, \ldots, r : 0 \le \pi \le 1 \right\} \tag{3.28}$$

for the data. This is identical to models (1.1) and (2.1) but in this section we omit the subscripts for simplicity.

3.7.1 Conservative Confidence Intervals

Suppose we want to construct an exact $100(1-\alpha)\%$ confidence interval for π. We need to find functions $a(\pi)$ and $b(\pi)$ such that for each fixed π

$$\mathrm{P}\{a(\pi) \leq Z \leq b(\pi); \pi\} = 1 - \alpha$$

and then obtain the confidence interval by inverting the inequalities to solve for π. One possibility is to use the so-called central interval for which

$$\mathrm{P}\{Z \leq a(\pi); \pi\} = \alpha/2 \quad \text{and} \quad \mathrm{P}\{Z \geq b(\pi); \pi\} = \alpha/2.$$

(For a likelihood-based interval, we require

$$a(\pi)^{26}(1 - a(\pi))^{149} = b(\pi)^{26}(1 - b(\pi))^{149}.$$

This problem is slightly more complex but is in principle treated in the same way as the central interval so we will consider only the simpler central interval.) For any fixed π, $\mathrm{P}\{Z < x; \pi\} = \mathrm{P}\{B(175, \pi) < x\}$ is a step function with jumps at $0, 1, \ldots, 175$ so there are either no solutions or infinitely many solutions to the equations $\mathrm{P}\{Z \leq a(\pi); \pi\} = \alpha/2$ and $\mathrm{P}\{Z \geq b(\pi); \pi\} = \alpha/2$.

We can overcome this problem by the practical and sensible strategy (adopted in the definition of a confidence set in (3.24)) of constructing a conservative interval which requires only that

$$\mathrm{P}\{Z \leq a(\pi); \pi\} \leq \alpha/2 \quad \text{and} \quad \mathrm{P}\{Z \geq b(\pi); \pi\} \geq \alpha/2 \tag{3.29}$$

which implies that

$$\mathrm{P}\{a(\pi) \leq Z \leq b(\pi); \pi\} \geq 1 - \alpha.$$

We solve the equations (3.29) for every π to obtain functions $a(\pi)$ and $b(\pi)$. If we plot the functions $z = a(\pi)$ and $z = b(\pi)$ against π, we obtain the graphical representation of the conservative confidence sets shown in Figure 3.8. For $z = 26$, we obtain the 95% confidence interval

$$[0.10, 0.21].$$

3.7.2 Approximate Confidence Intervals

The small dashed lines in Figure 3.8 represent the curves

$$175\pi \pm 1.96\{175\pi(1 - \pi)\}^{1/2}$$

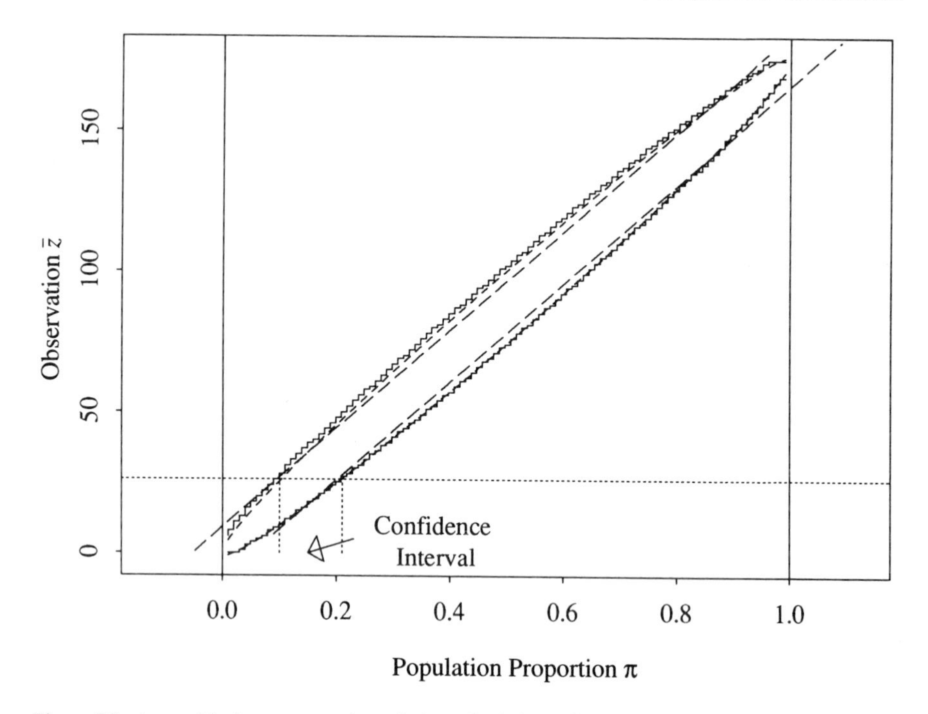

Figure 3.8. A graphical representation of the calculation of confidence limits for the binomial proportion π. The short dashed line (---) represents the calculation treating $(Z - r\pi)/\{r\pi(1-\pi)\}^{1/2}$ as a Gaussianly distributed pivotal quantity and the long dashed line (———) represents the calculation treating $(Z - r\pi)/\{Z(r-Z)/r\}^{1/2}$ as a Gaussianly distributed pivotal quantity.

which are the basis for confidence intervals derived using the Gaussian approximation to the binomial distribution. That is we treat $(Z - r\pi)/\{r\pi(1-\pi)\}^{1/2}$ as Gaussianly distributed so the 95% confidence set is

$$\left\{\pi\colon \frac{|z - r\pi|}{\{r\pi(1-\pi)\}^{1/2}} \leq 1.96\right\} \tag{3.30}$$

which is the interval with endpoints

$$\frac{(2z + 1.96^2)r \pm \{(2z + 1.96^2)^2 r^2 - 4rz^2(r + 1.96^2)\}^{1/2}}{2r(r + 1.96^2)}.$$

For the control eyes of subjects receiving argon treatment in the other eye, (3.30) yields [0.10, 0.21]. The larger dashed lines represent the more usual approximate confidence intervals based on treating $(Z - r\pi)/\{Z(r-Z)/r\}^{1/2}$

as Gaussianly distributed. The 95% confidence interval is

$$\left\{\frac{z}{r} - 1.96\left\{\frac{z(r-z)}{r^3}\right\}^{1/2} \leq \pi \leq \frac{z}{r} + 1.96\left\{\frac{z(r-z)}{r^3}\right\}^{1/2}\right\} \tag{3.31}$$

which is [0.10, 0.20] for the control eyes of subjects receiving argon treatment in the other eye.

The approximate confidence intervals are extremely accurate because $r = 175$ is quite large and z/r is moderate. They are less useful when z/r is near 0 or 1 and we obtain confidence intervals which contain values of π outside [0, 1]. This is a general problem with confidence intervals but in this example is due to the approximation rather than the methodology. Approximations of the kind presented here are developed in Chapter 4.

3.7.3 Randomized Confidence Intervals

An alternative approach is to smooth the distribution of Z by convolution with a continuous distribution. Let $U \sim \mathrm{U}(0, 1)$ be independent of Z and set $X = Z + U$. Then X has a continuous distribution with piecewise linear distribution function

$$F(x; \pi) = \sum_{s=0}^{[x]} \binom{r}{s} \pi^s (1-\pi)^{r-s} + (x - [x]) \binom{r}{[x]} \pi^{[x]} (1-\pi)^{r-[x]}, \qquad 0 \leq x \leq r+1,$$

where $[x]$ is the integer part of x. We solve the equations

$$F(a(\pi); \pi) = \alpha/2 \quad \text{and} \quad F(b(\pi); \pi) = 1 - \alpha/2$$

for $a(\pi)$ and $b(\pi)$ and proceed exactly as before. To obtain the value to read off the y-axis we generate a realization u of a U(0, 1) random variable and use $z + u$. The result is a *randomized confidence interval* of exact level $100(1 - \alpha)\%$. A randomized test of $\mathrm{H}_0\colon \pi = \pi_0$ can be carried out by checking whether π_0 lies in the randomized confidence interval.

Randomized intervals and tests are motivated by the desire to obtain inferences based on probability statements which attain the prescribed level exactly. This may be a useful theoretical property but it does mean that in practice there are cases when the (decision theoretic) conclusion to accept or reject a hypothesis is determined by the value of the randomization variable u. It seems preferable to relax strict adherence to a fixed level as we did implicitly when we defined the size of hypothesis tests in (3.16) and the level of confidence intervals in (3.24) by means of probability inequalities rather than exact equalities.

PROBLEMS

3.7.1. Let $Z_1, \ldots, Z_n$ be independent binary random variables which take on the value 0 or 1 with probability $1 - \pi$ and π respectively. Show that the $100(1 - \alpha)\%$ confidence interval obtained by treating the observations as independent Gaussian observations with unknown mean and variance is a reasonable approximation (in large samples) to a $100(1 - \alpha)\%$ confidence interval for π. For what values of π is the interval attractive? Explain.

3.7.2. Suppose that in the Diabetic Retinopathy Study (Section 1.1.1), we want to compare the incidence of severe visual loss within 2 years in the two sets of control eyes. Recall from (1.2) that these can be modeled by independent binomial(r_1, π_{11}) and binomial(r_2, π_{21}) distributions. Discuss ways of setting confidence intervals for $\eta = \pi_{11} - \pi_{21}$.

3.8 THE BEHRENS–FISHER AND FIELLER–CREASY PROBLEMS

In many simple problems (like the caffeine problem), confidence intervals coincide with fiducial intervals. Initially therefore, fiducial and confidence sets were thought to be essentially the same. However, differences emerged in the famous *Behrens–Fisher* and *Fieller–Creasy problems* and the two methods were recognized as having different interpretations and potentially providing numerically different intervals.

Both the Behrens–Fisher and Fieller–Creasy problems are concerned with the situation in which we observe realizations of $n_1 + n_2$ independent Gaussian random variables, n_1 of which have mean μ_1 and variance σ_1^2 and n_2 of which have mean μ_2 and variance σ_2^2. The Behrens–Fisher problem is the problem of making inferences about $\delta = \mu_1 - \mu_2$ when σ_1 and σ_2 are unequal and the Fieller–Creasy problem is that of making inferences about $\rho = \mu_1/\mu_2$. Since in both problems, we have two independent Gaussian samples, the likelihoods depend on the data only through the two sample means and sample variances. Let $\bar{z}_1$, s_1, $\bar{z}_2$, and s_2 denote the sample mean and standard deviation for the two sets of observations respectively.

3.8.1 The Behrens–Fisher Problem

Recall from Section 2.6.1 that the fiducial distribution of μ_1 can be written as $\mu_1 \mid \mathbf{Z}_1 \sim \bar{z}_1 + n^{-1/2} s_1 T_1$, where T_1 is distributed with the Student t distribution with $n_1 - 1$ degrees of freedom. Since the two samples are independent, the joint fiducial distribution of μ_1 and μ_2 is the product of their two distributions and the fiducial distribution of $\delta = \mu_1 - \mu_2$ is

$$\delta \mid \mathbf{Z} \sim \bar{z}_1 - \bar{z}_2 + n_1^{-1/2} s_1 T_1 - n_2^{-1/2} s_2 T_2$$

or

$$\frac{\delta - (\bar{z}_1 - \bar{z}_2)}{s} \Bigg| \mathbf{Z} \sim \frac{n_1^{-1/2} s_1}{s} T_1 - \frac{n_2^{-1/2} s_2}{s} T_2$$

$$\sim T_1 \cos(a) - T_2 \sin(a). \quad (3.32)$$

where $s^2 = n_1^{-1} s_1^2 + n_2^{-1} s_2^2$ and $\tan(a) = \{n_1 s_2^2 / n_2 s_1^2\}^{1/2}$, a result derived by Behrens (1929) and Fisher (1935b). Jeffreys (1940) showed that if all the parameters are a priori independent then the *Behrens–Fisher distribution* is also the posterior distribution of δ from the Jeffreys prior. The fiducial/Bayesian credibility interval for δ is then

$$[\bar{z}_1 - \bar{z}_2 - b(1 - \alpha/2)s, \bar{z}_1 - \bar{z}_2 + b(1 - \alpha/2)s], \quad (3.33)$$

where $b(1 - \alpha/2)$ is the $(1 - \alpha/2)$th quantile of the Behrens–Fisher distribution. The Behrens–Fisher distribution is not particularly tractable but Cochran (1964) derived a useful approximation in terms of the Student t distribution, namely that

$$b(\alpha) \approx \{n_1^{-1} s_1^2 G_{n_1 - 1}^{-1}(\alpha) + n_2^{-1} s_2^2 G_{n_2 - 1}^{-1}(\alpha)\}/s^2, \quad (3.34)$$

where G_ν is the distribution function of the Student t distribution with ν degrees of freedom.

The basic problem in constructing a confidence interval for δ is to find a pivotal quantity whose distribution does not depend on $\theta = \sigma_1^2/\sigma_2^2$. For example, the obvious candidate

$$\frac{\bar{z}_1 - \bar{z}_2 - \delta}{s} = \frac{(\bar{z}_1 - \bar{z}_2 - \delta)/\{n_1^{-1}\sigma_1^2 + n_2^{-1}\sigma_2^2\}^{1/2}}{s/\{n_1^{-1}\sigma_1^2 + n_2^{-1}\sigma_2^2\}^{1/2}}$$

$$\sim \frac{G}{\{\nu_1^{-1}\theta K_1 + \nu_2^{-1} N K_2\}^{1/2}/\{\theta + N\}^{1/2}},$$

where G, K_1, and K_2 are independent random variables with standard Gaussian, $\chi^2_{n_1 - 1}$ and $\chi^2_{n_2 - 1}$ distributions respectively, and $N = n_1/n_2$, has a distribution which depends on θ. Incidentally, this shows that the fiducial interval (3.33) is not a confidence interval. To obtain a confidence interval for δ, we need to find a function $h_\alpha(s_1^2, s_2^2)$ which does not depend on any of the unknown parameters such that

$$1 - \alpha = \mathrm{P}\{|\bar{Z}_1 - \bar{Z}_2 - \delta| \leq h_\alpha(S_1^2, S_2^2); \delta, \sigma_1, \sigma_2\} \quad (3.35)$$

for all δ, σ_1, and σ_2 but Linnik (1968, Theorem 8.3.1, p. 148) showed that no such function exists. This example therefore illustrates the potential difficulties in eliminating nuisance parameters in exact frequentist inference.

The best a frequentist can hope for are randomized confidence intervals (such as may be obtained by randomly pairing min (n_1, n_2) observations from each sample) or approximate confidence intervals. The most famous approximate confidence intervals are due to Welch (1937b; 1947). In his first paper, Welch suggested that we approximate the sampling distribution of the statistic $(\bar{z}_1 - \bar{z}_2 - \delta)/s$ by the Student t distribution with

$$\frac{(\theta + N)^2}{(n_1 - 1)^{-1}\theta^2 + N^2(n_2 - 1)^{-1}}$$

degrees of freedom. In practice, to make the approximation usable, we estimate θ by s_1^2/s_2^2. (This can also be viewed as a first order approximation to the Behrens–Fisher distribution but it turns out to be inferior to Cochran's approximation (3.34).) In his second paper, Welch showed that (3.35) holds approximately when $h_\alpha(s_1^2, s_2^2)$ is replaced by

$$h_\alpha^*(s_1^2, s_2^2) = s\Phi^{-1}(1 - \alpha)\left[1 + \{1 + \Phi^{-1}(1 - \alpha)\}^2 \frac{\{n_1^{-2}v_1^{-1}s_1^4 + n_2^{-2}v_2^{-1}s_2^4\}}{4s^4}\right.$$
$$\left. - \{1 + \Phi^{-1}(1 - \alpha)^2\}^2 \frac{\{n_1^{-2}v_1^{-2}s_1^4 + n_2^{-2}v_2^{-2}s_2^4\}}{2s^4}\right].$$

3.8.2 The Fieller–Creasy Problem

In the general Fieller–Creasy problem the variances are unknown but the essential issues are still brought out when the variances are known so we will take advantage of the considerable simplification obtained by assuming that $\sigma_1 = \sigma_2 = 1$ is known and $n_1 = n_2 = n$.

If we assume that μ_1 and μ_2 are a priori independent and assign to them uniform improper prior distributions then $\mu_1, \mu_2 \mid \mathbf{Z} = \mathbf{z}$ are independent normal random variables with means $\bar{z}_1$ and $\bar{z}_2$ and common variance $1/n$. To obtain the posterior distribution of (ρ, λ), we make the transformation $\rho = \mu_1/\mu_2$ and $\lambda = \mu_2$, integrate over λ to obtain the marginal posterior distribution of ρ and then derive the Bayesian credibility set for ρ. The last integral is not particularly tractable except in the case that $\bar{z}_1 = \bar{z}_2 = 0$ in which case

$$g(\rho \mid \mathbf{Z} = \mathbf{z}) = \frac{1}{\pi(1 + \rho^2)}$$

and we need to choose $[a, b]$ to satisfy

$$1 - \alpha = \int_a^b g(\rho \mid \mathbf{Z} = \mathbf{z})\, d\rho = \frac{1}{\pi}\{\arctan(b) - \arctan(a)\}.$$

The posterior distribution is symmetric about the origin so $a = -b$,

$$b = \tan(\pi(1-\alpha)/2)$$

and hence, the $100(1-\alpha)\%$ credibility interval for the case $\bar{z}_1 = \bar{z}_2 = 0$ is

$$[-\tan(\pi(1-\alpha)/2, \tan(\pi(1-\alpha)/2)]. \tag{3.36}$$

Creasy (1954) obtained (3.36) from the fiducial distribution of (μ_1, μ_2). An alternative fiducial distribution has been obtained by James et al. (1974) but we will not pursue this here.

Suppose that we seek a $100(1-\alpha)\%$ confidence set for the ratio $\rho = \mu_1/\mu_2$. It is clear that $\bar{Z}_1$ and $\bar{Z}_2$ are independent normal random variables with means μ_1 and μ_2 and common variance $1/n$. It follows that

$$\bar{Z}_1 - \rho\bar{Z}_2 \sim \mathrm{N}(\mu_1 - \rho\mu_2, (1+\rho^2)/n) = \mathrm{N}(0, (1+\rho^2)/n)$$

so that

$$\frac{\bar{Z}_1 - \rho\bar{Z}_2}{(1+\rho^2)^{1/2}/n^{1/2}} \sim \mathrm{N}(0, 1).$$

Note that $n^{1/2}(\bar{Z}_1 - \rho\bar{Z}_2)/(1+\rho^2)^{1/2}$ has a distribution which does not depend on any unknown parameters but it is not a monotone function of ρ so it is not a pivotal quantity. Following Fieller (1940), we can construct a $100(1-\alpha)\%$ confidence interval for ρ (which incidentally does have certain optimality properties) by inverting the relationship

$$1 - \alpha = \mathrm{P}\left\{\left|\frac{\bar{Z}_1 - \rho\bar{Z}_2}{(1+\rho^2)^{1/2}/n^{1/2}}\right| \le b\right\} = 2\Phi(b) - 1.$$

Since $b = \Phi^{-1}(1-\alpha/2)$, we need to solve the equation

$$\frac{|\bar{Z}_1 - \rho\bar{Z}_2|}{(1+\rho^2)^{1/2}/n^{1/2}} \le \Phi^{-1}(1-\alpha/2)$$

for ρ. The endpoints of the interval are the roots of the quadratic equation

$$\{\bar{Z}_2^2 - n^{-1}\Phi^{-1}(1-\alpha/2)^2\}\rho^2 - 2\bar{Z}_1\bar{Z}_2\rho + \{\bar{Z}_1^2 - n^{-1}\Phi^{-1}(1-\alpha/2)^2\} = 0. \tag{3.37}$$

The roots of (3.37) are both real if

$$\bar{Z}_2^2 + \bar{Z}_1^2 \ge n^{-1}\Phi^{-1}(1-\alpha/2)^2$$

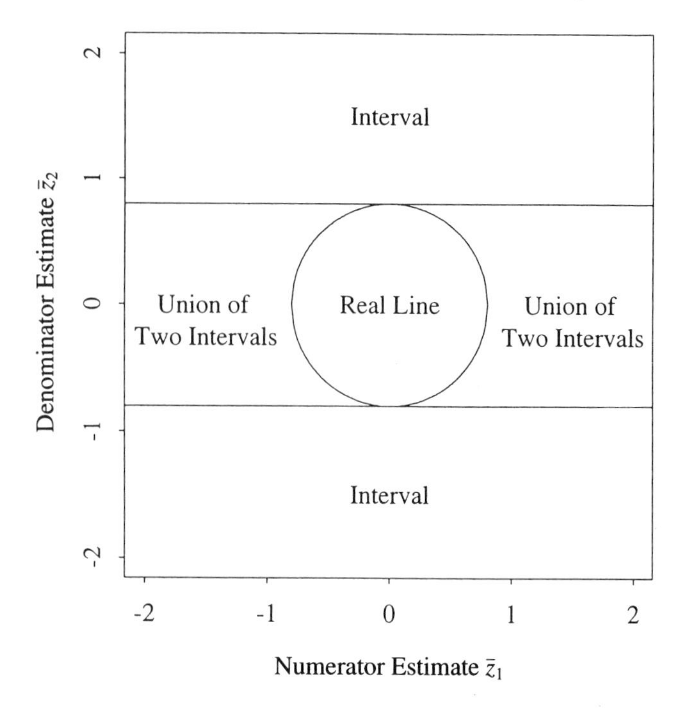

Figure 3.9. The regions in the $(\bar{z}_1, \bar{z}_2)$ space which result in different types of confidence sets for $\rho = \mu_1/\mu_2$ in the Fieller–Creasy problem. The circle has equation $\bar{z}_1^2 + \bar{z}_2^2 = n^{-1}\{\Phi^{-1}(1-\alpha/2)\}^2$ and the horizontal lines $\bar{z}_2^2 = n^{-1}\{\Phi^{-1}(1-\alpha/2)\}^2$.

and in this case the interval is the set of points between the roots if $\bar{Z}_2^2 \geq n^{-1}\Phi^{-1}(1-\alpha/2)^2$ and the set of points on either side of the roots if $\bar{Z}_2^2 < n^{-1}\Phi^{-1}(1-\alpha/2)^2$. The roots of (3.37) are both complex if

$$\bar{Z}_2^2 + \bar{Z}_1^2 < n^{-1}\Phi^{-1}(1-\alpha/2)^2$$

and in this case the confidence interval is the whole real line. The regions in $(\bar{z}_1, \bar{z}_2)$ space which produce the three kinds of set are shown in Figure 3.9. The region in which the confidence interval is the whole real line decreases as n and/or α increase but always contains the point $\bar{z}_1 = \bar{z}_2 = 0$.

Although the confidence sets for this problem may at first be disturbing, they do satisfy the requirement that in repeated samples they contain the parameter the specified proportion of times and there is no reason to expect such sets to be simple intervals. Empty sets and the whole space are acceptable and interpretable outcomes in that they mean that, at the given level, either no possible values of the parameter are consistent with the model and the data or, all possible values of the parameter are consistent with the data. Since confidence sets implicitly incorporate a test of $H_0\colon \theta \in \Omega$, an empty $100(1-\alpha)\%$ confidence or likelihood set means that an α level test would reject H_0. Thus an empty interval means that the model may be inappropriate. Sometimes, a different level set will be a nonempty proper subset of the parameter space. This suggests again that it is best to regard confidence sets as a collection of sets indexed by the level and to consider several choices of level. However, notice that changing the level does not avoid the problem for the confidence interval when $\bar{y} = \bar{z} = 0$.

PROBLEMS

3.8.1. Suppose we have observations $Z_1, \ldots, Z_n$ on the model

$$\mathscr{F} = \left\{ f(\mathbf{x}; \theta) = \prod_{i=1}^{n} \frac{\theta}{x_i^2} I(x_i > \theta) \colon \theta > 0 \right\}.$$

Show that the maximum likelihood estimator of θ is given by Z_{n1}, where $Z_{n1} \leq Z_{n2} \leq \cdots \leq Z_{nn}$. Then show that a $100(1-\alpha)\%$ confidence interval for θ is given by

$$[\alpha_1^{1/n} Z_{n1}, (1-\alpha_2)^{1/n} Z_{n1}], \qquad \alpha_1 + \alpha_2 = \alpha.$$

What values of α_1 and α_2 yield a likelihood-based interval? Use the Jeffreys prior for a scale parameter to obtain a $100(1-\alpha)\%$ Bayesian credibility interval for θ. What is the frequentist coverage probability of the credibility interval?

3.8.2. Suppose that $Z \mid \mu \sim \mathrm{N}(\mu, 1)$. For the improper prior $g(\mu) \propto e^{2\mu}$, find the posterior distribution of $\mu \mid Z = z$. Show that the set $[Z + 2, \infty)$ is a 97.72% Bayesian credibility set for μ but only a 2.28% confidence set for μ.

3.8.3. (Stein, 1959) Suppose that we observe $\{Z_1, \ldots, Z_n\}$ on the model

$$\mathscr{F} = \left\{ f(\mathbf{z}; \mu_1, \ldots, \mu_n) = \prod_{i=1}^{n} \frac{1}{(2\pi)^{1/2}} \exp\left\{ -\frac{(z_i - \mu_i)^2}{2} \right\}, -\infty < z_i < \infty \colon \mu_i \in \mathbb{R} \right\}.$$

Suppose that we are interested in obtaining a $100(1-\alpha)\%$ confidence set of the form $[\theta_\alpha, \infty)$ for $\theta = \sum_{i=1}^n \mu_i^2$. Show that with $\hat{\theta} = \sum_{i=1}^n Z_i^2$,

$$C_\alpha(\mathbf{z}) = \left[\hat{\theta} - n + 2\Phi^{-1}(1-\alpha)^2 - 2\Phi^{-1}(1-\alpha) \left\{ \Phi^{-1}(1-\alpha)^2 - \frac{n}{2} + \hat{\theta} \right\}^{1/2}, \infty \right).$$

is a $100(1-\alpha)\%$ confidence set for θ. Show that if $n^{-2}\hat{\theta} \to 0$ as $n \to \infty$, the coverage probability of the Bayesian/fiducial interval obtained in Problem 2.4.5 tends to 0 as $n \to \infty$.

3.9 CONDITIONAL INFERENCE

The frequentist methods we have considered in this chapter are justified by their repeated sampling properties – provided the model holds, they usually work well and fail only in a specified proportion of cases. Fisher felt that hypothesis testing and confidence interval theory overemphasized repeated sampling properties at the expense of considering other properties. In particular, he felt that methods with good repeated sampling properties do not necessarily use all the available information.

3.9.1 Relevant Subsets

To show that confidence sets do not necessarily use all the available information, Fisher introduced the concept of a relevant subset which was subsequently formalized by Buehler (1959). Suppose that we want to make an inference which asserts that for a set $C_\alpha(\mathbf{z})$, we have $P\{\theta \in C_\alpha(\mathbf{Z})\} = 1 - \alpha$. Then a subset $S(\mathbf{z})$ of the sample space $\mathscr{X}$ is a *relevant subset* if, for some $\varepsilon > 0$, either

$$P\{\theta \in C_\alpha(\mathbf{Z}) \mid S(\mathbf{Z})\} \geq 1 - \alpha + \varepsilon \text{ (positively biased)} \tag{3.38}$$

or

$$P\{\theta \in C_\alpha(\mathbf{Z}) \mid S(\mathbf{Z})\} \leq 1 - \alpha - \varepsilon \text{ (negatively biased)} \tag{3.39}$$

for all parameter values. Conditioning on a positively biased relevant subset makes the inference hold at a higher level than that stated, so by using the information in the relevant subset we can in principle make a more precise conditional statement and, in this sense, we have not used all the available information. The set

$$S(\mathbf{Z}) = \{\mathbf{Z}: \bar{Z}_2^2 + \bar{Z}_1^2 < n^{-1}\Phi^{-1}(1 - \alpha/2)^2\}$$

is a positively biased relevant subset for the confidence interval in the Fieller–Creasy problem because when the data lie in this set, the $100(1 - \alpha)\%$ confidence interval $(-\infty, \infty)$ is actually a 100% confidence interval. Conversely, conditioning on a negatively biased relevant subset shows that the claimed level is too high. In the Behrens–Fisher problem, Fisher (1956b) showed that the set

$$S(\mathbf{Z}) = \{\mathbf{Z}: S_2^2 = S_1^2\}$$

is a negatively biased relevant subset for Welch's (1947) approximate confidence interval because for $n_1 = n_2 = 7$ and $\alpha = 0.1$ we obtain

$$P\{\bar{Z}_1 - \bar{Z}_2 - \delta \leq h_{0.1}^*(S_1^2, S_2^2) \mid S_1^2 = S_2^2; \delta, \sigma_1, \sigma_2\} \leq 0.802.$$

The existence of relevant subsets does not contradict the confidence property which is a long run average property that holds over the whole sample space and can fail over regions of the sample space. However, Fisher argued that poor conditional properties are an argument for not using confidence sets.

3.9.2 Ancillary Statistics

Fisher's point that confidence intervals can have poor conditional properties raises the question of how to make inferences which have good conditional properties. Posing the question slightly differently, we can ask: What should we condition on in constructing inferences to ensure that they have good conditional properties? Fisher argued that any function of the data that has a distribution which does not depend on unknown parameters can be viewed as ancillary to the inference in the sense that it affects the precision of inferences but not their position. He called such functions *ancillary statistics* and argued that inference should be made conditionally on ancillary statistics. (The definition and use of ancillary statistics is discussed in more detail in Section 7.4.)

To illustrate Fisher's argument, consider the location model discussed by Pitman (1938) for which

$$\mathscr{F} = \left\{ f(\mathbf{z}; \mu) = \prod_{j=1}^{n} I(-\tfrac{1}{2} \le z_j - \mu \le \tfrac{1}{2}) \colon \mu \in \mathbb{R} \right\} \tag{3.40}$$

and suppose that we are interested in making inferences about μ. The likelihood is

$$f(\mathbf{z}; \mu) = \prod_{j=1}^{n} I(-\tfrac{1}{2} \le z_j - \mu \le \tfrac{1}{2}) = I(-\tfrac{1}{2} \le z_{n1} - \mu \le z_{nn} - \mu \le \tfrac{1}{2}), \qquad \mu \in \mathbb{R},$$

where $Z_{n1} \le Z_{n2} \le \cdots \le Z_{nn}$ are the order statistics of the data. The likelihood equals 1 for $t \in (z_{nn} - \frac{1}{2}, z_{n1} + \frac{1}{2})$ so is maximized at any point in this interval. In particular, we may take the midpoint or midrange $(z_{nn} + z_{n1})/2$ as the value at which the likelihood is maximized. Under (3.40), the midrange has distribution function (see 8b in the Appendix)

$$F(x) = \begin{cases} 2^{n-1}(x + \frac{1}{2} - \mu)^n & \mu - \frac{1}{2} \le x \le \mu \\ 1 - 2^{n-1}(\mu + \frac{1}{2} - x)^n & \mu \le x \le \mu + \frac{1}{2}, \end{cases}$$

so $(m - b, m - a)$ is a $100(1 - \alpha)\%$ confidence interval for μ for any $a \le b$ such that

$$\alpha = \mathrm{P}(a \le M - \mu \le b) = F(b + \mu) - F(a + \mu).$$

The central confidence interval is obtained from

$$\alpha/2 = 2^{n-1}(a + \tfrac{1}{2})^n \quad \Rightarrow \quad a = \frac{\alpha^{1/n} - 1}{2}$$

and

$$1 - \alpha/2 = 1 - 2^{n-1}(\tfrac{1}{2} - b)^n \quad \Rightarrow \quad b = \frac{1 - \alpha^{1/n}}{2}$$

whence we obtain

$$\left\{m - \frac{(1 - \alpha^{1/n})}{2}, \; m + \frac{(1 - \alpha^{1/n})}{2}\right\}.$$

In particular, a 100% confidence interval is

$$(m - \tfrac{1}{2}, m + \tfrac{1}{2}). \tag{3.41}$$

It is apparent that the sample range contains information which we have not used in constructing the confidence interval. If the range is 1, the maximum possible value, then we have the endpoints of the distribution exactly and consequently we have the midpoint μ of the distribution exactly. Thus for a range w near 1, we should have a small confidence interval. Conversely, for w near 0, we should have a large interval. That is, instead of averaging over all values of the range w, we should condition on the actually observed value of w in deriving our inference.

Under (3.40), the marginal distribution of the range is

$$f(w) = n(n - 1)w^{n-2}(1 - w), \qquad 0 \leq w \leq 1,$$

which does not depend on μ so the range is ancillary and, according to Fisher, we should base inference on the conditional distribution

$$f(m \mid w) = (1 - w)^{-1}, \qquad \mu - \tfrac{1}{2} + \frac{w}{2} \leq m \leq \mu + \tfrac{1}{2} - \frac{w}{2}.$$

Since $F(m \mid w) = \{m - \mu + \tfrac{1}{2} - w/2\}/(1 - w)$, $\mu - \tfrac{1}{2} + w/2 \leq m \leq \mu + \tfrac{1}{2} - w/2$, we find that $(m - b, m - a)$ is a $100(1 - \alpha)\%$ conditional confidence interval for μ for any $a \leq b$ such that

$$\alpha = \mathrm{P}(a \leq M - \mu \leq b \mid w) = F(b + \mu \mid w) - F(a + \mu \mid w).$$

The central confidence interval is obtained from

$$\alpha/2 = \frac{a + \frac{1}{2} - \frac{w}{2}}{(1-w)} \quad \Rightarrow \quad a = (1-w)\alpha/2 + \frac{w}{2} - \frac{1}{2}$$

$$1-\alpha/2 = \frac{b + \frac{1}{2} - \frac{w}{2}}{(1-w)} \quad \Rightarrow \quad b = (1-w)(1-\alpha/2) + \frac{w}{2} - \frac{1}{2} = -(1-w)\alpha/2 - \frac{w}{2} + \frac{1}{2}$$

whence we obtain

$$\left\{m + (1-w)\alpha/2 + \frac{w}{2} - \frac{1}{2},\, m - (1-w)\alpha/2 - \frac{w}{2} + \frac{1}{2}\right\}.$$

The likelihood equals 1 for $t \in (z_{(n)} - \frac{1}{2}, z_{(1)} + \frac{1}{2}) = (m - \frac{1}{2} + w/2, m + \frac{1}{2} - w/2)$ and 0 otherwise so the only nonempty likelihood based interval for μ is the 100% interval

$$\left(m - \frac{1}{2} + \frac{w}{2},\, m + \frac{1}{2} - \frac{w}{2}\right). \tag{3.42}$$

We recover the unconditional 100% interval (3.41) from (3.42) only when $w = 0$! Thus the unconditional interval acts as though we are always in the worst case.

The interpretation of a conditional interval is the same as that of an unconditional interview except that instead of $100(1-\alpha)\%$ of intervals containing μ, we have that $100(1-\alpha)\%$ of intervals calculated from samples with the same sample range contain μ. In other words, sampling is restricted to the subset of the sample space for which samples have the given range rather than applying to the whole sample space.

3.9.3 Conditional Inference in the Location–Scale Model

A surprising feature of inference conditional on ancillary statistics discovered by Fisher (1934) and Pitman (1938) is that it can be carried out in considerable generality and in particular for the general location–scale model

$$\mathscr{F} = \left\{ f(\mathbf{z}; \mu, \sigma) = \prod_{i=1}^{n} \frac{1}{\sigma} h\left(\frac{z_i - \mu}{\sigma}\right),\ -\infty < z_i < \infty : \mu \in \mathbb{R}, \sigma > 0 \right\}. \tag{3.43}$$

Consider any equivariant estimators $(\hat{\mu}, \hat{\sigma})$ which satisfy (3.5). It follows from (3.5) that the configuration $\{C_i = (Z_i - \hat{\mu})/\hat{\sigma}, i = 1, \ldots, n\}$ satisfies

$$\hat{\mu}(\mathbf{C}) = 0 \quad \text{and} \quad \hat{\sigma}(\mathbf{C}) = 1$$

so C_n and C_{n-1} can be expressed in terms of $C_1, \ldots, C_{n-2}$. Suppose for definiteness that $C_{n-1} = l_1(C_1, \ldots, C_{n-2})$ and $C_n = l_2(C_1, \ldots, C_{n-2})$. Also, from (3.5), for $a > 0$,

$$C_i(a\mathbf{Z} + b) = \frac{aZ_i + b - \hat{\mu}(a\mathbf{Z} + b)}{\hat{\sigma}(a\mathbf{Z} + b)} = \frac{Z_i - \hat{\mu}(\mathbf{Z})}{\hat{\sigma}(\mathbf{z})} = C_i(\mathbf{Z})$$

so the configuration is invariant to linear transformations and therefore has a distribution which does not depend on (μ, σ). That is, the configuration is an ancillary statistic and, according to Fisher, we should base inference on the conditional distribution of $(\hat{\mu}, \hat{\sigma})$ given the configuration.

The joint distribution of $\mathbf{Z}$ under (3.43) is given by

$$f(\mathbf{z}; \mu, \sigma) = \frac{1}{\sigma^n} \prod_{i=1}^{n} h\left(\frac{z_i - \mu}{\sigma}\right).$$

The Jacobian of the transformation

$$z_i = \hat{\sigma} c_i + \hat{\mu}, \qquad i = 1, \ldots, n-2$$

$$z_{n-1} = \hat{\sigma} l_1(c_1, \ldots, c_{n-2}) + \hat{\mu}$$

$$z_n = \hat{\sigma} l_2(c_1, \ldots, c_{n-2}) + \hat{\mu}$$

which maps $\mathbf{Z}$ to $(\hat{\mu}, \hat{\sigma}, C_1, \ldots, C_{n-2})$ is $\hat{\sigma}^{n-2} k(c, n)$, where k depends on the definition of the estimator $(\hat{\mu}, \hat{\sigma})$, so the joint density of $(\hat{\mu}, \hat{\sigma}, C_1, \ldots, C_{n-2})$ is (see 8 in the Appendix)

$$f(\hat{\mu}, \hat{\sigma}, c_1, \ldots, c_{n-2}) = k(c, n) \frac{\hat{\sigma}^{n-2}}{\sigma^n} \prod_{i=1}^{n} h\left(\frac{\hat{\sigma} c_i + \hat{\mu} - \mu}{\sigma}\right).$$

Next, the transformation

$$\hat{\mu} = \sigma v u + \mu$$

$$\hat{\sigma} = \sigma v$$

$$c_i = c_i$$

which maps $(\hat{\mu}, \hat{\sigma}, C_1, \ldots, C_{n-2})$ to $(U, V, C_1, \ldots, C_{n-2})$ where $U = (\hat{\mu} - \mu)/\hat{\sigma}$ and $V = \hat{\sigma}/\sigma$ are pivotal quantities, has Jacobian $v\sigma^2$, so the joint density of $(U, V, C_1, \ldots, C_{n-2})$ is

$$f(u, v, c_1, \ldots, c_{n-2}) = k(c, n) v^{n-1} \prod_{i=1}^{n} h(vc_i + vu).$$

The conditional density of (U, V) given the configuration is therefore

$$f(u, v \mid c_1, \ldots, c_{n-2}) = \frac{v^{n-1} \prod_{i=1}^{n} h(vc_i + vu)}{\int_{-\infty}^{\infty} \int_{0}^{\infty} v^{n-1} \prod_{i=1}^{n} h(vc_i + vu)\, dv\, du}. \tag{3.44}$$

The conditional distribution function of the pivotal quantities $U = (\hat{\mu} - \mu)/\hat{\sigma}$ and $V = \hat{\sigma}/\sigma$ is, by definition

$$\begin{aligned} K(u_0, v_0; \mathbf{c}) &= \mathrm{P}\left\{\frac{\hat{\mu} - \mu}{\hat{\sigma}} \le u_0, \frac{\hat{\sigma}}{\sigma} \le v_0 \mid C_1 = c_1, \ldots, C_{n-2} = c_{n-2}\right\} \\ &= \mathrm{P}\{U \le u_0, V \le v_0 \mid C_1 = c_1, \ldots, C_{n-2} = c_{n-2}\} \\ &= \frac{\int_{-\infty}^{u_0} \int_{0}^{v_0} v^{n-1} \prod_{i=1}^{n} h(vc_i + vu)\, dv\, du}{\int_{-\infty}^{\infty} \int_{0}^{\infty} v^{n-1} \prod_{i=1}^{n} h(vc_i + vu)\, dv\, du}. \end{aligned} \tag{3.45}$$

Making the change of variables

$$u = (\hat{\mu} - t)/\hat{\sigma}$$

$$v = \hat{\sigma}/s$$

which has Jacobian $1/s^2$, we have that

$$K(u_0, v_0; \mathbf{c}) = \frac{\int_{\hat{\mu} - \hat{\sigma} u_0}^{\infty} \int_{\hat{\sigma}/v_0}^{\infty} \frac{1}{s^{n+1}} \prod_{i=1}^{n} h\left(\frac{\hat{\sigma} c_i + \hat{\mu} - t}{s}\right) ds\, dt}{\int_{-\infty}^{\infty} \int_{0}^{\infty} \frac{1}{s^{n+1}} \prod_{i=1}^{n} h\left(\frac{\hat{\sigma} c_i + \hat{\mu} - t}{s}\right) ds\, dt}.$$

Since $\hat{\sigma} c_i + \hat{\mu} = z_i$ by definition, we can write

$$\begin{aligned} K(u_0, v_0; \mathbf{c}) &= \frac{\int_{\hat{\mu} - \hat{\sigma} u_0}^{\infty} \int_{\hat{\sigma}/v_0}^{\infty} \frac{1}{s^{n+1}} \prod_{i=1}^{n} h\left(\frac{z_i - t}{s}\right) ds\, dt}{\int_{-\infty}^{\infty} \int_{0}^{\infty} \frac{1}{s^{n+1}} \prod_{i=1}^{n} h\left(\frac{z_i - t}{s}\right) ds\, dt} \\ &= 1 - G\left(\hat{\mu} - \hat{\sigma} u_0, \frac{\hat{\sigma}}{v_0}; \mathbf{z}\right), \end{aligned} \tag{3.46}$$

say, where

$$G(u, v; \mathbf{z}) = 1 - \frac{\int_u^\infty \int_v^\infty \frac{1}{s^{n+1}} \prod_{i=1}^n h\left(\frac{z_i - t}{s}\right) ds\, dt}{\int_{-\infty}^\infty \int_0^\infty \frac{1}{s^{n+1}} \prod_{i=1}^n h\left(\frac{z_i - t}{s}\right) ds\, dt} \tag{3.47}$$

Although the joint conditional distribution function (3.45) depends on the choice of $(\hat{\mu}, \hat{\sigma})$ (and hence the configuration), conditional confidence intervals do not. Since

$$\begin{aligned} 1 - \alpha &= \mathrm{P}\{\hat{\mu} - \hat{\sigma}b(\mathbf{c}) \leq \mu \leq \hat{\mu} - \hat{\sigma}a(\mathbf{c}) \mid \mathbf{C} = \mathbf{c}\} \\ &= \mathrm{P}\left\{a(\mathbf{c}) \leq \frac{\hat{\mu} - \mu}{\hat{\sigma}} \leq b(\mathbf{c}) \mid \mathbf{C} = \mathbf{c}\right\} \\ &= K(b(\mathbf{c}), \infty; \mathbf{c}) - K(a(\mathbf{c}), \infty; \mathbf{c}), \end{aligned}$$

a $100(1 - \alpha)\%$ conditional confidence interval for μ is given by

$$[\hat{\mu} - \hat{\sigma}b(\mathbf{c}), \hat{\mu} - \hat{\sigma}a(\mathbf{c})], \tag{3.48}$$

where

$$1 - \alpha = K(b(\mathbf{c}), \infty; \mathbf{c}) - K(a(\mathbf{c}), \infty; \mathbf{c}). \tag{3.49}$$

Clearly, if $a(\mathbf{c}) = K^{-1}(q_1, \infty; \mathbf{c})$ and $b(\mathbf{c}) = K^{-1}(q_2, \infty; \mathbf{c})$, where it is understood that we are inverting the first argument only, then by (3.49) q_1 and q_2 must satisfy $0 < q_1, q_2 < 1$, $q_2 - q_1 = 1 - \alpha$. But (3.46) implies that

$$K^{-1}(q, \infty; \mathbf{c}) = \frac{\hat{\mu} - G^{-1}(1 - q, \infty; \mathbf{z})}{\hat{\sigma}},$$

so the $100(1 - \alpha)\%$ conditional confidence interval for μ in (3.43) can be written

$$[\hat{\mu} - \hat{\sigma}K^{-1}(q_2, \infty; \mathbf{c}), \hat{\mu} - \hat{\sigma}K^{-1}(q_1, \infty; \mathbf{c})] = [G^{-1}(1 - q_2, \infty; \mathbf{z}), G^{-1}(1 - q_1, \infty; \mathbf{z}].$$

This interval is invariant to the choice of $(\hat{\mu}, \hat{\sigma})$ (and hence the configuration) so will be the same for any pair of equivariant estimators $(\hat{\mu}, \hat{\sigma})$ because G defined in (3.47) is invariant to the choice of $(\hat{\mu}, \hat{\sigma})$.

3.9.4 The Relationship Between Conditional, Bayesian, and Fiducial Inference

If we think of G defined in (3.47) as a joint distribution function, it is numerically identical to the distribution function of the joint posterior distribution of

(μ, σ) based on the Jeffreys prior and therefore to the joint distribution function of the fiducial distribution for (μ, σ). Indeed, since by (3.46) and (3.45)

$$1 - G\left(\hat{\mu} - \hat{\sigma}u_0, \frac{\hat{\sigma}}{v_0}; \mathbf{z}\right) = P\left\{\frac{\hat{\mu} - \mu}{\hat{\sigma}} \leq u_0, \frac{\hat{\sigma}}{\sigma} \leq v_0 \,\middle|\, c_1, \ldots, c_{n-2}\right\}$$

$$= P\left\{\mu \geq \hat{\mu} - \hat{\sigma}u_0, \sigma \geq \frac{\hat{\sigma}}{v_0} \,\middle|\, c_1, \ldots, c_{n-2}\right\},$$

the fiducial argument can be applied directly to obtain the fiducial density

$$f_f(t, s) = \frac{\dfrac{1}{s^{n+1}} \displaystyle\prod_{i=1}^{n} h\left(\frac{z_i - t}{s}\right)}{\displaystyle\int_{-\infty}^{\infty} \int_0^{\infty} \frac{1}{s^{n+1}} \prod_{i=1}^{n} h\left(\frac{z_i - t}{s}\right) ds\, dt}.$$

Thus, the conditional inference calculation is numerically equivalent to the fiducial calculation and this entirely frequentist argument can be interpreted as a justification for fiducial inference.

3.9.5 The Relationship Between Conditional and Unconditional Inference

Conditional confidence intervals like (3.48) are also valid unconditional confidence intervals because

$$P\{\hat{\mu} - \hat{\sigma}b(\mathbf{c}) \leq \mu \leq \hat{\mu} - \hat{\sigma}a(\mathbf{c})\} = P\left\{a(\mathbf{c}) \leq \frac{\hat{\mu} - \mu}{\hat{\sigma}} \leq b(\mathbf{c})\right\}$$

$$= E\left[P\left\{a(\mathbf{c}) \leq \frac{\hat{\mu} - \mu}{\hat{\sigma}} \leq b(\mathbf{c}) \mid \mathbf{C} = \mathbf{c}\right\}\right]$$

$$= 1 - \alpha.$$

However, conditional intervals are not necessarily optimal unconditionally. Indeed, Neyman rejected conditional inference because the resulting procedures can be suboptimal in terms of unconditional power; see Welch (1939). Nonetheless, conditional inference is now seen to have a role to play in frequentist statistics; see Kiefer (1977b) and Lehmann (1983, Chapter 3).

3.9.6 Difficulties in Conditional Inference

Clearly, conditioning on ancillary statistics can lead to improved inferences. However, Basu (1964) showed that ancillary statistics are not necessarily unique and different choices of ancillary statistics can lead to different results. See

Section 7.4. Fortunately, this does not happen with the transformation group structure underlying location–scale (and regression) models – any configuration can be used.

Fisher believed that it is not possible to improve on inferences conditioned on ancillary statistics. Since this conditioning is implicit in fiducial theory, it should not be possible to do better without incorporating prior information. However, contrary to Fisher's belief, Buehler and Federson (1963) and later Brown (1967) showed that the simple confidence interval for the mean of a Gaussian model is vulnerable to relevant subsets. In particular, Brown (1967) showed that there exists a constant K such that

$$\mathrm{P}\left\{\bar{Z}-n^{-1/2}SG_{\nu}^{-1}(1-\alpha/2)\leq\mu\leq\bar{Z}+n^{-1/2}SG_{\nu}^{-1}(1-\alpha/2)\,\middle|\,\frac{|\bar{Z}|}{S}\leq K\right\}\geq 1-\alpha+\varepsilon,$$

for $\varepsilon>0$, so by (3.38) $S(\mathbf{Z})=\{\mathbf{Z}\colon |\bar{Z}|/S\leq K\}$ is a positively biased relevant subset which is not eliminated by conditioning on an ancillary statistic. On the other hand, Robinson (1975) constructed an elegant example in which there is a negatively biased relevant subset (see (3.39)) which is not an ancillary statistic. Thus Fisher's criticism of confidence intervals applies equally to his own fiducial intervals.

Pierce (1973) showed that any procedure which cannot be interpreted as a Bayesian procedure derived from a proper prior is vulnerable to relevant subsets. This means that good conditional properties can only be achieved from a Bayesian analysis based on a proper prior and that good frequentist properties can only hold for procedures which are in a sense limits of proper Bayesian procedures. Thus, we either have to restrict the class of procedures we consider or be satisfied with weaker conditional properties (Robinson, 1979a, b). While conditional inference remains attractive, in anything other than a fully conditional Bayesian analysis, the general question of what to condition on is quite delicate.

PROBLEMS

3.9.1. (Cox, 1958) Suppose that a random variable Y is equally like to be $\mathrm{N}(\mu, 1)$ or $\mathrm{N}(\mu, 100)$. The data is a realization of (Y, C) where C is an indicator variable taking the value 1 or 2 according to whether Y has the first or second distribution. Suppose we observe $(y, c) = (2.35, 1)$. Set a conditional $100(1-\alpha)\%$ confidence interval for μ given $c = 1$ and compare it to the unconditional $100(1-\alpha)\%$ confidence interval.

3.9.2. Suppose that we have observations $\{(Y_1, X_1), \ldots, (Y_n, X_n)\}$ on the Gaussian simple regression model

$$\mathscr{F} = \left\{ f(\mathbf{y}; \alpha, \beta) = f_X(\mathbf{x}) \prod_{i=1}^{n} \frac{1}{(2\pi)^{1/2}} \exp\left\{ -\frac{(y_i - \alpha - x_i\beta)^2}{2} \right\}, \right.$$
$$\left. -\infty \leq y_i \leq \infty : \alpha, \beta \in \mathbb{R} \right\}.$$

In this model, the explanatory variables $\mathbf{X} = (X_1, \ldots, X_n)$ are treated as random with joint density function $f_X(\mathbf{x})$. Show that the maximum likelihood estimators of α and β are the same as when $\mathbf{X}$ is treated as fixed. Show that

$$Y_j \mid \mathbf{X} = \mathbf{x} \sim \text{independent } \mathrm{N}(\alpha + x_j\beta, 1).$$

What is the conditional sampling distribution of the maximum likelihood estimator $\hat{\beta}$ given that $\mathbf{X} = \mathbf{x}$? Set a conditional $100(1 - \alpha)\%$ confidence interval for β given $\mathbf{X} = \mathbf{x}$. Find the unconditional distribution of $\hat{\beta}$ when $\mathbf{X}$ has a multivariate Gaussian distribution with mean 0 and variance $\sigma^2 I$, where I is the $n \times n$ identity matrix, so $\sum_{i=1}^{n} (X_i - \bar{X})^2 \sim \sigma^2 \chi^2_{n-1}$ and set an approximate unconditional $100(1 - \alpha)\%$ confidence interval for β. Compare the two results.

3.9.3. Consider the location model (3.40). For any equivariant estimator $\hat{\mu}$ of μ which satisfies $\hat{\mu}(\mathbf{z} + b) = \hat{\mu}(\mathbf{z}) + b$, show that the configuration $\{c_i = z_i - \hat{\mu}, i = 1, \ldots, n\}$ is an ancillary statistic. Show that the conditional confidence interval for μ derived from the conditional distribution of $\hat{\mu}$ given the configuration is the same as the interval obtained in Section 3.9.2 by conditioning on the sample range.

3.9.4. Consider the scale model

$$\mathscr{F} = \left\{ f(\mathbf{z}; \sigma) = \prod_{i=1}^{n} \frac{1}{\sigma} h\left(\frac{z_i}{\sigma}\right), -\infty < z_i < \infty : \sigma > 0 \right\}.$$

For any equivariant estimator $\hat{\sigma}$ of σ which satisfies $\hat{\sigma}(a\mathbf{z}) = |a|\hat{\sigma}(\mathbf{z})$, show that the configuration $\{c_i = z_i/\hat{\sigma}, i = 1, \ldots, n\}$ is an ancillary statistic. Find the conditional distribution of $\hat{\sigma}$ given the configuration and show that conditional confidence intervals for σ do not depend on the choice of $\hat{\sigma}$.

3.9.5. Suppose we adopt the exponential model discussed in Problem 3.1.5 for Proschan's (1963) air conditioning data which was presented in Problem 1.5.11. Find a $100(1 - \alpha)\%$ conditional confidence interval for λ and then for $\theta = (1/\lambda) \log 2$. Comment on the relationship between these intervals and the unconditional ones obtained in Problem 3.6.3.

3.9.6. Consider the regression model

$$\mathscr{F} = \left\{ f(\mathbf{y}; \alpha, \beta, \sigma) = \prod_{i=1}^{n} \frac{1}{\sigma} h\left(\frac{y_i - \alpha - x_i\beta}{\sigma}\right), \right.$$

$$\left. -\infty < y_i < \infty\colon \alpha, \beta \in \mathbb{R}, \sigma > 0 \right\}.$$

Let $\gamma = (\alpha, \beta)^{\mathrm{T}}$ and X be the $n \times 2$ matrix with ith row $(1, x_i)$. Consider any equivariant estimators $(\hat{\gamma}, \hat{\sigma})$ which satisfy

$$\hat{\gamma}(a\mathbf{y} + Xb) = a\hat{\gamma}(\mathbf{y}) + b \quad \text{and} \quad \hat{\sigma}(a\mathbf{y} + Xb) = |a|\hat{\sigma}(\mathbf{y}).$$

Show that the configuration $\{c_i = (y_i - \hat{\alpha} - x_i\hat{\beta})/\hat{\sigma},\ i = 1, \ldots, n\}$ is an ancillary statistic. Show that the conditional distribution of $(\hat{\alpha}, \hat{\beta}, \hat{\sigma})$ given the configuration is the same as the posterior distribution of (α, β, σ) using the Jeffreys prior.

3.10 SIMULATION

Repeated sampling or frequentist properties are often described as objective because they can in principle be verified empirically. That is, if we can repeat the data collection process to obtain new sets of data and calculate tests and/or confidence intervals for each data set, we can obtain a collection of tests and/or confidence intervals which we can use to check whether the desired outcomes occur in the required proportions. This is complicated by the fact that the required outcomes are expressed in terms of true parameter values which are usually unknown. However, if we carry out an artificial experiment or *simulation study*, we know the true parameter values and we can then explore the repeated sampling properties of the procedures.

An important point about quantities derived in a simulation is that they are and should be treated like any other data sets. We can model the characteristics they represent and then proceed to make inferences about these characteristics. The fact that the data are generated in an artificial statistical experiment does not give them an inherently inferior status to other data.

3.10.1 Simulating Coverage Probabilities

Suppose we want to investigate the coverage probability of the 95% confidence interval

$$C_{0.05}(\mathbf{z}) = \left\{\mu\colon \bar{z} - \frac{2.11s}{\sqrt{18}} \le \mu \le \bar{z} + \frac{2.11s}{\sqrt{18}}\right\}$$

for the mean derived from (3.27) when we have $n = 18$ observations from the Gaussian model (3.1). Most statistical packages have the capacity to generate pseudorandom numbers from standard distributions such as the Gaussian distribution (for details, see for example Devroye, 1986, Ripley, 1987, and the discussion in Sections 6.8.1–6.8.3) so we can proceed as follows: generate a large number, say N, independent data sets of size $n = 18$ from a Gaussian distribution with specified mean μ_0 and variance σ_0^2, compute the interval $C_{0.05}(\mathbf{z})$ from each data set, and record whether the interval contains μ_0 or not. If we denote each data set by $\mathbf{z}_j$, at the end of the simulation, we have N observations of the form $I\{C_{0.05}(\mathbf{z}_j)$ contains $\mu_0\}$, $j = 1, \ldots, N$, where I is the indicator function. Notice that once we have generated the data sets, we do not use our knowledge of the true value μ_0 until we evaluate the intervals. The observations $I\{C_{0.05}(\mathbf{z}_j)$ contains $\mu_0\}$ are realizations of independent random variables (by design) which take on only the values 0 or 1 and take the value 1 with constant probability π, the actual coverage probability of the interval $C_{0.05}(\mathbf{z})$ under the settings of the simulation. It follows that the number Y of intervals containing the true parameter value μ_0 has a binomial(N, π) distribution so we can make inferences about π and, in particular, set confidence intervals for π.

To implement the simulation, we need to choose the simulation settings, in this case N, μ_0, and σ_0, and obtain realizations of Gaussian random variables. If we use the Gaussian approximation to the binomial distribution (Section 3.7), we obtain an approximate 95% confidence interval for π from (3.30) as

$$\left\{\pi : \frac{|y/N - \pi|}{\{\pi(1-\pi)/N\}^{1/2}} \leq 1.96\right\}.$$

We hope that π will equal 0.95 (the supposed or nominal level of $C_{0.05}(\mathbf{z})$). In this case, the standard error $\{\pi(1-\pi)/N\}^{1/2}$ equals $0.2179/\mathrm{N}^{1/2}$, so the approximate 95% confidence interval is

$$\frac{y}{N} \pm \frac{1.96 \times 0.2179}{N^{1/2}}.$$

If we want to estimate π to within Δ with 95% confidence, we need to choose N so that

$$\frac{1.96 \times 0.2179}{N^{1/2}} \leq \Delta \quad \text{or} \quad N \geq \left(\frac{1.96 \times 0.2179}{\Delta}\right)^2.$$

For $\Delta = 0.02$, $N = 456$, for $\Delta = 0.01$, $N = 1825$, and so on. A common choice is to compromise and take $N = 1000$. We also need to specify μ_0 and σ_0. We can choose these to match their estimates from the caffeine data or we can note that $C_{0.05}(\mathbf{z})$ is equivariant to the choice of μ_0 and σ_0 in the sense that $C_{0.05}(\mu_0 + \sigma_0\mathbf{z}) = \mu_0 + \sigma_0 C_{0.05}(\mathbf{z})$ and therefore simply choose the most convenient values $\mu_0 = 0$ and $\sigma_0 = 1$.

An approximate 95% confidence interval for the coverage probability of $C_{0.05}(\mathbf{z})$ based on a simulation using $N = 1000$, $\mu_0 = 0$, and $\sigma_0 = 1$ is obtained from (3.31) as $0.941 \pm 1.96\sqrt{0.94 \times 0.06/1000} = 0.941 \pm 0.015$. This interval contains the nominal coverage probability 0.95, so the claimed repeated sampling property is plausible.

3.10.2 Simulating the Expected Length of an Interval

We can also examine the distribution of the lengths of the confidence intervals. The actual distribution of lengths can be derived from the distribution of s^2 which is $\sigma^2\chi^2_{17}$. However, the Gaussian qq-plot in Figure 3.10 shows that, in this problem, the distribution of lengths can be approximated by a Gaussian distribution. Using the Gaussian approximation to the distribution of the mean lengths, we see from (3.27) that a 95% confidence interval for the mean length of the intervals $C_{0.05}(\mathbf{z})$ is $0.977 \pm 1.96 \times 0.0053 = 0.977 \pm 0.010$.

3.10.3 Simulating the Level and Power of Tests

If we are studying tests, we can estimate the significance level (the proportion of tests which reject the null hypothesis when it is true) and the power (the

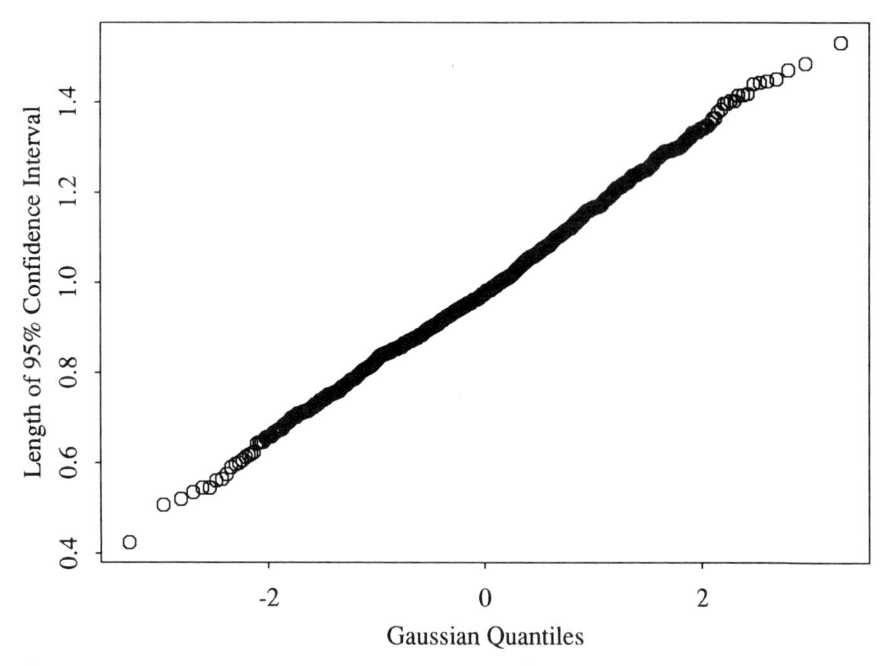

Figure 3.10. A Gaussian qq-plot for the lengths of simulated 95% confidence intervals for the Gaussian mean μ under the standard Gaussian model.

proportion of tests which reject the null hypothesis when it is false). The procedure is analogous to that for estimating the coverage probability.

3.10.4 Simulating Conditional Properties

We can also use a simulation to study conditional repeated sampling properties. The only difference from what we have described above is that instead of simply generating data from a model, we need to generate data from a model which also satisfies the condition under which the conditional properties are to be evaluated. For example, in the $U(\mu - \frac{1}{2}, \mu + \frac{1}{2})$ problem considered in Section 3.9.2, we can evaluate unconditional repeated sampling properties by generating data from the model $U(\mu - \frac{1}{2}, \mu + \frac{1}{2})$ or we can evaluate conditional repeated sampling properties by generating data from the model $U(\mu - \frac{1}{2}, \mu + \frac{1}{2})$ for which the sample range equals a specified value, say w_0.

3.10.5 Limitations of Simulation

Two difficulties with simulation are that there is variability between simulations and the results are often quite specific to the settings we have chosen. The first difficulty is intrinsic to the inference problem and is dealt with by choosing N large and by setting confidence intervals for the characteristics of interest but there is little we can do about the second. This means that simulations are less satisfactory than theoretical results but nonetheless provide a useful supplement to theoretical results and often can be used when theoretical results are unavailable.

PROBLEMS

3.10.1. In Section 3.7 we described the approximate confidence intervals (3.30) and (3.31) for a binomial proportion π. Carry out a simulation to evaluate the repeated sampling properties of these two intervals in small samples. Try the settings $r = 15$ and $r = 30$, $\pi = 0.5$ and $\pi = 0.2$. What do you conclude?

3.10.2. Suppose that the Gaussian setup described in Section 3.8 holds. Then when σ_1 and σ_2 are unequal, the fiducial interval for $\delta = \mu_1 - \mu_2$ is given by (3.33) and we can use Cochran's (1964) approximation (3.34) or Welch's (1937b) Student t approximation with

$$\hat{\nu} = \frac{(s_1^2/s_2^2 + n_1/n_2)^2}{(n_1 - 1)^{-1}[s_1^2/s_2^2]^2 + [n_1/n_2]^2(n_2 - 1)^{-1}}$$

degrees of freedom, for the quantiles of the Behrens–Fisher distribution.

Carry out a simulation to evaluate the repeated sampling properties of these two intervals. Try the settings $n_1 = 20$, $n_2 = 20$ and $n_1 = 18$, $n_2 = 25$, $\mu_1 = \mu_2 = 0$, $\sigma_1 = 1$, $\sigma_2 = 3$, and $\sigma_1 = 1$, $\sigma_2 = 5$. What do you conclude?

FURTHER READING

The frequentist approach to inference is developed at different levels of mathematical generality in numerous texts. Bickel and Doksum (1977) and Casella and Berger (1990) present the theory at an intermediate level, whereas Lehmann (1983, 1959/1991) gives mathematical treatments of point estimation and hypothesis testing respectively. Cox and Hinkley (1974) and Barnett (1982) contain useful material with plenty of discussion; Silvey (1970) is a suitable presentation for many purposes. Edgeworth's and Fisher's contributions to maximum likelihood estimation are discussed by Pratt (1976). The presentation of the material on statistical testing in this chapter owes much to Gigerenzer et al. (1989). The Behrens–Fisher and Fieller–Creasy problems have been widely discussed, and Wallace (1980) is an excellent reference. Conditional inference remains topical, and Casella (1992) is an accessible review.

CHAPTER 4

Large Sample Theory

In the pressure vessel failure problem (Section 1.1.2), the data $\mathbf{z}$ represent the failure times in hours of 20 pressure vessels and we are interested, amongst other things, in estimating the median failure time. We showed (Section 1.2.2) that the data can plausibly be treated as a realization of a random variable $\mathbf{Z}$ generated by the exponential model

$$\mathscr{F} = \left\{ f(\mathbf{y}; \lambda) = \prod_{i=1}^{n} \lambda \exp(-\lambda y_i), y_i > 0 : \lambda > 0 \right\}. \tag{4.1}$$

In this model, the median failure time is $\theta = \lambda^{-1} \log(2)$.

Suppose that we want to make frequentist inferences about θ. The sample mean $\bar{z} = 575.53$ is the maximum likelihood estimate of λ, so we may choose to base inference on a function of $\bar{z}$. In this case, we require the sampling distribution of $\bar{Z}$. It is a standard result of distribution theory that the sum S of n independent exponential(λ) random variables has a gamma(n, λ) distribution so the density of the normalized sum $\bar{Z} = S/n$ under $\mathscr{F}$ is

$$g(m; n, \lambda) = \frac{n^n \lambda}{\Gamma(n)} (\lambda m)^{n-1} e^{-\lambda n m}, \qquad m > 0.$$

The distribution function is obtained by integrating the density as

$$G(x; n, \lambda) = \frac{n^n \lambda}{\Gamma(n)} \int_0^x (\lambda m)^{n-1} e^{-\lambda n m}\, dm, \qquad x > 0.$$

We can show that $\lambda \bar{Z}$ has distribution function $G(x; n, 1)$ so $\lambda \bar{Z}$ is a pivotal quantity (Section 3.6.1) and a $100(1-\alpha)\%$ confidence interval for the median failure time is a realization of the random interval

$$\left(\frac{\bar{Z}\log(2)}{b}, \frac{\bar{Z}\log(2)}{a}\right),$$

where $a < b$ are any non-negative numbers satisfying

$$G(b; n, 1) - G(a; n, 1) = 1 - \alpha.$$

In particular, the central $100(1-\alpha)\%$ confidence interval is a realization of

$$\left(\frac{\bar{Z}\log(2)}{G^{-1}(1-\alpha/2; n, 1)}, \frac{\bar{Z}\log(2)}{G^{-1}(\alpha/2; n, 1)}\right), \tag{4.2}$$

where $G^{-1}(u; n, 1)$ is the quantile function defined by $G(G^{-1}(u; n, 1); n, 1) = u$, so the endpoints of the interval are obtained by multiplying $\bar{z}\log(2) = 398.93$ by $1/G^{-1}(1-\alpha/2; n, 1)$ and $1/G^{-1}(\alpha/2; n, 1)$ respectively.

4.1 APPROXIMATE CONFIDENCE INTERVALS

To implement the confidence interval (4.2), we require values of the quantile function $G^{-1}(u; n, 1)$ for selected u. Except in unusual cases (such as $n = 1$), we cannot obtain these exactly so we have to resort to approximations. Some of the approximations for the gamma distribution are of sufficiently high quality to be considered exact: Harter (1964) gives tables of $G^{-1}(u; n, 1)$ correct to 6 significant figures which yield the 95% confidence interval for θ

$$(268.90, 653.10).$$

Although we rarely require such accuracy in statistical work, the "exact" results do provide a baseline against which we can compare simpler approximations.

4.1.1 Approximation by Direct Expansion

Since we have an expression for the density $g(m; n, \lambda)$ of $\bar{Z}$, we can try to approximate $g(m; n, \lambda)$, integrate the approximation to obtain an approximation to the distribution function $G(x; n, \lambda)$ and then invert this to obtain an approximation to the quantile function $G^{-1}(u; n, \lambda)$.

4.1.2 Consistent Estimation

A reasonable first step is to approximate the density $g(m; n, \lambda)$ by its limit as the sample size $n \to \infty$ which is hopefully simpler than g itself. We can use

Stirling's formula

$$\Gamma(n) = (2\pi)^{1/2} e^{-n} n^{n-1/2}\{1 + O(n^{-1})\}$$

or equivalently

$$\Gamma(n)^{-1} = \frac{1}{(2\pi)^{1/2}} e^{n} n^{-n+1/2}\{1 + O(n^{-1})\}$$

to approximate the gamma function in $g(m; n, \lambda)$ so that

$$\begin{aligned} g(m; n, \lambda) &= \frac{n^{1/2}\lambda}{(2\pi)^{1/2}} (\lambda m)^{n-1} e^{n(1-\lambda m)}\{1 + O(n^{-1})\} \\ &= \frac{n^{1/2}\lambda}{(2\pi)^{1/2}} \exp\{1 - \lambda m + (n-1)(\log(\lambda m) + 1 - \lambda m)\}\{1 + O(n^{-1})\}. \end{aligned} \tag{4.3}$$

It follows from (4.3) that

$$g(m; n, \lambda) = O(n^{1/2} e^{-k(m,\lambda)n}),$$

where $k(m, \lambda) = \log(\lambda m) + 1 - \lambda m$. Since

$$k(m, \lambda) = \begin{cases} \log(1 + \lambda m - 1) - (\lambda m - 1) < 0 & \text{for } \lambda m \neq 1 \\ 0 & \text{for } \lambda m = 1, \end{cases}$$

$g(m; n, \lambda)$ converges to 0 as $n \to \infty$ whenever $m \neq \lambda^{-1}$ and diverges to ∞ as $n \to \infty$ when $m = \lambda^{-1}$. This implies that

$$G(x; n, \lambda) = \mathrm{P}\{\bar{Z} \leq x\} \to I\left(x \geq \frac{1}{\lambda}\right) \qquad \text{for all } x \neq \frac{1}{\lambda} \tag{4.4}$$

so, as $n \to \infty$, the sampling distribution of $\bar{Z}$ converges to the *degenerate distribution* with all its mass concentrated at λ^{-1}. We say that $\bar{Z}$ is *consistent* for λ^{-1}.

The consistency of $\bar{Z}$ for λ^{-1} implies that $\bar{Z}$ is close to λ^{-1} in the sense that

$$\mathrm{P}\left\{\left|\bar{Z} - \frac{1}{\lambda}\right| > \varepsilon\right\} = 1 - G\left(\frac{1}{\lambda} + \varepsilon; n, \lambda\right) + G\left(\frac{1}{\lambda} - \varepsilon; n, \lambda\right) \to 0 \tag{4.5}$$

as $n \to \infty$, and provides some justification for using $\bar{Z}$ to make inferences about λ^{-1}.

4.1.3 Approximation by a Nondegenerate Distribution

Degenerate distributions do not provide useful approximations to proper distributions because they do not describe the variability in a proper distribution. Thus we need to obtain a nondegenerate approximation to $G(x; n, \lambda)$. We can achieve this by approximating the distribution of the centered and scaled statistic $a_n(\bar{Z} - \lambda^{-1})$, where $a_n \to \infty$, rather than the distribution of $\bar{Z}$.

Let $t = a_n(m - \lambda^{-1})$ so $m = \lambda^{-1} + a_n^{-1}t$ and $dm = a_n^{-1}\,dt$, where $a_n \to \infty$. From approximation (4.3) we have that the density of $a_n(\bar{Z} - \lambda^{-1})$ satisfies

$$\frac{1}{a_n} g\left(\frac{1}{\lambda} + \frac{t}{a_n}; n, \lambda\right) = \frac{n^{1/2}\lambda}{(2\pi)^{1/2} a_n} \exp\left\{-\frac{\lambda t}{a_n} + (n-1)\left(\log\left(1 + \frac{\lambda t}{a_n}\right) - \frac{\lambda t}{a_n}\right)\right\}\{1 + O(n^{-1})\}.$$

Expanding the log function in the exponent in a Taylor series (see 2 in the Appendix) leads to

$$\frac{1}{a_n} g\left(\frac{1}{\lambda} + \frac{t}{a_n}; n, \lambda\right) = \frac{n^{1/2}\lambda}{(2\pi)^{1/2} a_n} \exp\left\{-\frac{\lambda t}{a_n} - (n-1)\left(\frac{(\lambda t)^2}{2a_n^2} - \frac{(\lambda t)^3}{3a_n^3} + O(a_n^{-4})\right)\right\}$$
$$\times \{1 + O(n^{-1})\}$$

and setting $a_n = n^{1/2}\lambda$ yields

$$\frac{1}{n^{1/2}\lambda} g\left(\frac{1}{\lambda} + \frac{t}{n^{1/2}\lambda}; n, \lambda\right) = \frac{1}{(2\pi)^{1/2}} \exp\left\{-\frac{t^2}{2} + \frac{1}{n^{1/2}}\left(\frac{t^3}{3} - t\right) + O(n^{-1})\right\}\{1 + O(n^{-1})\}$$
$$= \phi(t) \exp\left\{\frac{t^3 - 3t}{3n^{1/2}} + O(n^{-1})\right\}\{1 + O(n^{-1})\},$$

where $\phi(t) = (2\pi)^{-1/2} \exp(-t^2/2)$ is the standard Gaussian density. Expanding the exponential function in a Taylor series (see 2 in the Appendix) yields

$$\frac{1}{n^{1/2}\lambda} g\left(\frac{1}{\lambda} + \frac{t}{n^{1/2}\lambda}; n, \lambda\right) = \phi(t)\left\{1 + \frac{t^3 - 3t}{3n^{1/2}} + O(n^{-1})\right\}. \tag{4.6}$$

It follows from (4.6) that $n^{-1/2}\lambda^{-1}g(\lambda^{-1} + n^{-1/2}\lambda^{-1}t; n, \lambda) \to \phi(t)$ for every $t \in \mathbb{R}$ and hence

$$\mathrm{P}\left\{n^{1/2}\lambda\left(\bar{Z} - \frac{1}{\lambda}\right) \le t\right\} = \int_{-\infty}^{t} \frac{1}{n^{1/2}\lambda} g\left(\frac{1}{\lambda} + \frac{r}{n^{1/2}\lambda}; n, \lambda\right) dr \to \Phi(t),$$

for every $t \in \mathbb{R}$. (This argument can be made rigorous by applying the dominated convergence theorem.) Thus the limit of the distribution function of $n^{1/2}\lambda(\bar{Z} - \lambda^{-1})$ is a standard Gaussian distribution function and we can approximate the sampling distribution of $n^{1/2}\lambda(\bar{Z} - \lambda^{-1})$ by a standard Gaussian distribution. Since

$$G(t; n, \lambda) = \mathrm{P}\{\bar{Z} \leq x\} \sim \Phi\left(n^{1/2}\lambda\left(x - \frac{1}{\lambda}\right)\right), \tag{4.7}$$

an equivalent statement is that we can approximate the sampling distribution of $\bar{Z}$ by the $\mathrm{N}(\lambda^{-1}, n^{-1}\lambda^{-2})$ distribution. The corresponding approximation to the quantile function of $\bar{Z}$ is

$$G^{-1}(u; n, \lambda) \sim \frac{1}{\lambda} + \frac{1}{n^{1/2}\lambda}\Phi^{-1}(u).$$

We obtain an approximate $100(1 - \alpha)\%$ confidence interval for the median failure time $\theta = \lambda^{-1}\log(2)$ by substituting the approximation to $G^{-1}(u; n, \lambda)$ into (4.2). If we did not already have an explicit expression for the confidence interval in terms of G^{-1} (which requires us to know that $\lambda\bar{Z}$ is a pivotal quantity), we could derive such an interval from the Gaussian approximation. Notice that

$$1 - \alpha \approx \mathrm{P}\left\{-n^{-1/2}\Phi^{-1}(1 - \alpha/2) \leq \lambda\left(\bar{Z} - \frac{1}{\lambda}\right) \leq n^{-1/2}\Phi^{-1}(1 - \alpha/2)\right\},$$

so we can treat $\lambda(\bar{Z} - \lambda^{-1})$ as an approximate pivotal quantity and invert the inequalities to solve for λ^{-1}. We find that

$$1 - \alpha \approx \mathrm{P}\left\{\frac{\bar{Z}}{1 + n^{-1/2}\Phi^{-1}(1 - \alpha/2)} \leq \frac{1}{\lambda} \leq \frac{\bar{Z}}{1 - n^{-1/2}\Phi^{-1}(1 - \alpha/2)}\right\}$$

so that a realization of

$$\left\{\frac{\bar{Z}\log(2)}{1 + n^{-1/2}\Phi^{-1}(1 - \alpha/2)}, \frac{\bar{Z}\log(2)}{1 - n^{-1/2}\Phi^{-1}(1 - \alpha/2)}\right\}$$

is an approximate $100(1 - \alpha)\%$ confidence interval for θ. This is the same as the interval obtained by substituting $G^{-1}(u; n, 1) \sim 1 + n^{-1/2}\Phi^{-1}(u)$ into (4.2).

The Gaussian approximation (4.7) is compared to the exact sampling distribution in Figure 4.1. The approximation is not particularly good in the tails of the distribution for $n = 20$ and only becomes really good when n is quite large (see Table 4.1).

4.1.4 Edgeworth and Cornish–Fisher Expansions

The Gaussian approximation to the sampling distribution of $\bar{Z}$ is derived from the leading term in (4.6) so we may be able to improve the approximation in

Table 4.1. Comparison of Different Approximations to the Gamma Quantile Function

n	u	Gamma	Approximation			
			Studentized	Gaussian	Lognormal	Cube Root
10	0.005	0.3717	0.5511	0.1855	0.4428	0.3692
	0.010	0.4130	0.5762	0.2643	0.4792	0.4113
	0.025	0.4795	0.6174	0.3802	0.5381	0.4787
	0.050	0.5425	0.6578	0.4799	0.5944	0.5424
	0.950	1.5705	2.0840	1.5201	1.6823	1.5701
	0.975	1.7085	2.6302	1.6198	1.8585	1.7086
	0.990	1.8783	3.7829	1.7357	2.0869	1.8796
	0.995	1.9998	5.3923	1.8145	2.2582	2.0023
20	0.005	0.5177	0.6345	0.4240	0.5622	0.5167
	0.010	0.5541	0.6578	0.4798	0.5944	0.5535
	0.025	0.6108	0.6953	0.5617	0.6452	0.6106
	0.050	0.6627	0.7311	0.6322	0.6923	0.6627
	0.950	1.3940	1.5818	1.3678	1.4446	1.3938
	0.975	1.4835	1.7802	1.4383	1.5500	1.4836
	0.990	1.5923	2.0841	1.5202	1.6823	1.5928
	0.995	1.6691	2.3583	1.5760	1.7789	1.6701
50	0.005	0.6733	0.7330	0.6357	0.6947	0.6730
	0.010	0.7006	0.7524	0.6710	0.7196	0.7005
	0.025	0.7422	0.7830	0.7228	0.7579	0.7422
	0.050	0.7793	0.8113	0.7674	0.7925	0.7793
	0.950	1.2434	1.3031	1.2326	1.2619	1.2434
	0.975	1.2956	1.3835	1.2772	1.3194	1.2956
	0.990	1.3581	1.4903	1.3290	1.3896	1.3582
	0.995	1.4017	1.5730	1.3643	1.4395	1.4019
100	0.005	0.7612	0.7952	0.7424	0.7729	0.7611
	0.010	0.7822	0.8113	0.7674	0.7924	0.7821
	0.025	0.8136	0.8361	0.8040	0.8220	0.8136
	0.050	0.8414	0.8587	0.8355	0.8483	0.8414
	0.950	1.1700	1.1969	1.1645	1.1788	1.1700
	0.975	1.2053	1.2438	1.1960	1.2165	1.2053
	0.990	1.2472	1.3032	1.2326	1.2619	1.2473
	0.995	1.2763	1.3470	1.2576	1.2938	1.2764

The studentized (Gaussian) approximation is $\{1 - n^{-1/2}\Phi^{-1}(u)\}^{-1}$, the Gaussian approximation is $1 + n^{-1/2}\Phi^{-1}(u)$, the lognormal approximation is $\exp\{n^{-1/2}\Phi^{-1}(u)\}$, and the cube root approximation is $\{1 - 1/9n + \Phi^{-1}(u)/3n^{1/2}\}^3$.

small samples by including additional terms from (4.6). Each extra term involves a higher power of $n^{-1/2}$ so including more terms decreases the order of the remainder. The resulting approximation is called an *Edgeworth expansion* (Edgeworth, 1905). In particular, we find that the two-term Edgeworth expansion for the density (which contains the Gaussian approximation and first term expansion) is

$$\frac{1}{n^{1/2}\lambda} g\left(\frac{1}{\lambda} + \frac{t}{n^{1/2}\lambda}; n, \lambda\right)$$

$$= \phi(t)\left\{1 + \frac{t^3 - 3t}{3n^{1/2}} + \frac{2t^6 - 21t^4 + 36t^2 - 3}{36n} + O(n^{-3/2})\right\}. \quad (4.8)$$

The Edgeworth expansion for the distribution function is

$$\mathrm{P}\left\{n^{1/2}\lambda\left(\bar{Z} - \frac{1}{\lambda}\right) \leq t\right\}$$

$$= \int_{-\infty}^{t} \phi(r)\left\{1 + \frac{r^3 - 3r}{3n^{1/2}} + \frac{2r^6 - 21r^4 + 36r^2 - 3}{36n} + O(n^{-3/2})\right\} dr$$

$$= \Phi(t) + \phi(t)\frac{t^2 - 1}{3n^{1/2}} + \phi(t)\frac{2t^5 - 11t^3 + 3t}{36n} + O(n^{-3/2}). \quad (4.9)$$

The Edgeworth expansion for the distribution function can be inverted to obtain an expansion for the quantile function. Let $Q(u, n, 1)$ denote the quantile function of $n^{1/2}\lambda(\bar{Z} - \lambda^{-1})$. Then set $Q(u; n, 1) = \Phi^{-1}(u) + a/n^{1/2} + b/n \ldots$, use (4.9) to expand the right-hand side of the equation

$$u = \mathrm{P}\{n^{1/2}\lambda(\bar{Z} - \lambda^{-1}) \leq \Phi^{-1}(u) + a/n^{1/2} + b/n^{1/2} \ldots\}$$

in powers of $n^{-1/2}$, equate the coefficients of $n^{-1/2}, n^{-1} \ldots$ to zero and solve for $a, b \ldots$. We obtain the *Cornish–Fisher expansion* (Cornish and Fisher, 1937)

$$Q(u; n, 1) = \Phi^{-1}(u) + \frac{u^2 - 1}{3n^{1/2}} + \frac{u^3 - 7u}{36n} + O(n^{-3/2})$$

which can be used to construct approximate confidence intervals in (4.2).

4.1.5 Asymptotic Expansions

Asymptotic expansions like the Edgeworth and Cornish–Fisher expansions are widely used in statistics, so it is worth knowing some of their properties. First, asymptotic expansions are not unique because different functions can have the same expansion. Second, simple operations such as addition, multiplication, and term by term integration (such as we used in Section 4.1.3) can be carried out on asymptotic expansions. Term by term differentiation is, however, not generally permitted. Asymptotic expansions are usually not convergent when viewed as infinite series but, when terminated after a finite number of terms,

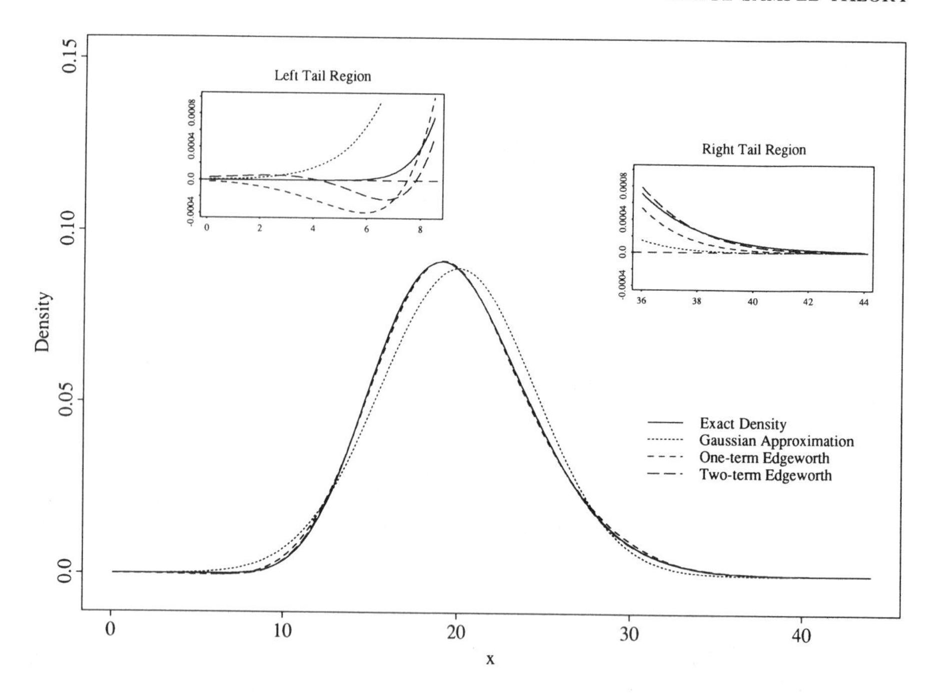

Figure 4.1. The sampling density of the standardized sample mean under the exponential model for the pressure vessel failure data and Edgeworth approximations of different orders.

the error term is less than the first omitted term which decreases as $n \to \infty$. Thus they have a natural measure of the magnitude of the error in the approximation.

Unfortunately, including further terms in an asymptotic expansion need not improve the quality of the approximation it provides. This is shown in Figure 4.1 which contains plots of the density function of $\lambda\bar{Z}$ and the approximations corresponding to the Gaussian approximation (4.7) and the one- and two-term Edgeworth expansions (4.8). Oscillatory effects introduced by the higher order terms make the lower tail of the approximations negative. Edgeworth expansions often fail in the tails which is typically where we want to use them. We will explore alternative ways of improving the Gaussian approximation in Sections 4.1.8 and 4.4 below.

4.1.6 Convergence in Distribution and Convergence in Probability

It is useful to simplify future discussion by introducing some formal terminology to describe the convergence of distribution functions.

The fundamental concept from the inference perspective is that of *convergence in distribution*.

A sequence of random variables $\{X_n\}$ converges in distribution to X and we write $X_n \xrightarrow{\mathscr{D}} X$ if the distribution function $G_n(x)$ of X_n converges to the distribution function $G(x)$ of X at every point x at which $G(x)$ is continuous.

This definition is general enough to apply to both (4.4) and the Gaussian approximation (4.7). We can abbreviate the statement of (4.4) that the distribution function of $\bar{Z}$ converges to that of the degenerate distribution with all its mass at λ^{-1} by saying that $\bar{Z}$ converges in distribution to λ^{-1} and write

$$\bar{Z} \xrightarrow{\mathscr{D}} \frac{1}{\lambda}.$$

Similarly, we can express the Gaussian approximation (4.7) as

$$n^{1/2}\lambda\left(\bar{Z} - \frac{1}{\lambda}\right) \xrightarrow{\mathscr{D}} \mathrm{N}(0, 1). \tag{4.10}$$

Here we call λ^{-1} the *asymptotic mean* (which means that $\bar{Z}$ is asymptotically unbiased for λ^{-1}) and $n^{-1}\lambda^{-2}$ the *asymptotic variance* of $\bar{Z}$.

Convergence in distribution to a constant is a special case both of convergence in distribution and another general mode of convergence known as *convergence in probability*.

A sequence of random variables $\{X_n\}$ converges in probability to X and we write $X_n = X + o_p(1)$ if for every $\varepsilon > 0$,

$$\mathrm{P}\{|X_n - X| > \varepsilon\} \to 0 \qquad \textit{as } n \to \infty.$$

Convergence in probability implies convergence in distribution and (4.5) shows that convergence in distribution to a constant implies convergence in probability to that constant. Thus (4.4) is identical to the assertion that $\bar{Z}$ converges in probability to λ^{-1} and we can write

$$\bar{Z} = \frac{1}{\lambda} + o_p(1).$$

In fact, we can make a stronger statement because (4.10) implies that $\bar{Z}$ converges in probability to λ^{-1} at the same rate as $n^{-1/2}$. We write

$$\bar{Z} - \frac{1}{\lambda} = O_p(n^{-1/2}).$$

The relationship between convergence in distribution and convergence in probability means that we can define a *consistent estimator* in terms of either mode of convergence.

A sequence of estimators $\{\hat{\theta}_n\}$ *is consistent for a (constant) parameter* θ *if either* $\hat{\theta}_n \xrightarrow{\mathscr{D}} \theta$ *or* $\hat{\theta}_n = \theta + o_p(1)$.

4.1.7 Establishing Convergence of Estimators

The approach used in Sections 4.1.1–4.1.5 of approximating the sampling distribution by direct expansion has two disadvantages. First, it quickly becomes tedious and second and more importantly, it depends on our being able to write down the exact density of the sampling distribution. There are many situations in which this is very difficult or impossible to do so we need another approach. Fortunately, one of the most important contributions of probability theory has been the development of general theorems which enable us to obtain approximations to the sampling distribution of a statistic by considering the distribution of the data $\mathbf{Z}_n = (Z_1, \ldots, Z_n)$ rather than the sampling distribution of the statistic of interest. These results simplify the calculations and can be applied even when the sampling distribution is unknown.

Let $Z_1, \ldots, Z_n$ be independent random variables with means $\mu_1, \ldots, \mu_n$ and variances $\sigma_1^2, \ldots, \sigma_n^2$. Set $V_n = \sum_{i=1}^n \sigma_i^2$. The basic tool for finding asymptotic approximations to sampling distributions is the central limit theorem which applies when the *Lindeberg condition* holds.

Lindeberg condition For each $\varepsilon > 0$,

$$V_n^{-1} \sum_{i=1}^n \mathrm{E}\{(Z_i - \mu_i)^2 I(|Z_i - \mu_i| > \varepsilon V_n^{1/2})\} \to 0 \qquad \text{as } n \to \infty.$$

Theorem 4.1 (Lindeberg, 1922; Feller, 1935) *Let* $Z_1, \ldots, Z_n$ *be independent random variables with means* $\mu_1, \ldots, \mu_n$ *and variances* $\sigma_1^2, \ldots, \sigma_n^2$. *Set* $V_n = \sum_{i=1}^n \sigma_i^2$ *and suppose that the Lindeberg condition holds. Then*

$$\mathrm{P}\left\{V_n^{-1/2} \sum_{i=1}^n (Z_i - \mu_i) \le x\right\} \to \Phi(x), \qquad \textit{for all } x \in \mathbb{R},$$

or, equivalently,

$$V_n^{-1/2} \sum_{i=1}^n (Z_i - \mu_i) \xrightarrow{\mathscr{D}} \mathrm{N}(0, 1) \qquad \textit{as } n \to \infty.$$

The Lindeberg condition can be difficult to check so we often use conditions which imply the Lindeberg condition. For example, the Lindeberg condition is implied by the *Lyapounov condition.*

Lyapounov condition For some $\delta > 0$,

$$V_n^{-\delta/2} \sum_{i=1}^n \mathrm{E}|Z_i - \mu_i|^{2+\delta} \to 0 \qquad \text{as } n \to \infty.$$

Even simpler conditions can be used if the Z_i have common mean $\mu = \mu_1 = \cdots = \mu_n$ and variance $\sigma^2 = \sigma_1^2 = \cdots = \sigma_n^2$ (in this case, the Lindberg condition holds if $\mathrm{E}|Z_i - \mu|^{2+\delta} \leq M < \infty$, for some $\delta > 0$) or if the Z_i are identically distributed. When the Z_i are independent and identically distributed with finite variance, the central limit theorem is known as the *Lindeberg–Levy central limit theorem.*

Corollary 4.1 *Let $Z_1, \ldots, Z_n$ be independent and identically distributed random variables with mean μ and variance σ^2. Then*

$$\mathrm{P}\left\{n^{-1/2} \sum_{i=1}^{n} \frac{(Z_i - \mu)}{\sigma} \leq x\right\} \to \Phi(x) \qquad \textit{for all } x \in \mathbb{R}.$$

Equations (4.7) and (4.10) follow immediately from the Lindeberg–Levy central limit theorem and the fact that $\mathrm{E}Z = \lambda^{-1}$ and $\mathrm{Var}\,(Z) = \lambda^{-2}$ under the exponential model (4.1).

Edgeworth expansions can also be obtained by an extension of the central limit theorem. If in addition to the conditions of Corollary 4.1, $Z_1, \ldots, Z_n$ have third cumulant $\kappa_3 = \mathrm{E}(Z_1 - \mu)^3/\sigma^{3/2}$ and fourth cumulant $\kappa_4 = \mathrm{E}(Z_1 - \mu)^4/\sigma^4 - 3$, then

$$\mathrm{P}\left\{n^{-1/2} \sum_{i=1}^{n} \frac{(Z_i - \mu)}{\sigma} \leq x\right\} = \Phi(x) - \phi(x)\frac{\kappa_3(x^2 - 1)}{6n^{1/2}}$$
$$- \phi(x)\frac{3\kappa_4(x^3 - 3x) + \kappa_3^2(x^5 - 10x^3 + 15x)}{72n} + \cdots.$$

The distribution function expansion (4.9) is obtained by substituting the fact that $\kappa_3 = 2$ and $\kappa_4 = 6$ for the standardized exponential distribution.

The analogue of the Lindeberg–Levy central limit theorem for convergence in probability is the *weak law of large numbers*. If $Z_1, \ldots, Z_n$ are independent and identically distributed random variables with mcan μ,

$$n^{-1} \sum_{i=1}^{n} Z_i = \mu + o_p(1).$$

The weak law of large numbers establishes the consistency of $\bar{Z}$ for λ^{-1} under (4.1) more simply than the argument leading to (4.5).

We often need to establish convergence in probability for a general sequence of random variables $\{X_n\}$ which is not necessarily a sequence of means. In this case we may be able to use *Chebychev's inequality*, which states that

$$\mathrm{P}\{|X_n - \eta| \geq \varepsilon\} \leq \frac{\mathrm{E}(X_n - \eta)^2}{\varepsilon^2}$$

so that if $E(X_n - \eta)^2 \to 0$ then $X_n = \eta + o_p(1)$. Indeed, we have the stronger result that

$$X_n - \eta = O_p\{(E(X_n - \eta)^2)^{1/2}\}$$

so that we can find the rate of convergence by computing moments.

Finally, if $X_n - \eta = o_p(1)$ and g is continuous at η, it follows that $g(X_n) - g(\eta) = o_p(1)$ and the limits in probability of sums, products, and continuous functions of sequences converging in probability are these functions of the limits.

4.1.8 Smooth Functions of Estimators

We noted in Section 4.1.5 that Edgeworth expansions often break down in the tails. This suggests that we consider expansions with better leading terms rather than expansions with more terms. A simple way to obtain expansions with better leading terms (based on good applied practice) is to approximate the sampling distribution of a nonlinear function h of $\bar{Z}$ rather than of $\bar{Z}$. (Another approach is presented in Section 4.4.)

Provided h is smooth enough, we can make a Taylor expansion (see 2 in the Appendix) of $h(\bar{Z})$ about μ to obtain

$$h(\bar{Z}) = h(\mu) + h'(\mu)(\bar{Z} - \mu) + h''(\mu)\frac{(\bar{Z} - \mu)^2}{2!} \dots,$$

an expansion in increasing powers of $\bar{Z} - \mu$. Since $\bar{Z} - \mu = O_p(n^{-1/2})$ by the central limit theorem, this is a stochastic expansion in increasing powers of terms which are of order $n^{-1/2}$ in probability. This asymptotic expansion of $h(\bar{Z})$ is useful in obtaining expansions of the moments and the sampling distribution of $h(\bar{Z})$. For example,

$$Eh(\bar{Z}) = h(\mu) + \frac{h''(\mu)\sigma^2}{n2!} \dots,$$

$$\operatorname{Var} h(\bar{Z}) = \frac{h'(\mu)^2\sigma^2}{n} \dots,$$

$$E\{h(\bar{Z}) - h(\mu)\}^3 = \frac{h'(\mu)^3\kappa_3 + 3h''(\mu)h'(\mu)^2\sigma^4}{n^2} + \cdots,$$

and so on. Notice that these are expansions in increasing powers of n^{-1} and not $n^{-1/2}$.

We obtain an approximation to the sampling distribution of $h(\bar{Z})$ by writing

$$n^{1/2}\{h(\bar{Z}) - h(\mu)\} = h'(\mu)n^{1/2}(\bar{Z} - \mu) + h''(\mu)n^{1/2}\frac{(\bar{Z} - \mu)^2}{2!} \dots. \quad (4.11)$$

Aside from the first term on the right-hand side of (4.11), we have terms of smaller order than $n^{1/2}O_p(n^{-1}) = O_p(n^{-1/2})$ which should vanish in the limit leaving us with only the first term to which we can apply the central limit theorem. That these terms vanish can be established by a result which is often referred to as *Slutsky's theorem.*

Theorem 4.2 (Slutsky, 1925; Cramer, 1946, p. 254) *Suppose that $X_n \xrightarrow{\mathscr{D}} X$ and that U_n and V_n converge in probability to u and v respectively. Then*

$$U_n + V_n X_n = u + vX_n + o_p(1) \xrightarrow{\mathscr{D}} u + vX.$$

In particular, when $n^{1/2}(T_n - \theta) \xrightarrow{\mathscr{D}} \mathrm{N}(0, \sigma^2)$,

$$U_n + V_n n^{1/2}(T_n - \theta) \xrightarrow{\mathscr{D}} \mathrm{N}(u, v^2\sigma^2).$$

On applying Theorem 4.2 to the right-hand side of (4.11), we have $T_n = \bar{Z}$, $\theta = \mu$, $V_n = 1$ and U_n represents the terms beyond the first. The central limit theorem establishes that $n^{1/2}(\bar{Z} - \mu)$ satisfies the required condition and that $U_n = o_p(n^{-1/2})$ converges in probability to $u = 0$. This argument is used so frequently that we state it explicitly as the *transformation theorem.*

Theorem 4.3 *Suppose that $n^{1/2}(T_n - \theta) \xrightarrow{\mathscr{D}} X$ and that $h: \mathbb{R} \to \mathbb{R}$ such that the derivative h' of h exists and is nonzero at θ. Then*

$$n^{1/2}\{h(T_n) - h(\theta)\} = n^{1/2}h'(\theta)(T_n - \theta) + o_p(1) \xrightarrow{\mathscr{D}} h'(\theta)X.$$

Thus when $n^{1/2}(T_n - \theta) \xrightarrow{\mathscr{D}} \mathrm{N}(0, \sigma^2)$,

$$n^{1/2}\{h(T_n) - h(\theta)\} \xrightarrow{\mathscr{D}} \mathrm{N}(0, h'(\theta)^2\sigma^2).$$

Theorems 4.2 and 4.3 expand the applicability of the central limit theorem by allowing us to convert in probability expansions of statistics into approximations to the sampling distributions of those statistics, and many large sample approximations to the sampling distributions of statistics are built up from these results.

4.1.9 Variance Stabilizing Transformations

From Theorem 4.3 and (4.10) we see that applying a smooth transformation h to the sample mean from the exponential model (4.1) leads to

$$n^{1/2}\left\{h(\bar{Z}) - h\left(\frac{1}{\lambda}\right)\right\} \xrightarrow{\mathscr{D}} \mathrm{N}\left(0, \frac{h'\left(\frac{1}{\lambda}\right)^2}{\lambda^2}\right)$$

but what h should we choose? The log transformation is widely used in applied statistics when we have non-negative data, so we can try $h(x) = \log(x)$. In this case $h'(x) = x^{-1}$ and $h'(\lambda^{-1}) = \lambda$ so by Theorem 4.3,

$$n^{1/2}(\log(\bar{Z}) + \log(\lambda)) = n^{1/2}\log(\lambda\bar{Z}) \xrightarrow{\mathscr{D}} \mathrm{N}(0,1)$$

The log transformation has removed the dependence of the asymptotic variance on λ so is called a *variance stabilizing transformation.*

Generally, if the asymptotic variance σ^2 of a statistic T_n is a function $\sigma^2(\mu)$ of the asymptotic mean μ, the variance stabilizing transformation for T_n is the function h satisfying

$$h'(\mu) \propto \frac{1}{\sigma(\mu)}$$

or

$$h(\mu) \propto \int \frac{1}{\sigma(\mu)}\, d\mu.$$

The Gaussian approximation on the log scale is equivalent to approximating the distribution of $\bar{Z}$ by a lognormal distribution. This should work better than our previous approximations because it is asymmetric. The lognormal approximation yields

$$G(x; n, \lambda) = \mathrm{P}\{\bar{Z} \le x\} \sim \Phi\{n^{1/2}\log(\lambda x)\}$$

so

$$G^{-1}(u; n, 1) \sim \exp\{n^{-1/2}\Phi^{-1}(u)\}$$

which can be substituted directly into (4.2). Alternatively, we can treat $\log(\lambda\bar{Z})$ as an approximate pivotal quantity and obtain the same confidence interval from first principles.

4.1.10 Symmetrizing Transformations

A second approach to choosing a transformation is to try to choose a transformation to make the transformed distribution nearly symmetric. An imperfect but simple measure of the asymmetry of the transformed distribution is $\mathrm{E}\{h(\bar{Z}) - h(\mu)\}^3$, which equals 0 for a symmetric distribution. Equating the first term in the expansion for this moment (Section 4.1.8) to 0, we see that we need to solve for h in the equation

$$h'(\mu)\kappa_3 + 3h''(\mu)\sigma^4 = 0.$$

If we take h to be a power transformation of the form $h(x) = x^q$, we have

$h'(x) = qx^{q-1}$ and $h''(x) = q(q-1)x^{q-2}$ so

$$\mu^{q-1}\kappa_3 + 3(q-1)\mu^{q-2}\sigma^4 = 0$$

or

$$q = 1 - \frac{\mu\kappa_3}{3\sigma^4}$$

For the gamma distribution, we obtain $q = 1/3$, the *Wilson–Hilferty* (1931) *transformation.* In this case, we have

$$n^{1/2}\left\{\bar{Z}^{1/3} - \frac{1}{\lambda^{1/3}}\right\} \xrightarrow{\mathscr{D}} \mathrm{N}\left(0, \frac{1}{9\lambda^{2/3}}\right).$$

Even better results are obtained by including the next term in the expansion of $\mathrm{E}h(\bar{Z})$ to obtain

$$n^{1/2}\left\{\bar{Z}^{1/3} - \frac{1}{\lambda^{1/3}}\left(1 - \frac{1}{9n}\right)\right\} \xrightarrow{\mathscr{D}} \mathrm{N}\left(0, \frac{1}{9\lambda^{2/3}}\right).$$

This cube root normal approximation yields

$$G^{-1}(u; n, 1) \sim \left\{1 - \frac{1}{9n} + \frac{\Phi^{-1}(u)}{3n^{1/2}}\right\}^3.$$

The densities corresponding to the lognormal and cube root normal approximations are compared with the actual sampling density in Figure 4.2 and the critical points corresponding to these approximations are compared with the "exact" critical points in Table 4.1. The cube root normal approximation is by far the best of those we have considered. An approximate 95% confidence interval for the median failure time based on this approximation is

$$(268.89, 653.34)$$

which is very close to the "exact" interval.

4.1.11 Studentization

The variance stabilizing transformation served the purpose both of introducing a better scale for the approximation and of stabilizing the variance. If we are only interested in the latter aim, Theorem 4.2 provides an alternative way to achieve it: since $\bar{Z}$ converges in probability to λ^{-1}, Theorem 4.2 ensures

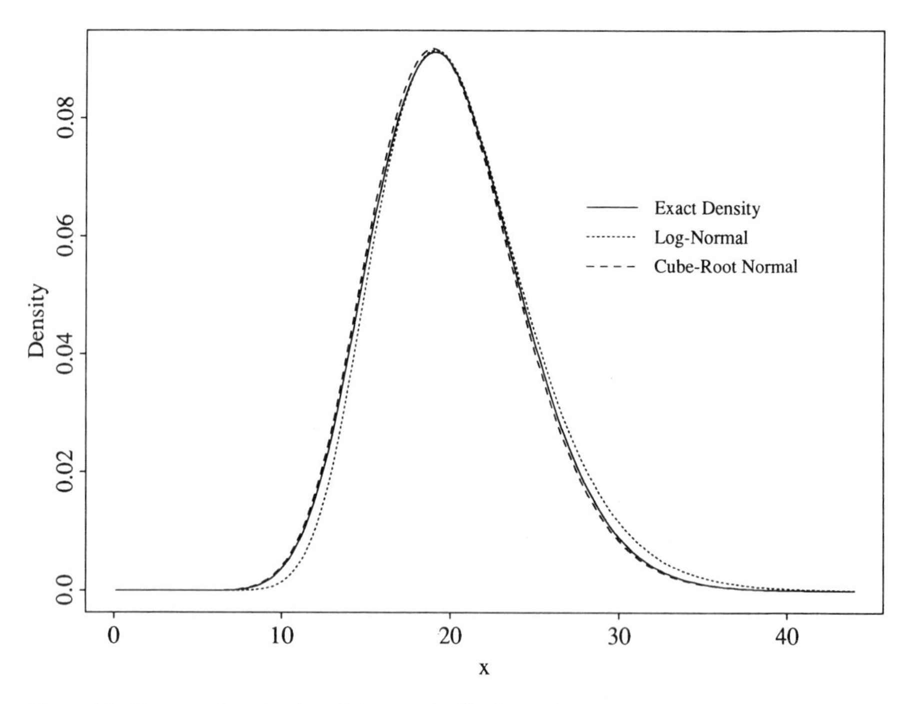

Figure 4.2. The sampling density of the standardized sample mean under the exponential model for the pressure vessel failure data and Gaussian approximation on different scales.

that

$$n^{1/2}\frac{\left(\bar{Z}-\frac{1}{\lambda}\right)}{\bar{Z}} = \frac{1}{\bar{Z}\lambda}n^{1/2}\lambda\left(\bar{Z}-\frac{1}{\lambda}\right) \xrightarrow{\mathscr{D}} \mathrm{N}(0, 1)$$

so we have eliminated the dependence of the variance on the mean. This process is known as *studentization.*

Treating $\bar{Z}^{-1}(\bar{Z} - \lambda^{-1})$ as an approximate pivotal quantity, we find that

$$\{\bar{Z}(1 - n^{-1/2}\Phi^{-1}(1 - \alpha/2)), \bar{Z}(1 + n^{-1/2}\Phi^{-1}(1 - \alpha/2))\}$$

is an approximate $100(1 - \alpha)\%$ confidence interval for λ^{-1}. Comparing this interval to (4.2), we see that studentization is like making the approximation

$$G^{-1}(u; n, 1) = \frac{1}{\{1 - n^{-1/2}\Phi^{-1}(u)\}}.$$

This approximation is better than our original normal approximation but inferior to those obtained by transformation.

PROBLEMS

4.1.1. Suppose that we have observations $\mathbf{Z}$ on the uniform model

$$\mathscr{F} = \left\{ f(\mathbf{y}; \theta) = \prod_{i=1}^{n} \frac{1}{\theta} I(0 \le y_i \le \theta); \theta \ge 0 \right\}$$

where $I(\cdot)$ is the indicator function. What happens to $\mathrm{P}\{Z_{nn} \le x\}$ as $n \to \infty$? Show that this implies that for all $\varepsilon > 0$,

$$\mathrm{P}\{Z_{nn} < \theta - \varepsilon\} \to 0 \qquad \text{as } n \to \infty.$$

Interpret this result. Then show that as $n \to \infty$

$$\mathrm{P}\{n(Z_{nn} - \theta) \le y\} \to \exp(y/\theta), \qquad y < 0.$$

Use this result to obtain a large sample confidence interval for θ and compare this to the exact likelihood based confidence interval.

4.1.2. In Problem 4.1.1, obtain the next two terms in the expansion of $\mathrm{P}\{n(Y_{nn} - \theta) \le y\}$. On the same set of axes, plot the exact probability function, the leading term in the expansion, the approximation based on the first two terms in the expansion and that based on the first three terms. Comment.

4.1.3. Suppose that we have observations $\mathbf{Z}$ on the model

$$\mathscr{F} = \left\{ f(\mathbf{y}; \theta) = \prod_{i=1}^{n} \left(\frac{\theta}{y_i^2} \right) I(y_i > \theta) \colon \theta > 0 \right\}.$$

Show that as $n \to \infty$,

$$\mathrm{P}\{n(Z_{n1} - \theta) \le x\} \to 1 - \exp\left(-\frac{x}{\theta} \right), \qquad x > 0.$$

Construct a $100(1 - \alpha)\%$ large sample confidence interval for θ and compare it to the exact interval obtained in Problem 3.8.1.

4.1.4. Suppose that we have observations $\mathbf{Z}$ on the Poisson model

$$\mathscr{F} = \left\{ f(\mathbf{y}; \lambda) = \prod_{i=1}^{n} \frac{\lambda^{y_i} \exp(-\lambda)}{y_i!}, \; y_i = 0, 1, 2 \ldots \colon \lambda > 0 \right\}.$$

Find the asymptotic distribution of $\sqrt{\bar{Z}}$ and show how to use it to set an approximate $100(1-\alpha)\%$ confidence interval for λ. Use Rutherford and Geiger's (1910) alpha particle data from Problem 1.5.1 to set a 95% confidence interval for the mean number of scintillations due to alpha particles from the radioactive decay of polonium samples in 7.5 seconds.

4.1.5. Suppose that we make an observation Z on the binomial model

$$\mathscr{F} = \left\{ f(z; \theta) = \binom{n}{z} \theta^z (1-\theta)^{n-z}, z = 0, 1, \ldots, n: 0 \leq \theta \leq 1 \right\}$$

and that the log-odds ratio $\psi = \log(\theta/(1-\theta))$ is the function of interest. Find the maximum likelihood estimator of ψ and then find its asymptotic sampling distribution. Set an approximate $100(1-\alpha)\%$ confidence interval for ψ. Hence obtain an approximate $100(1-\alpha)\%$ confidence interval for θ. Does this interval have any advantage over the usual approximate confidence interval based on the maximum likelihood estimator of θ? Explain.

4.1.6. Suppose that $R \sim \text{binomial}(n, \theta)$. Find the asymptotic distribution of R and then use Theorem 4.3 to construct an approximate $100(1-\alpha)\%$ confidence interval for $1/\theta$. Now suppose that N is the number of independent binomial$(1, \theta)$ trials till the rth success and so has a negative binomial distribution

$$P(N = n) = \binom{n-1}{r-1} \theta^r (1-\theta)^{n-r}, \qquad n = r, r+1, \ldots, \quad 0 \leq \theta \leq 1.$$

Find the asymptotic distribution of N and then construct an approximate $100(1-\alpha)\%$ confidence interval for $1/\theta$. Compare the two intervals.

4.1.7. Suppose that we have n independent observations $\mathbf{Z}$ on the Gaussian model

$$\mathscr{F} = \left\{ f(\mathbf{z}; \mu, \sigma) = \prod_{i=1}^{n} \frac{1}{(2\pi\sigma^2)^{1/2}} \exp\left\{ -\frac{(z_i - \mu)^2}{2\sigma^2} \right\}, z_i \in \mathbb{R}: \mu \in \mathbb{R} \backslash \{0\}, \sigma > 0 \right\}$$

and that $\psi = \mu^2$ is the functional of interest. Then an unbiased estimator of ψ is given by

$$T_n = \bar{Z}^2 - \frac{S^2}{n}, \quad \text{where } \bar{Z} = n^{-1} \sum_{i=1}^{n} Y_i \text{ and } S^2 = (n-1)^{-1} \sum_{i=1}^{n} (Z_i - \bar{Z})^2.$$

Find the asymptotic sampling distribution of T_n and then write down a $100(1-\alpha)\%$ confidence interval for ψ.

4.1.8. Suppose we have observations $\mathbf{Z}$ on the variance component model in which

$$Z_{ij} = \mu + a_i + u_{ij}, \qquad j = 1, \ldots, m, i = 1, \ldots, g,$$

where $\{a_i\}$ are unobserved independent $\mathrm{N}(0, \sigma_a^2)$ random variables which are independent of $\{u_{ij}\}$ which are independent $\mathrm{N}(0, \sigma_u^2)$ random variables, and μ is an unknown parameter. Find the exact distribution of $\hat{\tau}_a/m = \sum_{i=1}^{g} (\bar{Z}_i - \bar{Z})^2/g$, where $\bar{Z}_i = m^{-1} \sum_{j=1}^{m} Z_{ij}$, and then show that it is a consistent estimator of $\tau_a/m = \sigma_a^2 + \sigma_u^2/m$ when $g \to \infty$ with m fixed, but not when $m \to \infty$ with g fixed. What happens when $m, g \to \infty$? Interpret the results.

4.2 MULTIPARAMETER PROBLEMS

In Section 4.1, we treated the pressure vessel failure time data as though the observations were actually generated by the exponential model (4.1). As we noted in Section 1.2.2, the gamma model is often used to explore the appropriateness of the exponential model. That is, we treat the data as a realization of $\mathbf{Z}$ generated by the model

$$\mathscr{F} = \left\{ f(\mathbf{y}, \lambda, \kappa) = \prod_{i=1}^{n} \frac{1}{\Gamma(\kappa)} \lambda(\lambda y_i)^{\kappa - 1} \exp(-\lambda y_i), y_i > 0 \colon \lambda > 0 \right\} \tag{4.12}$$

and explore whether the nonexponentiality parameter κ is close to 1.

4.2.1 Maximum Likelihood Estimation Under the Gamma Model

The log-likelihood under the gamma model is

$$\ell(\lambda, \kappa) \propto n\kappa \log(\lambda) + \kappa \sum_{i=1}^{n} \log(z_i) - \sum_{i=1}^{n} \lambda z_i - n \log\{\Gamma(\kappa)\}$$

which is maximized at (λ, κ) satisfying

$$0 = \frac{n\kappa}{\lambda} - \sum_{i=1}^{n} z_i$$

$$0 = n \log(\lambda) + \sum_{i=1}^{n} \log(z_i) - n\psi(\kappa),$$

where $\psi(x) = \partial \log\{\Gamma(x)\}/\partial x$ is the digamma function. Using the first equation to eliminate λ from the second, we obtain after some manipulation

$$\lambda = \frac{\kappa}{\bar{z}}$$

and

$$0 = n^{-1} \sum_{i=1}^{n} \log(z_i) - \log(\bar{z}) - \psi(\kappa) + \log(\kappa). \tag{4.13}$$

Noting that the right-hand side of (4.13) is continuous in $\kappa > 0$, that

$$n^{-1} \sum_{i=1}^{n} \log(z_i) - \log(\bar{z}) < 0$$

by Jensen's inequality (see 1 in the Appendix), and that $\psi(\kappa) - \log(\kappa) < 0$,

$$\begin{aligned} \psi(\kappa) - \log(\kappa) &\to 0 \qquad \text{as } \kappa \to \infty \\ &\to -\infty \qquad \text{as } \kappa \to 0, \end{aligned}$$

we see that (4.13) always has at least one solution. Since the derivative of the right-hand side of (4.13) is

$$-\psi'(\kappa) + \frac{1}{\kappa} < 0,$$

(4.13) has precisely one solution and there is a unique $(\hat{\lambda}, \hat{\kappa})$ which maximizes the likelihood.

To find the maximum likelihood estimates $(\hat{\lambda}, \hat{\kappa})$, we need to solve (4.13) for κ and then set $\lambda = \kappa/\bar{z}$. Greenwood and Durand (1960) suggested an approximation to the solution of this equation: let $y = |n^{-1} \sum_{i=1}^{n} \log(z_i) - \log(\bar{z})|$ and then set

$$\hat{\kappa} = \begin{cases} \dfrac{0.5000876 + 0.1648852y - 0.0544274y^2}{y} & 0 < y < 0.5772 \\[2ex] \dfrac{8.898919 + 9.059950y + 0.9775373y^2}{y(17.79728 + 11.968477y + y^2)} & 0.5772 \le y \le 17. \end{cases}$$

We find that $y = 1.072$ so we should use the second approximation. Solving the equations for our data, we find that $(\hat{\lambda}, \hat{\kappa}) = (0.001, 0.579)$. The right-hand side of the estimating equation (4.13) evaluated at this value equals 0.0002.

The next step is to find an approximation to the sampling distribution of the maximum likelihood estimator $(\hat{\lambda}, \hat{\kappa})$. This is complicated by the fact that these estimators are only implicitly defined. However, we can obtain an asymptotic expansion for $(\hat{\lambda}, \hat{\kappa})$ and then approximate the asymptotic distribution of the terms in this expansion. The approach we use is much more transparent if, instead of restricting attention to the maximum likelihood estimators under the gamma model, we work with a general class of estimators and a general model.

4.2.2 Estimating Equations

Suppose that we treat the data as a realization of $\mathbf{Z}$ generated by the model

$$\mathscr{F} = \left\{ f(\mathbf{y}; \theta) = \prod_{i=1}^{n} f(y_i; \theta): \theta \in \Omega \right\}. \tag{4.14}$$

Consider the class of estimators of θ which are solutions $\hat{\theta}$ of a general *estimating equation* of the form

$$\sum_{i=1}^{n} \eta(Z_i, \theta) = 0. \tag{4.15}$$

We call $\sum_{i=1}^{n} \eta(Z_i, \theta)$ an *estimating function* (Edgeworth, 1908–9; Godambe, 1960; 1991) and $\hat{\theta}$ a *maximum likelihood type* or *M-estimator* (Huber, 1964).

Maximum likelihood estimators $\hat{\theta}$ for the model $\mathscr{F}$ correspond to setting $\eta(x, \theta) = \partial \log \{f(x; \theta)\}/\partial\theta$ but we obtain a useful generalization of maximum likelihood estimation if we allow flexibility in the choice of η in (4.15).

If the expected value of the estimating function is 0 under $\mathscr{F}$ so

$$\int_{-\infty}^{\infty} \eta(z, \theta) f(z; \theta)\, dz = 0,$$

we say that the estimating equation is *unbiased* for θ under $\mathscr{F}$. Unbiasedness of the estimating function implies that when the estimating equation procedure is applied to the population represented by $\mathscr{F}$, the estimator is the parameter we are trying to estimate. This property is called *Fisher consistency*.

More formally, define a function $\theta(\cdot)$ from the set of all distribution functions to the parameter space Ω as a solution of the equation

$$\int_{-\infty}^{\infty} \eta(z, \theta)\, dF(z) = 0, \tag{4.16}$$

where F is an arbitrary distribution function. Replacing F in (4.16) by the empirical distribution function F_n (Section 1.5.2) produces (4.15) so we can write $\hat{\theta} = \theta(F_n)$. Let F_0 denote the distribution which generated $\mathbf{Z}$ so that $F_0(x) = F(x; \theta_0)$ for some θ_0 (called the *true parameter value*) whenever $F_0 \in \mathscr{F}$. Solving (4.16) at F_0 produces $\theta(F_0)$ and $\hat{\theta}$ is Fisher consistent for θ_0 if $\theta(F_0) = \theta_0$.

The formalization in terms of the function $\theta(\cdot)$ is useful because it tells us that $\hat{\theta}$ is estimating $\theta(F_0)$ when $Z_i, \ldots, Z_n$ are independent and identically distributed random variables with common distribution function ($F_0 \notin \mathscr{F}$).

Fisher consistency is not the same as consistency defined in Section 4.1.6 but it is closely related because F_n is consistent for F_0 and, when $\theta(\cdot)$ is continuous at F_0, it follows that $\hat{\theta}$ is consistent for $\theta(F_0)$. We will show in Section 4.2.4 that, under further conditions, $\theta(F_0)$ is the asymptotic mean of $\hat{\theta}$.

We derive the properties of M-estimators in Sections 4.2.4–4.2.6, specialize the results to maximum likelihood estimators in Sections 4.2.7–4.2.10, and then apply them to make inferences about the parameters in the gamma model (4.12) in Section 4.2.11.

4.2.3 Establishing Convergence for Random Vectors

In multiparameter problems, we have to establish approximations to the sampling distribution of a vector estimator. Although the calculations become more complicated, the manipulation of vectors raises no substantive difficulties. The techniques we use are very similar to those we would use in the case $p = 1$ and the results include $p = 1$ as a special case.

A simple extension of the central limit theorem can be established using the *Cramer–Wold device.*

Theorem 4.4 (Cramer and Wold, 1936) *The random p-vector X_n converges in distribution to X if and only if for each fixed p-vector a, $a^{\mathrm{T}}X_n$ converges in distribution to $a^{\mathrm{T}}X$.*

For our purposes we require only the multivariate version of the Lindeberg–Levy central limit theorem which is readily established from Corollary 4.1 and Theorem 4.4.

Theorem 4.5 *Let $Z_1, \ldots, Z_n$ be independent and identically distributed random p-vectors with mean p-vector μ and $p \times p$ variance matrix Σ. Then*

$$n^{-1/2} \sum_{i=1}^{n} (Z_i - \mu) \xrightarrow{\mathscr{D}} \mathrm{N}_p(0, \Sigma) \qquad \text{as } n \to \infty,$$

where N_p denotes the p-dimensional multivariate Gaussian distribution.

Convergence in probability is even easier to extend to the multiparameter case because a vector or matrix converges in probability if and only if its components do so.

4.2.4 The Approximate Sampling Distribution of an M-Estimator

The basic procedure for approximating the sampling distribution of an M-estimator is to expand the estimating equation (4.15) in a Taylor series (see 2 in the Appendix) about the point $\theta(F_0)$ to produce an expansion in increasing powers of $\hat{\theta} - \theta(F_0)$ which is of the form

$$0 = n^{-1} \sum_{i=1}^{n} \eta(Z_i, \hat{\theta}) = n^{-1} \sum_{i=1}^{n} \eta(Z_i, \theta(F_0)) + n^{-1} \sum_{i=1}^{n} \eta'(Z_i, \theta(F_0))(\hat{\theta} - \theta(F_0)) + \cdots,$$

where $\eta'(z, t)$ denotes the matrix with (i, j)th component $\partial\eta_i(z, t)/\partial t_j$. We then invert this expansion to obtain an expansion for $\hat{\theta} - \theta(F_0)$ in increasing powers of $n^{-1} \sum_{i=1}^{n} \eta(Z_i, \theta(F_0))$ which is typically of order $n^{-1/2}$ in probability and can be used to obtain approximations to the sampling distribution of $\hat{\theta} - \theta(F_0)$.

The procedure is particularly straightforward if we are only interested in the leading term. In this case, we have the expansion

$$n^{1/2}(\hat{\theta} - \theta(F_0)) = -\left\{n^{-1} \sum_{i=1}^{n} \eta'(Z_i, \theta^*)\right\}^{-1} n^{-1/2} \sum_{i=1}^{n} \eta(Z_i, \theta(F_0)), \quad (4.17)$$

where θ^* is between $\hat{\theta}$ and $\theta(F_0)$ in the sense that $|\theta^* - \theta(F_0)| \leq |\hat{\theta} - \theta(F_0)|$.

Provided $Z_1, \ldots, Z_n$ are independent and identically distributed random variables, $n^{-1/2} \sum_{i=1}^{n} \eta(Z_i, \theta(F_0))$ is a sum of independent and identically distributed random variables. If in addition

$$\mathrm{E}_{F_0}\eta(Z, \theta(F_0)) = 0,$$

and

$$\mathrm{E}_{F_0}\eta(Z, \theta(F_0))\eta(Z, \theta(F_0))^{\mathrm{T}} = A_{F_0}(\theta(F_0)) < \infty,$$

it follows from the central limit theorem that

$$n^{-1/2} \sum_{i=1}^{n} \eta(Z_i, \theta(F_0)) \xrightarrow{\mathscr{D}} \mathrm{N}(0, A_{F_0}(\theta(F_0))). \quad (4.18)$$

Next, write

$$n^{-1} \sum_{i=1}^{n} \eta'(Z_i, \theta^*) = n^{-1} \sum_{i=1}^{n} \eta'(Z_i, \theta(F_0)) + n^{-1} \sum_{i=1}^{n} \{\eta'(Z_i, \theta^*) - \eta'(Z_i, \theta(F_0))\}. \quad (4.19)$$

The first term is the mean of independent and identically distributed random variables so, provided $-\mathrm{E}_{F_0}\eta'(Z, \theta(F_0)) = B_{F_0}(\theta(F_0)) < \infty$, the weak law of large numbers ensures that

$$n^{-1} \sum_{i=1}^{n} \eta'(Z_i, \theta(F_0)) = -B_{F_0}((\theta(F_0)) + o_p(1). \quad (4.20)$$

Also, provided $\hat{\theta} = \theta(F_0) + o_p(1)$ and $\eta'(y, t)$ is continuous at $t = \theta(F_0)$ uniformly in y, we can show that

$$n^{-1} \sum_{i=1}^{n} \{\eta'(Z_i, \theta^*) - \eta'(Z_i, \theta(F_0))\} = o_p(1). \tag{4.21}$$

Finally, if $B_{F_0}(\theta(F_0))$ is nonsingular, we can apply Theorem 4.2 to (4.17)–(4.21) to obtain

$$n^{1/2}(\hat{\theta} - \theta(F_0)) = B_{F_0}(\theta(F_0))^{-1} n^{-1/2} \sum_{i=1}^{n} \eta(Z_i, \theta(F_0)) + o_p(1)$$

from which it follows that

$$n^{1/2}(\hat{\theta} - \theta(F_0)) \xrightarrow{\mathscr{D}} \mathrm{N}(0, B_{F_0}(\theta(F_0))^{-1} A_{F_0}(\theta(F_0)) B_{F_0}(\theta(F_0))^{-\mathrm{T}}),$$

where $M^{-\mathrm{T}} = (M^{-1})^{\mathrm{T}}$.

Collecting all the conditions, we have proved the following result.

Theorem 4.6 *Let $Z_1, \ldots, Z_n$ be independent and identically distributed random variables with common distribution function F_0. Suppose that*

1. *$\theta(F_0)$ is an interior point of the parameter space Ω and an isolated root of the equation $\mathrm{E}_{F_0}\eta(Z_i, \theta) = 0$,*
2. *$\eta'(y, t)$ is continuous at $t = \theta(F_0)$ uniformly in y,*
3. *$\mathrm{E}_{F_0}\eta(Z_i, \theta(F_0))\eta(Z_i, \theta(F_0))^{\mathrm{T}} = A_{F_0}(\theta(F_0)) < \infty$*

and

4. *$\mathrm{E}_{F_0}\eta'(Z_i, \theta(F_0)) = B_{F_0}(\theta(F_0)) < \infty$ and nonsingular.*

Then if $\hat{\theta} = \theta(F_0) + o_p(1)$ and $n^{-1/2} \sum_{i=1}^{n} \eta(Z_i, \hat{\theta}) = o_p(1)$,

$$n^{1/2}(\hat{\theta} - \theta(F_0)) = -B_{F_0}(\theta(F_0))^{-1} n^{-1/2} \sum_{i=1}^{n} \eta(Z_i, \theta(F_0)) + o_p(1)$$

which implies that

$$n^{1/2}(\hat{\theta} - \theta(F_0)) \xrightarrow{\mathscr{D}} \mathrm{N}(0, B_{F_0}(\theta(F_0))^{-1} A_{F_0}(\theta(F_0)) B_{F_0}(\theta(F_0))^{-\mathrm{T}}).$$

The last two (unnumbered) conditions ensure that $\hat{\theta}$ is a consistent estimator which satisfies the estimating equations. Condition 1 ensures that $\theta(F_0)$ is not on the boundary of Ω so that a Gaussian approximation to the sampling distribution is plausible (see Moran, 1971). Condition 2 justifies the Taylor series expansion of the estimating equations and ensures that the remainder vanishes. Conditions 3 and 4 justify the application of the weak law of large numbers to the denominator and then the central limit theorem to the numerator in the rearranged Taylor expansion expressing $n^{1/2}(\hat{\theta} - \theta(F_0))$ as the ratio of two terms.

A more general version of Theorem 4.6 has been given by Huber (1967).

4.2.5 Approximate Standard Errors for Estimating *M*-Estimators

Theorem 4.6 establishes that the asymptotic variance of an M-estimator $\hat{\theta}$ is

$$V_{F_0}(\theta(F_0)) = n^{-1}B_{F_0}(\theta(F_0))^{-1}A_{F_0}(\theta(F_0))B_{F_0}(\theta(F_0))^{-\mathrm{T}}.$$

A natural estimator of $V_{F_0}(\theta(F_0))$ is

$$\hat{V}(\hat{\theta}) = n^{-1}\hat{B}(\hat{\theta})^{-1}\hat{A}(\hat{\theta})\hat{B}(\hat{\theta})^{-\mathrm{T}}, \tag{4.22}$$

where

$$\hat{A}(\theta) = n^{-1}\sum_{i=1}^{n}\eta(Z_i;\theta)\eta(Z_i;\theta)^{\mathrm{T}} \quad \text{and} \quad \hat{B}(\theta) = n^{-1}\sum_{i=1}^{n}\eta'(Z_i;\theta).$$

The estimator (4.22) is consistent for $V_{F_0}(\theta(F_0))$ under the conditions of Theorem 4.6 so $\{\operatorname{diag}\hat{V}(\hat{\theta})\}^{1/2}$ gives approximate standard errors for $\hat{\theta}$.

4.2.6 Solving Estimating Equations

The Newton–Raphson method described in Section 2.7.7 as an algorithm for obtaining the maximum likelihood estimates can be applied to solve the more general (4.15). The algorithm is based on a linear expansion of the estimating function in (4.15) instead of a quadratic expansion of the log-likelihood function but the end result is the same.

We can modify the Newton–Raphson method by replacing the normalized Hessian matrix $n^{-1}\sum_{i=1}^{n}\eta'(Z_i, \theta_{(m)})$ by the estimate $-B_{F(\cdot;\theta_{(m)})}(\theta_{(m)})$ of its limit under the model $\mathscr{F}$. At least when the estimating function is the derivative of the log-likelihood, the resulting algorithm is known as *Fisher's method of scoring.* As we saw in Section 4.2.1, the form of η may suggest additional alternative algorithms.

4.2.7 Why Maximum Likelihood Estimates the True Parameter

Suppose that the model $\mathscr{F} = \{f(\mathbf{y};\theta) = \prod_{i=1}^{n} f(y_i;\theta) : \theta \in \Omega\}$ holds so that $F_0 \in \mathscr{F}$. Let θ_0 denote the true parameter value which identifies the distribution in the model which actually generated the data so F_0 denotes the distribution with density $f(\mathbf{y}; \theta_0)$. By Jensen's inequality (see 1 in the Appendix)

$$\mathrm{E}_{F_0} \log \left\{ \frac{f(Z; \theta)}{f(Z; \theta_0)} \right\} < \log \mathrm{E}_{F_0} \left\{ \frac{f(Z; \theta)}{f(Z; \theta_0)} \right\}$$
$$= \log \left\{ \int_{\text{support}\{f(\cdot; \theta_0)\}} f(y; \theta)\, dy \right\}$$

If the densities in $\mathscr{F}$ have the same support and the densities in the model are distinct in the sense that $f(y; \theta_0) \neq f(y; \theta)$ whenever $\theta \neq \theta_0$,

$$\mathrm{E}_{F_0} \log \left\{ \frac{f(Z; \theta)}{f(Z; \theta_0)} \right\} < 0 \tag{4.23}$$

and θ_0 maximizes $\mathrm{E}_{F_0} \log \{f(Z; \theta)\}$.

Since $\mathrm{E}_{F_0} \log \{f(Z_i; \theta)\}$ is maximized at $\theta = \theta_0$, the true value θ_0 satisfies the equation

$$\left. \frac{\partial \mathrm{E}_{F_0} \log f(Z; \theta)}{\partial \theta} \right|_{\theta = \theta_0} = 0.$$

If we can interchange the order of expectation (i.e., integration) and differentiation we have that

$$\mathrm{E}_{F_0} \eta(Z, \theta_0) = \mathrm{E}_{F_0} \left. \frac{\partial \log f(Z; \theta)}{\partial \theta} \right|_{\theta = \theta_0} = 0$$

so the estimating equation is unbiased for θ_0 when $\mathscr{F}$ holds and $\hat{\theta}$ is Fisher consistent for θ_0.

4.2.8 The Approximate Sampling Distribution of Maximum Likelihood Estimators

A simplification to Theorem 4.6 is often available for maximum likelihood estimators. If we can interchange the order of integration and differentiation twice,

$$\begin{aligned} B_{F_0}(\theta_0) &= -\mathrm{E}_{F_0} \eta'(Z, \theta_0) \\ &= -\int \frac{\partial^2 \log f(x, \theta_0)}{\partial \theta_0\, \partial \theta_0^{\mathrm{T}}} f(x, \theta_0)\, dx \\ &= -\int \left[\frac{1}{f(x, \theta_0)} \frac{\partial^2 f(x, \theta_0)}{\partial \theta_0\, \partial \theta_0^{\mathrm{T}}} - \frac{1}{f(x, \theta_0)^2} \frac{\partial f(x, \theta_0)}{\partial \theta_0} \frac{\partial f(x, \theta_0)^{\mathrm{T}}}{\partial \theta_0} \right] f(x, \theta_0)\, dx \end{aligned}$$

$$= -\int \frac{\partial^2 f(x, \theta_0)}{\partial \theta_0 \, \partial \theta_0^{\mathrm{T}}} \, dx + \int \frac{\partial \log \{f(x, \theta_0)\}}{\partial \theta_0} \frac{\partial \log \{f(x, \theta_0)\}^{\mathrm{T}}}{\partial \theta_0} f(x, \theta_0) \, dx$$

$$= -\frac{\partial^2}{\partial \theta_0 \, \partial \theta_0^{\mathrm{T}}} \int f(x, \theta_0) \, dx + \mathrm{E}_{F_0} \eta(Z, \theta_0) \eta(Z, \theta_0)^{\mathrm{T}}$$

$$= A_{F_0}(\theta_0).$$

The common value of $A_{F_0}(\theta) = B_{F_0}(\theta)$ is denoted by $I(\theta)$ and is the Fisher information matrix defined in Section 2.3.2.

Specializing the statement of Theorem 4.6, we have the following result.

Corollary 4.6 *Let $Z_1, \ldots, Z_n$ be observations on a model*

$$\mathscr{F} = \left\{ f(\mathbf{y}; \theta) = \prod_{i=1}^{n} f(y_i; \theta) : \theta \in \Omega \right\}$$

which satisfies

1. *θ_0 is an interior point of the parameter space Ω*
2. *the support of $f(\cdot, \theta)$ does not depend on θ and $f(y; \theta_0) \neq f(y; \theta)$ for $\theta_0 \neq \theta$*
3. *$\eta(x, \theta) = \partial \log \{f(x, \theta)\}/\partial \theta$ and the second derivative of $f(x, \theta)$ with respect to θ is finite for each x in the support of $f(x, \theta)$ and continuous at θ_0 uniformly in x*
4. *the integral $\int f(x, \theta) \, dx$ can be differentiated twice under the integral sign*
5. *the Fisher information*

$$I(\theta) = \int_{-\infty}^{\infty} \frac{\partial \log f(x, \theta)}{\partial \theta} \frac{\partial \log f(x, \theta)^{\mathrm{T}}}{\partial \theta} f(x, \theta) \, dx$$

is finite and nonsingular at $\theta = \theta_0$.

Then if $\hat{\theta} = \theta_0 + o_p(1)$ and $n^{-1/2} \sum_{i=1}^{n} \eta(Z_i, \hat{\theta}) = o_p(1)$,

$$n^{1/2}(\hat{\theta} - \theta_0) = -I(\theta_0)^{-1} n^{-1/2} \sum_{i=1}^{n} \eta(Z_i, \theta_0) + o_p(1)$$

which implies that

$$n^{1/2}(\hat{\theta} - \theta_0) \xrightarrow{\mathscr{D}} \mathrm{N}(0, I(\theta_0)^{-1}).$$

4.2.9 Approximate Standard Errors for Maximum Likelihood Estimators

To use Corollary 4.6 to make inferences about θ_0, we need to construct an estimator of the Fisher information matrix $I(\theta_0)$. We could use the estimator $\hat{V}(\hat{\theta})$ defined in (4.22) but, under the assumed model, we usually use the *observed information*

$$\mathscr{I} = -n^{-1} \sum_{i=1}^{n} \left. \frac{\partial^2 \log f(Z_i, \theta)}{\partial \theta^2} \right|_{\theta = \hat{\theta}}$$

or the *expected information*

$$I(\hat{\theta}) = -\left. \left\{ \mathrm{E}_\theta \frac{\partial^2 \log f(Z, \theta)}{\partial \theta^2} \right\} \right|_{\theta = \hat{\theta}}.$$

These estimators are generally different except in the case of distributions in the exponential family (Section 1.3.1) for which they are identical. In particular, they are identical for the gamma model (4.12). Both of these estimators are consistent for $I(\theta_0)$ under the conditions of Corollary 4.6 so either estimator can be used to make approximate inferences about θ_0. Both estimators require $\hat{\theta}$ but $\mathscr{I}$ is typically simpler to obtain than $I(\hat{\theta})$ because it does not require the expectation of the second derivative matrix.

Using the observed Fisher information, an approximate $100(1 - \alpha)\%$ confidence interval for θ_{01} is given by

$$[\hat{\theta}_1 - n^{-1/2}(\mathscr{I}^{11})^{1/2}\Phi^{-1}(1 - \alpha/2), \hat{\theta}_1 + n^{-1/2}(\mathscr{I}^{11})^{1/2}\Phi^{-1}(1 - \alpha/2)], \quad (4.24)$$

where $\mathscr{I}^{11}$ denotes the $(1, 1)$th element of $\mathscr{I}^{-1}$. Confidence intervals for the other components of θ_0 are easily obtained.

4.2.10 Consistency of Maximum Likelihood Estimators

To apply Corollary 4.6, we still have to show that $\hat{\theta}$ is a consistent estimator of θ_0. This is surprisingly difficult to do and requires rather technical conditions.

Since we are trying to maximize the log-likelihood $\sum_{i=1}^{n} \log \{f(Z_i, \theta)\}$ to estimate θ_0, we can base a consistency proof on the likelihood function. This has been done very elegantly by Wald (1949) who first proved consistency for the case that the parameter space Ω contains only a finite number of points and then extended the result to more general sets Ω satisfying compactness conditions which enable them to be approximated by finite sets.

An alternative approach due to Cramer (1946, pp. 500–4) is to show that a root of the likelihood equation is consistent. The difficulty with this approach is that if the equtions have multiple roots, it is impossible to tell which of these are consistent. This difficulty can be overcome by specifying an algorithm for choosing a single root and then showing that this root is consistent.

4.2.11 Maximum Likelihood Inference Under the Gamma Model

Technical arguments can be used to show that distributions in the exponential family satisfy the conditions of Corollary 4.6; see for example Lehmann (1959/1991, pp. 57–60; 1983, pp. 438–42). For our problem with the gamma model (4.12), $\theta = (\lambda, \kappa)^{\mathrm{T}}$ and

$$\eta(x, \theta) = \left(\frac{\kappa}{\lambda} - x, \log(\lambda) + \log(x) - \psi(\kappa)\right)^{\mathrm{T}}.$$

It follows that

$$\begin{aligned} I(\theta) &= -\mathrm{E}\eta'(Z, \theta) \\ &= \begin{pmatrix} \kappa/\lambda^2 & -\lambda^{-1} \\ -\lambda^{-1} & \psi'(\kappa) \end{pmatrix} \end{aligned}$$

and inverting this matrix, we obtain

$$I(\theta)^{-1} = \frac{\lambda^2}{\kappa\psi'(\kappa) - 1}\begin{pmatrix} \psi'(\kappa) & \lambda^{-1} \\ \lambda^{-1} & \kappa/\lambda^2 \end{pmatrix}.$$

Solving the estimating equations for our data, we find that $\hat{\theta} = (0.001, 0.579)$ and

$$\left(\frac{\mathscr{I}}{20}\right)^{-1} = \begin{pmatrix} 1.58 \times 10^{-7} & 4.08 \times 10^{-5} \\ 4.08 \times 10^{-5} & 0.023 \end{pmatrix}$$

An approximate 95% confidence interval for κ is obtained from (4.24) as

$$(0.27, 0.88).$$

This interval does not contain $\kappa = 1$ so provides evidence against the adoption of an exponential model for the pressure vessel failure data.

The fact that κ is a non-negative parameter suggests that we should consider a log-normal approximation to the sampling distribution of $\hat{\kappa}$. From Theorem 4.3, we obtain

$$\operatorname{Var}(\log(\hat{\kappa})) \sim \frac{1}{n\kappa(\kappa\psi'(\kappa) - 1)}$$

which produces the standard error 0.265 and hence the 95% confidence interval

$$(0.34, 0.97).$$

Although both approximations lead to similar conclusions in this case, we may still seek to investigate which scale provides the better approximation. Ideally, we would like to compare the approximations to the exact results but we do not know the exact distribution of $\hat{\kappa}$. Nonetheless, we can make some progress through a computer simulation (see Section 3.10). We can generate 1000 data sets of size $n = 20$ from a gamma distribution with $\lambda = 0.001$ and $\kappa_0 = 0.5$, compute a nominal 95% confidence interval for κ from each data set using the two approximations, and then compare the estimated coverage probabilities of the intervals (the proportion of intervals containing the actual value of κ_0) and the distributions of the lengths of the intervals.

Using the Gaussian approximation to the binomial distribution to set confidence intervals for the actual coverage probabilities (see Section 3.7.2), we find that the approximation on the raw scale produces nominal 95% confidence intervals for κ_0 which have an estimated coverage probability of $0.97 \pm 1.96\sqrt{0.97 \times 0.03/1000} = 0.97 \pm 0.01$ whereas that on the log scale produces nominal 95% confidence intervals for κ_0 which have an estimated coverage probability of $0.92 \pm 1.96\sqrt{0.92 \times 0.08/1000} = 0.92 \pm 0.01$. The distributions of the lengths of the confidence intervals are shown in Figure 4.3. Using the Gaussian approximation to the distribution of the mean lengths, we see

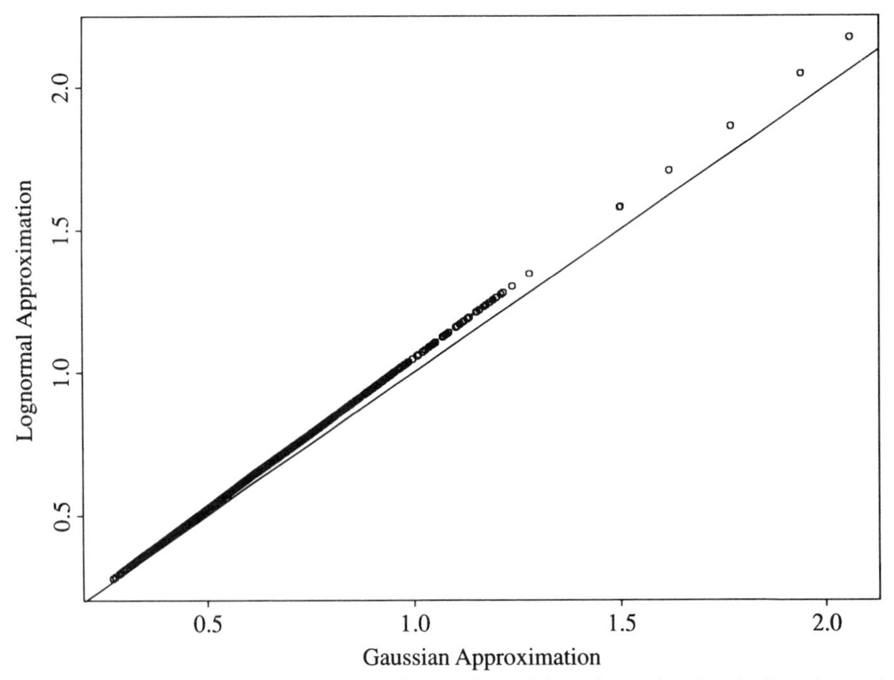

Figure 4.3. A qq-plot of the lengths of simulated 95% confidence intervals using the Gaussian and lognormal approximations to the sampling distribution of the maximum likelihood estimator of the gamma shape parameter κ under the $\Gamma(0, 5, 0.001)$ model.

that the approximation on the raw scale produces nominal 95% confidence intervals for κ_0 with mean length $0.60 \pm 1.96 \times 0.006 = 0.60 \pm 0.01$ whereas that on the log scale produces nominal 95% confidence intervals for κ_0 with mean length $0.63 \pm 1.96 \times 0.007 = 0.63 \pm 0.01$. There is not a great difference between the results produced by the two methods but the confidence intervals produced on the raw scale have slightly better coverage and are typically slightly shorter than those produced on the log scale. More detailed comparisons can be made by extending the range of n, λ, κ, and nominal levels considered.

PROBLEMS

4.2.1. Suppose that we have observations $\mathbf{Z}$ on the multivariate Student t model

$$\mathscr{F} = \left\{ f(\mathbf{y}; \mu, \sigma) = \frac{\Gamma\left(\frac{v+n}{2}\right)}{(v\pi\sigma^2)^{n/2}\Gamma\left(\frac{v}{2}\right)} \frac{1}{\left\{1 + \sum_{i=1}^{n} \frac{(y_i - \mu)^2}{2v}\right\}^{(v+n)/2}} \right.$$

$$\left. -\infty \leq y_i \leq \infty : \mu \in \mathbb{R}, \sigma > 0 \right\},$$

where $v > 0$ is known. Use the representation $Z_i = \mu + \sigma Y_i / h^{1/2}$, where Y_i are independent standard Gaussian random variables which are independent of $h \sim \chi_v^2 / v$ to find the sampling distribution of the maximum likelihood estimator $\hat{\mu}$ of μ as $n \to \infty$. Show that $\hat{\sigma}^2$ is inconsistent by showing that it converges in probability to the random variable σ^2 / h.

4.2.2. Suppose that we observe $\mathbf{Z}$ on the model

$$\mathscr{F} = \left\{ f(\mathbf{z}; \theta) = \prod_{i=1}^{n} \theta z_i^{\theta - 1} \exp(-z_i^{\theta}), z_i > 0 : \theta > 0 \right\}.$$

The maximum likelihood estimator of θ cannot be written down explicitly. Nonetheless, show that the likelihood equations have a unique root which equals the maximum likelihood estimator. (Hint: show that the likelihood equation is a continuous function of θ, takes positive and negative values and crosses the zero axis once.) Find the asymptotic sampling distribution of the maximum likelihood estimator and show how to use it to set an approximate $100(1 - \alpha)\%$ confidence interval for θ.

4.2.3. *Pareto's distribution* is sometimes used to represent the distribution of incomes over a population. Suppose that we observe $\mathbf{Z}$ on the Pareto model

$$\mathscr{F} = \left\{ f(\mathbf{y}; \lambda) = \prod_{i=1}^{n} \frac{\kappa\lambda^{\kappa}}{y_i^{\kappa+1}},\ y_i > \lambda\colon \kappa, \lambda > 0 \right\}.$$

Suppose initially that λ is known. Obtain the maximum likelihood estimator $\hat{\kappa}$ of κ. Find the asymptotic distribution of $\hat{\kappa}$ and hence of the maximum likelihood estimator of the median $\lambda 2^{1/\kappa}$. Construct a $100(1-\alpha)\%$ large sample confidence interval for the median income in the population.

4.2.4. In the context of Problem 4.2.3, suppose now that both κ and λ are unknown. Obtain the maximum likelihood estimators of κ and λ. Show that $\hat{\lambda} - \lambda = O_p(n^{-1})$. Hence or otherwise show that the asymptotic distribution of the maximum likelihood estimator of the median when λ is unknown is the same as when λ is known. (Malik, 1970, has obtained the exact distribution theory for this problem.)

4.2.5. We noted in Section 1.3.2 that the Weibull model is often used in place of the gamma model to explore the applicability of the exponential model. In the Weibull model we treat the data as a realization of $\mathbf{Z}$ generated by

$$\mathscr{F} = \left\{ f(\mathbf{y}, \lambda, \kappa) = \prod_{i=1}^{n} \kappa\lambda(\lambda y_i)^{\kappa-1} \exp\{-(\lambda y_i)^{\kappa}\},\ y_i > 0\colon \lambda > 0 \right\}.$$

Show that the log-likelihood is maximized at (λ, κ) satisfying

$$\lambda = \frac{1}{\{n^{-1}\sum_{i=1}^{n} z_i^{\kappa}\}^{1/\kappa}}$$

$$0 = \frac{1}{\kappa} + n^{-1}\sum_{i=1}^{n} \log(z_i) - \frac{\sum_{i=1}^{n} z_i^{\kappa}\log(z_i)}{\sum_{i=1}^{n} z_i^{\kappa}}$$

and show that there is a unique $(\hat{\lambda}, \hat{\kappa})$ which maximizes the likelihood. Write down an approximation to the sampling distribution of the maximum likelihood estimator.

4.2.6. Fit the Weibull model of Problem 4.2.5 to the pressure vessel failure time data presented in Table 1.2 and use it to make inferences about κ and then the median of the failure time distribution. Carry out a simulation to explore the quality of the Gausian approximation and the repeated sampling properties of the inferences about κ at the estimated parameter values.

4.2.7. Suppose that the conditions of Theorem 4.6 hold and that $\theta^* - \theta(F_0) = O_p(n^{-1/2})$. Show that the one-step estimator

$$\hat{\theta} = \theta^* - \left\{ \sum_{i=1}^{n} \eta'(Z_i, \theta^*) \right\}^{-1} \sum_{i=1}^{n} \eta(Z_i, \theta^*)$$

is asymptotically equivalent to a root of the estimating equations (4.15).

4.3 THE CHOICE OF INFERENCE PROCEDURE

We showed in Sections 4.2.6–4.2.11 that maximizing the likelihood for the gamma model (4.12) is a reasonable method of estimating the parameters of the model. However, the implementation of the procedure requires us to solve an implicit equation and to approximate the sampling distribution of the implicitly defined estimator so other approaches may be simpler to implement.

4.3.1 Method of Moments Estimation for the Gamma Model

If we compute the sample moments $m_k = n^{-1} \sum_{j=1}^{n} Z_j^k$, $k = 1, 2$, their expectations under the gamma model (4.12), which are $Em_1 = \kappa/\lambda$ and $Em_2 = \kappa(1 + \kappa)/\lambda^2$, and then solve the system of equations

$$m_1 = \frac{\hat{\kappa}_{\mathrm{m}}}{\hat{\lambda}_{\mathrm{m}}}$$

$$m_2 = \frac{\hat{\kappa}_{\mathrm{m}}(1 + \hat{\kappa}_{\mathrm{m}})}{\hat{\lambda}_{\mathrm{m}}^2},$$

we obtain the explicit method of moments estimators (Section 3.1.1)

$$\hat{\lambda}_{\mathrm{m}} = \frac{\hat{\kappa}_{\mathrm{m}}}{m_1}$$

$$\hat{\kappa}_{\mathrm{m}} = \frac{m_1^2}{m_2 - m_1^2}$$

of λ and κ.

4.3.2 The Sampling Distribution of the Method of Moments Estimators

We can apply Theorem 4.6 with $\theta = (\lambda, \kappa)^{\mathrm{T}}$ and

$$\eta(y, \theta) = (y - \kappa/\lambda, y^2 - \kappa(1 + \kappa)/\lambda^2)^{\mathrm{T}}$$

to obtain the approximate sampling distribution of the method of moments estimators but it is straightforward and instructive to obtain the approximate sampling distribution directly by considering the random vector $(m_1, m_2)^T$.

The vector $(m_1, m_2)^T$ is a sum of independent and identically distributed random vectors $(Z, Z^2)^T$ which, under the gamma model (4.12) have mean $(\kappa/\lambda, \kappa(\kappa+1)/\lambda^2)$ and variance matrix

$$V = \begin{pmatrix} \kappa/\lambda^2 & 2\kappa(\kappa+1)/\lambda^3 \\ 2\kappa(\kappa+1)/\lambda^3 & 2\kappa(\kappa+1)(2\kappa+3)/\lambda^4 \end{pmatrix}.$$

By the multivariate central limit theorem,

$$n^{1/2}\left(m_1 - \frac{\kappa}{\lambda}, m_2 - \frac{\kappa(\kappa+1)}{\lambda^2}\right) \xrightarrow{\mathscr{D}} \mathrm{N}_2(0, V) \qquad \text{as } n \to \infty.$$

Since

$$\hat{\kappa}_{\mathrm{m}} = h(m_1, m_2) = \frac{m_1^2}{m_2 - m_1^2}$$

and

$$h'(m_1, m_2) = \frac{m_1}{(m_2 - m_1^2)^2}(2m_2, -m_1)^T,$$

by a Taylor expansion (see 2 in the Appendix)

$$\begin{aligned} \hat{\kappa}_{\mathrm{m}} - \kappa &= h(m_1, m_2) - h\left(\frac{\kappa}{\lambda}, \frac{\kappa(1+\kappa)}{\lambda^2}\right) \\ &= \left(m_1 - \frac{\kappa}{\lambda}, m_2 - \frac{\kappa(\kappa+1)}{\lambda^2}\right)h'\left(\frac{\kappa}{\lambda}, \frac{\kappa(\kappa+1)}{\lambda^2}\right) + o_p(n^{-1/2}) \end{aligned}$$

whence, as $n \to \infty$,

$$\begin{aligned} n^{1/2}(\hat{\kappa}_{\mathrm{m}} - \kappa) &\xrightarrow{\mathscr{D}} \mathrm{N}\left(0, h'\left(\frac{\kappa}{\lambda}, \frac{\kappa(\kappa+1)}{\lambda^2}\right)^T V h'\left(\frac{\kappa}{\lambda}, \frac{\kappa(\kappa+1)}{\lambda^2}\right)\right) \\ &= \mathrm{N}(0, 2\kappa(\kappa+1)). \end{aligned}$$

For the pressure vessel data, we obtain $\hat{\kappa}_{\mathrm{m}} = 1.056$ with standard error $\sqrt{2\hat{\kappa}_{\mathrm{m}}(\hat{\kappa}_{\mathrm{m}} + 1)/n} = 0.466$ so that an approximate 95% confidence interval for κ is

$$(0.14, 1.97).$$

In contrast to the maximum likelihood analysis of Section 4.2.11, there is now no evidence against the exponential model.

4.3.3 Asymptotic Relative Efficiency

If we compare the analyses of the gamma model (4.12) in Sections 4.2.11 and 4.3.2, we see that the confidence intervals are based on different point estimates which have different standard errors. Comparisons based on a single realization of **Z** may be misleading so we need a more sophisticated approach. Since the maximum likelihood estimator $\hat{\kappa}$ satisfies

$$n^{1/2}(\hat{\kappa} - \kappa) \xrightarrow{\mathscr{D}} \mathrm{N}\left(0, \frac{\kappa}{\kappa\psi'(\kappa) - 1}\right) \qquad \text{as } n \to \infty$$

and the method of moments estimator $\hat{\kappa}_{\mathrm{m}}$ satisfies

$$n^{1/2}(\hat{\kappa}_{\mathrm{m}} - \kappa) \xrightarrow{\mathscr{D}} \mathrm{N}(0, 2\kappa(\kappa + 1)) \qquad \text{as } n \to \infty,$$

asymptotically, the centers of the confidence intervals are the same and the lengths are determined by the asymptotic standard deviations of the estimators. That is, as noted by Wilks (1938a), asymptotic comparisons of confidence intervals derived from asymptotically unbiased and Gaussian estimators are comparisons of asymptotic standard deviations or, equivalently, variances.

Asymptotic variances can be compared by computing the *asymptotic relative efficiency* of $\hat{\kappa}_{\mathrm{m}}$ with respect to $\hat{\kappa}$ which is defined to be

$$\operatorname{are}(\hat{\kappa}_{\mathrm{m}}, \hat{\kappa}) = \frac{\operatorname{Var}(\hat{\kappa})}{\operatorname{Var}(\hat{\kappa}_{\mathrm{m}})} = \frac{\dfrac{\kappa}{(\kappa\psi'(\kappa - 1)}}{2\kappa(\kappa + 1)} = \frac{1}{2(\kappa + 1)(\kappa\psi'(\kappa) - 1)}.$$

From the transformation theorem, for any smooth transformation h,

$$\operatorname{are}(h(\hat{\kappa}_{\mathrm{m}}), h(\hat{\kappa})) = \frac{1}{2(\kappa + 1)(\kappa\psi'(\kappa) - 1)}$$

so the asymptotic relative efficiency does not change when we change the scale on which we are comparing the estimators. The asymptotic relative efficiency is less than 1 for all κ so the maximum likelihood estimator has a smaller asymptotic variance (and hence on average produces shorter confidence intervals) than the method of moments estimator. We say that the maximum likelihood estimator is more efficient than the method of moments estimator.

4.3.4 Asymptotic Efficiency

Suppose that **Z** is generated by the model $\mathscr{F} = \{f(\mathbf{y}; \theta) = \prod_{i=1}^{n} f(y_i; \theta) : \theta \in \Omega\}$ which satisfies regularity conditions like those of Corollary 4.6: see Bahadur

(1964) for details. Then Bahadur (1964) showed that if $T_n = t(\mathbf{Z})$ is any estimator satisfying $n^{1/2}(T_n - \theta) \xrightarrow{\mathscr{D}} \mathrm{N}_p(0, \Sigma(\theta))$, we have the inequality

$$\Sigma(\theta)_{jj} \geq I(\theta)^{jj}, \qquad j = 1, \ldots, p, \quad \text{for almost all } \theta \in \Omega,$$

where $I(\theta)^{jj}$ denotes the (j, j)th component (i.e., jth diagonal element) of the matrix $I(\theta)^{-1}$. When θ contains a single parameter, $I(\theta)^{jj} = 1/I(\theta)$. The inequality can fail for isolated values of θ but holds for all $\theta \in \Omega$ if $\Sigma(\theta)$ is continuous.

Bahadur's (1964) result shows that there is a limit to how efficient an estimator can be and makes it sensible to define the *asymptotic efficiency* of the jth component of any estimator satisfying $n^{1/2}(T_n - \theta) \xrightarrow{\mathscr{D}} \mathrm{N}(0, \Sigma(\theta))$ to be

$$\text{ae}\,(T_{nj}) = \frac{I(\theta)^{jj}}{\Sigma(\theta)_{jj}} \leq 1.$$

Any estimator T_n of θ which satisfies

$$n^{1/2}(T_n - \theta) \xrightarrow{\mathscr{D}} \mathrm{N}(0, \mathrm{I}(\theta)^{-1}) \qquad \text{for all } \theta \in \Omega$$

is *componentwise asymptotically efficient* for θ. Asymptotically efficient estimators are not unique because we can always add a sequence converging to 0 faster than $n^{-1/2}$ to an asymptotically efficient estimator without changing its limiting distribution.

It follows from Corollary 4.6 that, under appropriate conditions, maximum likelihood estimators are asymptotically efficient and therefore on average produce confidence intervals which are as short as possibe. Thus $\log(2)/\bar{Z}$ is an efficient estimator of the median failure time under the exponential model (4.1) and the maximum likelihood estimator $(\hat{\lambda}, \hat{\kappa})$ is an efficient estimator of (λ, κ) under the gamma model (4.12). This is the original basis on which Fisher (1922) advocated the use of maximum likelihood estimators over method of moments estimators.

4.3.5 Adaptive Estimation

In general, asymptotic efficiency depends both on the model and on what we know about the model. If we "know" $\kappa = 1$ in the gamma model, the exponential model holds and $1/\bar{Z}$ is an asymptotically efficient estimator of λ with asymptotic variance $I(\lambda)^{-1} = I_{11}(\lambda, \kappa)^{-1} = \lambda^2$. However, if κ is unknown and estimated in the gamma model, $\hat{\kappa}/\bar{Z}$ is asymptotically efficient for λ with asymptotic variance

$$I(\lambda, \kappa)^{11} = \frac{\lambda^2 \psi'(\kappa)}{\kappa\psi'(\kappa) - 1}.$$

We have the inequality

$$I(\lambda)^{-1} \leq I(\lambda, \kappa)^{11}$$

so there is a price to pay for having to estimate κ. In fact, in general,

$$I_{jj}(\theta)^{-1} \leq I(\theta)^{jj}.$$

We have equality whenever $I(\theta)$ is a diagonal matrix and in this case there is no loss of efficiency for having to estimate the other components of θ. If we can construct an efficient estimator T_{nj} of θ_j that depends on the other components of θ but having to estimate these other components does not affect the efficiency of T_{nj} in the sense that the asymptotic variance of T_{nj} equals $I_{jj}(\theta)^{-1}$, we call T_{nj} an *adaptive estimator* of θ_j. (For an example, see Problem 4.2.4.)

4.3.6 Mean Squared Error

It is not necessary to restrict the estimators on which confidence intervals are based to be asymptotically unbiased. For example, if we allow an asymptotic bias $b(\theta)$ so the estimator T_n is required to satisfy

$$n^{1/2}(T_n - \theta) \xrightarrow{\mathcal{D}} \mathrm{N}_p(b(\theta), \Sigma(\theta)) \qquad \text{as } n \to \infty,$$

comparisons based on asymptotic variances are less meaningful than those based on the componentwise asymptotic mean squared errors

$$\Sigma(\theta)_{jj} + b_j(\theta)^2.$$

Biased estimators can be useful, but it is important to recognize that confidence intervals which ignore the bias can have very poor coverage properties when the bias is large.

4.3.7 Other Criteria Which Impact on the Choice of Procedure

The case for using maximum likelihood estimators is based on their general applicability, the sufficiency of the likelihood function (Section 2.6.2), the observation that the expected log-likelihood is maximized at the true parameter value (Section 4.2.6), and the fact that maximum likelihood estimators are often asymptotically efficient (Section 4.3.4). This suggests that we should use the maximum likelihood analysis of the gamma model. However, asymptotic efficiency is not entirely compelling as an optimality criterion because it neglects concepts like simplicity: we may decide to use the method of moments estimator of the shape parameter κ in the gamma model on the grounds that it is simpler than the maximum likelihood estimator. A stronger reason for using the method of moments analysis for the gamma model is that it is more stable than the maximum likelihood analysis. The maximum likelihood estimator depends on the data through $y = |n^{-1} \sum_{i=1}^{n} \log(z_i) - \log(\bar{z})|$ so is more sensitive to observations close to 0 than the method of moments estimator. In fact, if we exclude the two smallest observations (0.75 and 1.70), the maximum likelihood estimator of κ increases from 0.56 to 0.96 (which changes the inference) and the method of moments estimator increases from 1.06 to 1.33 (which does not affect the inference). Thus the method of moments analysis is preferable in this case.

All of the comparisons we have made in this section have assumed that the underlying model holds exactly. Since models rarely hold exactly, we may ask about the value of our comparisons when the model holds only approximately. This issue is addressed in Chapter 5.

PROBLEMS

4.3.1. Suppose that we observe $\bar{Z}$ on the gamma model (4.12). For $n = 20$, $\lambda = 1$, $\kappa = 0.5$, 1, and 1.5, simulate the sampling distribution of the method of moments estimator $\hat{\kappa}_{\mathrm{m}}$. Compare the simulated sampling distribution to the Gaussian approximation to the sampling distribution in each case.

4.3.2. Suppose that we have observations $\mathbf{Z}$ on the multivariate Student t model of Problem 4.2.1. Use the fact that $\lambda_4 = \mathrm{E}(Z_i - \mu)^4/\{\mathrm{E}(Z_i - \mu)^2\}^2$ satisfies

$$\lambda_4 = 3 + \frac{6}{\nu - 4}, \qquad \text{provided } \nu > 4,$$

to construct a method of moments estimator of ν. Show that the method of moments estimator $\hat{\nu}$ diverges to infinity in probability as $n \to \infty$. If on the other hand, the Student t model

$$\mathscr{F} = \left\{ f(\mathbf{y}; \mu, \sigma) = \prod_{i=1}^{n} \frac{\Gamma\left(\dfrac{\nu+1}{2}\right)}{(\nu\pi\sigma^2)^{1/2}\Gamma\left(\dfrac{\nu}{2}\right)} \frac{1}{\left\{1 + \dfrac{(y_i - \mu)^2}{\nu\sigma^2}\right\}^{(\nu+1)/2}}, \right.$$

$$\left. -\infty \le y_i \le \infty \colon \mu \in \mathbb{R}, \sigma > 0 \right\},$$

holds, show that the estimator $\hat{\nu}$ constructed above converges in probability to ν.

4.3.3. The negative binomial model

$$\mathscr{F} = \left\{ f(\mathbf{z}; \lambda, \delta) = \prod_{i=1}^{n} \left\{ z_i! \Gamma\left(\frac{\lambda}{\delta}\right) \right\}^{-1} \Gamma\left(\frac{z_i + \lambda}{\delta}\right) \delta^{z_i} (1 + \delta)^{-(z_i + \lambda/\delta)}, \right.$$

$$\left. z_i = 0, 1, 2, \ldots, \lambda > 0, \delta > 0 \right\}$$

was presented in Problem 3.2.3 as a possible model for Mosteller and Wallace's (1964, 1984, p. 33) data (presented in Problem 3.2.2) on the number of occurrences of the word "may" in 262 blocks of text written by James Madison. Here $EZ = \lambda$ is the mean of the distribution, $\operatorname{Var}(Z) = \lambda(1 + \delta)$ is the variance of the distribution, and $\delta = 0$ corresponds to the Poisson distribution. As an alternative to the χ^2 goodness of fit test, find a large sample approximation to the sampling distribution of the method of moments estimator of δ and set an approximate confidence interval for δ. What do you conclude? If we do a formal test of H_0: $\delta = 0$, we need to approximate the sampling distribution of the method of moments estimator under H_0. What difficulties are caused by the fact that δ is on the boundary of the parameter space? See Lawless (1987) for further discussion.

(Hint:

$$\operatorname{Var}\begin{pmatrix} Z \\ Z^2 \end{pmatrix} = \lambda(1 + \delta)\begin{pmatrix} 1 & 1 + 2\delta + 2\lambda \\ 1 + 2\delta + 2\lambda & 1 + 6\delta + 6\lambda + 6\delta^2 + 10\lambda\delta + 4\lambda^2 \end{pmatrix}).$$

4.3.4. Suppose that we observe $\{(Y_1, x_1), \ldots, (Y_n, x_n)\}$ on the model

$$\mathscr{F} = \left\{ f(y; \beta, \sigma) = \prod_{i=1}^{n} \frac{1}{(2\pi\sigma^2 v(x_i))^{1/2}} \exp\left\{ -\frac{(y_i - x_i\beta)^2}{2\sigma^2 v(x_i)} \right\}, \right.$$

$$\left. y_i \in \mathbb{R}; \beta \in \mathbb{R}, \sigma > 0 \right\}.$$

Compare the maximum likelihood estimator $\hat{\beta}$ of β to the least squares estimator

$$\hat{\beta}_{\mathrm{LS}} = \frac{\sum_{i=1}^{n} x_i y_i}{\sum_{i=1}^{n} x_i^2}.$$

4.4 IMPROVING THE GAUSSIAN APPROXIMATION

In Sections 4.2.2–4.2.5, we assumed that we have independent observations $\mathbf{Z}$ with underlying distribution F_0 and we established a Gaussian approximation to the sampling distribution of the M-estimator $\hat{\theta}$ satisfying

$$\sum_{i=1}^{n} \eta(Z_i, \hat{\theta}) = 0.$$

In particular, in Theorem 4.6, we showed that

$$n^{1/2}(\hat{\theta} - \theta(F_0)) \xrightarrow{\mathscr{D}} \mathrm{N}(0, V_{F_0}(\theta(F_0))),$$

where $V_{F_0}(\theta(F_0)) = B_{F_0}(\theta(F_0))^{-1} A_{F_0}(\theta(F_0)) B_{F_0}(\theta(F_0))^{-\mathrm{T}}$, with

$$A_{F_0}(\theta) = \mathrm{E}_{F_0}\eta(Z, \theta)\eta(Z, \theta)^{\mathrm{T}} \quad \text{and} \quad B_{F_0}(\theta) = -\mathrm{E}_{F_0}\eta'(Z, \theta),$$

and $\theta(F_0)$ satisfies

$$\mathrm{E}_{F_0}\eta(Z, \theta(F_0)) = 0.$$

This approximation is adequate for many purposes but it can be important when the sample size is small to use more accurate approximations to the sampling distribution of $\hat{\theta}$.

4.4.1 Choice of Scale and Parameterization

As we noted in Sections 4.1.8–4.1.10, we can often improve an asymptotic approximation by changing the scale on which we make the approximation. For example, in Section 4.2.11 we considered approximating the distribution of $\log(\hat{\kappa})$ rather than $\hat{\kappa}$. Although the change of scale did not lead to improvement in that case, it may be helpful in other problems.

In multidimensional problems, we can consider reparameterization as well as transformation. Cox and Reid (1987) suggested reparameterizing so that the nuisance parameters are orthogonal to the parameters of interest in the sense that the corresponding Fisher information matrix is block diagonal. This is a sensible suggestion but it is generally difficult to implement and can lead to uninterpretable parameters. In our gamma problem, the parameterization in terms of $\mu = \kappa/\lambda$ and κ is orthogonal (Problem 4.4.2) so we could consider using this parameterization. It does not, however, change the approximations to the marginal sampling distribution of the estimators of κ.

4.2.2 Multivariate Edgeworth Expansions

Since the Gaussian approximation was obtained from a linear expansion of the estimating equations, we can try to improve the approximation by extending the expansion and then developing a *multivariate Edgeworth expansion* for the sampling distribution. This is unattractive because it involves tedious moment calculations (the multivariate Edgeworth expansion is of a similar form to the general Edgeworth expansion given in (4.8) and (4.9) but involves multivariate moments) and because, as we saw in Section 4.1.5, the Edgeworth expansion can perform poorly in the tails.

The Edgeworth expansion does show that the Gaussian approximation can be applied to the density $f_{\hat{\theta}}(x)$ and that this Gaussian density approximation performs extremely well in the center of the distribution. That is, if the dimension of the parameter space Ω is p, the Gaussian density approximation

$$f_{\hat{\theta}}(x) \sim \frac{n^{p/2}|B_{F_0}(\theta(F_0))|}{(2\pi)^{p/2}|A_{F_0}(\theta(F_0))|^{1/2}} \exp\left[-\frac{n}{2}(x - \theta(F_0))^{\mathrm{T}} V_{F_0}(\theta(F_0))^{-1}(x - \theta(F_0)) \right]$$

is of $O(n^{-1/2})$ but, when $x = \theta(F_0)$, the asymptotic mean of the sampling distribution, the Gaussian approximation reduces to

$$f_{\hat{\theta}}(x) \sim \frac{n^{p/2}|B_{F_0}(\theta(F_0))|}{(2\pi)^{p/2}|A_{F_0}(\theta(F_0))|^{1/2}} \tag{4.25}$$

with a relative error of $O(n^{-1})$. (This corresponds to setting $x = 0$ in the multivariate version of the Edgeworth expansion (4.8).)

If we can construct a density $f_{\hat{\theta},x}(x)$ by recentering $f_{\hat{\theta}}(x)$ at x so that the asymptotic mean of $\hat{\theta}$ under $f_{\hat{\theta},x}(x)$ equals x, then we can use the Edgeworth expansion (4.25) to construct a good Gaussian density approximation at x and, by repeating this at each x, we may be able to construct an approximation to $f_{\hat{\theta}}(x)$ with a relative error of $O(n^{-1})$. The key step is to recenter the sampling distribution of $\hat{\theta}$ so that it has asymptotic mean x.

4.4.3 Saddlepoint Approximation

If we examine the discussion in Section 4.2.2, we see that recentering the sampling distribution of $\hat{\theta}$ so that it has asymptotic mean x requires us to construct a distribution F_x such that

$$\mathrm{E}_{F_x}\eta(Z, x) = 0.$$

This can be achieved by a process known as *exponential tilting* which was used to approximate sampling distributions by Esscher (1932): for $\alpha \in \mathbb{R}^p$, define

$$\begin{aligned} f_x(y) &= \frac{\exp\{\alpha^{\mathrm{T}}\eta(y, x)\}}{M_x(\alpha)} f_0(y) \\ &= \exp\{\alpha^{\mathrm{T}}\eta(y, x) - K_x(\alpha)\} f_0(y), \end{aligned}$$

where $M_x(\alpha) = \int \exp\{\alpha^{\mathrm{T}}\eta(y, x)\} f_0(y)\, dy$ and $K_x(\alpha) = \log\{M_x(\alpha)\}$. The function

$M_x(\cdot)$, which is chosen to ensure that $f_x(y)$ is properly normalized, is the moment generating function of $\eta(Z; x)$ and $K_x(\cdot)$ is the cumulant generating function. The distribution with density $f_x(y)$ is sometimes called the *conjugate distribution* of $f_0(y)$. (This should not be confused with the conjugate prior density introduced in Section 2.2.5.)

Differentiating both sides of the identity

$$1 = \int \exp\{\alpha^{\mathrm{T}}\eta(y, x) - K_x(\alpha)\} f_0(y)\, dy$$

with respect to α, we obtain

$$\begin{aligned} 0 &= \int \{\eta(y, x) - K_x'(\alpha)\} \exp\{\alpha^{\mathrm{T}}\eta(y, x) - K_x(\alpha)\} f_0(y)\, dy \\ &= \mathrm{E}_{F_x}\eta(Z, x) - K_x'(\alpha), \end{aligned} \tag{4.26}$$

and differentiating both sides with respect to α again,

$$0 = \mathrm{E}_{F_x}\{\eta(Z, x) - K_x'(\alpha)\}\{\eta(Z, x) - K_x'(\alpha)\}^{\mathrm{T}} - K_x''(\alpha). \tag{4.27}$$

Hence, if we choose $\alpha(x)$ so that

$$0 = \int \eta(y, x) \exp\{\alpha(x)^{\mathrm{T}}\eta(y, x)\} f_0(y)\, dy, \tag{4.28}$$

we have from (4.26) that

$$\mathrm{E}_{F_x}\eta(Z, x) = K_x'(\alpha(x)) = 0$$

and from (4.27) that

$$A_{F_x}(x) = \mathrm{E}_{F_x}\eta(Z, x)\eta(Z, x)^{\mathrm{T}} = K_x''(\alpha(x)).$$

It follows that the density $f_{\hat{\theta}, x}(x)$ of $\hat{\theta}$ when the shifted distribution F_x with density $f_x(\mathbf{y}) = \prod_{i=1}^{n} f_x(y_i)$ holds can be approximated at x by the Gaussian density approximation

$$f_{\hat{\theta}, x}(x) \sim \frac{n^{p/2}|B_{F_x}(x)|}{(2\pi)^{p/2}|A_{F_x}(x)|^{1/2}} \tag{4.29}$$

with a relative error of $O(1/n)$, where $B_{F_x}(x) = -\mathrm{E}_{F_x}\eta'(Z, x)$.

The last step is to relate the density $f_{\hat{\theta}}(x)$ of $\hat{\theta}$ under the model density F_0 to the density $f_{\hat{\theta}, x}(x)$ of $\hat{\theta}$ under the shifted distribution F_x. Field (1982) showed that

$$f_\theta(x) = f_{\theta,x}(x) \exp\{nK_x(\alpha(x))\}$$

which, together with (4.29) yields the approximation

$$f_\theta(x) \sim \frac{n^{p/2}|B_{F_x}(x)|}{(2\pi)^{p/2}|A_{F_x}(x)|^{1/2}} \exp\{nK_x(\alpha(x))\}. \tag{4.30}$$

This approximation is called the *saddlepoint approximation* (and $\alpha(x)$ the *saddlepoint*) from its alternative derivation by means of complex contour integration; for the case of the sample mean, see Daniels (1954). It is also sometimes called a *small sample asymptotic approximation*, following Hampel (1973), because of its remarkable accuracy in small samples. This accuracy is rather better than suggested by the relative error of the approximation, which Field (1982) showed is $O(n^{-1})$ for x in a compact set. Numerical integration to obtain the normalizing constant generally improves the approximation and can reduce the relative error in the approximation to $O(n^{-3/2})$. As the saddlepoint approximation is non-negative, it does not suffer the tail problems of the Edgeworth expansion.

Much of the work on saddlepoint approximation has focused on the sample mean, maximum likelihood estimators in exponential family models or estimation in location–scale models. These are problems where the method generally works well. A number of different approximation formulae – corresponding to different ways of approaching the problem – are available, but there are close relationships between the different methods. For discussion of the connections between the different approaches and their use in different contexts, see the monographs by Barndorff-Nielsen and Cox (1989) and Field and Ronchetti (1990), and the review paper by Reid (1988).

4.4.4 Saddlepoint Approximation in the Exponential Model

We can use the saddlepoint approximation to approximate the sampling distribution of the sample mean under the exponential model and compare the approximation to those we obtained in Section 4.1. In this case, $\eta(y, x) = y - x$ so the moment generating function of η is

$$M_x(\alpha) = \int \exp\{\alpha(y - x)\}\lambda \exp(-\lambda y)\, dy = \exp(-x\alpha)\frac{\lambda}{(\lambda - \alpha)}, \qquad \alpha < \lambda.$$

It follows that $K_x(\alpha) = -x\alpha + \log(\lambda/(\lambda - \alpha))$, $K_x'(\alpha) = -x + 1/(\lambda - \alpha)$ and $K_x''(\alpha) = 1/(\lambda - \alpha)^2$, so from (4.28)

$$\alpha(x) = \lambda - \frac{1}{x}.$$

We therefore have $K_x(\alpha(x)) = \log(\lambda x) + 1 - \lambda x$, $A_{F_x}(x) = K_x''(\alpha(x)) = x^2$ and $B_{F_x}(x) = 1$ and hence, substituting into (4.30),

$$f_\theta(x; \theta_0) \sim \frac{n^{1/2}}{(2\pi)^{1/2} x} \exp\{n(1 - \lambda x)\}(\lambda x)^n = \frac{n^{1/2}}{(2\pi)^{1/2}} \exp(n) x^{n-1} \lambda^n \exp(-nx\lambda)$$

which is exact after renormalization! Thus the saddlepoint approximation does not achieve our original goal of simplifying the form of the exact distribution, but it certainly cannot be improved on as an approximation. The saddlepoint approximation for the sampling distribution of the mean is exact only for the Gaussian, gamma, and inverse Gaussian models (Daniels, 1980) but, as noted above, it is remarkably accurate in many other problems.

4.4.5 Saddlepoint Approximation in the Gamma Model

Now suppose that we want to construct the saddlepoint approximation to the sampling distribution of the method of moments estimator $\hat{\kappa}_m$ of κ when the gamma model

$$\mathscr{F} = \left\{ f(\mathbf{y}; \lambda, \kappa) = \prod_{i=1}^{n} \frac{1}{\Gamma(\kappa)} \lambda(\lambda y_i)^{\kappa - 1} \exp(-\lambda y_i),\ y_i > 0: \lambda > 0 \right\}$$

holds. The method of moments estimator of Section 4.3.1 corresponds to choosing $\eta(y, x) = (y - x_2/x_1, y^2 - x_2(1 + x_2)/x_1^2)^{\mathrm{T}}$ so that the moment generating function of η is

$$\begin{aligned} M_x(\alpha) &= \frac{1}{\Gamma(\kappa)} \int \exp\left\{ \alpha_1\left(y - \frac{x_2}{x_1}\right) + \alpha_2\left(y^2 - \frac{x_2(1 + x_2)}{x_1^2}\right) \right\} \lambda(\lambda y)^{\kappa - 1} \exp(-\lambda y)\, dy \\ &= \frac{\lambda^\kappa}{\Gamma(\kappa)} \exp\left\{ -\frac{\alpha_1 x_2}{x_1} - \frac{\alpha_2 x_2(1 + x_2)}{x_1^2} \right\} \int y^{\kappa - 1} \exp\{-(\lambda - \alpha_1)y + \alpha_2 y^2\}\, dy. \end{aligned}$$

Writing

$$I_r = \int y^r \exp\{-(\lambda - \alpha_1)y + \alpha_2 y^2\}\, dy = (\lambda - \alpha_1)^{-(r+1)} \int z^r \exp\left\{ -z + \frac{\alpha_2 z^2}{(\lambda - \alpha_1)^2} \right\} dz,$$

it follows that

$$K_x(\alpha) = \kappa \log(\lambda) - \log\{\Gamma(\kappa)\} - \frac{\alpha_1 x_2}{x_1} - \frac{\alpha_2 x_2(1 + x_2)}{x_1^2} + \log\{I_{\kappa - 1}\},$$

$$K_x'(\alpha) = \begin{pmatrix} I_\kappa / I_{\kappa - 1} - x_2/x_1 \\ I_{\kappa + 1}/I_{\kappa - 1} - x_2(1 + x_2)/x_1^2 \end{pmatrix}$$

and

$$K_x''(\alpha) = \begin{pmatrix} I_{\kappa+1}I_{\kappa-1} - I_\kappa^2 & I_{\kappa+2}I_{\kappa-1} - I_{\kappa+1}I_\kappa \\ I_{\kappa+2}I_{\kappa-1} - I_{\kappa+1}I_\kappa & I_{\kappa+3}I_{\kappa-1} - I_{\kappa+1}^2 \end{pmatrix} \Big/ I_{\kappa-1}^2.$$

Also, since

$$\eta'(y, x) = \begin{pmatrix} x_2/x_1^2 & -1/x_1 \\ 2x_2(1 + x_2)/x_1^3 & (-1 + 2x_2)/x_1^2 \end{pmatrix}$$

does not depend on y, we do not need to integrate to obtain $B_{F_x}(x) = -\eta'(y, x)$.

To compute the above saddlepoint approximation, we set up a fine grid of x values over the support of the sampling distribution of $\hat{\theta}$ and then compute the saddlepoint approximation at each x. Most of the burden in the computation is in solving the equation

$$0 = K_x'(\alpha(x))$$

for the saddlepoint $\alpha(x)$. Once we have the saddlepoint, we can calculate the remaining terms and hence the saddlepoint approximation at x. The integrals I_r for $r = \kappa - 1, \kappa + 1, \kappa + 2, \kappa + 3$, need to be done numerically (see Section 2.2), for example by the Gaussian quadrature formula based on the Laguerre polynomials (Abramowitz and Stegun, 1970, p. 890). Once we have the saddlepoint approximation on a grid of x values, we can do further numerical integration to obtain the normalizing constant and approximations to the marginal densities.

The computations are in principle straightforward but are surprisingly difficult in practice. Experience reported in Field and Ronchetti (1990, pp. 108–9) is that it is important to use good initial values when solving the saddlepoint equation. We can start at $x = \theta_0$ for which $\alpha(\theta_0) = 0$ and then use a very fine grid so that each step is small. However, even this may not be enough to avoid the difficult numerical problems. Even though $\hat{\kappa}_m$ is scale invariant so its distribution does not depend on λ, the choice of λ greatly effects the numerical properties of the procedure.

4.4.6 Approximations to the Distribution Function

The saddlepoint approximation is an approximation to the sampling density of a statistic. To construct tests and confidence intervals, we need to obtain an approximation to the marginal sampling distribution function. These can in principle be obtained by numerical integration though this is obviously unattractive in higher dimensional problems. For one dimensional estimators, we need to evaluate the integral

$$\mathrm{P}(\hat{\theta} \le t) = 1 - \int_t^\infty f_\theta(x; \theta_0)\, dx \sim 1 - \int_t^\infty \frac{n^{1/2} B_{F_x}(x)}{(2\pi)^{1/2} A_{F_x}(x)^{1/2}} \exp\{nK_x(\alpha(x))\}\, dx.$$

An analytic approximation given by Field and Ronchetti (1990, p. 115) which is derived using Laplace's method (Problem 4.4.6) is

$$P(\hat{\theta} \leq t) \sim 1 - \frac{1}{(2\pi n)^{1/2} A_{F_t}(t)^{1/2} \alpha(t)} \exp\{nK_t(\alpha(t))\}.$$

A different approximation for the case $\hat{\theta} = \bar{Z}$ was obtained by Lugannani and Rice (1980). These explicit approximations are very accurate and greatly simplify the calculation of tail probabilities. However, it is far less clear how to proceed in higher dimensional problems.

4.4.7 Saddlepoint Approximations in Statistical Inference

With notable exceptions, much of the effort to date has been directed to obtaining saddlepoint approximations rather than to using them in inference. In location–scale and other problems in which we can construct pivotal quantities, we can use approximations to the marginal tail probabilities to make inferences. This is not possible for the gamma shape parameter κ but we may be able to proceed by replacing κ by $\hat{\kappa}$ in the approximation. The approach of Tingley and Field (1990) is applicable to this problem but its presentation would take us too far afield. Other uses of the saddlepoint approximation in inference are mentioned in Sections 6.6.2 and 6.7.6.

PROBLEMS

4.4.1. Suppose that we observe $\{(Y_1, x_1), \ldots, (Y_n, x_n)\}$ on the model

$$\mathscr{F} = \left\{ f(y; \beta, \sigma) = \prod_{i=1}^{n} \frac{1}{(2\pi\sigma^2 v(x_i))^{1/2}} \exp\left\{ -\frac{(y_i - x_i\beta)^2}{2\sigma^2 v(x_i)} \right\}, \right.$$

$$\left. y_i \in \mathbb{R};\ \beta \in \mathbb{R},\ \sigma > 0 \right\}.$$

Find the Fisher information matrix and show that the β, σ parameterization is orthogonal.

4.4.2. Suppose that we observe $\mathbf{Z}$ on the gamma model (4.12). Reparameterize the model in terms of $\mu = \kappa/\lambda$ and $\tau = \kappa$. Show that the parameterization is orthogonal.

4.4.3. Suppose that we observe Z_1, Z_2 on the model

$$\mathscr{F} = \left\{ f(y_1, y_2; \mu_1, \mu_2) = \frac{1}{2\pi} \exp\left\{ -\frac{(y_1 - \mu_1)^2}{2} - \frac{(y_2 - \mu_2)^2}{2} \right\}, \right.$$

$$\left. y_1, y_2 \in \mathbb{R}; \mu_1, \mu_2 \in \mathbb{R} \right\}.$$

Reparameterize the model in terms of $\rho = \mu_1/\mu_2$ and $\tau = (\mu_1^2 + \mu_2^2)^{1/2}$. Show that the parameterization is orthogonal.

4.4.4. Suppose that we observe **Z** on the Gaussian model

$$\mathscr{F} = \left\{ f(\mathbf{z}; \mu, \sigma) = \prod_{i=1}^{n} \frac{1}{(2\pi\sigma^2)^{1/2}} \exp\left\{ -\frac{(z_i - \mu)^2}{2\sigma^2} \right\}, z_i \in \mathbb{R}: \mu \in \mathbb{R}, \sigma > 0 \right\}.$$

Show that the saddlepoint approximation to the density of $\bar{Z}$ is exact.

4.4.5. For the gamma model (4.12), show that the moment generating function of $\eta(Z, x)$, where $\eta(y, x) = (x_2 x_1^{-1} - y, \log(x_1) + \log(y) - \psi(x_2))^{\mathrm{T}}$ is

$$M_x(\alpha) = \frac{\lambda^\kappa \Gamma(\alpha_2 + \kappa)}{(\alpha_1 + \lambda)^{\alpha_2 + \kappa} \Gamma(\kappa)} \exp\left\{ \frac{\alpha_1 x_2}{x_1} + \alpha_2 (\log(x_1) - \psi(x_2)) \right\}.$$

Hence or otherwise show that the saddlepoint approximation to the sampling distribution of the maximum likelihood estimator $\hat{\theta}$ of $\theta = (\lambda, \kappa)^{\mathrm{T}}$ is

$$f_{\hat{\theta}}(x; \theta) \sim \frac{n|\psi'(x_2)x_2 - 1|(\alpha_1(x) + \lambda)}{2\pi|\psi'(\alpha_2(x) + \kappa)(\alpha_2(x) + \kappa) - 1|^{1/2} x_1^2} M_x(\alpha(x))^n,$$

where $\alpha_1(x) = (\alpha_2(x) + \kappa)x_1/x_2 - \lambda$ and $\alpha_2(x)$ satisfies

$$\psi(\alpha_2 + \kappa) - \log\left\{ (\alpha_2(x) + \kappa)\left(\frac{x_1}{x_2} - \lambda + \kappa \right) \right\} = \psi(x_2) - \log(x_1).$$

4.4.6. Consider the integral

$$f(v) = \int_t^s g(x)\, e^{vh(x)}\, dx,$$

where g and h are real, continuous functions, h' is continuous, and h is minimized at $x = t$. By approximating the integral over $[t, s]$ by the integral over $[t, t + \delta)$ for $\delta > 0$, making the change of variable

$y = h(t) - h(x)$, approximating the nonexponential terms by their values at α, and evaluating the resulting simple integral, show that

$$f(v) \sim -\frac{g(t)}{vh'(t)} e^{vh(t)}, \qquad \text{as } v \to \infty.$$

Apply the above result to obtain the Laplace approximation to the tail probability of the saddlepoint approximation

$$\mathrm{P}(\hat{\theta} \geq t) \sim \int_t^\infty \frac{n^{1/2} B_{F_x}(x)}{(2\pi)^{1/2} A_{F_x}(x)^{1/2}} \exp\{nK_x(\alpha(x))\}\, dx.$$

(Hint: Show that $\partial K_x(\alpha(x))/\partial x = -\alpha(x) B_{F_x}(x)$.)

4.5 HYPOTHESIS TESTING

In our exploration of the quality of the exponential model (4.1) in Sections 4.2 and 4.3, we fitted the gamma model (4.12) and used a confidence interval to assess the magnitude of the shape parameter κ. An alternative approach is to construct a formal test of the null hypothesis $\mathrm{H}_0: \kappa = \kappa_0$ against $\mathrm{H}_1: \kappa \neq \kappa_0$ and then apply it with $\kappa_0 = 1$.

4.5.1 Tests for a Single Parameter

Consider any estimator $\bar{\kappa}$ of κ which satisfies

$$n^{1/2}(\bar{\kappa} - \kappa) \xrightarrow{\mathscr{D}} \mathrm{N}(0, \tau(\kappa)^2).$$

Recall that for the maximum likelihood estimator $\tau(\kappa)^2 = \kappa/\{\kappa\psi'(\kappa) - 1\}$ (Section 4.2.11) and for the method of moments estimator $\tau(\kappa)^2 = 2\kappa(\kappa + 1)$ (Section 4.3.2). An approximate α level Neyman–Pearson test can be based on the critical region $\{\mathbf{z}: n^{1/2}|\hat{\kappa} - \kappa_0|/\tau(\kappa_0) \geq c\}$, where α satisfies

$$\alpha = \mathrm{P}\left\{n^{1/2} \frac{|\hat{\kappa} - \kappa_0|}{\tau(\kappa_0)} \geq c; \mathrm{H}_0\right\} \to 2\{1 - \Phi(c)\}$$

and $\Phi(\cdot)$ is the standard Gaussian distribution function.

4.5.2 Consistency in Testing

Under $\mathrm{H}_1: \kappa = \kappa_1$, the power of the test with critical region

$$\{\mathbf{z}: n^{1/2}|\hat{\kappa} - \kappa_0|/\tau(\kappa_0) \geq c\}$$

is

$$\begin{aligned}
\mathrm{P}\left\{n^{1/2}\frac{|\hat{\kappa}-\kappa_0|}{\tau(\kappa_0)} \geq c; \mathrm{H}_1\right\}
&= \mathrm{P}\{\hat{\kappa} \geq \kappa_0 + n^{-1/2}\tau(\kappa_0)c; \mathrm{H}_1\} + \mathrm{P}\{\hat{\kappa} \leq \kappa_0 - n^{-1/2}\tau(\kappa_0)c; \mathrm{H}_1\} \\
&= \mathrm{P}\left\{n^{1/2}\frac{(\hat{\kappa}-\kappa_1)}{\tau(\kappa_1)} \geq n^{1/2}\frac{(\kappa_0-\kappa_1)}{\tau(\kappa_1)} + \frac{\tau(\kappa_0)c}{\tau(\kappa_1)}; \mathrm{H}_1\right\} \\
&= \mathrm{P}\left\{n^{1/2}\frac{(\hat{\kappa}-\kappa_1)}{\tau(\kappa_1)} \leq n^{1/2}\frac{(\kappa_0-\kappa_1)}{\tau(\kappa_1)} - \frac{\tau(\kappa_0)c}{\tau(\kappa_1)}; \mathrm{H}_1\right\} \\
&\to 1
\end{aligned}$$

as $n \to \infty$. This implies that the test will reject any fixed alternative hypothesis as $n \to \infty$. We say that the test is *consistent*.

4.5.3 Local Alternative Hypotheses and Local Power

To obtain a nontrivial limit result for the power of the test, we follow Pitman (1949) and consider local alternative hypotheses of the form $\mathrm{H}_1: \kappa = \kappa_n = \kappa_0 + \xi/n^{1/2}$. In this case, the *asymptotic power function* is

$$\begin{aligned}
\mathrm{P}\left\{n^{1/2}\frac{|\hat{\kappa}-\kappa_0|}{\tau(\kappa_0)} \geq c; \mathrm{H}_1\right\} &= \mathrm{P}\left\{n^{1/2}\frac{(\hat{\kappa}-\kappa_n)}{\tau(\kappa_n)} \geq -\frac{\xi}{\tau(\kappa_n)} + \frac{\tau(\kappa_0)c}{\tau(\kappa_n)}; \mathrm{H}_1\right\} \\
&\quad + \mathrm{P}\left\{n^{1/2}\frac{(\hat{\kappa}-\kappa_n)}{\tau(\kappa_n)} \leq -\frac{\xi}{\tau(\kappa_n)} - \frac{\tau(\kappa_0)c}{\tau(\kappa_n)}; \mathrm{H}_1\right\} \\
&\to 1 - \Phi\left\{-\frac{\xi}{\tau(\kappa_0)} + c\right\} + \Phi\left\{-\frac{\xi}{\tau(\kappa_0)} - c\right\},
\end{aligned}$$

provided $n^{1/2}(\hat{\kappa}-\kappa_n)/\tau(\kappa_n)$ is asymptotically normal under H_1 and τ is continuous and positive at κ_0. The quantity $1/\tau(\kappa_0)$ is called the *efficacy* of the test and determines its power.

It is convenient when dealing with local alternatives to consider the *asymptotic local power* at ξ which is the derivative of the asymptotic power function with respect to ξ,

$$\left[\phi\left\{-\frac{\xi}{\tau(\kappa_0)} + c\right\} - \phi\left\{-\frac{\xi}{\tau(\kappa_0)} - c\right\}\right] \Big/ \tau(\kappa_0),$$

where $\phi(\cdot)$ is the standard Gaussian density. The asymptotic local power at

$\xi = 0$ is maximized by

$$\xi^* = \frac{2c\phi(c)}{\tau(\kappa_0)^2},$$

and ξ^* is maximized by making $\tau(\kappa_0)^2$ as small as possible.

4.5.4 Pitman Efficiency

The Pitman efficiency of two tests is defined to be the squared ratio of their efficacies which is the same as the asymptotic relative efficiency of the underlying estimators and hence is often referred to as the *asymptotic relative efficiency* of the tests. It is reasonable to use tests with maximum asymptotic local power (i.e., tests based on the maximum likelihood estimators) but, as we saw in Section 4.3.7, the choice of test generally requires more than the consideration of asymptotic local power.

4.5.5 Other Approaches to Comparing Tests

In Pitman's approach to comparing the properties of tests, the alternatives are local in the sense that they converge to the null hypothesis at the rate $n^{-1/2}$. If we want to consider fixed alternatives, then we need to let the size converge to 0 and/or the power converge to 1 as $n \to \infty$ in order to get nontrivial results. Approaches of this type have been proposed by Chernoff (1952), Hodges and Lehmann (1956), Bahadur (1960), and Hoeffding (1965). Rubin and Sethuraman (1965) developed an approach to comparing tests in which the alternatives, the size, and the power depend on the sample size. A comparison of these approaches is given by Serfling (1980, Chapter 10).

4.5.6 Tests of Several Parameters

Now suppose that we want to test a hypothesis which constrains several parameters. As in our development in Section 4.2, the development is more transparent if we suppose we have independent observations $\mathbf{Z}$ with underlying distribution F_0. Consider testing the null hypothesis that q specified linear functions of $\theta(F_0)$ equal 0 against the general alternative. There is considerable simplification in the formulae we obtain with no real loss of generality if we suppose that the model is parameterized so that the null hypothesis can be expressed as $\mathrm{H}_0\colon \theta_1(F_0) = 0$ where $\theta(F_0) = (\theta_1(F_0)^{\mathrm{T}}, \theta_2(F_0)^{\mathrm{T}})^{\mathrm{T}}$ or, equivalently, as $\mathrm{H}_0\colon H^{\mathrm{T}}\theta(F_0) = 0$, where $H^{\mathrm{T}} = (I_q, 0)$.

4.5.7 The Wald Test

Consider an estimator $\hat{\theta}$ for which

$$n^{1/2}(\hat{\theta} - \theta(F_0)) \xrightarrow{\mathscr{D}} \mathrm{N}(0, \Sigma_{F_0}).$$

It follows that

$$n^{1/2}H^{\mathrm{T}}(\hat{\theta} - \theta(F_0)) \sim \mathrm{N}(0, H^{\mathrm{T}}\Sigma_{F_0}H),$$

or, under H_0,

$$n^{1/2}\{H^{\mathrm{T}}\Sigma_{F_0}H\}^{-1/2}H^{T}\hat{\theta} \sim \mathrm{N}_q(0, I_q),$$

where I_q is the $q \times q$ identity matrix. If $\hat{\Sigma}$ is a consistent estimator of Σ_{F_0}, we can apply Theorem 4.2 and take the cross-product of both sides to obtain

$$W = n\hat{\theta}^{\mathrm{T}}H\{H^{\mathrm{T}}\hat{\Sigma}H\}^{-1}H^{\mathrm{T}}\hat{\theta} \sim \chi_q^2,$$

where χ_q^2 denotes the *chi-squared distribution* with q degrees of freedom. (We will justify this result in Section 4.5.8.) Note that

$$W = n\hat{\theta}_1^{\mathrm{T}}\{\hat{\Sigma}_{11}\}^{-1}\hat{\theta}_1, \tag{4.31}$$

where $\hat{\theta}_1$ is the vector of the first q components of θ and $\hat{\Sigma}_{11}$ is the $(1, 1)$ block of $\hat{\Sigma}$. An obvious approximate α level test of H_0 known as the *Wald test* (Wald, 1943) is $\{\mathbf{z}: W \geq K_q^{-1}(1 - \alpha)\}$, where K_q^{-1} is the quantile function of the chi-squared distribution with q degrees of freedom.

While we do not require the distribution of the test statistic under an alternative hypothesis to implement the test, it may still be useful for power calculations. If we use the fact that $H^{\mathrm{T}}\theta(F_0) = \xi \neq 0$ under the alternative hypothesis, we have

$$n^{1/2}\{H^{\mathrm{T}}\hat{\Sigma}H\}^{-1/2}H^{\mathrm{T}}\hat{\theta} = N + n^{1/2}\{H^{\mathrm{T}}\Sigma_{F_0}H\}^{-1/2}\xi + o_p(1),$$

where $\mathrm{N} \sim \mathrm{N}_q(0, I_q)$. Hence W diverges to infinity for every fixed $\xi \neq 0$ and the test is consistent. To obtain a nontrivial limit result, consider the local alternative hypothesis H_n: $H^{\mathrm{T}}\theta(F_0) = n^{-1/2}\xi$ as in Section 4.5.3. Under H_n, we obtain

$$\begin{aligned} W &= (N + \{H^{\mathrm{T}}\Sigma_{F_0}H\}^{-1/2}\xi)^{\mathrm{T}}(N + \{H^{\mathrm{T}}\Sigma_{F_0}H\}^{-1/2}\xi) + o_p(1) \\ &\sim \chi_q^2(\xi^{\mathrm{T}}\{H^{\mathrm{T}}\Sigma_{F_0}H\}^{-1}\xi), \end{aligned}$$

where $\chi_q^2(\delta)$ denotes the *noncentral chi-squared distribution* with q degrees of freedom and noncentrality parameter δ.

4.5.8 The Continuous Mapping Theorem

The large sample approximations to the sampling distribution of W obtained in Section 4.5.7 can be justified by the continuous mapping theorem.

Theorem 4.7 *Suppose that $X_n \xrightarrow{\mathscr{D}} X$ and that $g: \mathbb{R} \to \mathbb{R}$ is continuous. Then*

$$g\{X_n\} \xrightarrow{\mathscr{D}} g\{X\}.$$

In particular, if $X_n = n^{1/2}(T_n - \theta) \xrightarrow{\mathscr{D}} \mathrm{N}(0, \Sigma)$, we obtain that

$$g\{n^{1/2}(T_n - \theta)\} \xrightarrow{\mathscr{D}} g\{\mathrm{N}(0, \Sigma)\}.$$

Under the null hypothesis, the sampling distribution of the Wald statistic is obtained by applying the continuous mapping theorem to $g(v) = v^{\mathrm{T}}\Sigma^{-1}v$ with $v \sim \mathrm{N}_q(0, \Sigma)$. More generally, under local alternative hypotheses, we apply the continuous mapping theorem to $g(v) = v^{\mathrm{T}}Cv$ for some C with $v \sim \mathrm{N}_q(\mu, \Sigma)$. In this case,

$$g\{\mathrm{N}(\mu, \Sigma)\} = \sum_{i=1}^{q} w_i(N_i + [Q^{\mathrm{T}}\Sigma^{-1/2}\mu]_i)^2$$

where $q = \operatorname{rank}(\Sigma C)$, w_i are the eigenvalues of the matrix ΣC, N_i are independent standard Gaussian random variables $1 \le i \le q$, and Q is the matrix of eigenvectors of $\Sigma^{1/2}C\Sigma^{1/2}$. When $w_i = w$, the distribution is w times that of a noncentral χ^2 random variable with q degrees of freedom and noncentrality parameter $\mu\Sigma^{-1}\mu$. In particular, if $\Sigma = C^{-1}$ and $\mu = 0$, we obtain

$$g\{\mathrm{N}(0, \Sigma^{-1})\} = \chi_q^2.$$

4.5.9 The Score Test

If $\hat{\theta}$ is a root of the estimating equations

$$\sum_{i=1}^{n} \eta(Z_i, \theta) = 0$$

and the conditions of Theorem 4.6 apply, then we have $\Sigma_{F_0} = V_{F_0}(\theta(F_0))$. In this case, an alternative to the Wald test is the *score test* which is based on examining whether the estimator $\hat{\theta}_R$ computed under H_0 and which therefore satisfies $H^{\mathrm{T}}\hat{\theta}_R = 0$ also satisfies the unrestricted estimating equations $\sum_{i=1}^{n} \eta(Z_i, \hat{\theta}_R) = 0$. Now, exactly as in Section 4.2.4, we can show that

$$n^{-1} \sum_{i=1}^{n} \eta(Z_i, \hat{\theta}_R) = n^{-1} \sum_{i=1}^{n} \eta(Z_i, \theta(F_0)) - B_{F_0}(\theta_0(F_0))(\hat{\theta}_R - \theta(F_0)) + o_p(n^{-1/2})$$

so, under local alternatives H_n: $H^{\mathrm{T}}\theta(F_0) = n^{-1/2}\xi$,

$$\begin{aligned}
H^{\mathrm{T}}B_{F_0}(\theta_0(F_0))^{-1}n^{-1/2}\sum_{i=1}^{n}\eta(Z_i,\hat{\theta}_R) \\
&= H^{\mathrm{T}}B_{F_0}(\theta_0(F_0))^{-1}n^{-1/2}\sum_{i=1}^{n}\eta(Z_i,\theta(F_0)) + \xi + o_p(1) \\
&\xrightarrow{\mathscr{D}} \mathrm{N}_q(\xi, H^{\mathrm{T}}V_{F_0}(\theta(F_0))H),
\end{aligned}$$

as in (4.18). If $\hat{B}(\hat{\theta}_R)$ and $\hat{A}(\hat{\theta}_R)$ are consistent estimators of $B_{F_0}(\theta_0)$ and $A_{F_0}(\theta_0)$ respectively and $\hat{V}(\hat{\theta}_R) = \hat{B}(\hat{\theta}_R)^{-1}\hat{A}(\hat{\theta}_R)\hat{B}(\hat{\theta}_R)^{-\mathrm{T}}$, then

$$\begin{aligned}
S &= n^{-1}\sum_{i=1}^{n}\eta(z_i,\hat{\theta}_R)^{\mathrm{T}}\hat{B}(\hat{\theta}_R)^{-1}H\{H^{\mathrm{T}}\hat{V}(\hat{\theta}_R)H\}^{-1}H^{\mathrm{T}}\hat{B}(\hat{\theta}_R)^{-1}\sum_{i=1}^{n}\eta(z_i,\hat{\theta}_R) \\
&= n^{-1}\sum_{i=1}^{n}\eta_1(z_i,\hat{\theta}_R)^{\mathrm{T}}\hat{B}(\hat{\theta}_R)^{11}\{[\hat{V}(\hat{\theta}_R)]_{11}\}^{-1}\hat{B}(\hat{\theta}_R)^{11}\sum_{i=1}^{n}\eta_1(z_i,\hat{\theta}_R),
\end{aligned} \tag{4.32}$$

because $\sum_{i=1}^{n}\eta_2(z_i,\hat{\theta}_R) = 0$. It follows from Theorem 4.7 that S has an asymptotic $\chi_q^2(\delta)$ distribution, with $\delta = \xi^{\mathrm{T}}\{[V_{F_0}(\theta(F_0))]_{11}\}^{-1}\xi$.

We can in fact show that when $\Sigma_{F_0} = V_{F_0}(\theta(F_0))$, W and S are asymptotically equivalent in the sense that $W - S = o_p(1)$ under local alternatives. Nonetheless, in small samples, they may give different results and in this case S will often be preferable to W because the Gaussian approximation to the sampling distribution of the estimating equation is usually better than the Gaussian approximation to the sampling distribution of the estimator.

4.5.10 The Lagrange Multiplier Form of the Score Test

The estimator $\hat{\theta}_R$ computed under H_0, can be computed from the estimating equations

$$\begin{aligned}
0 &= n^{-1}\sum_{i=1}^{n}\eta(Z_i,\hat{\theta}_R) - H\hat{\lambda} \\
0 &= -H^{\mathrm{T}}\hat{\theta}_R,
\end{aligned} \tag{4.33}$$

where $\hat{\lambda}$ is a q-vector Lagrange multiplier so the score test can be re-expressed in terms of the Lagrange multiplier as

$$\begin{aligned}
S &= n\hat{\lambda}^{\mathrm{T}}H^{\mathrm{T}}\hat{B}(\hat{\theta}_R)^{-1}H\{H^{\mathrm{T}}\hat{V}(\hat{\theta}_R)H\}^{-1}H^{\mathrm{T}}\hat{B}(\hat{\theta}_R)^{-1}H\hat{\lambda} \\
&= n\hat{\lambda}^{\mathrm{T}}\hat{B}(\hat{\theta}_R)^{11}\{[\hat{V}(\hat{\theta}_R)]_{11}\}^{-1}\hat{B}(\hat{\theta}_R)^{11}\hat{\lambda}.
\end{aligned}$$

In this form, the score test is also referred to as the *Lagrange multiplier test* (Silvey, 1959).

4.5.11 The Likelihood Ratio Test

As we noted in Section 3.5.1, we can also test H_0 using the likelihood ratio or, equivalently, twice the logarithm of the likelihood ratio

$$\Delta = 2\left[\sum_{i=1}^{n} \log\{f(z_i; \hat{\theta}_R)\} - \sum_{i=1}^{n} \log\{f(z_i; \hat{\theta})\}\right], \tag{4.34}$$

where $\hat{\theta}$ and $\hat{\theta}_R$ are obtained by maximizing the likelihood without restriction and under H_0 respectively.

To implement the test, we need to approximate the sampling distribution of Δ under H_0. We proceed in two stages: We obtain an expansion for $\hat{\theta} - \hat{\theta}_R$ (which enables us to obtain a large sample approximation to the sampling distribution of $\hat{\theta} - \hat{\theta}_R$) and then expand Δ in powers of $\hat{\theta} - \hat{\theta}_R$ which allows us to express Δ as a function of $\hat{\theta} - \hat{\theta}_R$. The argument is due to Wilks (1938b).

Let $\eta(z, \theta) = \partial \log\{f(z; \theta)\}/\partial\theta$. In conformity with our development of the Wald test, it is instructive to develop the approximation without assuming that $A_{F_0}(\theta(F_0)) = B_{F_0}(\theta_0(F_0))$. From the proof of Theorem 4.6 we have that

$$\hat{\theta} - \theta(F_0) = B_{F_0}(\theta_0(F_0))^{-1} n^{-1} \sum_{i=1}^{n} \eta(Z_i, \theta(F_0)) + o_p(n^{-1/2}). \tag{4.35}$$

Similarly, expanding the restricted estimating equations (4.33), we obtain

$$0 = n^{-1} \sum_{i=1}^{n} \eta(Z_i, \theta(F_0)) - B_{F_0}(\theta(F_0))(\hat{\theta}_R - \theta(F_0)) - H\hat{\lambda} + o_p(n^{-1/2})$$

$$0 = -H^{\mathrm{T}}(\hat{\theta}_R - \theta(F_0)) - H^{\mathrm{T}}\theta(F_0)$$

or, equivalently

$$\begin{pmatrix} \hat{\theta}_R - \theta(F_0) \\ \hat{\lambda} \end{pmatrix} = \begin{pmatrix} B_{F_0}(\theta(F_0)) & H \\ H^{\mathrm{T}} & 0 \end{pmatrix}^{-1} \begin{pmatrix} n^{-1} \sum_{i=1}^{n} \eta(Z_i, \theta(F_0)) \\ n^{1/2} H^{\mathrm{T}}\theta(F_0) \end{pmatrix} + o_p(n^{-1/2}).$$

It follows from standard results on inverting patterned matrices (see 3d in the Appendix) and the fact that under H_0 we have $H^{\mathrm{T}}\theta(F_0) = 0$ that

$$\hat{\theta}_R - \theta(F_0) = \{B_{F_0}(\theta(F_0))^{-1} - R_{F_0} H^{\mathrm{T}} B_{F_0}(\theta(F_0))^{-1}\} n^{-1} \sum_{i=1}^{n} \eta(Z_i, \theta(F_0)) + o_p(n^{-1/2}), \tag{4.36}$$

where $R_{F_0} = B_{F_0}(\theta(F_0))^{-1} H\{H^{\mathrm{T}} B_{F_0}(\theta_0(F_0))^{-1} H\}^{-1}$. Subtracting (4.36) from (4.35), we obtain

$$\hat{\theta} - \hat{\theta}_R = R_{F_0} H^{\mathrm{T}} B_{F_0}(\theta(F_0))^{-1} n^{-1} \sum_{i=1}^{n} \eta(Z_i, \theta(F_0)) + o_p(n^{-1/2})$$

$$= n^{-1/2} R_{F_0} \zeta + o_p(n^{-1/2}), \tag{4.37}$$

say, where $\zeta = H^{\mathrm{T}} B_{F_0}(\theta(F_0))^{-1} n^{-1/2} \sum_{i=1}^{n} \eta(Z_i, \theta(F_0))$. From (4.18), we have that

$$\zeta \sim \mathrm{N}_q(0, H^{\mathrm{T}} V_{F_0}(\theta(F_0)) H).$$

Under the conditions of the asymptotic normality theorem, we can expand Δ in a Taylor expansion so that for some $|\tilde{\theta} - \hat{\theta}_R| \leq |\hat{\theta} - \hat{\theta}_R|$,

$$\Delta = 2\left[\sum_{i=1}^{n} \log\{f(z_i; \hat{\theta})\} - \sum_{i=1}^{n} \log\{f(z_i; \hat{\theta}_R)\}\right]$$

$$= -2(\hat{\theta} - \hat{\theta}_R)^{\mathrm{T}} \sum_{i=1}^{n} \eta(z_i; \hat{\theta}) - (\hat{\theta} - \hat{\theta}_R)^{\mathrm{T}} \sum_{i=1}^{n} \eta'(z_i, \tilde{\theta})(\hat{\theta}_F - \hat{\theta}_R)$$

$$= n(\hat{\theta} - \hat{\theta}_R)^{\mathrm{T}} B_{F_0}(\theta(F_0))(\hat{\theta} - \hat{\theta}_R) + o_p(1).$$

Substituting (4.37) for $\hat{\theta} - \hat{\theta}_R$, we have that under H_0,

$$\Delta = \zeta^{\mathrm{T}} R_{F_0}^{\mathrm{T}} B_{F_0}(\theta(F_0)) R_{F_0}(\theta(F_0)) \zeta + o_p(1)$$

$$= \zeta^{\mathrm{T}} \{H^{\mathrm{T}} B_{F_0}(\theta(F_0))^{-1} H\}^{-1} \zeta + o_p(1)$$

$$\sim \sum_{i=1}^{q} w_i N_i^2$$

where $q = \text{rank}\,(V_{F_0}(\theta(F_0))_{11}\{B_{F_0}(\theta(F_0))^{11}\}^{-1})$, w_i are the eigenvalues of the matrix $V_{F_0}(\theta(F_0))_{11}\{B_{F_0}(\theta(F_0))^{11}\}^{-1}$, and N_i are independent standard Gaussian random variables $1 \leq i \leq q$, by Theorem 4.7. If $F_0 \in \mathscr{F}$ holds, $B_{F_0}(\theta(F_0))^{11} = I_{F_0}(\theta_0)^{11} = V_{F_0}(\theta(F_0))_{11}$ and we have

$$\Delta \sim \chi_q^2.$$

We can extend this result to local alternatives. If the model $\mathscr{F}$ holds and $\Sigma_{F_0} = V_{F_0}(\theta(F_0))$, we have the stronger result that W, S, and Δ are asymptotically equivalent under local alternatives.

The results for the Wald, score, and likelihood ratio tests presented in Sections 4.5.7–4.5.11 were obtained at different levels of generality by Foutz and Srivastava (1977), Kent (1982), Ronchetti (1982), and Heritier and Ronchetti (1994).

4.5.12 Tests for Parameters on the Boundary of the Parameter Space

The approximations developed in Sections 4.5.7, 4.5.9, and 4.5.11 break down if $\theta(F_0)$ is on the boundary of the parameter space Ω because then the underlying Gaussian approximations are inappropriate. Results which are applicable to this problem have been obtained for the likelihood ratio test by Chernoff (1954) and the Wald test by Moran (1971).

4.5.13 Tests Under the Gamma Model

For the gamma model, the likelihood ratio test of $H_0: \kappa = \kappa_0$ is based on

$$\Delta = 2\left[\sum_{i=1}^{n} \log\left\{f\left(z_i; \frac{\kappa_0}{\bar{z}}, \kappa_0\right)\right\} - \prod_{i=1}^{n} \log\left\{f\left(z_i; \frac{\hat{\kappa}}{\bar{z}}, \hat{\kappa}\right)\right\}\right],$$

where $\hat{\kappa}$ is the unrestricted maximum likelihood estimator while the score test is based on

$$n^{-1}\left\{\sum_{i=1}^{n} \log(z_i) - n\log(\bar{z}) - n\psi(\kappa_0) + n\log(\kappa_0)\right\}^2 \frac{\kappa_0}{\kappa_0\psi'(\kappa_0) - 1}$$

which conveniently saves us from having to estimate κ. The Wald test is based on the square of (and hence is equivalent to) the test we considered in Section 4.5.1.

4.5.14 Confidence Intervals Derived from Tests

The Wald, score, and likelihood ratio tests can also be used to obtain approximate confidence intervals. Suppose we partition θ into the first component θ_1 and the $(p-1)$-vector θ_2 of the remaining components so $\theta = (\theta_1, \theta_2)$ and θ_1 is the parameter of interest. By reordering and relabeling the components of θ if necessary, θ_1 can be taken to be any of the components of θ. Approximate $100(1-\alpha)\%$ confidence intervals for $\theta_1(F_0)$ are given by:

$$\text{Wald: } \left\{\theta_1 : \frac{n(\hat{\theta}_1 - \theta_1)^2}{\hat{\Sigma}_{11}} \le K_1^{-1}(1-\alpha)\right\}$$

$$\text{Score: } \left\{\theta_1 : n^{-1}\left\{\sum_{i=1}^{n} \eta_1(Z_i; \theta_1)\right\}^2 \frac{\{B_{F_0}(\hat{\theta}(\theta_1))^{11}\}^2}{[V_{F_0}(\hat{\theta}(\theta_1))]_{11}} \le K_1^{-1}(1-\alpha)\right\}$$

$$\text{Likelihood: } \left\{\theta_1 : 2\left[\sum_{i=1}^{n} \log\{f(z_i; \hat{\theta}(\theta_1))\} - \sum_{i=1}^{n} \log\{f(z_i; \hat{\theta})\}\right] \le K_*^{-1}(1-\alpha)\right\}$$

where $V_{F_0}(\theta) = B_{F_0}(\theta)^{-1}A_{F_0}(\theta)B_{F_0}(\theta)^{-\mathrm{T}}$, $\hat{\theta}(\theta_1)^{\mathrm{T}} = (\theta_1, \hat{\theta}_2(\theta_1)^{\mathrm{T}})$, $\hat{\theta}_2(\theta_1)$ maximizes the likelihood for fixed θ_1, $\hat{\theta}$ is the maximum likelihood estimator, K_1 is the distribution function of the χ_1^2 distribution, and K_* is the distribution of twice the log-likelihood ratio under H_0. We can use $\hat{B}(\hat{\theta}(\theta_1))$ and $\hat{V}(\hat{\theta}(\theta_1))$ instead of $B_{F_0}(\hat{\theta}(\theta_1))$ and $V_{F_0}(\hat{\theta}(\theta_1))$ in the score statistic to simplify the computations. The interval derived from the Wald test reduces to

$$[\hat{\theta}_1 - n^{-1/2}\{\hat{\Sigma}_{11}\}^{1/2}\Phi^{-1}(1 - \alpha/2), \hat{\theta}_1 + n^{-1/2}\{\hat{\Sigma}_{11}\}^{1/2}\Phi^{-1}(1 - \alpha/2)]$$

which, if the model $\mathscr{F}$ holds, is the same as the interval (4.24) obtained directly from the Gaussian approximation to the sampling distribution of $\hat{\theta}_1$ and Theorem 4.2. In large samples, the three intervals will be similar but in finite samples can be quite different. The score and likelihood based intervals will often be preferable to the interval based on the Wald test because they do not enforce a symmetric interval.

4.5.15 Modified Profile Likelihood

We see from Sections 4.5.11 and 4.5.14 that the profile log-likelihood for θ_1 introduced in Section 2.7.2 and defined in general by

$$m(\theta_1) = \sum_{i=1}^{n} \log\{f(z_i; \hat{\theta}(\theta_1))\},$$

where $\hat{\theta}(\theta_1)^{\mathrm{T}} = (\theta_1^{\mathrm{T}}, \hat{\theta}_2(\theta_1)^{\mathrm{T}})$ and $\hat{\theta}_2(\theta_1)$ maximizes the likelihood for fixed θ_1, is fundamental for likelihood based inferences about θ_1. However, the profile likelihood may not behave like a likelihood. For example, the estimating equations for θ_1 derived from $m(\theta_1)$ can be biased in the sense that

$$\mathrm{E}m'(\theta_1) = B(\theta_1) \neq 0.$$

In principle, the bias can be removed by solving the unbiased estimating equation

$$m'(\theta_1) - B(\theta_1) = 0.$$

If we only have an expansion for the bias, we can make an adjustment based on the leading terms to reduce the order of the bias. If we can find a function $b(\theta_1)$ such that $b'(\theta_1) = B(\theta_1)$, then we can construct the *modified profile log-likelihood*

$$m(\theta_1) - b(\theta_1).$$

The most important adjustment of this type is that for the profile likelihood of the variance parameters in Gaussian regression models proposed by Patterson and Thompson (1971) and discussed by Harville (1977). (A simple illustration of the method is given in Problem 4.5.5.) In this case, the modified profile likelihood is variously called the *reduced*, *restricted*, or *residual likelihood*.

When the estimating equation is biased under the model, the maximum likelihood estimator $\hat{\theta}_1$ is also biased for θ_1. Thus the modified profile likelihood provides a method of reducing the bias in the maximum likelihood estimator.

Alternative adjustments to the profile likelihood can also be made; see Barndorff-Nielsen (1983), Cox and Reid (1987), and McCullagh and Tibshirani (1990) for details.

4.5.16 Bartlett Adjustment

We can also consider trying to improve the χ^2 approximations to the sampling distributions of the test statistics. A simple approach considered by Bartlett (1937) which often works surprisingly well for statistics S satisfying

$$S \xrightarrow{\mathscr{D}} \chi_q^2$$

is to consider applying the approximation to νS, where $\nu \to 1$ as $n \to \infty$ is chosen so that the mean of νS equals q, the mean of χ_q^2. Even if we cannot make the mean of νS exactly equal to q, we may be able to make it closer to q. In particular, Lawley (1956) noted that if

$$\mathrm{E}S = q\left\{1 + \frac{a}{n} + O(n^{-2})\right\},$$

we can take $\nu = 1/(1 + a/n)$ or $\nu = 1 - a/n$ to obtain

$$\nu \mathrm{E}S = q + O(n^{-2}).$$

This kind of adjustment is often referred to as *Bartlett adjustment*. In practice, a may depend on unknown parameters and in this case, we replace a by an estimate $\hat{a}$.

When $F_0 \in \mathscr{F}$, the χ^2 approximation to the sampling distribution of the log-likelihood ratio statistic is typically of order n^{-1}. The Bartlett adjustment improves the approximation to order n^{-2} provided $\hat{a} - a = O_p(n^{-1/2})$ (Barndorff-Nielsen and Hall, 1988). This is established by showing that the Bartlett adjustment which explicitly improves the order of the approximation of the mean by the asymptotic mean q actually improves the approximation of all the cumulants by the asymptotic cumulants and hence of the approximation to the sampling distribution. This is intuitively plausible because the asymptotic cumulants are all simple functions of q. Bartlett adjustments can also be applied to modified profile log likelihoods (DiCiccio and Stern, 1994).

PROBLEMS

4.5.1. Suppose we have observations $(Z_1, \ldots, Z_s)$ on the multinomial model

$$\mathscr{F} = \left\{ f(\mathbf{z}, \lambda, \kappa) = \frac{n!}{z_1! \cdots z_s!} \theta_1^{z_1} \cdots \theta_s^{z_s},\ z_i = 0, 1, \ldots, n, \right.$$

$$\left. \sum_{i=1}^{s} z_i = n\colon 0 \leq \theta_i \leq 1,\ \sum_{i=1}^{s} \theta_i = 1 \right\}.$$

Find the maximum likelihood estimators of $\theta_1, \ldots, \theta_s$. Show that the score test of $\mathrm{H}_0\colon H^{\mathrm{T}}\theta = 0$ is the χ^2 goodness of fit test introduced in Problem 3.2.1.

4.5.2. Suppose that we have observations $\mathbf{Z}$ on the model

$$\mathscr{F} = \left\{ f(\mathbf{y}; \theta) = \prod_{i=1}^{n} f(y_i; \theta)\colon \theta \in \Omega \right\},$$

where Ω is a subset of the real line. Consider a level α test of $\mathrm{H}_0\colon \theta = \theta_0$ against the general alternative based on $C_\alpha(\mathbf{Z}) = \{\mathbf{Z} n^{1/2} |T_n - \mu(\theta_0)|/\tau(\theta_0) \geq k\}$ for some $k > 0$, where T_n is an estimator of θ satisfying $n^{1/2}(T_n - \mu(\theta)) \to \mathrm{N}(0, \tau(\theta)^2)$. Find k. Show that the power function of the test against the sequence of local alternatives $\mathrm{H}_n\colon \theta = \theta_0 + n^{-1/2}\xi$ is approximately

$$1 - \Phi\left(k - \frac{\xi\mu'(\theta_0)}{\tau(\theta_0)}\right) + \Phi\left(-k - \frac{\xi\mu'(\theta_0)}{\tau(\theta_0)}\right).$$

Show that the local power at $\xi = 0$ is 0 and apply Bahadur's theorem (Section 4.3.4) to show that the maximum local power is $2k\phi(k)\mu'(\theta_0)^2 I(\theta_0)$ which is achieved if the estimator T_n is efficient.

4.5.3. Suppose that we have observations $\mathbf{Z}$ on the gamma model (4.12). Show how to use the score test (based on the maximum likelihood estimator of (λ, κ)) to construct approximate confidence intervals for κ and λ.

4.5.4. In the context of Problem 4.5.3, show how to construct approximate confidence intervals for κ and λ using the score test based on the method of moments estimator.

4.5.5. For the Gaussian model of Problem 4.3.8, show that the estimating equation for σ^2 derived from the profile likelihood for σ^2 is biased. Show how to construct an unbiased estimating equation from the profile likehood and construct the associated modified profile likelihood. Show that the maximum modified profile likelihood or REML estimator is unbiased for σ^2.

4.5.6. Consider the variance component model presented in Section 1.5.6:

$$\mathscr{F} = \left\{ f(\mathbf{z}; \mu, \sigma_a, \sigma_u) = \frac{1}{(2\pi)^{mg/2}|\Sigma|^{1/2}} \exp\{-(\mathbf{z} - \mu)^{\mathrm{T}}\Sigma^{-1}(\mathbf{z} - \mu)/2\}, \right.$$

$$\left. -\infty < z_{ij} < \infty \colon \mu \in \mathbb{R}, \sigma_a > 0, \sigma_u > 0 \right\},$$

where Σ is the block diagonal matrix with blocks $\sigma_a^2 J + \sigma_u^2 I$, where J is the $m \times m$ matrix will all elements equal to 1 and I is the $m \times m$ identity matrix. Obtain the joint profile likelihood of $\tau_u = \sigma_u^2$ and $\tau_a = m\sigma_a^2 + \sigma_u^2$. Explore the bias of the estimating equations and construct a modified joint profile likelihood for which the estimating equations are both unbiased.

4.5.7. (Welch, 1937b). Suppose that we have observations $\mathbf{Z}$ on the model

$$\mathscr{F} = \left\{ f(\mathbf{y}; \mu_1, \mu_2, \sigma_1, \sigma_2) = \prod_{j=1}^{2} \prod_{i=1}^{n_j} \frac{1}{(2\pi\sigma_j^2)^{1/2}} \exp\left\{ -\frac{(y_{ij} - \mu_j)^2}{2\sigma_j^2} \right\}, \right.$$

$$\left. -\infty \leq y_{ij} \leq \infty \colon \mu_1, \mu_2 \in \mathbb{R}, \sigma_1, \sigma_2 > 0 \right\}.$$

Let $\bar{z}_1$, s_1, $\bar{z}_2$, and s_2 denote the sample mean and standard deviation for the two sets of observations. Obtain the mean and variance of

$$W = \frac{s_1^2/n_1 + s_2^2/n_2}{\sigma_1^2/n_1 + \sigma_2^2/n_2}.$$

Find g and v such that $W \sim g\chi_v^2$ approximately by equating the mean and variance of W to the mean and variance of gX, where $X \sim \chi_v^2$. Then use the fact that W is independent of

$$V = \frac{\bar{z}_1 - \bar{z}_2 - (\mu_1 - \mu_2)}{\sqrt{\sigma_1^2/n_1 + \sigma_2^2/n_2}}$$

to show that the $V/W^{1/2} \sim t_v$ approximately. (Hint: Under the Gaussian model, Var $(s_j^2) = 2\sigma_j^4/(n_j - 1)$.)

4.5.8. Suppose that we have observations $\mathbf{Z}$ on the exponential model

$$\mathscr{F} = \left\{ f(\mathbf{y}; \lambda) = \prod_{j=1}^{2} \prod_{i=1}^{n_j} \lambda_j \exp(-\lambda_j y_{ij}), 0 \le y_{ij} \le \infty : \lambda_1, \lambda_2 > 0 \right\}$$

Show that twice the log-likelihood ratio for testing $H_0: \lambda_1 = \lambda_2$ against the general alternative is

$$\Delta = 2n \log(\bar{y}) - 2n_1 \log(\bar{y}_1) - 2n_2 \log(\bar{y}_2),$$

where $n = n_1 + n_2$, $\bar{y}_j = n_j^{-1} \sum_{i=1}^{n_j} y_{ij}$, and $\bar{y} = n^{-1} \sum_{j=1}^{2} \sum_{i=1}^{n_j} y_{ij}$. Show that if $X \sim \Gamma(\kappa, \lambda)$, then $\mathrm{E} \log(X) = \psi(\kappa) - \log(\lambda)$, where ψ is the digamma function. Use the fact that

$$\psi(\kappa) = \log(\kappa) - \frac{1}{2\kappa} - \frac{1}{12\kappa^2} + O(\kappa^{-3}) \qquad \text{as } \kappa \to \infty,$$

to show that, under H_0,

$$\mathrm{E}\Delta = 1 + \frac{1}{6}\left(\sum_{i=1}^{2} \frac{1}{n_i} - \frac{1}{n} \right) + O(N^{-2}),$$

where N refers indifferently to n, n_1, or n_2. Show how to make a Bartlett adjustment to Δ.

4.5.9. Suppose that we have observations $\mathbf{Z}$ on the model

$$\mathscr{F} = \left\{ f(\mathbf{y}; \theta) = \prod_{j=1}^{k} \prod_{i=1}^{n_j} \frac{1}{(2\pi\sigma_j^2)^{1/2}} \exp\left\{ -\frac{(y_{ij} - \mu_j)^2}{2\sigma_j^2} \right\}, \right.$$

$$\left. -\infty \le y_{ij} \le \infty : \theta = (\mu_1, \ldots, \mu_k, \sigma_1, \ldots, \sigma_k)^{\mathrm{T}}, \mu_i \in \mathbb{R}, \sigma_i > 0 \right\}.$$

Bartlett (1937) suggested that a test of the hypothesis $H_0: \sigma_1 = \cdots = \sigma_k$ be based on the modified log-likelihood ratio test statistic

$$\Delta = \nu \log(s^2) - \sum_{j=1}^{k} \nu_j \log(s_j^2),$$

where

$$\nu = \sum_{j=1}^{k} n_j - k,\ \nu_j = n_j - 1,\ s_j^2 = \nu_j^{-1} \sum_{i=1}^{n_j} (y_{ij} - \bar{y}_j)^2,\ s^2 = \nu^{-1} \sum_{j=1}^{k} \sum_{i=1}^{n_j} (y_{ij} - \bar{y}_j)^2$$

and

$$\bar{y}_j = n_j^{-1} \sum_{i=1}^{n_j} y_{ij}.$$

Show that

$$\mathrm{E}\Delta = (k-1)\left\{1 + \frac{1}{3(k-1)}\left(\sum_{j=1}^{k} \frac{1}{v_j} - \frac{1}{v}\right)\right\} + O(N^{-2}),$$

where N refers indifferently to v or v_j, and show how to make a Bartlett adjustment to Δ.

4.6 LIKELIHOOD AND BAYESIAN THEORY

Likelihood and posterior density functions can be intractable (particularly in multidimensional problems) so we need general approximations to these functions too. Throughout this section, we adopt the general model

$$\mathscr{F} = \left\{ f(\mathbf{y}; \theta) = \prod_{i=1}^{n} f(y_i; \theta) \colon \theta \in \Omega \right\}$$

and develop approximations under it.

4.6.1 Approximating the Likelihood Function

To approximate the likelihood function, we proceed as in Section 2.7.7 to construct a quadratic approximation to the log-likelihood. Suppose that conditions like those of the corollary of Theorem 4.6 hold so that we can expand the likelihood function about its mode, the maximum likelihood estimate $\hat{\theta}$. Then for θ in a neighbourhood of $\hat{\theta}$, we have the expansion

$$\log\left\{\frac{f(\mathbf{z}; \theta)}{f(\mathbf{z}; \hat{\theta})}\right\} = \sum_{i=1}^{n} \log f(z_i; \theta) - \sum_{i=1}^{n} \log f(z_i; \hat{\theta})$$

$$\sim (\theta - \hat{\theta})^{\mathrm{T}} \sum_{i=1}^{n} \eta(z_i, \hat{\theta}) - \frac{n}{2}(\theta - \hat{\theta})^{\mathrm{T}} \mathscr{I} (\theta - \hat{\theta}),$$

where $\eta(z, \theta) = \partial \log\{f(z; \theta)\}/\partial\theta$ and $\mathscr{I} = -n^{-1} \sum_{i=1}^{n} \eta'(z_i, \hat{\theta})$. Exponentiating both sides and using the fact that $\hat{\theta}$ satisfies the likelihood equations, we obtain the asymptotic expansion

$$\frac{f(\mathbf{z}; \theta)}{f(\mathbf{z}; \hat{\theta})} \sim \exp\left\{-\frac{n}{2}(\theta - \hat{\theta})^{\mathrm{T}} \mathscr{I} (\theta - \hat{\theta})\right\}. \tag{4.38}$$

That is, we can approximate the log-likelihood in the region of its mode $\hat{\theta}$ by a quadratic function with an error which is generally of order $n^{-1/2}$. In contrast to the preceding sections, $\hat{\theta}$ is not regarded as an estimator but rather as a sensible point about which to center the approximating quadratic function.

Suppose we partition θ into the first component θ_1 and the $(p-1)$-vector θ_2 of the remaining components so $\theta = (\theta_1, \theta_2)$. By reordering and relabeling the components of θ if necessary, there is no loss of generality in treating θ_1 as the parameter of interest. If we take the Gaussian approximation (4.3.8) and compute the profile likelihood for θ_1, we obtain an approximation to the profile likelihood for θ_1 of the form

$$\frac{f(\mathbf{z}; \theta_1, \hat{\theta}_2(\theta_1))}{f(\mathbf{z}; \hat{\theta})} \sim \exp\left\{-\frac{n}{2}\frac{(\theta_1 - \hat{\theta}_1)^2}{\mathcal{I}^{11}}\right\}.$$

(See Problem 4.6.1.) We can then use the approach of Section 2.7.3 to obtain approximate likelihood intervals for θ_1.

4.6.2 Laplace's Method for Approximating Posterior Distributions When the Likelihood Dominates the Prior

For a prior distribution with density $g(\theta)$, the posterior distribution of $\theta \mid \mathbf{Z} = \mathbf{z}$ is

$$g(\theta \mid \mathbf{z}) = \frac{f(\mathbf{z}; \theta)g(\theta)}{\int f(\mathbf{z}; \theta)g(\theta)\, d\theta}.$$

Provided the likelihood dominates the prior, we can approximate the prior in a neighbourhood of $\hat{\theta}$ by $g(\hat{\theta})$ and use the approximation (4.38) to the likelihood to obtain

$$f(\mathbf{z}; \theta)g(\theta) \sim f(\mathbf{z}; \hat{\theta})g(\hat{\theta}) \exp\left[-\frac{n}{2}(\theta - \hat{\theta})^{\mathrm{T}}\mathcal{I}(\theta - \hat{\theta})\right].$$

On integrating,

$$\int f(\mathbf{z}; \theta)g(\theta)\, d\theta \sim f(\mathbf{z}; \hat{\theta})g(\hat{\theta}) \int \exp\left[-\frac{n}{2}(\theta - \hat{\theta})^{\mathrm{T}}\mathcal{I}(\theta - \hat{\theta})\right] d\theta$$
$$= f(\mathbf{z}; \hat{\theta})g(\hat{\theta})\left(\frac{2\pi}{n}\right)^{p/2}\frac{1}{|\mathcal{I}|^{1/2}}.$$

This method of approximating integrals by the integral of the integrand expanded about its modal value is known as *Laplace's method*. It works well when the integrand has a dominant mode because then most of the contribution of the integrand to the integral is in the neighbourhood of its mode and the underlying Gaussian approximation captures this contribution. Using Laplace's method, we obtain the approximation

$$g(\theta \mid \mathbf{z}) \sim \frac{n^{p/2}|\mathscr{I}|^{1/2}}{(2\pi)^{1/2}} \exp\left[-\frac{n}{2}(\theta - \hat{\theta})^{\mathrm{T}}\mathscr{I}(\theta - \hat{\theta})\right] \tag{4.39}$$

in a neighborhood of $\hat{\theta}$. That is, up to the first term, $\theta \mid \mathbf{Z} = \mathbf{z} \sim \mathrm{N}(\hat{\theta}, n^{-1}\mathscr{I}^{-1})$ with an error which is typically $O(n^{-1/2})$. A formal argument is given by Walker (1969).

From the properties of the multivariate Gaussian distribution, it follows that an approximation to the marginal posterior distribution of the first component θ_1 of θ is $\mathrm{N}(\hat{\theta}_1, \mathscr{I}^{11})$ and hence that an approximate $100(1-\alpha)\%$ Bayesian credibility interval for θ_1 is given by

$$[\hat{\theta}_1 - n^{-1/2}(\mathscr{I}^{11})^{1/2}\Phi^{-1}(1-\alpha/2), \hat{\theta}_1 + n^{-1/2}(\mathscr{I}^{11})^{1/2}\Phi^{-1}(1-\alpha/2)]. \tag{4.40}$$

4.6.3 Approximating Conditional Sampling Distributions

In the case of a location–scale or regression problem, the conditional density of the data given the configuration (Section 3.9.3) is the same as the posterior density using the Jeffreys prior. This means that the above argument applies equally to the conditional density of the data given the configuration and the interval (4.40) is also a large sample approximation to the conditional confidence interval for θ_{01}. See Efron and Hinkley (1978) and Hinkley (1978). The interval (4.40) is also numerically the same as the large sample confidence interval (4.24) based on the maximum likelihood estimator and the observed Fisher information so that, to first order, the three types of intervals are in numerical agreement in large samples.

4.6.4 Laplace's Method for Approximating Posterior Distributions When the Likelihood Does Not Dominate the Prior

A more natural approximation to the posterior distribution than (4.39) when the likelihood does not dominate the prior is obtained if we expand about the posterior mode $\tilde{\theta}$ (which maximizes $f(\mathbf{z}; \theta)g(\theta)$ rather than the mode of the likelihood $\hat{\theta}$ (which maximizes $f(\mathbf{z}; \theta)$). If we let

$$\mathscr{H} = -n^{-1} \left.\frac{\partial^2 \log\{f(\mathbf{z}; \theta)g(\theta)\}}{\partial\theta\, \partial\theta^{\mathrm{T}}}\right|_{\theta=\tilde{\theta}}$$

we obtain

$$f(\mathbf{z}; \theta)g(\theta) \sim f(\mathbf{z}; \tilde{\theta})g(\tilde{\theta}) \exp\left[-\frac{n}{2}(\theta - \tilde{\theta})^{\mathrm{T}}\mathscr{H}(\theta - \tilde{\theta})\right],$$

so on integrating

$$\int f(\mathbf{z};\theta)g(\theta)\,d\theta \sim f(\mathbf{z};\tilde{\theta})g(\tilde{\theta})\int \exp\left[-\frac{n}{2}(\theta-\tilde{\theta})^{\mathrm{T}}\mathscr{H}(\theta-\tilde{\theta})\right]d\theta$$

$$= f(\mathbf{z};\tilde{\theta})g(\tilde{\theta})\left(\frac{2\pi}{n}\right)^{p/2}\frac{1}{|\mathscr{H}|^{1/2}}.$$

This application of Laplace's method leads to the approximation

$$g(\theta\mid\mathbf{z}) \sim \frac{n^{p/2}|\mathscr{H}|^{1/2}}{(2\pi)^{p/2}}\exp\left[-\frac{n}{2}(\theta-\tilde{\theta})^{\mathrm{T}}\mathscr{H}(\theta-\tilde{\theta})\right]. \tag{4.41}$$

That is $\theta \mid \mathbf{Z}=\mathbf{z} \sim \mathrm{N}(\tilde{\theta}, n^{-1}\mathscr{H}^{-1})$. This approximation has an error of $O(n^{-1/2})$ and is similar in large samples to (4.39) but may be preferable when the likelihood does not dominate the prior.

4.6.5 Laplace's Method for Approximating Marginal Posterior Distributions

As pointed out in Section 2.1.6, the quantities of interest in a Bayesian analysis are integrals of the posterior distribution which are in fact ratios of integrals. If we apply Laplace's method to each integral separately and then take the ratio of the approximations, we often obtain very accurate approximations.

Suppose for example that we want to approximate the marginal posterior distribution of θ_1 which is

$$g_1(\theta_1\mid\mathbf{z}) = \frac{\int f(\mathbf{z};\theta_1,\theta_2)g(\theta_1,\theta_2)\,d\theta_2}{\int f(\mathbf{z};\theta)g(\theta)\,d\theta},$$

where we have partitioned θ into the first component θ_1 and the $(p-1)$-vector θ_2 of the remaining components so $\theta=(\theta_1,\theta_2)$. Let $\tilde{\theta}$ denote the posterior mode (the maximum of $f(\mathbf{z};\theta)g(\theta)$) as before and $\theta_2^*(\theta_1)$ the maximum of $f(\mathbf{z};\theta_1,\theta_2)g(\theta_1,\theta_2)$ with θ_1 held fixed. Also, let

$$\mathscr{H}_1^* = -n^{-1}\left.\frac{\partial^2 \log\{f(\mathbf{z};\theta_1,\theta_2)g(\theta_1,\theta_2)\}}{\partial\theta_2\,\partial\theta_2^{\mathrm{T}}}\right|_{\theta_2=\theta_2^*(\theta_1)}.$$

The denominator is the same as before but the numerator becomes

$$\int f(\mathbf{z};\theta_1,\theta_2)g(\theta_1,\theta_2)\,d\theta_2 \sim f(\mathbf{z};\theta_1,\theta_2^*(\theta_1))g(\theta_1,\theta_2^*(\theta_1))\left(\frac{2\pi}{n}\right)^{(p-1)/2}\frac{1}{|\mathscr{H}_1^*|^{1/2}}.$$

It follows that

$$g_1(\theta_1 \mid \mathbf{z}) \sim n^{1/2} \frac{|\mathscr{H}|^{1/2}}{(2\pi)^{1/2}|\mathscr{H}_1^*|^{1/2}} \frac{f(\mathbf{z}; \theta_1, \theta_2^*(\theta_1))g(\theta_1, \theta_2^*(\theta_1))}{f(\mathbf{z}; \tilde{\theta})g(\tilde{\theta})}.$$

This approximation was suggested by Leonard (1982). It generally has a relative error of $O(n^{-1})$ and, as with the saddlepoint approximation, the relative error can be reduced further to $O(n^{-3/2})$ by renormalizing the density by numerical integration; see Tierney and Kadane (1986) for details. This approximation should be compared to $\theta_1 \mid \mathbf{Z} = \mathbf{z} \sim \mathrm{N}(\tilde{\theta}_1, n^{-1}\mathscr{H}^{11})$, obtained from (4.41) with an error of $O(n^{-1/2})$.

The computation of the approximate marginal density requires maximizing $f(\mathbf{z}; \theta)g(\theta)$ and computing $f(\mathbf{z}; \tilde{\theta})g(\tilde{\theta})$ and $|\mathscr{H}|$, and for a grid of θ_1 values, maximizing $f(\mathbf{z}; \theta_1, \theta_2)g(\theta_1, \theta_2)$ over θ_2 and computing $f(\mathbf{z}; \theta_1, \theta_2^*))g(\theta_1, \theta_2^*(\theta_1))$ and $|\mathscr{H}_1^*|$. This is considerably simpler than what is required for saddlepoint approximation. See Section 4.4.5.

Laplace's method can also be applied in a straightforward way to the computation of posterior moments. (Problem 4.6.6.)

PROBLEMS

4.6.1. Suppose that we have an observation $Z \in \mathbb{R}^p$ on the model

$$\mathscr{F} = \left\{ f(z; \mu, \sigma) = \frac{1}{(2\pi)^{p/2}|\Sigma|^{1/2}} \exp\left\{-\tfrac{1}{2}(z-\mu)^{\mathrm{T}}\Sigma(z-\mu)\right\}, \right.$$

$$\left. -\infty \le z_i \le \infty \colon \mu \in \mathbb{R}^p \right\},$$

where Σ is known. Show that the profile likelihood of μ_1, the first component of μ, is proportional to

$$\exp\left\{-\frac{(z_1-\mu_1)^2}{2\Sigma^{11}}\right\}.$$

Apply this result to the large sample approximation to the likelihood $f(\mathbf{z}; \theta)$ to obtain a large sample approximation to the profile likelihood of θ_1 and show how to obtain a likelihood set for θ_1.

4.6.2. Apply Laplace's method to the gamma function $\Gamma(v+1) = \int_0^\infty u^v e^{-u}\,du$ to show that

$$\Gamma(v+1) \sim (2\pi)^{1/2} v^{v+1/2} e^{-v}.$$

(Hint: Make the change of variables $u = vt$.)

4.6.3. Suppose that we have observations $\mathbf{Z}$ on the model

$$\mathscr{F} = \left\{ f(\mathbf{y}; \mu, \sigma) = \prod_{i=1}^{n} \frac{1}{(2\pi\sigma^2)^{1/2}} \exp\left\{ -\frac{(y_i - \mu)^2}{2\sigma^2}, \right. \right.$$

$$\left. -\infty \leq y_i \leq \infty \colon \mu \in \mathbb{R}, \sigma > 0 \right\}$$

and we use the Jeffreys prior (Section 2.3.2) in our analysis. Use Laplace's method to obtain analytic approximations to the marginal posterior distributions of μ and σ. How do these relate to the exact marginal posterior distributions?

4.6.4. Suppose that we observe $\mathbf{Z}$ on the gamma model (4.12). Find the posterior distribution for the prior $g(\lambda, \kappa) = \lambda^{-1}$, $\kappa, \lambda > 0$. Use Laplace's method to obtain analytic approximations to the marginal posterior distributions of λ and κ. For the pressure vessel failure data presented in Table 1.2, plot the marginal posterior distribution of κ. What do you conclude?

4.6.5. Use Laplace's method to obtain the approximate posterior odds ratios given in Subsections 2.5.3 and 2.5.4 for the problem of testing $H_0\colon \mu = 0$ against $H_1\colon \mu \neq 0$ under a Gaussian model and the Jeffreys Cauchy prior.

4.6.6. Suppose that we have observed $\mathbf{Z}$ on the model

$$\mathscr{F} = \left\{ f(\mathbf{y}; \theta) = \prod_{i=1}^{n} f(y_i; \theta) \colon \theta \in \Omega \right\}.$$

Let $g(\theta)$ denote the prior density of θ and $g(\theta \mid \mathbf{z})$ the posterior density from this prior. We want to approximate

$$\int m(\theta) g(\theta \mid \mathbf{z})\, d\theta,$$

the posterior mean of m, where $m\colon \mathbb{R}^p \to \mathbb{R}$ is a real function. Use Laplace's method to obtain an analytic approximation to the posterior mean of $m(\theta)$. Specialize the result to the marginal posterior mean and the marginal posterior variance of the first component θ_1 of θ.

FURTHER READING

Large sample theory is presented in many textbooks. Silvey (1970) and Bickel and Doksum (1977) contain useful material at an introductory level. Large sample theory for maximum likelihood estimators is widely discussed; Bahadur (1971) and Chernoff (1972) give general results. Serfling (1980) and Sen and Singer (1983) are general references to large sample theory. Saddlepoint approximation is discussed in detail in Barndorff–Nielsen and Cox (1989) and Field and Ronchetti (1990). These texts give a number of useful references. The application of Laplace's method in Bayesian analysis together with a number of references is presented in Bernado and Smith (1994).

CHAPTER 5

Robust Inference

In the stellar velocity problem presented in Section 1.1.4, the data $\mathbf{z}$ are the components of velocity (in km/s) orthogonal to the galactic plane of stars in three different groups. The standard model treats the groups as independent and the data in each group as a realization of $\mathbf{Z}$ generated by the Gaussian model

$$\mathscr{F} = \left\{ f(\mathbf{z}; \mu, \sigma) = \prod_{i=1}^{n} \frac{1}{\sigma} \phi\left(\frac{z_i - \mu}{\sigma}\right), \quad -\infty \leq z_i \leq \infty; -\infty \leq \mu \leq \infty, \sigma > 0 \right\} \tag{5.1}$$

with $\mu = 0$ as the core model for the bulk of the data. (We adopt the more general model in which μ is an unspecified nuisance parameter to make the discussion in this chapter more generally applicable.) We observed in Section 1.2.4 that the model $\mathscr{F}$ does not hold exactly because there are outliers in the data and noted that the contamination model

$$\mathscr{G} = \left\{ g(\mathbf{z}; \mu, \sigma, \varepsilon, g) = \prod_{i=1}^{n} \left\{ (1 - \varepsilon) \frac{1}{\sigma} \phi\left(\frac{z_i - \mu}{\sigma}\right) + \varepsilon c(z_i) \right\}, \right.$$
$$\left. -\infty \leq z_i \leq \infty; -\infty \leq \mu \leq \infty, \sigma > 0, \varepsilon > 0, c \in \mathscr{C} \right\} \tag{5.2}$$

is a useful model to focus our thinking about outliers.

5.1 THE STANDARD DEVIATION

When the Gaussian model (5.1) holds exactly, we often base inference about σ on the maximum likelihood estimator

$$\hat{\sigma} = \left\{ \frac{1}{n} \sum_{i=1}^{n} (Z_i - \bar{Z})^2 \right\}^{1/2}.$$

Under (5.1),

$$n\hat{\sigma}^2 \sim \sigma^2 \chi^2_{n-1}$$

so that if K_ν is the distribution function of the χ^2_ν distribution, we have that

$$1 - \alpha = \mathrm{P}\left\{K_{n-1}^{-1}(\alpha/2) \le \frac{n\hat{\sigma}^2}{\sigma^2} \le K_{n-1}^{-1}(1 - \alpha/2)\right\}$$

$$= \mathrm{P}\left\{\frac{n\hat{\sigma}^2}{K_{n-1}^{-1}(1 - \alpha/2)} \le \sigma^2 \le \frac{n\hat{\sigma}^2}{K_{n-1}^{-1}(\alpha/2)}\right\}$$

from which it follows that a realization of

$$\{n^{1/2}\hat{\sigma}/K_{n-1}^{-1}(1 - \alpha/2)^{1/2}, n^{1/2}\hat{\sigma}/K_{n-1}^{-1}(\alpha/2)^{1/2}\}$$

is an exact $100(1 - \alpha)\%$ confidence interval for σ.

For later comparison, it is convenient to develop a large sample approximation to the exact confidence interval. Under the model $\mathscr{F}$, we have $\mathrm{E}\hat{\sigma}^2 = (n - 1)\sigma^2/n$ and $\mathrm{Var}\, \hat{\sigma}^2 = 2(n - 1)\sigma^4/n^2$ so we obtain the large sample approximation

$$n^{1/2}(\hat{\sigma}^2 - \sigma^2) \xrightarrow{\mathscr{D}} \mathrm{N}(0, 2\sigma^4).$$

It makes considerable sense to treat scale estimators on the log scale (because they are non-negative and the log scale is appropriate for comparisons of relative magnitudes) and we will do so throughout this section. Applying Theorem 4.3 with $h(x) = \log(x^{1/2})$, we have

$$n^{1/2}(\log(\hat{\sigma}) - \log(\sigma)) \xrightarrow{\mathscr{D}} \mathrm{N}(0, 1/2) \tag{5.3}$$

so that an approximate $100(1 - \alpha)\%$ confidence level for $\log(\sigma)$ is

$$\log(\hat{\sigma}) \pm \frac{1}{n^{1/2}2^{1/2}} \Phi^{-1}(1 - \alpha/2)$$

and hence an approximate $100(1 - \alpha)\%$ confidence interval for σ is

$$[\hat{\sigma} \exp\{-n^{-1/2}2^{-1/2}\Phi^{-1}(1 - \alpha/2)\}, \hat{\sigma} \exp\{n^{-1/2}2^{-1/2}\Phi^{-1}(1 - \alpha/2)\}]. \tag{5.4}$$

Exponentiating the approximate 95% confidence intervals for $\log(\sigma)$ as in (5.4), we obtain the following results for the three sets of stars:

Disk:	[33.45, 65.53]
Intermediate:	[69.64, 110.54]
Halo:	[69.41, 131.11].

These intervals suggest that the intermediate stars are halo stars.

5.1.1 Nonparametric Standard Errors

Now suppose that the components of $\mathbf{Z}$ are in fact independent with common distribution F_0 and that F_0 is not necessarily in $\mathscr{F}$. Then the sample standard deviation $\hat{\sigma}$ is estimating $\sigma(F_0)$, the standard deviation of the underlying distribution. To obtain a large sample approximation to the sampling distribution of $\hat{\sigma}$, we use the identity $a^2 - b^2 = 2b(a - b) + (b - a)^2$ to write

$$\begin{aligned}\hat{\sigma}^2 - \sigma^2(F_0) &= n^{-1} \sum_{i=1}^{n} \left\{Z_i^2 - \int_{-\infty}^{\infty} x^2\, dF_0(x)\right\} - \left\{n^{-1} \sum_{i=1}^{n} Z_i\right\}^2 + \left\{\int_{-\infty}^{\infty} x\, dF_0(x)\right\}^2 \\ &= n^{-1} \sum_{i=1}^{n} \left\{Z_i^2 - \int_{-\infty}^{\infty} x^2\, dF_0(x) - 2\mu(F_0)(Z_i - \mu(F_0))\right\} \\ &\qquad - \left\{n^{-1} \sum_{i=1}^{n} Z_i - \mu(F_0)\right\}^2,\end{aligned}$$

where $\mu(F_0) = \int_{-\infty}^{\infty} x\, dF_0(x)$. We can then write

$$n^{1/2}\{\hat{\sigma}^2 - \sigma^2(F_0)\} = n^{-1/2} \sum_{i=1}^{n} \mathrm{IF}(Z_i; F_0, \hat{\sigma}^2) + O_p(n^{-1/2}),$$

where

$$\begin{aligned}\mathrm{IF}(z; F_0, \hat{\sigma}^2) &= z^2 - \int_{-\infty}^{\infty} x^2\, dF_0(x) - 2\mu(F_0)(z - \mu(F_0)) \\ &= (z - \mu(F_0))^2 - \sigma^2(F_0)\end{aligned} \tag{5.5}$$

is called the *influence function* of $\hat{\sigma}^2$ at F_0 (Hampel 1968; 1974). This function is plotted in Figure 5.2b. Provided $\mathrm{E}_{F_0}\mathrm{IF}(Z; F_0, \hat{\sigma}^2)^2 < \infty$, it follows from the central limit theorem (Theorem 4.1) and Theorem 4.2 that

$$n^{1/2}\{\hat{\sigma}^2 - \sigma^2(F_0)\} \xrightarrow{\mathscr{D}} \mathrm{N}(0, \mathrm{E}_{F_0}\mathrm{IF}(Z; F_0, \hat{\sigma}^2)^2),$$

where

$$\mathrm{E}_{F_0}\mathrm{IF}(Z; F_0, \hat{\sigma}^2)^2 = \mathrm{E}_{F_0}(Z - \mu(F_0))^4 - \sigma^4(F_0).$$

On the log scale, we obtain from Theorem 4.3

$$\begin{aligned}n^{1/2}\{\log \hat{\sigma} - \log \sigma(F_0)\} &= n^{-1/2} \sum_{i=1}^{n} \frac{\mathrm{IF}(Z_i; F_0, \hat{\sigma}^2)}{2\sigma^2(F_0)} + O_p(n^{-1/2}) \\ &\xrightarrow{\mathscr{D}} \mathrm{N}\left(0, \mathrm{E}_{F_0} \frac{\mathrm{IF}(Z; F_0, \hat{\sigma}^2)^2}{4\sigma^4(F_0)}\right).\end{aligned} \tag{5.6}$$

Notice that we can sensibly write $\mathrm{IF}(z; F_0, \log(\hat{\sigma})) = \mathrm{IF}(z; F_0, \hat{\sigma}^2)/2\sigma^2(F_0)$ but it is just as convenient to use the latter expression.

Under the Gaussian model (5.1), $\mathrm{E}(Z - \mu(F_0))^4 = 3\sigma^4$ so (5.6) includes (5.3) as a special case. For long tailed distributions $\mathrm{E}(Z - \mu(F_0))^4 > 3\sigma^4(F_0)$ so in this case, the standard error increases to preserve the level of the confidence interval for $\sigma(F_0)$. This suggests that we use an estimate of $\{\mathrm{E}_{F_0}(Z - \mu(F_0))^4 - \sigma^4(F_0)\}^{1/2}/(4n)^{1/2}\sigma^2(F_0)$, rather than $1/(2n)^{1/2}$ as the standard error of $\log(\hat{\sigma})$ since we lose nothing (asymptotically) under the model and preserve the level otherwise. This is an illustration of *nonparametric variance estimation*: we get a valid inference about $\sigma(F_0)$ provided $\mathrm{E}_{F_0}Z^4 < \infty$, both when $F_0 \in \mathscr{F}$ and $F_0 \notin \mathscr{F}$.

Using a nonparametric variance estimate, the approximate $100(1 - \alpha)\%$ confidence interval for $\sigma(F_0)$ is

$$[\hat{\sigma} \exp\{-n^{-1/2}\hat{\tau}^{1/2}\Phi^{-1}(1 - \alpha/2)\}, \hat{\sigma} \exp\{n^{-1/2}\hat{\tau}^{1/2}\Phi^{-1}(1 - \alpha/2)\}], \quad (5.7)$$

where $\hat{\tau} = 4^{-1}\hat{\sigma}^{-4}\{n^{-1}\sum_{i=1}^{n}(Z_i - \bar{Z})^4 - \hat{\sigma}^4\}$. For the three sets of stars, we obtain the following approximate 95% confidence intervals:

Disk:	[28.54, 76.80]
Intermediate:	[63.57, 121.09]
Halo:	[42.26, 215.32].

These confidence intervals are wider than before and therefore make the conclusion far less clear.

5.1.2 Robustness of Validity and Robustness of Efficiency

The confidence interval (5.7) based on an estimate of

$$\frac{\{\mathrm{E}_{F_0}(Z - \mu(F_0))^4 - \sigma^4(F_0)\}^{1/2}}{(4n)^{1/2}\sigma^2(F_0)}$$

achieves its intended level asymptotically for any distribution which has finite fourth moment. This property is known as *robustness of validity* and was the focus of attention in the early studies of robustness by Pearson (1929; 1931), Geary (1936), Gayen (1950), Box (1953), Box and Andersen (1955) and Scheffé (1959, Chapter 10). (Although these studies all concerned testing, the consequences for confidence intervals are easily extracted.) However, robustness of validity is a weak property which says nothing about the effect of non-normality on the length of the confidence intervals. Indeed, (5.7) widens as the tails on the distribution lengthen and the width can easily become infinite if the underlying distribution does not have a finite fourth moment. In fact, in his critique of the method of moments, Fisher (1922) showed that, even if the underlying distribution does have a finite fourth moment, the sample variance can be a very inefficient estimator of the underlying variance. Intervals which maintain their length as well as their level in a neighbourhood of the model are said to have *robustness of efficiency*, a more valuable property than robustness of validity alone.

5.1.3 Misspecification Bias

Suppose that $F_0 \in \mathscr{G}$, where $\mathscr{G}$ is a contamination model like (5.2) with core F. As we noted in Section 1.2.4, we may want to estimate the standard deviation $\sigma(F)$ of the core of $\mathscr{G}$ rather than the standard deviation $\sigma(F_0)$ of the underlying distribution. In this case, there is a bias due to departures from the model $\mathscr{F}$ which is represented by the difference $\sigma(F_0) - \sigma(F)$. Using a Taylor series approximation (see 2 in the Appendix)

$$\log\{\sigma(F_0)\} - \log\{\sigma(F)\} \approx \frac{\sigma^2(F_0) - \sigma^2(F)}{2\sigma^2(F_0)}.$$

Arguing exactly as in Section 5.1.1,

$$\sigma^2(F_0) - \sigma^2(F) = \int_{-\infty}^{\infty} \mathrm{IF}(z; F, \hat{\sigma}^2)\, dF_0(z) - \left\{\int_{-\infty}^{\infty} z\, d(F_0 - F)(z)\right\}^2,$$

where $\mathrm{IF}(z; F, \hat{\sigma}^2)$ is given by (5.5). Ignoring the second term, we have

$$\begin{aligned}\log\{\sigma(F_0)\} - \log\{\sigma(F)\} &\approx \int_{-\infty}^{\infty} \frac{\mathrm{IF}(z; F, \hat{\sigma}^2)}{2\sigma^2(F)}\, dF_0(z) \\ &= \mathrm{E}_{F_0} \frac{\mathrm{IF}(Z; F, \hat{\sigma}^2)}{2\sigma^2(F)}.\end{aligned} \tag{5.8}$$

We can justify omitting the second term in the bias if we replace F_0 by a sequence of contaminated distributions which converges to F, as $n \to \infty$, because then the second term is of smaller order than the first. However, F_0 is usually a fixed distribution and in this case there is no real basis other than simplicity for ignoring the second term in the bias. The approximation to the bias should therefore be treated as suggestive. If F_0 is longer tailed than F, the bias will typically be positive; if F_0 is a distribution for which the variance is infinite, the bias will be infinite. That is, the bias can be arbitrarily large.

5.1.4 The Approximate Mean Squared Error

Combining (5.6) and (5.8), we see that the effect of heavy tailed contamination (and hence outliers) is to shift our confidence interval in the positive direction from where it should be (bias) and to make it wider than it should be (variance). Moreover, these changes can be arbitrarily large. Formally, the mean squared error can be written as

$$\begin{aligned}\mathrm{MSE}(\log(\hat{\sigma}); F_0) &\approx \frac{\{\mathrm{E}_{F_0}\,\mathrm{IF}(Z; \Phi, \hat{\sigma}^2)\}^2}{4\sigma^4} + n^{-1}\frac{\mathrm{E}_{F_0}\mathrm{IF}(Z; \Phi, \hat{\sigma}^2)^2}{4\sigma^4(F_0)} \\ &= \frac{\{\mathrm{E}_{F_0}(Z^2 - \sigma^2)\}^2}{4\sigma^4} + n^{-1}\frac{\{\mathrm{E}_{F_0}(Z - \mu(F_0))^4 - \sigma(F_0)\}}{4\sigma^4(F_0)}\end{aligned}$$

which can be arbitrarily large compared to

$$\mathrm{MSE}(\log(\hat{\sigma}); F) \sim \frac{1}{2n},$$

the value under (5.1).

5.1.5 Stability and Breakdown

A natural way to study the stability of an estimator is to add an additional observation x to the data $\mathbf{z}$ and then explore the effect of changing x on the estimator. Writing

$$\begin{aligned}\bar{z}(x) &= \frac{x}{n+1} + \frac{1}{n+1}\sum_{i=1}^{n} z_i \\ &= \frac{x}{n+1} + \frac{n}{n+1}\bar{z} \\ &= \bar{z} + \frac{1}{(n+1)}(x - \bar{z}),\end{aligned}$$

where $\bar{z} = n^{-1}\sum_{i=1}^{n} z_i$, we find for the variance estimator that

$$\begin{aligned}\hat{\sigma}^2(x) &= \frac{1}{n+1}\sum_{j=1}^{n}(z_j - \bar{z}(x))^2 + \frac{1}{n+1}(x - \bar{z}(x))^2 \\ &= \frac{1}{n+1}\sum_{j=1}^{n}\left(z_j - \bar{z} - \frac{1}{(n+1)}(x - \bar{z})\right)^2 + \frac{n^2(x - \bar{z})^2}{(n+1)^3}\end{aligned}$$

$$= \frac{1}{n+1} \sum_{j=1}^{n} (z_j - \bar{z})^2 + \frac{n}{(n+1)^3} (x - \bar{z})^2 + \frac{n^2 (x - \bar{z})^2}{(n+1)^3}$$

$$= \frac{n}{n+1} \hat{\sigma}^2 + \frac{n(x - \bar{z})^2}{(n+1)^2}.$$

The effect of changes in x on the estimate $\hat{\sigma}^2$ is, not surprisingly, quadratic.

Small changes to x do not change the estimate much, but the estimate can be made arbitrarily large by large changes to x. In fact, the estimator can break down in the sense of becoming infinite by moving a single observation to infinity. We define the *breakdown point* of an estimator to be the largest fraction of the data that can be moved arbitrarily without perturbing the estimator to the boundary of the parameter space. Thus the higher the breakdown point, the more robust the estimator against extreme outliers. The breakdown point of the sample variance is 0.

The normalized effect of observation x is called the *sensitivity curve* (Tukey, 1960). It is

$$(n+1)\{\hat{\sigma}^2(x) - \hat{\sigma}^2\} = \frac{n}{n+1} (x - \bar{z})^2 - \hat{\sigma}^2 \rightarrow (x - \mu(F_0))^2 - \sigma^2(F_0)$$

and by a Taylor series argument

$$(n+1)\{\log(\hat{\sigma}(x)) - \log(\hat{\sigma})\} \rightarrow \frac{(x - \mu(F_0))^2 - \sigma^2(F_0)}{2\sigma^2(F_0)}.$$

The limit of the sensitivity curve is the influence function of $\log(\hat{\sigma})$ at F_0. The supremum of the influence function is a measure of the maximum effect an observation can have on the estimator. It is called the *gross error sensitivity* of $\hat{\sigma}$ at F_0 (Hampel, 1968). For the sample standard deviation, the gross error sensitivity is infinite.

5.1.6 Formal Robustness Calculations

We can refine the calculations of Sections 5.1.1–5.1.4 for the standard deviation by extracting their common aspects. Basically, we have two distributions, say G and H, and we need to compare $\sigma^2(G)$ to $\sigma^2(H)$. The key idea is to expand $\sigma^2(H)$ about $\sigma^2(G)$ and then use the first (linear) term in this expansion to approximate the difference $\sigma^2(H) - \sigma^2(G)$. The fact that the argument is a function rather than a simple number complicates the expansion. Nonetheless, there is an appropriate calculus for this situation which yields the Taylor series expansion

$$\sigma^2(H) - \sigma^2(G) = d_1(G, H - G) + \tfrac{1}{2} d_2(G, H - G) + \cdots$$

expansions for $\sigma(H)$ and $\log \sigma(H)$ follow directly from a further Taylor expansion, where d_k is the kth order differential of σ^2 at G in the direction of H. See von Mises (1947), Serfling (1980, Chapter 6) and Fernholz (1983). The terms in the expansion are actually obtained using no more than the regular calculus as

$$d_k(G, H - G) = \frac{d^k}{d\lambda^k} \sigma^2(G + \lambda(H - G))|_{\lambda = 0}.$$

For the variance, we obtain

$$\begin{aligned} d_1(G, H - G) &= \int_{-\infty}^{\infty} \left\{x^2 - 2x \int_{-\infty}^{\infty} y \, dG(y)\right\} d(H - G)(x) \\ &= \int_{-\infty}^{\infty} \{x^2 - 2x\mu(G)\} \, dH(x) - \int_{-\infty}^{\infty} \{x^2 - 2x\mu(G)\} \, dG(x) \\ &= \int_{-\infty}^{\infty} \left\{z^2 - 2z\mu(G) - \int_{-\infty}^{\infty} \{x^2 - 2x\mu(G)\} \, dG(x)\right\} dH(z) \\ &= \int_{-\infty}^{\infty} \left\{\{z - \mu(G)\}^2 - \sigma^2(G)\right\} \mathrm{d}H(z), \\ d_2(G, H - G) &= -2\left\{\int_{-\infty}^{\infty} x \, d(H - G)(x)\right\}^2 \end{aligned}$$

and

$$d_k(G, H - G) = 0 \qquad \text{for } k \geq 3.$$

Hence, using the influence function

$$\mathrm{IF}(x; G, \hat{\sigma}^2) = \{x - \mu(G)\}^2 - \sigma^2(G)$$

as in (5.5), we have

$$\sigma^2(H) - \sigma^2(G) \approx d_1(G, H - G) = \int \mathrm{IF}(x, G, \hat{\sigma}^2) \, dH(x).$$

We can now see why the influence function is so fundamental.

- We obtain the approximate bias (Section 5.1.3) of the estimator as the expectation with respect to the actual distribution F_0 of the influence function at the core model F. That is, for $G = F$ and $H = F_0$.

- The asymptotic effect of a single observation at x (Section 5.1.5) on the estimator is obtained by setting $G = F_0$ and $H(z) = I(z \leq x)$, the degenerate distribution with all its mass at x. This is just the influence function itself which we can think of as the approximate bias when the core distribution is contaminated by a pointmass at x.
- If $F_0 = (1 - \varepsilon)F + \varepsilon C$ is a contaminated version of F with contaminating distribution $C \in \mathscr{C}$, we have

$$\begin{aligned}\sup_{C \in \mathscr{C}} |\sigma(F_0) - \sigma(F)| &\approx \sup_{F_0 \in \mathscr{C}} \left| \int_{-\infty}^{\infty} \mathrm{IF}(x, F, \hat{\sigma})\, d(F_0 - F)(x) \right| \\ &= \sup_{C \in \mathscr{C}} \left| \varepsilon \int_{-\infty}^{\infty} \mathrm{IF}(x, F, \hat{\sigma})\, dC(x) \right| \\ &= \varepsilon \sup_{x \in \mathbb{R}} |\mathrm{IF}(x, F, \hat{\sigma})|\end{aligned}$$

  so the maximum bias is determined by the maximum influence function at F, provided the interchange of supremum and integral is legitimate. This bound is called the gross error sensitivity (Section 5.1.5) because it reflects the largest effect a single gross outlier can have on the estimator.
- Finally, the estimator $\hat{\sigma}^2$ can be written as $\sigma^2(F_n)$, where F_n is the empirical distribution function (Section 1.5.2) so the asymptotic distribution (Section 5.1.1) is obtained from the leading term when $G = F_0$ and $H = F_n$.

The one important quantity we cannot compute directly from the influence function is the breakdown point (Section 5.1.5).

5.1.7 Distributional Robustness

The basic requirements for inferences designed for a core model $\mathscr{F}$ to be robust against departures from the assumed distributional shape are that they

- are valid and reasonably efficient under $\mathscr{F}$
- have close to their assumed properties under $\mathscr{F}$ when applied to distributions close to those in $\mathscr{F}$
- are not rendered useless by a small proportion of extreme deviations from $\mathscr{F}$.

The first requirement ensures that good inferences are produced when $\mathscr{F}$ holds, an important and often overlooked property of robust methods. The two robustness conditions can usefully be regarded as stability conditions; small perturbation to $\mathscr{F}$ should not change the properties of the estimator much, and a few extreme perturbations should not be disastrous.

These requirements can be interpreted in terms of the influence function and stability properties of an estimator. For an estimator to be robust, we require the influence function to be continuous so that small changes (including rounding and grouping) to the observations have only a small effect on the estimator and bounded for distributions in a neighborhood of the model so that the mean squared error of the estimator is bounded. In addition, we want an estimator to not be rendered uninformative by a small proportion of extreme outliers, so we want the estimator to have a nonzero breakdown point.

The sample standard deviation $\hat{\sigma}$ is asymptotically efficient under (5.1) so it meets the first requirement. However, as we saw in Section 5.1.4 and 5.1.5, it fails the second and third requirement, so we conclude that $\hat{\sigma}$ is not a robust estimator of $\sigma(F)$.

Interestingly, Fisher (1920) recommended the use of $\hat{\sigma}$ for stellar velocity data rather than the mean absolute deviation recommended by Eddington (1914) on the basis of its efficiency under the Gaussian model. Although the mean absolute deviation is more efficient than the standard deviation in a neighborhood of the Gaussian model (5.1) and therefore more robust (see Section 5.7 for some evidence of this), it is not really robust and we will not pursue it in detail. We will construct a class of distributionally robust estimators of $\sigma(F)$ in Section 5.4.

PROBLEMS

5.1.1. Let $T(\mathbf{Z}) = n^{1/2}(\bar{Z} - \mu)/S$ denote the t ratio. Plot $T(\mathbf{Z}, x)$ with $\mu = 0$ against x for the caffeine data (Section 1.1.3) to explore the effect of adding an additional observation at x. What happens as $x \to \infty$? Compare the result to that for the unstudentized sample mean. What are the implications of the results in terms of robustness of validity and robustness of efficiency?

5.1.2. Compute the sensitivity curves of the median and median absolute deviation from the median (Section 5.4.8) based on the caffeine data (Section 1.1.3) and compare these to those obtained for the mean and sample standard deviation in Section 5.1.

5.1.3. Compare the sensitivity surface of $\hat{\sigma}_1/\hat{\sigma}_2$, when $\hat{\sigma}_j$ is the sample standard deviation, the mean absolute deviation from the mean, and the median absolute deviation from the median based on the intermediate and halo stars (Section 1.1.4) by adding an extra observation to each sample.

5.1.4. Show that the sensitivity surface of the M-estimator $\hat{\theta}$ which satisfies

$$0 = \sum_{i=1}^{n} \eta(z_i; \hat{\theta})$$

can be written as

$$(n+1)(\hat{\theta}(x)-\hat{\theta}) \approx -(n+1)\left\{\sum_{i=1}^{n} \eta'(z_i;\hat{\theta}) + \eta'(x,\hat{\theta})\right\}^{-1} \eta(x,\hat{\theta}).$$

5.1.5. Cushny and Peebles (1905) published the results of a study of the effect of two optical isomers of hyoscyamine hydrobromide in increasing the sleeping time of 10 patients. They measured the average number of hours of sleep gained over several nights under the two versions of the drug. The paired differences between the results for the second drug (laevo rotatory) and the first (dextro rotatory) are

1.2 2.4 1.3 1.3 0.0 1.0 1.8 0.8 4.6 1.4

Data reprinted with permission from the *Journal of Physiology* (1905).

The sample mean equals 1.58 with a standard error of 0.39 and the $\alpha = 0.1$ trimmed mean (Section 6.9.2) equals 1.40 with a standard error of 0.2. The problem of interest is to make inferences about the difference in the mean effects of the two drugs.

What advantages are there to computing appropriate confidence intervals rather than p-values? Construct approximate 95% confidence interval for the difference in the mean effects of the two drugs using firstly the sample mean and then the trimmed mean. Compare the two intervals and comment on the use of the intervals to test H_0: The difference in the mean effects of the two drugs equals zero.

(Hint: Tukey and McLaughlin (1963) showed that the approximate pivotal quantity constructed from the trimmed mean and its standard error has approximately the Student t distribution with $n - 2[n\alpha]$ degrees of freedom.)

5.2 DEPARTURES FROM INDEPENDENCE

Departures from the assumed dependence structure (which in the stellar velocity example means independence) constitute an important deviation from the assumed model. These departures are typically rather harder to deal with than departures from the assumed distributional form because we need to specify the dependence structure quite closely.

5.2.1 Autoregressive Dependence

As an illustrative example, suppose that the data follow an autoregressive process (see Section 1.3.5) induced by the measurement process. This is appropriate when the observations are made sequentially in time and an observation is dependent on the immediately preceding observations. Explicitly, for $|\rho| < 1$, we assume that

$$Z_1 \sim \mu + \frac{\sigma U_1}{(1-\rho^2)^{1/2}}$$

and

$$Z_i - \mu = \rho(Z_{i-1} - \mu) + \sigma U_i, \qquad i = 2, \ldots, n,$$

with $\{U_i\}$ a sequence of independent random variables with mean 0 and variance 1. When $\rho = 0$ and $\{U_i\}$ are Gaussianly distributed, $Z_1, \ldots, Z_n$ are independent Gaussian random variables with mean μ and variance σ^2 so this contains our model $\mathscr{F}$ as a special case.

Let $\gamma(h) = \mathrm{E_G}(Z_i - \mu)(Z_{i-h} - \mu)$, where G indicates expectation under the autoregressive model, denote the autocovariance function of the autoregressive process. We obtain simple expressions for the bias and variance of the sample standard deviation $\hat{\sigma}$ under G in terms of $\gamma(h)$ in Section 5.2.2. The autocovariance function $\gamma(h)$ does not characterize all aspects of the dependence structure of a stochastic process, but it is a measure of dependence which allows us to obtain suggestive results relatively easily.

5.2.2 The Effect of Autoregressive Dependence

Under the autoregressive model,

$$\begin{aligned}
\mathrm{E_G}(\bar{Z} - \mu)^2 &= n^{-2} \sum_{i=1}^{n} \sum_{j=1}^{n} \mathrm{E_G}(Z_i - \mu)(Z_j - \mu) \\
&= n^{-2} \sum_{i=1}^{n} \sum_{j=1}^{n} \gamma(i-j) \\
&= n^{-1}\gamma(0) + 2n^{-2} \sum_{i=2}^{n} \sum_{j=1}^{i-1} \gamma(i-j) \\
&= n^{-1}\gamma(0) + 2n^{-2} \sum_{h=1}^{n-1} (n - |h|)\gamma(h)
\end{aligned}$$

so we have

$$\begin{aligned}
\mathrm{E_G}\hat{\sigma}^2 &= n^{-1} \sum_{i=1}^{n} \mathrm{E_G}(Z_i - \mu)^2 - \mathrm{E_G}(\bar{Z} - \mu)^2 \\
&= \gamma(0) - n^{-1}\gamma(0) - 2n^{-2} \sum_{h=1}^{n-1} (n - |h|)\gamma(h) \\
&= n^{-1}(n-1)\gamma(0) - 2n^{-2} \sum_{h=1}^{n-1} (n - |h|)\gamma(h) \\
&= \gamma(0) + O(n^{-1}).
\end{aligned}$$

Thus $\hat{\sigma}^2$ is estimating $\gamma(0)$ rather than $\sigma^2(G)$. Since $\gamma(h) = \sigma^2(G)\rho^{|h|}/(1-\rho^2)$, we have $\gamma(0) = \sigma^2(G)/(1-\rho^2)$ and hence

$$\log\{\gamma(0)^{1/2}\} - \log\{\sigma(G)\} \sim \frac{\gamma(0) - \sigma^2(G)}{2\sigma^2(G)} = \frac{\rho^2}{2(1-\rho^2)}.$$

The asymptotic bias of $\log(\hat{\sigma})$ is therefore

$$\log\{\gamma(0)^{1/2}\} - \log\{\sigma(\Phi)\} = \frac{\rho^2}{2(1-\rho^2)} + \frac{\mathrm{E_G IF}(Z; \Phi, \hat{\sigma}^2)}{2\sigma^2(\Phi)}.$$

The bias due to the unsuspected serial correlation is of order ρ^2 as $\rho \to 0$ so may be considered small.

The variance is more complicated to derive. However, suppressing the dependence of μ and σ on G, we see from Fuller (1976, p. 239) that

$$\begin{aligned}
\mathrm{Var_G}(\hat{\sigma}^2) &\sim n^{-1}\left\{\left[\frac{\mathrm{E_G}(Z-\mu)^4}{\sigma^4} - 3\right]\gamma(0)^2 + 2\sum_{h=-\infty}^{\infty}\gamma(h)^2\right\} \\
&= n^{-1}\left\{\left[\frac{\mathrm{E_G}(Z-\mu)^4}{\sigma^4} - 3\right]\gamma(0)^2 + 2\gamma(0)^2 + 4\sum_{h=1}^{\infty}\gamma(h)^2\right\} \\
&= n^{-1}\left\{\left[\frac{\mathrm{E_G}(Z-\mu)^4}{\sigma^4} - 1\right] + 4\sum_{h=1}^{\infty}\rho^{2h}\right\}\frac{\sigma^4}{(1-\rho)^2} \\
&= n^{-1}\left\{\left[\frac{\mathrm{E_G}(Z-\mu)^4}{\sigma^4} - 1\right] + \frac{4\rho^2}{(1-\rho^2)}\right\}\frac{\sigma^4}{(1-\rho)^2} \\
&= n^{-1}\left\{\left[\frac{\mathrm{E_G}(Z-\mu)^4}{\sigma^4} - 1\right] + \frac{4\rho^2}{(1-\rho)^2}\right\}\sigma^4\{1 + 2\rho + O(\rho^2)\} \\
&= n^{-1}[\mathrm{E_G}(Z-\mu)^4 - \sigma^4](1 + 2\rho + O(\rho^2)) \\
&= n^{-1}\mathrm{E_G IF}(Z; G, \hat{\sigma}^2)^2(1 + 2\rho + O(\rho^2)).
\end{aligned}$$

For negative ρ, treating $n^{-1}\mathrm{E_G IF}(Z; G, \hat{\sigma}^2)^2$ as the asymptotic variance $\hat{\sigma}^2$ is conservative while for positive ρ it is overoptimistic. Unfortunately, positive correlation ($\rho > 0$) is common in practice so we can conclude that ignoring serial dependence generally results in a slight positive bias in our estimate and confidence intervals which are narrower than they should be. More general results of this type have been obtained by Gastwirth and Rubin (1975), Portnoy (1977; 1979), Bickel and Herzberg (1979), and Kariya (1980).

5.2.3 Others Forms of Dependence

Other forms of unsuspected dependence may also be relevant. For example, some form of *spatial dependence* may be appropriate for the stellar velocity data. This would occur if observations on stars that are close to each other in space are dependent, which could again arise through the measurement process.

Finally, most simple models for serial and spatial dependence specify that the dependence decreases rapidly as the observations become further apart. These are effectively therefore models for *local* or *short range dependence.* Alternative models which allow *long range dependence* are also available. Under these models, the standard errors based on the independence assumption are much too small. Unfortunately, fairly large samples are required to fit these models in practice. See Hampel et al. (1986, pp. 387–97) and Beran (1991).

PROBLEMS

5.2.1. Use the results of Section 5.2.2 to explore the effects of unsuspected autoregressive dependence on inference for the mean based on the sample mean $\bar{Z}$.

5.2.2. A moving average process $\{Z_t\}$ is defined by the relationship

$$Z_t = e_t + \alpha e_{t-1},$$

where $\{e_t\}$ are independent and identically distributed. Show that

$$\gamma(h) = \begin{cases} (1+\alpha^2)\sigma^2 & h = 0 \\ \alpha\sigma^2 & h = 1 \\ 0 & \text{else} \end{cases}$$

and explore the effect of unsuspected moving average dependence on inference for the mean based on the sample mean $\bar{Z}$.

5.2.3. Explore the effect of unsuspected moving average dependence on inference for the variance based on the sample variance.

5.3 ROBUSTNESS THEORY

Suppose that we have a model $\mathscr{F}$ which is approximately correct in the sense that the actual underlying distribution F_0 is either in $\mathscr{F}$ or close to $\mathscr{F}$. Then the objectives of robustness theory are twofold:

1. Evaluate how procedures derived under $\mathscr{F}$ perform when the underlying distribution F_0 is close to $\mathscr{F}$ and
2. Find procedures for making inference which perform well both when F_0 is in $\mathscr{F}$ and F_0 is close to $\mathscr{F}$.

It is often straightforward to achieve objective (1) (see for example Sections 5.1–5.2), but objective (2) is usually achievable only if the departures from $\mathscr{F}$ are specific and limited.

The two main departures from a model $\mathscr{F}$ considered in robustness theory to date have been departures from the assumed distributional shape (usually manifested as outliers) and, to a lesser extent, departures from the assumed dependence structure. In the context of the stellar velocity problem, these are the important departures but, in other contexts, other departures may be of interest. Generally, in robustness theory, we are concerned with departures involving the nuisance aspects of the model because the analysis often depends on our being able to detect departures involving the aspects of the model of primary interest. Thus, in a regression problem (see for example Section 5.5) we may seek robustness against outliers in the error distribution but not against curvature in the regression relationship. There are data configurations for which it is difficult to distinguish between these situations but the conceptual distinction is still useful.

Formal theories of robustness need to take into account the fact that even apparently good fit does not ensure that an assumed model is correct. Figure 5.1 shows a plot of the Gaussian density and the density of the mixture distribution (like the one in the contamination model (5.2)) in which $\varepsilon = 0.05$ and c is the Cauchy density. Even though the Gaussian distribution has all its moments finite and the mixture has no finite moments, it is difficult to distinguish between the densities. It is even more difficult with samples from the two distributions unless extreme outliers are present.

The simplest way to describe departures from an assumed model $\mathscr{F}$ is to introduce a parametric model $\mathscr{G}$ which contains $\mathscr{F}$ and captures the departures from $\mathscr{F}$. That is, $\mathscr{G}$ describes a neighborhood of the core model $\mathscr{F}$. The introduction of $\mathscr{G}$ is a natural way to achieve objective 1 and can be used to achieve objective 2 if we fit $\mathscr{G}$ instead of $\mathscr{F}$. Indeed, the introduction of $\mathscr{G}$ may be the only way to achieve objective 2: the only way to achieve robustness against dependence seems to be by incorporating dependence explicitly into the analysis. However, this approach is really only satisfactory if we consider very specific, limited departures from $\mathscr{F}$. In practice, departures from $\mathscr{F}$ can take so many different forms that we run the risk of ending up with a very complicated (nonparsimonious) model without any guarantee that it describes the actual departures from $\mathscr{F}$ and there is often little information on the additional nuisance parameters in $\mathscr{G}$. This is particularly the case if we have a small number of outliers in the data. Fortunately, in this case, as we noted in Section 1.2.4, we can be fairly vague about $\mathscr{G}$, even to the extent of making $\mathscr{G}$ semiparametric (Section 1.3.3) and, as we saw in Section 5.1.7, distributional

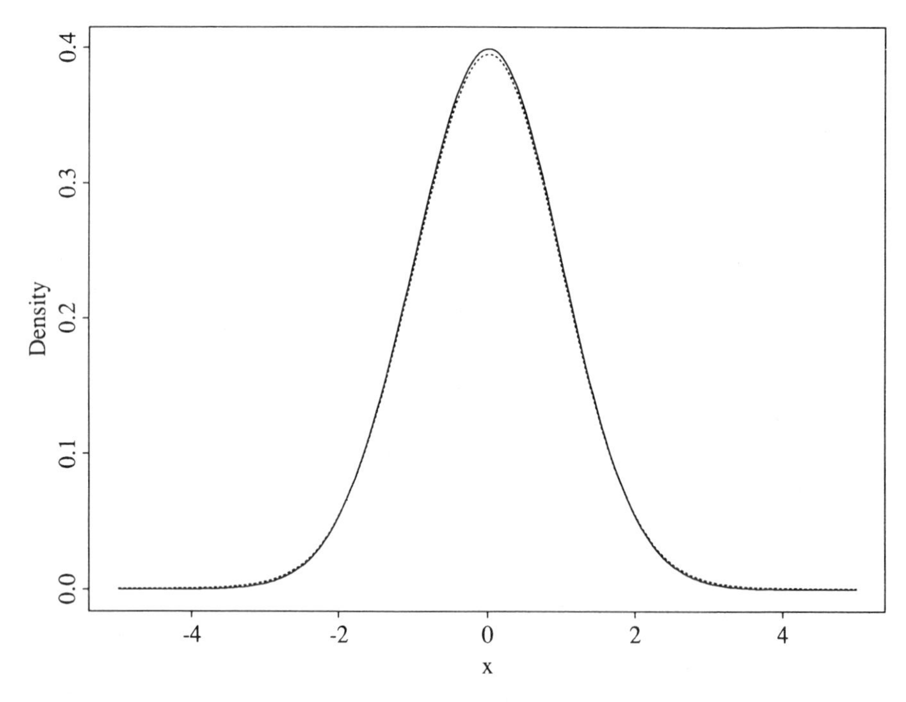

Figure 5.1. The densities of the Gaussian distribution and the 95% Gaussian and 5% Cauchy mixture distribution. After a plot by R. Koenker (personal communication). Used with permission.

robustness can be discussed in terms of the influence function and breakdown point. If we either cannot or decide not to represent the departures from $\mathscr{F}$ by a semiparametric model, we should still try to make a clear distinction between the aspects of the model which are of interest and those which are not and then avoid trying to estimate and make inferences about the additional nuisance parameters.

The core model concept produces an important distinction between robust and nonparametric inference. In robust inference, even if we adopt a semiparametric model, we can still exploit the core model $\mathscr{F}$ to provide compact descriptions of the data (including the variability in the data) and to clarify the objectives of the analysis. To be meaningful, $\mathscr{F}$ should reflect the aspects of the substantive problem of primary interest, incorporate the known and suspected structures in the data, and (in the outlier context) be a genuinely plausible model for the bulk of the data. Thus, for example, in formulating $\mathscr{F}$, we should not ignore information about grouping, regression relationships, and heteroscedasticity because to do so would result in an inappropriately simple model. In this sense at least, robustness is not a panacea for bad modeling and poor statistical practice.

5.4 BOUNDED INFLUENCE ESTIMATION

Suppose that we have observations **z** which, in the absence of outliers, we represent as realizations of **Z** generated by the Gaussian model (5.1). The maximum likelihood estimator of (μ, σ) satisfies the estimating equations

$$\sum_{i=1}^{n} \eta\left(\frac{Z_i - \mu}{\sigma}\right) = 0,$$

where $\eta(x) = (x, x^2 - 1)^{\mathrm{T}}$. The influence functions of the estimators defined by these estimating equations are shown in Figure 5.2; they are clearly continuous but unbounded so the maximum likelihood estimator is not distributionally robust (Section 5.1.7.).

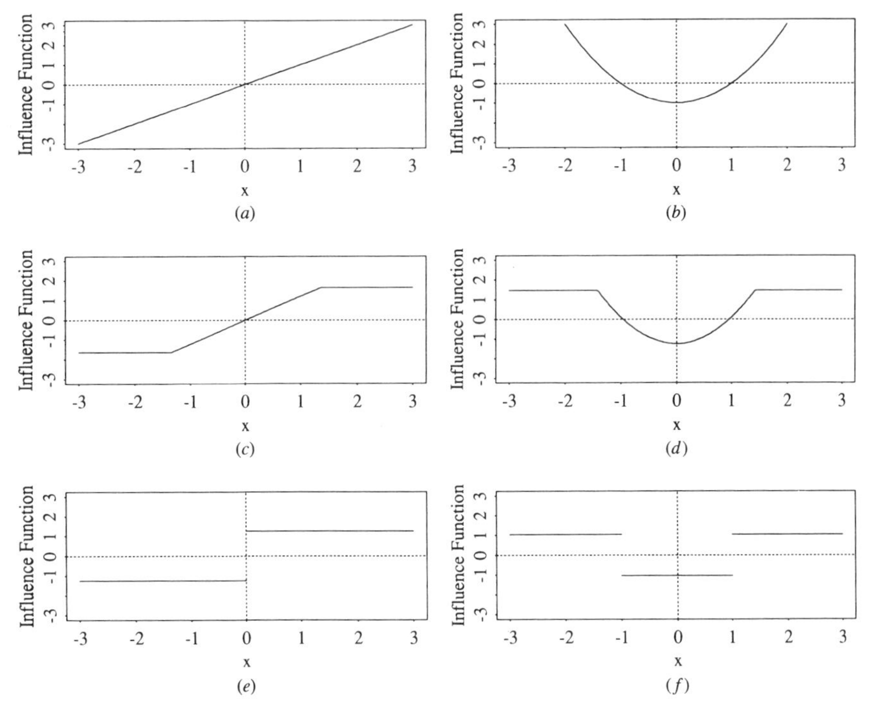

Figure 5.2. Influence functions for Huber's M-estimators at the standard Gaussian distribution: (a) location estimator with $c = \infty$ (sample mean); (b) scale estimator with $k = \infty$ (sample standard deviation); (c) location estimator with $c = 1.35$; (d) scale estimator with $k = 2$; (e) location estimator with $c = 0$ (sample median); (f) scale estimator with $k = 0$ (median absolute deviation from the median).

5.4.1 Huber's M-Estimators

An obvious modification to the Gaussian maximum likelihood estimating equations which produces estimators with continuous, bounded influence functions is obtained by truncating the two components (i.e. modify η so $n(x) = (\psi(x), X(x))^{T}$). The truncated version of the linear function is $\psi(x) = \max[-c, \min(c, x)]$ which is called the *Huber ψ-function.* The second component can be truncated in various ways but an attractively simple method is to set $\chi(x) = \min(k^2, x^2) - \beta(k)$. For further simplicity, we sometimes take $k = c$, in which case we can write $\chi(x) = \psi(x)^2 - \beta(c)$ and the method is called *Huber's proposal 2* since it was the second proposal of Huber (1964).

5.4.2 Fisher Consistency Under the Model

To ensure that we are actually estimating the parameters of the Gaussian model (5.1) when it holds (the first requirement for distributional robustness in Section 5.1.7), we require the estimators to be Fisher consistent under (5.1). Let $\phi(x) = (2\pi)^{-1/2}\exp(-x^2/2)$ denote the standard Gaussian density. Then

$$\int \psi\left(\frac{x-\mu}{\sigma}\right)\frac{1}{\sigma}\phi\left(\frac{x-\mu}{\sigma}\right)dx = \int \psi(x)\phi(x)\,dx = 0$$

because the product of the odd function ψ and the even function ϕ is odd and integrates to 0. Notice that this is true for any bounded, odd ψ and a symmetric core model. It follows that the estimator of μ is Fisher consistent under the Gaussian model. For the estimator of σ to be Fisher consistent, we require

$$\int \chi\left(\frac{x-\mu}{\sigma}\right)\frac{1}{\sigma}\phi\left(\frac{x-\mu}{\sigma}\right)dx = \int \chi(x)\phi(x)\,dx = 0.$$

This obviously holds for the scale estimator if we set

$$\begin{aligned}
\beta(k) &= \int \min(k^2, x^2)\phi(x)\,dx \\
&= \int_{-\infty}^{-k} k^2\phi(x)\,dx + \int_{-k}^{k} x^2\phi(x)\,dx + \int_{k}^{\infty} k^2\phi(x)\,dx \\
&= k^2\{\Phi(-k) + 1 - \Phi(k)\} + 2\Phi(k) - 1 - 2k\phi(k) \\
&= 2k^2\{1 - \Phi(k)\} + 2\Phi(k) - 1 - 2k\phi(k),
\end{aligned}$$

using (4b) in the Appendix.

If the data are actually generated by F_0, the Huber M-estimator $\hat{\theta} = (\hat{\mu}, \hat{\sigma})^{T}$ is estimating $\theta(F_0)$ which is a solution of the equation

$$E_{F_0}\eta(Z, \theta(F_0)) = 0.$$

See Section 4.2.2. We would like to estimate the core model parameter $\theta_0 = \theta(F)$ but we can only estimate $\theta(F_0)$. The bias $\theta(F_0) - \theta_0$ is unknown and cannot be estimated without making the kind of assumptions we want to avoid. The best we can do is to use estimators which estimate θ_0 under the core model, have small mean squared error near the core model and which limit the worst case bias and variance. As we noted in Section 5.1.7, this is achieved by choosing η so $\hat{\theta}$ is Fisher consistent for θ_0 under the core model, the influence function is continuous and bounded near the core model, and by ensuring that the estimator does not break down.

5.4.3 The Influence Function

The influence function of Huber's estimators at an arbitrary distribution G can be obtained from Theorem 4.6. It is

$$\mathrm{IF}(x; G, \hat{\theta}) = B_G(\theta(G))^{-1}\eta(x; \theta(G)), \tag{5.9}$$

where, with $y = \{x - \mu(G)\}/\sigma(G)$

$$B_G(\theta(G)) = -\frac{1}{\sigma(G)}\begin{pmatrix} \int \psi'(y)\, dG(x) & \int y\psi'(y)\, dG(x) \\ \int \chi'(y)\, dG(x) & \int y\chi'(y)\, dG(x) \end{pmatrix}.$$

The asymptotic bias of the estimator at G is

$$\theta(F_0) - \theta_0 \approx B_{F_0}(\theta_0)^{-1}\mathrm{E}_{F_0}\eta(Z, \theta_0)$$

and the asymptotic variance of the estimator at G is

$$V_G(\theta(G)) = n^{-1}B_G(\theta(G))^{-1}A_G(\theta(G))B_G(\theta(G))^{-\mathrm{T}},$$

where, with $y = \{x - \mu(G)\}/\sigma(G)$

$$A_G(\theta(G)) = \begin{pmatrix} \int \psi(y)^2\, dG(x) & \int \psi(y)\chi(y)\, dG(x) \\ \int \psi(y)\chi(y)\, dG(x) & \int \chi(y)^2\, dG(x) \end{pmatrix}.$$

5.4.4 Asymptotic Efficiency Under the Gaussian Model

Under the Gaussian model (5.1), we obtain

$$B_{\Phi}(\theta_0) = -\frac{1}{\sigma}\begin{pmatrix} 2\Phi(c) - 1 & 0 \\ 0 & 2\Phi(k) - 1 - 2k\phi(k) \end{pmatrix},$$

so from (5.9)

$$\mathrm{IF}(x; \Phi, \hat{\theta}) = \left(\frac{\psi(x)}{2\Phi(c) - 1}, \frac{\psi(x)^2 - \beta(k)}{2\Phi(k) - 1 - 2k\phi(k)}\right)^{\mathrm{T}}.$$

Moreover, under the Gaussian model (5.1),

$$A_{\Phi}(\theta_0) = \begin{pmatrix} \beta(c) & 0 \\ 0 & \gamma(k) - \beta(k)^2 \end{pmatrix},$$

where, by (4b) in the Appendix,

$$\begin{aligned} \gamma(k) &= \int_{-\infty}^{k} k^4\phi(x)\,dx + \int_{-k}^{k} x^4\phi(x)\,dx + \int_{k}^{\infty} k^4\phi(x)\,dx \\ &= k^4\{\Phi(-k) + 1 - \Phi(k)\} - x^3\phi(x)|_{-k}^{k} + 3\int_{-k}^{k} x^2\phi(x)\,dx \\ &= 2k^4\{1 - \Phi(k)\} - 2k^3\phi(k) + 3\{2\Phi(k) - 1 - 2k\phi(k)\}, \end{aligned}$$

so the asymptotic variance of Huber's M-estimators under the Gaussian model is

$$n^{-1}\sigma^2\begin{pmatrix} \beta(c)/\{2\Phi(c) - 1\}^2 & 0 \\ 0 & \{\gamma(k) - \beta(k)^2\}/\{2\Phi(k) - 1 - 2k\phi(k)\}^2 \end{pmatrix} \tag{5.10}$$

Since the asymptotic variance of the maximum likelihood estimator of σ under the Gaussian model (5.1) is $\sigma^2/2n$ (see Section 5.1.1), the asymptotic efficiency of the Huber estimator scale under the Gaussian model (5.1) is

$$\frac{2\{2\Phi(k) - 1 - 2k\phi(k)\}^2}{\gamma(k) - \beta(k)^2}.$$

We can compute the asymptotic efficiency for a range of values of k. The function is plotted in Figure 5.3 and selected results of this computation are given in Table 5.1. Thus we see that increasing c and k increases the efficiency of the estimator with full efficiency being achieved at $c = k = \infty$.

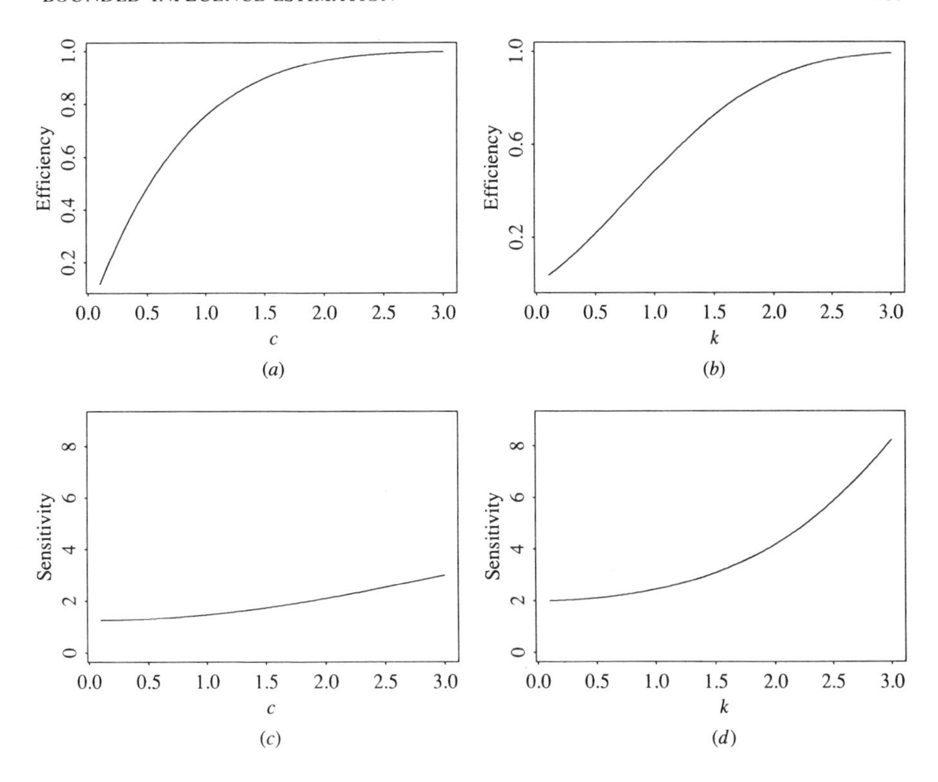

Figure 5.3. Asymptotic efficiencies and gross error sensitivities for Huber's M-estimators at the standard Gaussian distribution as a function of c and k: (*a*) efficiency of the location estimator; (*b*) efficiency of the scale estimator; (*c*) gross error sensitivity of the location estimator; (*d*) gross error sensitivity of the scale estimator.

Table 5.1. Asymptotic Efficiency of the Huber Proposal 2 M-Estimator at the Gaussian Model

c, k	0	0.35	0.7	1	1.35	1.7	2	3	∞
Location	0.67	0.75	0.84	0.90	0.95	0.98	0.99	1	1
Scale	0.37	0.14	0.32	0.48	0.66	0.80	0.89	0.99	1

5.4.5 Gross Error Sensitivity

Next, we need to evaluate the robustness of different choices of c and k. The gross error sensitivity (Section 5.1.6) of $\hat{\sigma}$ under (5.1) is

$$\sup_{x \in \mathbb{R}} |\mathrm{IF}(x; \Phi, \hat{\sigma})| = \frac{k^2 - \beta(k)}{2\Phi(k) - 1 - 2k\phi(k)}$$

$$= \frac{k^2\{2\Phi(k) - 1\}}{2\Phi(k) - 1 - 2k\phi(k)} - 1.$$

Table 5.2. Gross Error Sensitivity of the Huber Proposal 2 M-Estimator at the Gaussian Model

c, k	0	0.35	0.7	1	1.35	1.7	2	3	∞
Location	1.25	1.28	1.36	1.46	1.64	1.87	2.10	3.01	∞
Scale	1.03	2.05	2.20	2.43	2.85	3.45	4.17	8.25	∞

This function is plotted in Figure 5.3 and selected results of this computation are given in Table 5.2. Since the gross error sensitivity increases with c and k, the smaller c and k, the more robust the estimator.

5.4.6 Breakdown Point

We can also consider using the breakdown point to evaluate the robustness of estimators. The breakdown point is unfortunately not simple to calculate but in this instance tells the same story as the gross error sensitivity. Huber (1981, p. 143) shows that the breakdown point for the joint estimation of location and scale is the solution to

$$0 = \left\{\frac{\varepsilon}{1-\varepsilon}\right\}^2 c^2 - \beta(k) + (k^2 - \beta(k))\frac{\varepsilon}{1-\varepsilon}$$

or

$$\begin{aligned} 0 &= \varepsilon^2 c^2 - (1-\varepsilon)^2\beta(k) + (k^2 - \beta(k))\varepsilon(1-\varepsilon) \\ &= \varepsilon^2(c^2 - k^2) + \varepsilon(k^2 + \beta(k)) - \beta(k) \end{aligned}$$

so

$$\varepsilon = \begin{cases} \dfrac{\beta(k)}{\beta(k) + k^2} & \text{if } c = k \\[2ex] \dfrac{-(k^2 + \beta(k)) + \{(k^2 + \beta(k))^2 + 4(c^2 - k^2)\beta(k)\}^{1/2}}{2(c^2 - k^2)} & \text{if } c \neq k. \end{cases}$$

The breakdown point increases as c and k decrease so, in this problem, the information from the breakdown point is similar to that from the gross error sensitivity.

5.4.7 Compromise Estimators

We obviously need to make a tradeoff between efficiency at the model and robustness when the model does not hold. For the Gaussian mean, the usual choice is to take $c = 1.35$ which yields 95% efficiency at the Gaussian model. If we choose $k \neq c$, we would need $k > 2$ to achieve the same efficiency in the scale estimator. The choice $k = c = 1.35$ is relatively inefficient but fairly robust. Since the scale parameter is of interest, we will use $c = 1.35$ for location and $k = 2$ for scale.

There are two obvious approaches to obtaining approximate standard errors for the estimators. First, F_0 is assumed to be nearly Gaussian and our procedure is intended to be insensitive to small departures from the Gaussian model so we can use an estimate of (5.10). Second, we can use the nonparametric estimator introduced in (4.22). The nonparametric nature of this approach makes it the method of choice for most statisticians.

The standard deviations of the three stellar velocity data sets are 47, 88, and 95 while the M-estimators are 37, 67, and 93. We proceed to set approximate 95% confidence intervals for σ by exponentiating the large sample confidence intervals set on the log scale. The intervals produced using the Gaussian core distribution are substantially wider than those produced using the nonparametric variance estimate, which are given below:

Disk:	[28.28, 47.28]
Intermediate:	[57,43, 78.16]
Halo:	[73.40, 118.68].

This analysis suggests that, as shown in Welsh and Morrison (1990), the intermediate stars are more like halo stars than disk stars. However, the robust analysis shows that the evidence for the conclusion is weaker than appears from the analysis based on the sample standard deviation under the assumption of a Gaussian model.

5.4.8 Most Robust Estimators

We obtain estimators with the minimum gross error sensitivity and the maximum breakdown point (equal to 0.5) if we let $c, k \to 0$. The resulting estimators are the *median* and the *median absolute deviation from the median*:

$$\tilde{\mu} = \text{median}(Z_i)$$

$$\tilde{\sigma} = \frac{\text{mad}(Z_i)}{0.6745} = \frac{\text{median}\{|Z_i - \text{median}(Z_j)|\}}{0.6745}.$$

The efficiency of these estimators at the Gaussian model $\mathscr{F}$ is 0.67 and 0.37 respectively. The influence functions of these estimators

$$\text{IF}(x; G, \tilde{\mu}) = \frac{\text{sign}(x - \mu(G))}{2g\{\mu(G)\}}$$

$$\text{IF}(x; G, \tilde{\sigma}) =$$

$$\frac{\frac{1}{2}\,\text{sign}\,(|x - \mu(G)| - \sigma(G)) - [g\{\mu(G) + \sigma(G)\} - g\{\mu(G) - \sigma(G)\}]\,\text{IF}(x; G, \hat{\mu})}{0.6745[g\{\mu(G) + \sigma(G)\} + g\{\mu(G) - \sigma(G)\}]}$$

where $\mu(G) = G^{-1}(\frac{1}{2})$ and $\sigma(G) = H^{-1}(\frac{1}{2})$, with $H(x) = G(\mu(G)+x) - G(\mu(G)-x)$, are plotted at the Gaussian model in Figure 5.2. The estimating equations for the median and the median absolute deviation from the median are not continuous, so we cannot simply derive the influence functions by applying the results of Section 5.4.3. However, if G is smooth and strictly increasing near its median and quartiles, the arguments can be modified to produce the above results. The problem is that rounding or grouping effects mean that G has flat regions over which g equals 0 and the influence functions can be unbounded. This is the primary reason for requiring the influence functions to be smooth, and also illustrates the importance of considering the influence function at more than the core model.

5.4.9 Model Generated Estimators

Instead of modifying the maximum likelihood estimating equations from the core model to construct robust estimators, a different approach is to adopt a third model and then use the maximum likelihood estimators from this model as estimators of the parameters in $\mathscr{F}$. Thus, for a general location-scale model (1.18), we have the log-likelihood

$$\log\{f(z;\mu,\sigma)\} = -n\log(\sigma) + \sum_{i=1}^{n}\log\left\{h\left(\frac{Z_i-\mu}{\sigma}\right)\right\}$$

and hence the estimating equations

$$\sum_{i=1}^{n}\eta\left(\frac{Z_i-\mu}{\sigma}\right) = 0,$$

where $\eta(x) = -(h'(x)/h(x), xh'(x)/h(x)+1)^{\mathrm{T}}$. For the Gaussian model, $h'(x)/h(x) = -x$ and we recover the maximum likelihood estimators. However, if $h'(x)/h(x)$ and $xh'(x)/h(x)$ have smooth derivatives and are bounded, then after modifying the equations if necessary to ensure Fisher consistency at $\mathscr{F}$, the resulting estimators are robust estimators for $\mathscr{F}$. Notice that the requirement that $xh'(x)/h(x)$ be bounded means that $h'(x)/h(x) = O(x^{-1})$ as $|x| \to \infty$. Such estimators are called *redescending M-estimators.*

A classical choice is to use the Student t model with fixed, small degrees of freedom v. Here $h(x) \propto \{1 + x^2/v\}^{-(v+1)/2}$ so that $h'(x)/h(x) = -(v+1)x/\{v+x^2\}$ and $\eta(x) = ((v+1)x/\{v+x^2\},\ (v+1)x^2/\{v+x^2\} - (v+1)\delta)^{\mathrm{T}}$, where

$$\delta = \int \frac{x^2}{v+x^2}\,\phi(x)\,dx.$$

Numerical integration for $v = 3$ yields $\delta \approx 0.1899$. The influence functions are plotted in Figure 5.4. The location estimator has a redescending influence function while the scale estimator does not.

Redescending estimators are treated in exactly the same way as other M-estimators except that more care has to be taken when solving the estimating equations. It is important to note here that the Student t model is not a model we believe in; it is simply used to generate estimating equations.

5.4.10 Other Classes of Estimators

Finally, it is worth noting that we can explore the robustness properties of any estimator or class of estimators. Useful classes of estimators which have been considered are *L-estimators* (estimators which are linear combinations of order statistics—see Section 6.9.2), *R-estimators* (estimators based on ranks – see Section 6.9.1), *P-estimators* (Pitman estimators – see Section 3.1.4 – Johns, 1979; Huber, 1984), *minimum distance estimators* (see for example Knüsel, 1969; Parr and Schucany, 1980; Donoho and Liu, 1988), and *trimmed likelihood estimators* (Bednarski and Clarke, 1993), etc.

PROBLEMS

5.4.1. Suppose that we have independent observations $\mathbf{Z}$ for which we entertain the Gaussian model $\mathscr{F} = \{f(\mathbf{y};\theta) = \prod_{i=1}^{n} \phi(y_i - \theta) \colon \theta \in \mathbb{R}\}$. It is obviously a simplification to treat the variance as known but this avoids some extra complications without reducing the central issues. Consider the estimator θ^* of θ which is obtained by solving the equation

$$\sum_{i=1}^{n} \{\Phi(Z_i - \theta) - \tfrac{1}{2}\} = 0,$$

where $\Phi(x) = (2\pi)^{-1/2} \int_{-\infty}^{x} \exp(-y^2/2)\, dy$ is the standard Gaussian distribution function. Obtain the influence function of the estimator and make a rough sketch of it. Discuss whether or not the estimator is robust. Show that the asymptotic variance of the estimator θ^* at the model $\mathscr{F}$ is $1/12(E\phi(Z - \theta_0))^2$, where $\phi(x) = \Phi'(x)$ is the standard normal density. (Hint: $\Phi(Z - \theta_0) \sim$ uniform(0, 1) under $\mathscr{F}$.) Show that the asymptotic efficiency of the estimator θ^* under $\mathscr{F}$ is $3/\pi$. Would you use the estimator θ^* to estimate the mean of a normal model? Explain.

5.4.2. Suppose that we have independent observations $\mathbf{Z}$ which we assume follow the Gaussian model of Problem 5.4.1. In 1888, R.H. Smith (see Stigler, 1980) proposed the estimator θ^* of θ which is obtained by solving

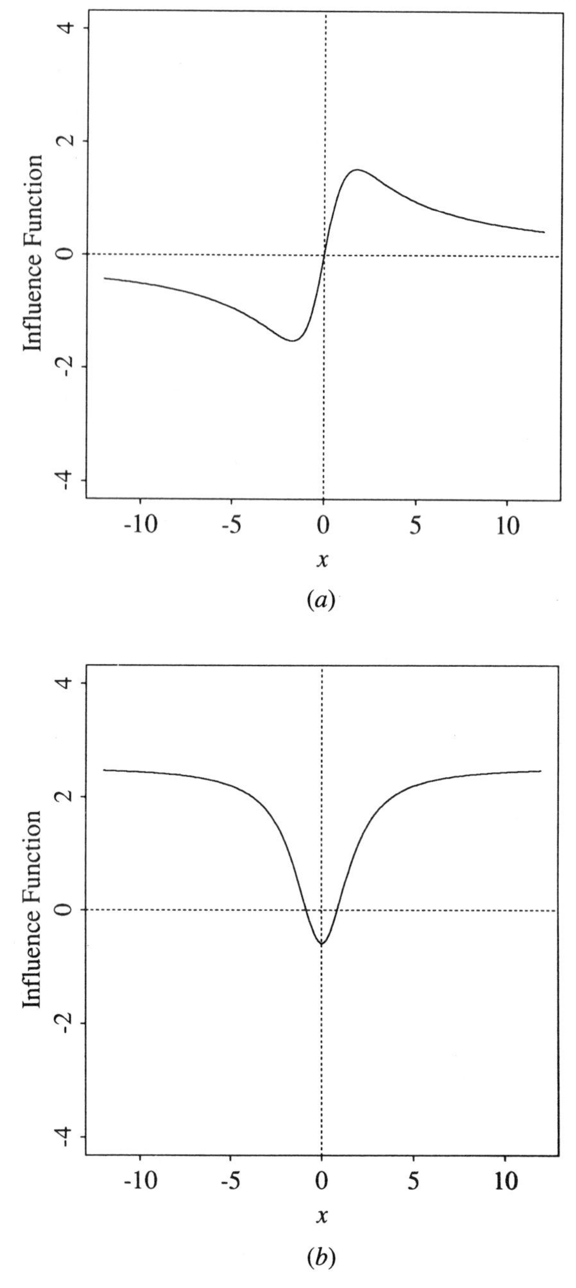

Figure 5.4. Influence functions for Student's t_3 M-estimators at the standard Gaussian distribution: (a) location estimator; (b) scale estimator.

the equation

$$\sum_{i=1}^{n} \psi(Z_i - \theta) = 0,$$

where $\psi(x) = x(r^2 - x^2)I(|x| \leq r)$, for some fixed $r > 0$. Discuss whether you would use Smith's estimator θ^* (for some value of r) to estimate the mean of a normal model. Be sure to include a discussion of the effect of different choices of r and to comment on the way the estimator treats extreme outliers. Write down an estimator of the asymptotic variance of Smith's estimator and use it to construct an approximate 95% confidence interval for θ.

5.4.3. Compute the asymptotic efficiency and the gross error sensitivity at the Gaussian model of the location M-estimator based on $\psi_c(x) = x(c^2 - x^2)I(|x| \leq c)$.

5.4.4. Suppose that we have independent observations $\mathbf{Z}$ with distribution function G and that we entertain the model

$$\mathscr{F} = \left\{ f_\sigma(\mathbf{y}) = \prod_{i=1}^{n} \frac{1}{\sigma} \exp\left(-\frac{y_i}{\sigma}\right), y_i > 0: \sigma > 0 \right\}$$

for the data. Find the influence function of the maximum likelihood estimator of σ under $\mathscr{F}$ and then discuss the robustnesss properties of this estimator. Now consider the estimator obtained by solving

$$\sum_{i=1}^{n} \chi\left(\frac{Z_i}{\sigma}\right) = 0,$$

where, for $k > 0$,

$$\chi(x) = \begin{cases} x^2 - 2 + 2(k+1)\,e^{-k} & x \leq k \\ k^2 - 2 + 2(k+1)\,e^{-k} & x > k. \end{cases}$$

The influence function of this estimator is

$$\mathrm{IF}(x, G) = \frac{\sigma\chi\left(\dfrac{x}{\sigma}\right)}{\mathrm{E}_G\left\{\dfrac{Z_1}{\sigma}\chi'\left(\dfrac{Z_1}{\sigma}\right)\right\}}.$$

Make a rough sketch of this function and discuss whether or not the estimator is robust. Show that $\mathrm{E}_G\chi(Z_1/\sigma) = 0$ at the model. Write down the asymptotic variance of the estimator and construct an estimator of this variance. Show how to set an approximate $100(1-\alpha)\%$ confidence interval for σ.

5.4.5. Apply the estimator of Problem 5.4.4 to the pressure vessel data from Section 1.1.2 and Problem 1.6. Compare the results to those obtained using exponential theory.

5.4.6. Suppose that we have n independent observations $Z_1, \ldots, Z_n$ which we believe follow the Poisson model

$$\mathscr{F} = \left\{ f_\theta(\mathbf{y}) = \prod_{i=1}^{n} \theta^{y_i} \exp(-\theta)/y_i!, \, y_i = 0, 1, \ldots : \theta > 0 \right\}.$$

Find the influence function of the maximum likelihood estimator and show that the estimator is not robust. Now consider the estimator $\hat{\theta}$ which satisfies

$$\sum_{i=1}^{n} \eta_\theta\left(\frac{Z_i}{\theta} - 1\right) = 0,$$

where, for $k > 0$ and some function $b(k, \theta)$,

$$\eta_\theta(x) = \begin{cases} x - b(k, \theta) & |x| \le k \\ k - b(k, \theta) & |x| > k. \end{cases}$$

Write down the equation which defines $b(k, \theta)$ so that $\hat{\theta}$ is consistent at the model. Outline an algorithm which could be used to compute $\hat{\theta}$.

5.4.7. We can calculate M-estimates of location and scale by alternating between calculating scale for a given location and location for a given scale. If we write $\chi(x) = \chi^*(x) - \mathrm{E}_\Phi \chi^*(Z)$, we can implement the scale step which updates from $(\mu_{(m+1)}, \sigma_{(m)})$ to $\sigma_{(m+1)}$ by computing

$$\sigma_{(m+1)} = \frac{\sigma_{(m)}}{n\mathrm{E}_\Phi \chi^*(e)} \sum_{i=1}^{n} \chi^*\left(\frac{z_i - \mu_{(m+1)}}{\sigma_{(m)}}\right).$$

Show that the solution satisfies the estimating equations. Describe the Newton–Raphson algorithm for the same step. Which method is preferable? Explain.

5.4.8. Suppose that we have independent observations $\mathbf{Z}$ which we assume follow the Gaussian model (5.1). Modify Smith's estimator (Problem 5.4.2) to allow for the fact that σ in unknown and construct a Huber proposal 2 estimator of σ. Draw a rough sketch of the influence functions. How does this estimator treat extreme outliers? What happens to the asymptotic efficiency as r increases? Discuss briefly the considerations involved in choosing a value of r. Write down an estimator of the asymptotic variance of Smith's estimator and use it to construct an approximate 95% confidence interval for μ.

5.4.9. Use the Huber M-estimator and the Tukey bisquare estimator for which $\psi(r) = r(c^2 - r^2)^2 I(|r| \le c)$ to analyse the effect of caffeine on the volume of urine data of Section 1.1.3. Compare your inferences about the mean change in the volume of urine to those obtained assuming the Gaussian model holds. Interpret your results.

5.5 CORROSION RESISTANCE OF STEEL PLATES

In Section 1.2.5 we discussed modeling the relationship between the weight loss of enamel covered steel plates subjected to 10% hydrochloric acid at a set temperature for a set time. We noted that in the absence of outliers, we might consider an initial plausible model of the form

$$\mathscr{F} = \left\{ f(\mathbf{y}; \gamma, \sigma) = \prod_{i=1}^{n} \frac{1}{\sigma} \phi\left(\frac{y_i - x_i^{\mathrm{T}}\gamma}{\sigma}\right), -\infty < y_i < \infty : \gamma \in \mathbb{R}^p, \sigma > 0 \right\}, \quad (5.11)$$

where $x_i = (1, z_i^{\mathrm{T}})^{\mathrm{T}}$, and $\gamma = (\alpha, \beta^{\mathrm{T}})^{\mathrm{T}}$. In the steel plate problem, the response y_i is the log weight loss, x_{1i} is the time, x_{2i} is the temperature, $z_i = (x_{1i}, x_{2i})^{\mathrm{T}}$, and $\beta = (\beta_1, \beta_2)^{\mathrm{T}}$. The slope parameters β are the parameters of interest. To preserve notation, let $\theta = (\gamma^{\mathrm{T}}, \sigma)^{\mathrm{T}}$. We assume without loss of generality that we have centered the covariates so that $\bar{z} = 0$.

5.5.1 The Gaussian Model Analysis

The log-likelihood function for the Gaussian model (5.11) is

$$\log f(\mathbf{z}; \theta) = -n \log(\sigma) - \sum_{i=1}^{n} \frac{(y_i - x_i^{\mathrm{T}}\gamma)^2}{2\sigma^2}$$

so the maximum likelihood estimator of θ satisfies the estimating equations

$$\sum_{i=1}^{n} \eta\left(x_i, \frac{y_i - x_i^{\mathrm{T}}\gamma}{\sigma}\right) = 0,$$

where $\eta(x, r) = (x^{\mathrm{T}}r, r^2 - 1)^{\mathrm{T}}$. The maximum likelihood estimator of γ is the same as the least squares estimator of γ (Section 1.5.4).

Residual and qq-plots of the residuals from the Gaussian maximum likelihood/least squares fit to the corrosion resistance data are shown in Figure 5.5. As expected from the preliminary analysis of Section 1.2.5, these plots show that there are a number of outliers in the data. The outliers correspond to plates which lost more weight than the other plates subjected to the same conditions of time and temperature. The residual distribution is asymmetric with a long right tail. There may also be some evidence of curvature and/or nonconstant

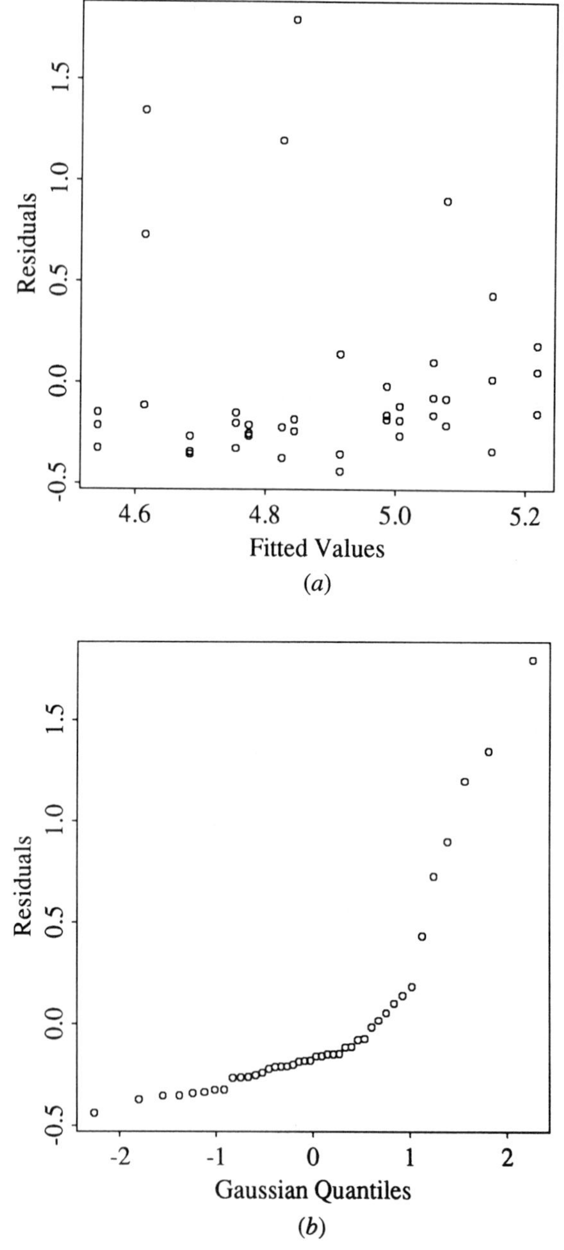

Figure 5.5. Least squares diagnostics for the corrosion resistance data: (*a*) residual plot; (*b*) Gaussian qq-plot.

Table 5.3. Least Squares Fit to the Corrosion Data

	Estimate	Standard Error	Ratio
Time	0.0354	0.0289	1.2243
Temperature	0.0116	0.0049	2.3602

variation for the bulk of the data in the residual plot though this is difficult to judge in the presence of outliers.

For later comparison, the least squares slope estimates, standard errors and the ratios of the parameter estimates to their standard errors are given in Table 5.3.

Notice that, if the Gaussian model (5.11) held (as we might mistakenly believe if we were to ignore the outliers), the ratios would have Student t distributions with $42 - 3 = 39$ degrees of freedom under the null hypothesis that the corresponding coefficient in the model equals 0. This means that in this analysis, the coefficient of temperature would be significant but that of time would not.

5.2.2 Huber's *M*-estimators for Regression

The functions $\eta(x, r) = (x^{\mathrm{T}}r, r^2 - 1)^{\mathrm{T}}$ are clearly continuous but unbounded in x and r. As in Section 5.4.1, we can obtain continuous, bounded versions of these functions by truncating the components of η. In the present example, the values of the explanatory variable are known so we do not need to bound the estimating equations in x. We can therefore adopt the approach of Section 5.4 directly and consider the estimator based on $\eta(x, r) = (x^{\mathrm{T}}\psi(r), \chi(r))^{\mathrm{T}}$, where $\psi(r) = \max[-c, \min(c, r)]$ is the Huber ψ-function and $\chi(x) = \min(k^2, x^2) - \beta(k)$.

The arguments applied in Section 5.4.2 establish that the estimator based on $\eta(x, r) = (x^{\mathrm{T}}\psi(r), \chi(r))^{\mathrm{T}}$ is consistent at the Gaussian model. The influence function of the estimator at a distribution G is

$$\mathrm{IF}(y, x); G, (\hat{\gamma}, \hat{\sigma})) = B_G(\theta(G))^{-1}\eta\left(x, \frac{y - x^{\mathrm{T}}\gamma(G)}{\sigma(G)}\right),$$

where, with $u = \{y - \alpha(G)\}/\sigma(G)$,

$$B_G(\theta(G)) = -\frac{1}{\sigma(G)}\begin{pmatrix} \int \psi'(u)\, dG(y) & 0 & \int u\psi'(u)\, dG(y) \\ 0 & \int \psi'(u)\, dG(y) n^{-1} \sum_{i=1}^{n} z_i z_i^{\mathrm{T}} & 0 \\ \int \chi'(u)\, dG(y) & 0 & \int u\chi'(u)\, dG(y) \end{pmatrix}$$

because we have centered the covariates so that $\bar{z} = 0$. It follows that the asymptotic variance of the estimator under G is

$$\mathrm{n}^{-1} B_G(\theta(G))^{-1} A_G(\theta(G)) B_G(\theta(G))^{-\mathrm{T}},$$

where, with $u = \{y - \alpha(G)\}/\sigma(G)$,

$$A_G(\theta(G)) = \begin{pmatrix} \int \psi(u)^2 \, dG(y) & 0 & \int \psi(u)\chi(u) \, dG(y) \\ 0 & \int \psi(u)^2 \, dG(y) n^{-1} \sum_{i=1}^{n} z_i z_i^{\mathrm{T}} & 0 \\ \int \psi(u)\chi(u) \, dG(y) & 0 & \int \chi(u)^2 \, dG(y) \end{pmatrix}$$

Centering so that $\bar{z} = 0$ ensures that the slope parameters β are orthogonal to (α, σ) and that the M-estimators of the slope parameters are Fisher consistent for the slope parameters in the core model even when G is an asymmetric distribution. That is, $\theta(G) = (\alpha(G), \beta_0^{\mathrm{T}}, \sigma(G))^{\mathrm{T}}$ for any G (Carroll and Welsh, 1988). The asymptotic covariance matrix $n^{-1} B_G(\theta(G))^{-1} A_G(\theta(G)) B_G(\theta(G))^{-\mathrm{T}}$ is block-diagonal only if $\int u\psi'(u) \, dG(y) = \int \chi'(u) \, dG(y) = \int \psi(u)\chi(u) \, dG(y) = 0$ (see for example Ruppert and Aldershof, 1989) which holds for the usual choices of ψ and χ when G is a symmetric distribution. Otherwise, the distribution of the intercept and the scale are asymptotically correlated.

Under the Gaussian model, the asymptotic variances of $\hat{\gamma} = (\hat{\alpha}, \hat{\beta}^{\mathrm{T}})^{\mathrm{T}}$ and $\hat{\sigma}$ are

$$\sigma^2 \frac{\beta(c)}{(2\Phi(c) - 1)} \left(n^{-1} \sum_{i=1}^{n} x_i x_i^{\mathrm{T}} \right)^{-1} \quad \text{and} \quad \frac{\sigma^2(\gamma(k) - \beta(k))^2}{(2\Phi(k) - 1 - 2k\phi(k))^2}$$

respectively, so the asymptotic efficiency calculations of Section 5.4.4 are applicable with the location parameter μ replaced by the regression parameter θ. The influence functions of $\hat{\gamma}$ and $\hat{\sigma}$ at the Gaussian model are

$$\frac{1}{2\Phi(c) - 1} \left(n^{-1} \sum_{i=1}^{n} x_i x_i^{\mathrm{T}} \right)^{-1} x \psi \left(\frac{y - x^{\mathrm{T}} \gamma_0}{\sigma_0} \right)$$

and

$$\frac{1}{2\Phi(k) - 1 - 2k\phi(k)} \chi \left(\frac{y - x^{\mathrm{T}} \gamma_0}{\sigma_0} \right)$$

respectively, so the gross error sensitivity calculations of Section 5.4.5 need to be modified to incorporate the (constant) vector $(n^{-1} \sum_{i=1}^{n} x_i x_i^{\mathrm{T}})^{-1} x$ but are otherwise also applicable.

The choice of c and k involves a tradeoff between efficiency at the model and robustness when the model does not hold. For the regression parameter γ, the usual choice is to take $c = 1.35$ which yields 95% efficiency at the Gaussian model. Since σ is a nuisance parameter in this problem, we may choose to make the efficiency/robustness tradeoff differently. A number of studies (Holland and Welsch, 1977; Hill and Holland, 1977; Denby and Mallows, 1977) have shown that good results for estimating θ are obtained if we estimate the scale using the standardized median absolute deviation which corresponds to $\chi(r) = \text{sign}\{|r| - 0.6745\}$. Essentially, the low efficiency of the scale estimate at the Gaussian model is compensated for by its maximal robustness against deviations from the Gaussian model. This choice is only attractive if we are interested in β, because then we do not have to estimate the asymptotic variance of $\hat{\sigma}$ in order to make inferences about $\hat{\beta}$ and the inefficiency of $\hat{\sigma}$ does not affect the efficiency of $\hat{\beta}$. This is not the case in general if we are interested in α but this problem is less common.

5.5.3 Iteratively Reweighted Least Squares

As pointed out in Section 4.2.6, estimating equations can be solved using general methods such as the Newton–Raphson method. However, the structure of the regression problem suggests an alternative *iteratively reweighted least squares* approach. If we rewrite the estimating equations for the regression parameter as

$$0 = \sum_{i=1}^{n} x_i \psi\left(\frac{y_i - x_i^{\mathrm{T}} \gamma}{\sigma}\right) = \sum_{i=1}^{n} x_i w_i(\theta) \frac{(y_i - x_i^{\mathrm{T}} \gamma)}{\sigma},$$

where $w_i(\theta) = \psi\{(y_i - x_i^{\mathrm{T}} \gamma)/\sigma\}/\{(y_i - x_i^T \gamma)/\sigma\}$, and use estimated weights based on $\theta_{(m)} = (\gamma_{(m)}^{\mathrm{T}}, \sigma_{(m)})^{\mathrm{T}}$, we can update $\gamma_{(m)}$ using

$$\gamma_{(m+1)} = \left\{\sum_{i=1}^{n} x_i w_i(\theta_{(m)}) x_i^{\mathrm{T}}\right\}^{-1} \sum_{i=1}^{n} x_i w_i(\theta_{(m)}) y_i$$

and $\sigma_{(m)}$ by recomputing the median absolute deviation explicitly for each $\gamma_{(m)}$. Iteration of this procedure until $n^{-1} \sum_{i=1}^{n} \eta\{x_i, (y_i - x_i^{\mathrm{T}} \gamma_{(m)})/\sigma_{(m)}\} \approx 0$ defines a sequence $\{\gamma_{(m)}, \sigma_{(m)}\}$ which converges to a root of the estimating equations.

5.5.4 Analysis Based on Huber's *M*-estimators

The residual and qq-plot from the fit based on the Huber estimator with the median absolute deviation are shown in Figure 5.6. All the residuals, including the outliers, are shown in these plots. The residuals which are truncated in the fit are truncated to lie on the dotted horizontal line nearest to them and this is indicated by the arrows pointing to these lines. The residuals from the bulk of the data lie between these two lines. To make it easier to study, the region between the two lines in these plots is replotted on a larger scale in Figure 5.7.

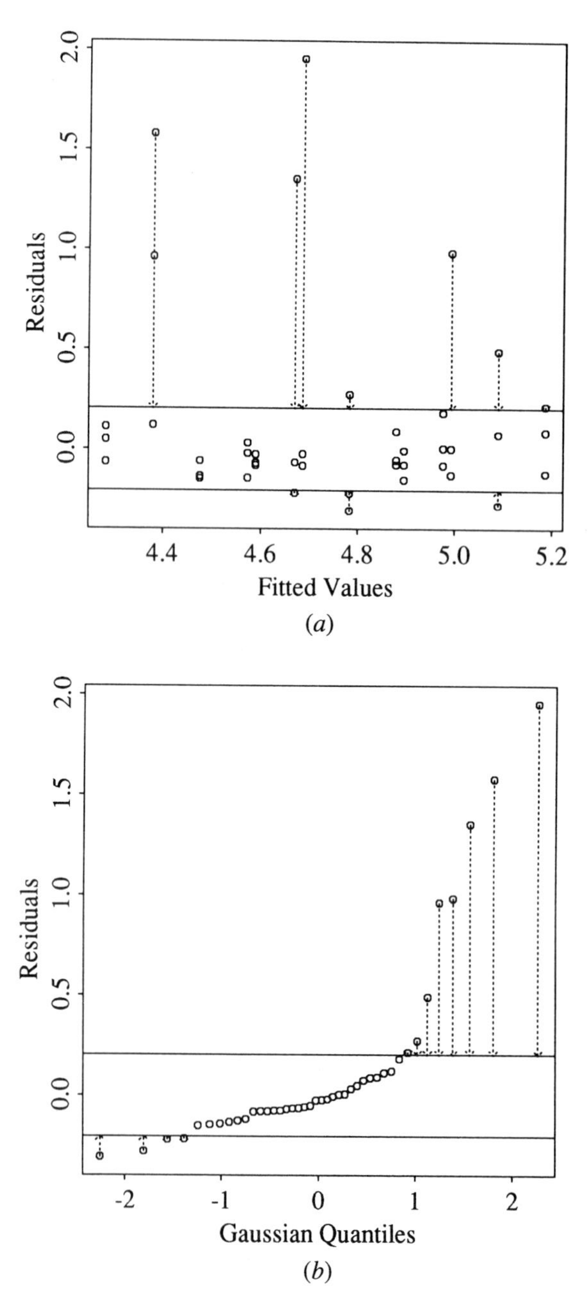

Figure 5.6. Huber M-estimator diagnostics for the corrosion resistance data: (*a*) residual plot; (*b*) Gaussian qq-plot.

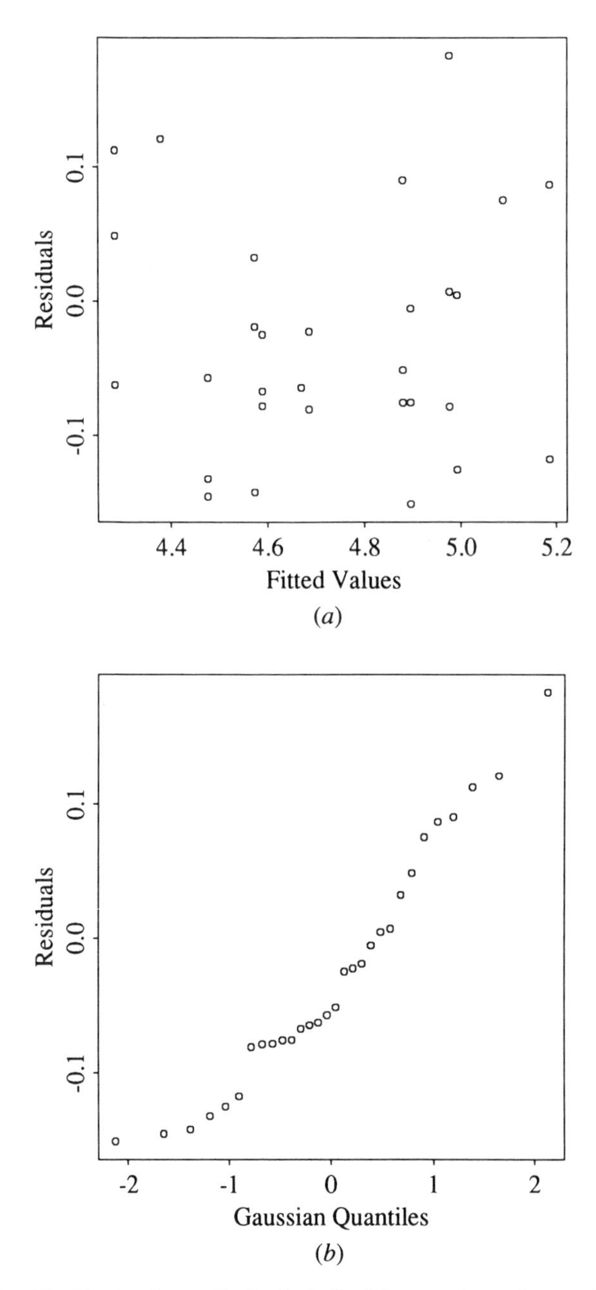

Figure 5.7. Huber M-estimator diagnostics for the bulk of the corrosion resistance data: (*a*) residual plot; (*b*) Gaussian qq-plot.

Table 5.4. Huber M-estimator Fit to the Corrosion Data

	Estimate	Standard Error	Ratio
Time	0.0484	0.0111	4.3810
Temperature	0.0153	0.0019	8.1748

The plots in Figure 5.7 could be obtained directly by plotting $(\hat{y}_i, \psi(r_i))$ instead of $(\hat{y}_i, r_i)$ and the qq-plot of $\psi(r_i)$ instead of r_i except that the inclusion of the truncated points in the qq-plot introduces some distortion so it is also useful to examine a qq-plot of the untruncated bulk of the data. There is perhaps a hint of curvature in the residual plot, but otherwise the plots suggest that the fit is reasonable. The use and interpretation of residuals from robust fits have recently been discussed by Cook and Weisberg (1989) and McKean et al. (1993).

The M-estimator slope parameter estimates, standard errors and the ratios of the parameter estimates to their standard errors are given in Table 5.3.

The slope parameter estimates are slightly larger than in the least squares fit but the standard errors are substantially smaller so the ratios have increased. If we use the large sample Gaussian approximation or the more conservative Student t approximation, we now find that the slope parameters are both significantly different from zero. Thus the results of this analysis are different from those obtained from the least squares analysis.

5.5.5 Analysis Based on Tukey's Bisquare Estimator

In the analysis of the steel plate data in Section 5.5.4 based on the Huber M-estimator, it is notable that the outliers are downweighted but not to zero. In contrast, redescending M-estimators such as that based on *Tukey's bisquare* for which $\psi(r) = r(c^2 - r^2)^2 I(|r| \leq c)$ give extreme observations zero weight. This is illustrated in the residual and qq-plots for this fit with the usual values of $c = 4.685$ shown in Figure 5.8. The standard errors are slightly smaller than in the Huber fit but the results are qualitatively similar.

Table 5.5. Tukey Biweight-estimator Fit to the Corrosion Data

	Estimate	Standard Error	Ratio
Time	0.0505	0.0086	5.8950
Temperature	0.0159	0.0015	10.9559

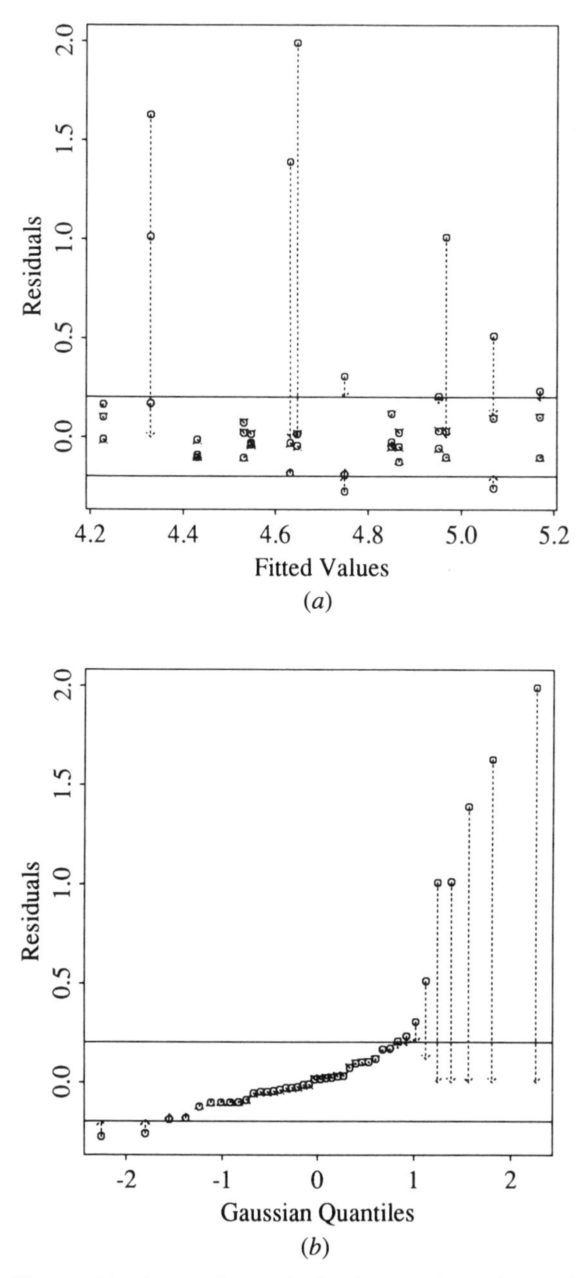

Figure 5.8. Tukey bisquare M-estimator diagnostics for the corrosion resistance data: (*a*) residual plot; (*b*) Gaussian qq-plot.

5.5.6 Generalized *M*-estimators

If the covariates are stochastic, we may encounter moderate errors in all the covariates (the errors-in-variables problem, see for example Fuller, 1987), gross errors in the covariates at some points, or both. Both of these problems are difficult to treat because they are underspecified and because in many respects they are more naturally multivariate than regression problems. The gross error problem is arguably closer to the classical framework, but is complicated by the difficulty of specifying realistic core models for the covariates, particularly in high dimensions.

The available approaches (Hampel et al., 1986, Chapter 6) have ignored this difficulty and simply tried to ensure that the estimating equation is bounded in both x and r by taking $\eta(x, r) = (u(x)x^{\mathrm{T}}\psi(r/v(x)), \chi(r))^{\mathrm{T}}$, for appropriate choices of $u(\cdot)$ and $v(\cdot)$. We obtain Huber type estimators with $u(x) = v(x) = 1$, Mallow's estimators with $u(x) = w(x)$, $v(x) = 1$, Andrews' estimators with $u(x) = 1$, $v(x) = w(x)$, Hill and Ryan's estimators with $u(x) = w(x)$, $v(x) = 1/w(x)$, and Schweppe's estimators with $u(x) = w(x)$, $v(x) = w(x)$.

The weights $w(x)$ can be obtained as $w(x) = \min\{1, b/\{z - \hat{\mu}_z)^{\mathrm{T}}\hat{C}_z^{-1}(z - \hat{\mu}_z)\}^{1/2}\}$, where $x = (1, z^{\mathrm{T}})^{\mathrm{T}}$, b is a tuning constant, $\hat{\mu}_z$ and $\hat{C}_z$ are estimates of the location and variance matrix of z, possibly but not necessarily defined as M-estimators (see Hampel et al., 1986, pp. 319–22 and Simpson et al., 1992, for some of the possibilities). Less satisfactory proposals include $w(x_i) = (1 - h_{ii})^{1/2}$, where $h_{ii} = x_i^{\mathrm{T}}\{\sum_{i=1}^{n} x_i x_i^{\mathrm{T}}\}^{-1} x_i$ is the leverage of the ith observation, and even simpler componentwise methods. In practice, these methods often need to be modified to allow different treatment of fixed and stochastic covariates in the same model.

5.5.7 High Breakdown Estimators

We have avoided discussing the breakdown properties of the bounded influence estimators partly because they depend on assumptions about the covariates and partly because they are disappointingly poor. In particular, the breakdown point of an equivariant M-estimator is bounded above by $1/(p + 1)$ (Hampel et al., 1986, pp. 296–99). This difficulty can be overcome by the class of *S-estimators* (Rousseeuw and Yohai, 1984) which are solutions to

$$\min s(\gamma),$$

where $s(\gamma)$ satisfies the equation $n^{-1}\sum_{i=1}^{n} \rho\{(y_i - x_i^{\mathrm{T}}\gamma)/s\} - K = 0$, with $K = \int \rho(z)\phi(z)\,dz$. Under regularity conditions, S-estimators have the same asymptotic distribution as M-estimators based on $\psi = \rho'$ but very different breakdown properties. To achieve high breakdown, $\psi = \rho'$ must redescend as in the Tukey bisquare estimator. In this case, $\rho'(r) = r(c^2 - r^2)^2 I(|r| \leq c)$ and we achieve 50% breakdown (and 28.7% efficiency at the Gaussian model) with $c = 1.547$ and $K = 0.1995$. More reasonably we achieve 25% breakdown (and 75.9% efficiency at the Gaussian model) with $c = 2.937$ and $\mathrm{K} = 0.3593$. The fact that the stochastic properties of S-estimators and M-estimators are the same but the breakdown properties are not (a phenomenon which also occurs with M-estimators and one-step M-estimators, see Simpson et al., 1992) emphasizes that breakdown is not a stochastic property.

PROBLEMS

5.5.1. Use robust methods to explore the relationship between the change in total catecholamine and the change in the volume of urine produced by the ingestion of caffeine (Section 1.1.3) and compare your results to those obtained using least squares methodology. What do you conclude?

5.5.2. Compare the analysis of the data from the second enamel covered steel plate experiment in Problem 1.5.9 using least squares and robust methods.

5.5.3 Suppose that we have independent observations $Z_1, \ldots, Z_n$ which we assume follow the Pareto model

$$\mathscr{F} = \left\{ f(\mathbf{y}; \lambda) = \prod_{i=1}^{n} \frac{\kappa \lambda^{\kappa}}{y_i^{\kappa+1}}, y_i > \lambda: \kappa, \lambda > 0 \right\}.$$

Suppose initially that λ is known. Find the influence function of the maximum likelihood estimator of κ and show that the estimator is not robust. Define a robust estimator of κ and show how you would use it to obtain a 95% confidence interval for κ. What modifications should we make if λ is unknown?

5.5.4. Suppose that we have independent observations $\mathbf{Z}$ with distribution function F_0 and that we entertain the model

$$\mathscr{F} = \left\{ f(\mathbf{z}; \kappa, \lambda) = \prod_{i=1}^{n} \frac{1}{\Gamma(\kappa)} \lambda^{\kappa} z_i^{\kappa-1} \exp(-\lambda z_i), z_i > 0: \lambda > 0 \right\}$$

for the data. Show how to modify the method of moments estimators to construct bounded influence estimators of λ and κ. Discuss the choice of any tuning constants you need to specify and show how to set an approximate confidence interval for κ.

5.5.5. Suppose that

$$y_i = x_i^{\mathrm{T}} \gamma + \sigma e_i, \qquad \gamma \in \mathbb{R}^p, \quad \sigma > 0,$$

with $\{e_i\}$ independent and identically distributed random variables with common distribution F_0. Assuming that the estimator $\hat{\gamma}$ defined by

$$\min s(\gamma),$$

where $s(\gamma)$ satisfies the equation $n^{-1}\sum_{i=1}^{n}\rho\{(y_i - x_i^{\mathrm{T}}\gamma)/s\} - K = 0$, with $K = \int \rho(z)\phi(z)\,dz$, is consistent for γ_0, obtain a large sample approximation to the sampling distribution of $\hat{\gamma}$.

5.6 TESTS BASED ON *M*-ESTIMATORS

One way to explore the possibility of curvature in the relationship between log weight loss, time and temperature is to expand the model to include quadratic terms so that $x_i = (1, (x_{1i} - \bar{x}_1)^2, (x_{2i} - \bar{x}_2)^2, (x_{1i} - \bar{x}_1)(x_{2i} - \bar{x}_2), x_{1i} - \bar{x}_1, x_{2i} - \bar{x}_2)^{\mathrm{T}}$, $\gamma = (\alpha, \beta_1, \beta_2, \beta_3, \beta_4, \beta_5)^{\mathrm{T}}$ and then test the hypothesis $\mathrm{H}_0\colon \beta_1 = \beta_2 = \beta_3 = 0$ against the general alternative.

In terms of the general multiple regression model, the null hypothesis can be expressed as $\mathrm{H}_0\colon H^{\mathrm{T}}\theta(F_0) = 0$, where $H^{\mathrm{T}} = (0, I_q, 0)$ (with $q = 3$) and $\theta(F_0) = (\alpha(F_0), \beta_0^{\mathrm{T}}, \sigma(F_0)^2)^{\mathrm{T}}$. If we reparameterize by reordering the parameters so that the intercept α is the last rather than the first component of γ (and making the analogous change to x_i), the hypothesis is in the form used in Section 4.5. However, if we are careful and keep track of terms, we do not have to make the reparameterization explicitly.

5.6.1 Wald, Score, and Likelihood Ratio Tests

If we fit the model using an M-estimator, the general results of Section 4.5 are applicable. In the particular case of Huber-type M-estimators for which $\eta(x, r) = (x^{\mathrm{T}}\psi(r), \chi(r))^{\mathrm{T}}$, the structure of the asymptotic covariance matrix and the fact that the hypothesis involves only slope parameters which are orthogonal to the intercept and scale parameter, produces important simplifications. First, the hypothesis involves only slope parameters β_0 which are always estimable. This means that we do not have to deal with bias in thinking about the hypothesis. Second, in the notation of Section 4.5, we find that with $u = (y - \mu(F_0))/\sigma(F_0)$, we have that

$$B_{F_0}(\theta(F_0))^{11} = \frac{1}{\int \psi'(u)\,dF_0(y)}\left[n^{-1}\sum_{i=1}^{n} z_i z_i^{\mathrm{T}}\right]^{11}$$

and

$$V_{F_0}(\theta(F_0))_{11} = \frac{\int \psi(u)^2\,dF_0(y)}{\{\int \psi'(u)\,dF_0(y)\}^2}\left[n^{-1}\sum_{i=1}^{n} z_i z_i^{\mathrm{T}}\right]^{11}.$$

Thus the Wald, score, and likelihood ratio test statistics are straightforward to compute and the asymptotic approximations to their sampling distributions are very simple.

The *Wald test statistic* for testing H_0 (Markatou et al., 1991, p. 205) is from (4.31)

$$W = \frac{\left\{n^{-1}\sum_{i=1}^{n}\psi'\left(\frac{y_i - x_i^{\mathrm{T}}\hat{\gamma}}{\hat{\sigma}}\right)\right\}^2}{n^{-1}\sum_{i=1}^{n}\psi\left(\frac{y_i - x_i^{\mathrm{T}}\hat{\gamma}}{\hat{\sigma}}\right)^2} n\hat{\gamma}_1^{\mathrm{T}}\left\{\left[n^{-1}\sum_{i=1}^{n} z_i z_i^{\mathrm{T}}\right]^{11}\right\}^{-1}\hat{\gamma}_1,$$

where $\hat{\gamma}_1$ is the vector of the first q components of $\hat{\gamma}$, and the *score test statistic* (Markatou and Hettmansperger, 1990) is from (4.32)

$$S = \frac{n^{-1}\sum_{i=1}^{n} z_{i1}^{\mathrm{T}}\psi(r_{Ri})\,[n^{-1}\sum_{i=1}^{n} z_i z_i^{\mathrm{T}}]^{11}\sum_{i=1}^{n} z_{i1}\psi(r_{Ri})}{n^{-1}\sum_{i=1}^{n}\psi(r_{Ri})^2},$$

where $r_{Ri} = (y_i - x_i^{\mathrm{T}}\hat{\gamma}_R)/\hat{\sigma}_R$ and z_{i1} is the vector of the first q components of z_i. If $\eta(x;\theta) = \partial\rho(x,\theta)/\partial\theta$ (so we maximize ρ to estimate θ), the *likelihood ratio test statistic* is

$$\Delta = 2\left\{\sum_{i=1}^{n}\rho\left(\frac{y_i - x_i^{\mathrm{T}}\hat{\gamma}_R}{\hat{\sigma}_R}\right) - \sum_{i=1}^{n}\rho\left(\frac{y_i - x_i^{\mathrm{T}}\hat{\gamma}}{\hat{\sigma}}\right)\right\},$$

Under local alternatives H_n: $H^{\mathrm{T}}\theta_0 = n^{-1/2}\xi$, it follows directly from Sections 4.5.7 and 4.5.9 that the Wald and score test statistics have asymptotic noncentral $\chi_q^2(\delta)$ distributions with noncentrality parameter

$$\delta = \frac{\{\int\psi'(u)\,dF_0(y)\}^2}{\int\psi(u)^2\,dF_0(y)}\xi^{\mathrm{T}}\left\{\left[n^{-1}\sum_{i=1}^{n} z_i z_i^{\mathrm{T}}\right]^{11}\right\}^{-1}\xi,$$

and hence that they have χ_q^2 distributions under H_0. From Section 4.5.11, the likelihood ratio test statistic Δ has asymptotically the same distribution as

$$\sum_{i=1}^{q} w_i(N_i + [Q^{\mathrm{T}}V_{F_0}(\theta(F_0))^{-1/2}\xi]_i)^2$$

where N_i are independent N(0, 1) random variables $1 \le i \le q$, w_i are the eigenvalues and Q is the matrix of eigenvectors of the matrix

$$\{B_{F_0}(\theta(F_0))^{11}\}^{-1/2}V_{F_0}(\theta(F_0))_{11}\{B_G(\theta(G))^{11}\}^{-\mathrm{T}/2}.$$

However,

$$\{B_{F_0}(\theta(F_0))^{11}\}^{-1}V_{F_0}(\theta(F_0))_{11} = \frac{\int \psi(u)^2\, dF_0(y) I_q}{\int \psi'(u)\, dF_0(y)}$$

so $w_i = \{\int \psi'(u)\, dF_0(y)\}^{-1} \int \psi(u)^2\, dF_0(y)$ and $Q = I$. Hence

$$\Delta \sim \frac{\int \psi(u)^2\, dF_0(y)}{\int \psi'(u)\, dF_0(y)} \chi_q^2(\delta)$$

or, equivalently, by Theorem 4.2

$$\frac{n^{-1} \sum_{i=1}^{n} \psi'\left(\dfrac{y_i - x_i^{\mathrm{T}}\hat{\gamma}}{\hat{\sigma}}\right)}{n^{-1} \sum_{i=1}^{n} \psi\left(\dfrac{y_i - x_i^{\mathrm{T}}\hat{\gamma}}{\hat{\sigma}}\right)^2} \Delta \sim \chi_q^2(\delta),$$

which is the result of Schrader and Hettmansperger (1980).

The Wald and score tests can be applied immediately to the problem of evaluating curvature in the steel plate example. The score test has the advantage that the large sample approximation to its sampling distribution is usually better than for the Wald test. (Also it does not involve ψ' so can be used straightforwardly even when ψ is not smooth.) The likelihood ratio test cannot be applied directly in the steel plate example because η is not constructed by differentiating a single criterion function ρ.

5.6.2 Modified Likelihood Ratio Tests

Richardson and Welsh (1996) show that we can modify the likelihood ratio test to

$$\Delta^* = 2\left\{\sum_{i=1}^{n} \rho^*\left(\frac{y_i - x_i^{\mathrm{T}}\hat{\gamma}_R}{\hat{\sigma}}\right) - \sum_{i=1}^{n} \rho^*\left(\frac{y_i - x_i^{\mathrm{T}}\hat{\gamma}}{\hat{\sigma}}\right)\right\},$$

where $\psi(x, \theta) = \partial\rho^*(x, \theta)/\partial\gamma$. The advantage of this statistic over the likelihood ratio statistic is that it is usually much simpler to obtain ρ^* than ρ. We can show that under local alternatives H_n,

$$\frac{n^{-1} \sum_{i=1}^{n} \psi'\left(\dfrac{y_i - x_i^{\mathrm{T}}\hat{\gamma}}{\hat{\sigma}}\right)}{n^{-1} \sum_{i=1}^{n} \psi\left(\dfrac{y_i - x_i^{\mathrm{T}}\hat{\gamma}}{\hat{\sigma}}\right)^2} \Delta^* \sim \chi_q^2(\delta),$$

so the asymptotic properties of Δ^* are the same as those of the other test statistics (Problem 5.6.3).

5.6.3 Application to the Steel Plate Data

The diagnostics for the expanded model are very similar to those obtained in Section 5.5. The test statistic, the normalized version of Δ^* based on the Huber ψ function, is 1.699 which is not significant when compared to a χ_3^2 (or more conservatively a $F(3, 36)$) distribution. There is therefore no strong evidence of curvature and our simpler initial model is an adequate description of the relationship between weight loss, time, and temperature.

5.6.4 Other Approaches to Model Selection

A different approach to model selection is to try to minimize a predictive criterion. The criteria of Akaike (1973), Mallows (1973), and Stone (1977) are distinguished by the different penalties they impose for including excess terms in the model but are otherwise very similar. Ronchetti and Staudte (1994) have recently developed a robust version of Mallows' proposal.

However we arrive at the final model, and whether we use robust methods or not, we need to keep in mind that the standard errors computed from the final model do not take the selection process into account and are therefore smaller than they should in fact be. While the standard errors can be interpreted conditionally, it is not always clear that this is appropriate. See the discussion in Section 1.5.

5.6.5 Robustness of Tests

The test statistics of Section 5.6.1–5.6.2 were based on robust estimators of the parameters, so intuitively they should be robust. That this is the case has been shown in considerable generality by Heritier and Ronchetti (1994) by extending earlier work of Ronchetti (1982) and Hampel et al. (1986, Chapter 7). Heritier and Ronchetti explored the effect of contamination on the level and power of tests by defining level and power influence functions. They showed that these are functions of the influence function of the underlying estimator and so inherit the properties of the underlying estimator. This is intuitively reasonable and means that we do not have to develop a new theory for tests.

Finally, as discussed in Section 4.5, we can also use the tests to construct robust approximate confidence intervals for the parameters.

PROBLEMS

5.6.1. Test for curvature in the second enamel covered steel plate experiment of Problem 1.5.9 using least squares and robust methods.

5.6.2. Analyse the steel plate corrosion data (Section 1.1.5) using the (incomplete) two-way classification model

$$y_{ijk} = \mu + \alpha_i + \beta_j + (\alpha\beta)_{ij} + e_{ijk},$$
$$\mathrm{i} = 1, \ldots, 5, \quad j = 1, 2, 3, \quad (i, j) \neq (5, 3), \quad k = 1, 2, 3,$$

where $\{e_{ijk}\}$ are independent $\mathrm{N}(0, \sigma^2)$ random variables at the core model, discussed in Section 1.2.5. Compare your conclusions to those obtained by fitting the regression model in Section 5.5.

5.6.3. Under the model (5.11) consider the statistic

$$\Delta^* = 2\left\{\sum_{i=1}^{n} \rho^*\left(\frac{y_i - z_i^{\mathrm{T}}\hat{\gamma}_R}{\hat{\sigma}}\right) - \sum_{i=1}^{n} \rho^*\left(\frac{y_i - z_i^{\mathrm{T}}\hat{\gamma}}{\hat{\sigma}}\right)\right\},$$

where $\hat{\gamma}$ is the M-estimator defined by $\sum_{i=1}^{n} x_i \psi\{(y_i - x_i^{\mathrm{T}}\hat{\gamma})/\hat{\sigma}\} = 0$ with $\psi(x) = \partial\rho^*(x)/\partial x$ and $\hat{\gamma}_R$ is the same M-estimator under H_0: $H^{\mathrm{T}}\gamma_0 = 0$. Assuming that $\hat{\gamma}$ and $\hat{\gamma}_R$ are consistent for γ_0, show that under local alternatives H_n: $H^{\mathrm{T}}\gamma_0 = n^{-12}\xi$,

$$\frac{n^{-1} \sum_{i=1}^{n} \psi'\left(\frac{y_i - x_i^{\mathrm{T}}\hat{\gamma}}{\hat{\sigma}}\right)}{n^{-1} \sum_{i=1}^{n} \psi\left(\frac{y_i - x_i^{\mathrm{T}}\hat{\gamma}}{\hat{\sigma}}\right)^2} \Delta^* \sim \chi_q^2(\delta),$$

where

$$\delta = \frac{\int \psi(u)^2 \, dG(y)}{\{\int \psi'(u) \, dG(y)\}^2} \xi^{\mathrm{T}} \left\{\left[n^{-1} \sum_{i=1}^{n} z_i z_i^{\mathrm{T}}\right]^{11}\right\}^{-1} \xi.$$

5.7 OTHER APPROACHES TO DISTRIBUTIONAL ROBUSTNESS

We have argued that for an estimator to be robust, it should have a continuous, bounded influence function in a neighborhood of the model and a reasonable breakdown point. This approach is essentially the influence function or the *infinitesimal approach* developed by Hampel (1968; 1971; 1974) and expounded in Hampel et al. (1986). Hampel placed the focus more firmly on the core model $\mathscr{F}$ and emphasized the importance of boundedness of the influence function or, equivalently, finiteness of the gross error sensitivity at the model $\mathscr{F}$. Estimators for which the gross error sensitivity is finite are called *B-robust* because their approximate bias is bounded. These estimators do not necessarily have a continuous influence function but insisting on this sensible requirement causes no particular difficulty.

5.7.1 Qualitative and Quantitative Robustness

Hampel (1968; 1971) also introduced the concept of *qualitative robustness* to formalize the requirement that an estimator have close to its assumed properties when applied to distributions close to those in $\mathscr{F}$. Essentially, this requirement is satisfied if the distribution of the estimator changes only very little when the model changes a little and this can be viewed as a continuity requirement on the estimator. Qualitative robustness is not easy to check because it requires us to treat the estimator as a function on the space of distribution functions and to deal with appropriate norms on this space. Moreover, qualitative robustness can identify some procedures as nonrobust (such as the sample mean for the Gaussian model) but it does not help distinguish between procedures which are qualitatively robust. Quantitative measures like the gross error sensitivity and the breakdown point are required for this purpose. Qualitative and B-robustness turn out to be closely related but are not identical. See for example Staudte and Sheather (1990, Section 3.2.4), Hampel et al. (1986, Sections 1.3d and 2.2b) and Huber (1981, Section 1.3 and Chapter 2).

5.7.2 Optimal B-Robust Estimators

As we saw in Section 5.4, robustness requirements typically lead to some loss of efficiency. This can be viewed as a form of insurance policy – we trade off optimality under the assumed model for protection against deviations from that model. However, it is desirable to achieve a balance between the two requirements and insist that the estimator be consistent and reasonably efficient for distributions in the model $\mathscr{F}$. Subject to a bound on the gross error sensitivity, Hampel and Krasker (see Hampel et al., 1986, p. 40) proved that the most efficient estimator at the model (the *optimal B-robust estimator*) is obtained by truncating the maximum likelihood estimating equations while making sure that the estimator is consistent at the core model. See Hampel et al. (1986, Section 4.3).

5.7.3 V-Robustness

Just as we derived the influence function as a linear approximation to an estimator, we can obtain linear approximations to both the gross error sensitivity and the asymptotic variance and use them to explore the effects of infinitesimal contamination on these functions. These linear approximations are called the *change of bias* and *change of variance functions* respectively. An estimator with a bounded influence function (finite gross error sensitivity) is called B-robust and one with a change of variance function which is bounded above (finite change of variance sensitivity) is called *V-robust* (Rousseeuw, 1981; Hampel et al., 1981; Hampel et al., 1986, p. 131). The requirements for V-robustness are slightly stronger than those for B-robustness, but they are closely related concepts.

5.7.4 High Breakdown Estimators

For structured data, B-robust M-estimators often have poor breakdown properties. For this reason, Rousseeuw (1984) and Rousseeuw and Yohai (1984) have argued that we should place the emphasis on finding *high breakdown estimators* rather than B-robust estimators. The concept of breakdown can be quite elusive, particularly in structured problems. For example, the highest possible breakdown point for estimating location in a single sample problem is 50% but for estimating both locations (or the difference in location) in a two sample problem is only 25%. Thus structure generally seems to decrease the highest attainable breakdown point because breakdown occurs more easily. Often the biggest problem with high dimensional data is that of collapse to lower dimension. This is important for variance estimation and for the large number of procedures (including regression) which involve standardization by an estimate of the variance matrix. Although we have defined breakdown to occur when an estimator moves to the boundary of the parameter space (Section 5.1.5), serious consequences may occur before the boundary is attained. Finally, high breakdown estimators can have unexpected properties (Rousseeuw and van Zomeren, 1990, which includes discussion; Hettmansperger and Sheather, 1992 and letters in response in 1993).

5.7.5 Minimax Robustness

The infinitesimal approach derives the influence function by considering an infinitesimal neighborhood of the model $\mathscr{F}$ but uses the influence function to extrapolate beyond this neighborhood. In some circumstances, it is possible to consider a *fixed neighborhood* of $\mathscr{F}$. Huber (1964) defined a neighborhood of the model in which the parameters of interest are identifiable and then found the distribution in this neighborhood (known as the *least favorable distribution*) such that an observation from this distribution has the least possible information about the parameters of interest. Any estimator (such as the maximum likelihood estimator) which is asymptotically efficient for the least favorable distribution minimizes the maximum possible variance, so is called a *minimax estimator*. Although we use the least favorable distribution to generate estimators, we do not believe that it describes the data generating process. Huber's approach turns out to be related to V-robustness rather than B-robustness. (The minimax bias problem leads to the most B-robust estimator; see for example Martin and Zamar, 1989.) Minimax theory is often regarded as pessimistic but is surprisingly optimistic in this problem in the sense that the least favorable distribution is less extreme than might be expected. One disadvantage of the approach is that it does not generalize nicely to arbitrary models. In addition, the fact that the parameters need to be identifiable over the whole neighborhood of $\mathscr{F}$ restricts the nature of the neighborhoods we can consider. This is unrealistic insofar as we have no control over the contamination and cannot reasonably insist that it is only of a particular form.

5.7.6 Local Contamination

Huber-Carol (1970), Bickel (1978), and Rieder (1978; 1980) considered *shrinking neighborhoods* of the model $\mathscr{F}$ which reduce to $\mathscr{F}$ as $n \to \infty$. Essentially, this means that we adopt a model which holds exactly asymptotically but not necessarily for finite n. Thus, the robustness problem is viewed as a finite sample problem. However, in practice, model fit tends to become poorer rather than better in large samples (see the discussion on testing Section 3.4.5) so the practical value of the approach is unclear.

5.7.7 Tukey's Approach

The end result of the Huber minimax approach is that we use the least favorable model to generate estimators of the parameters in $\mathscr{F}$. In a sense, we use a single model to represent the whole neighborhood of $\mathscr{F}$. Tukey (1981) has suggested that letting one distribution represent a neighborhood can be improved on by letting a small, finite set of distributions (called *pencils*) represent the neighborhood. Tukey calculates Pitman estimators (Section 3.1.5) for each of the distributions and produces a single estimate by taking linear combinations of these. These estimators are not necessarily robust in the sense of Hampel or Huber but seem to work quite well. Accessible presentations of these ideas are given by Morgenthaler (1986; 1987).

5.7.8 Likelihood Ratio Approaches

Huber (1965; 1968) explored bounding log-likelihood ratio test statistics and trying to maximize the minimum power over all alternatives. This approach leads to an exact finite sample theory but involves quite deep mathematics and seems difficult to extend beyond the location problem.

5.7.9 Nonparametric Approaches

All of the approaches discussed in Sections 5.7.1–5.7.8 involve a central parametric model $\mathscr{F}$. Parametric models are useful for simplifying the treatment of complex structural data, providing compact descriptions, making predictions, guiding statistical analyses, and shedding light on physical mechanisms. One response to the recognition that models do not hold exactly is to argue for nonparametric methods based on weaker, more complicated models. However, nonparametric methods may or may not be robust. In particular, the sample mean and sample variance are nonparametric estimators of the underlying mean and variance but are not robust (Section 5.1.7). At best, procedures based on these estimators enjoy robustness of validity but not robustness of efficiency (Section 5.1.2).

Generally, the focus in nonparametric methods is on validity rather than efficiency. Even when there is a focus on efficiency, as in the Gauss–Markov theorem, the result needs to be interpreted carefully. The Gauss–Markov theorem states that among all linear, unbiased estimators of the population mean, the sample mean has smallest variance. It omits to point out that, except at distributions which are very close to Gaussian, all linear unbiased estimators are poor so that the sample mean is the best of a bad lot. See Section 3.1.6. Nonparametric methods still require assumptions such as symmetry and independence, and departures from these are of interest.

Two approaches to robustness for nonparametric methods have been adopted. Bickel and Lehmann (1975a,b; 1976) based their approach on the asymptotic relative efficiency of procedures over nonparametric neighborhoods of nonparametric models. The asymptotic relative efficiency depends on the underlying distribution F_0, so they tried to find lower bounds over the whole neighborhood. These bounds tell us the worst loss but give no idea of the greatest gain. Some idea of the magnitude of gains can be obtained by considering specific distributions F_0. In the second approach reviewed recently by Rieder (1991), robustness is a property of the parameter being estimated rather than of the method of estimating the parameter. The idea is that the value of the parameter viewed as a function of the underlying distribution should change little as the distribution changes. This leads to a continuity requirement on the parameter.

5.7.10 Data Analytic Approaches

Finally, the oldest informal approach to robustness (which is still a common practice) is to examine the data for obvious outliers, delete these outliers, and then apply the optimal inference procedure for the assumed model to the cleaned data set. The idea behind this approach is to avoid having to deal with robust estimators. However, note that:

- It can be difficult to formalize this process so that its properties can be studied.
- It can be difficult to identify outliers.
- It is difficult to construct procedures which have high power.
- It is difficult to apply in complicated problems (such as multiple regression) where we cannot examine the relationships in the data without first fitting a model. In this case, we usually need to use robust estimators in the initial model fitting stage.
- Inference based on applying a standard procedure to the cleaned data will be based on distribution theory which ignores the cleaning process and hence will be inapplicable and possibly misleading.

While outlier deletion is arguably better than not doing anything at all, it is better to use robust methods which overcome the difficulties outlier deletion methods face.

5.8 LIKELIHOOD AND BAYESIAN THEORY

Consider again the stellar velocity data and the problem of making inferences about σ in the Gaussian model (5.1.). Recall that the likelihood of (μ, σ) under (5.1) is

$$f(\mathbf{z}; \mu, \sigma) = \frac{1}{(2\pi\sigma^2)^{n/2}} \exp\left\{-\frac{n\hat{\sigma}^2}{2\sigma^2} - \frac{n(\bar{z}-\mu)^2}{2\sigma^2}\right\}, \qquad (\mu, \sigma) \in \mathbb{R} \times [0, \infty).$$

For each fixed σ, the likelihood is maximized at $\tilde{\mu}(\sigma) = \bar{z}$ so replacing μ by $\tilde{\mu}(\sigma) = \bar{z}$, we obtain the profile likelihood for σ

$$f(\mathbf{z}; \tilde{\mu}(\sigma), \sigma) = \frac{1}{(2\pi\sigma^2)^{n/2}} \exp\left\{-\frac{n\hat{\sigma}^2}{2\sigma^2}\right\}, \qquad \sigma > 0.$$

The scaled profile likelihood is the likelihood ratio

$$\frac{f(\mathbf{z}; \tilde{\mu}(\sigma), \sigma)}{f(\mathbf{z}; \hat{\mu}(\hat{\sigma})} = \left\{\frac{\hat{\sigma}}{\sigma}\right\}^n \exp\left\{-\frac{n\hat{\sigma}^2}{2\sigma^2} + \frac{n}{2}\right\}, \qquad \sigma > 0. \tag{5.12}$$

5.8.1 Sensitivity of Profile Likelihoods

We can explore the sensitivity of the profile likelihood to an additional observation at a point x exactly as we did in Section 5.1.5. Using the results from Section 5.1.5, the effect on the likelihood of a single observation at x is

$$\frac{f(\mathbf{z}; x; \bar{z}(x), \sigma)}{f(\mathbf{z}; x; \hat{\mu}(x), \hat{\sigma}(x))} = \left\{\frac{\hat{\sigma}(x)}{\sigma}\right\}^n \exp\left\{-\frac{n\hat{\sigma}(x)^2}{2\sigma^2} + \frac{n}{2}\right\}$$

where $\hat{\sigma}^2(x) = \{n/(n+1)\}\hat{\sigma}^2 + (n+1)^{-2}n(x-\bar{z})^2$. This function is plotted for the disk data and selected x in Figure 5.9. We observe that as x increases, the location and spread of the profile likelihood increase.

5.8.2 Robust Likelihoods

Just as we obtained robust parameter estimators in Section 5.4 by modifying the maximum likelihood estimating equations, we can consider modifying the likelihood to obtain a robust likelihood. In particular, if the estimating function $\sum_{i=1}^n \eta(Z_i; \theta)$ satisfies $\eta(x; \theta) = \partial\rho(x; \theta)/\partial\theta$ for some ρ with appropriate properties, we obtain a likelihood when we take the likelihood proportional to $\exp\{-\sum_{i=1}^n \rho(Z_i; \theta)\}$. It is a minimal requirement that the mode of the

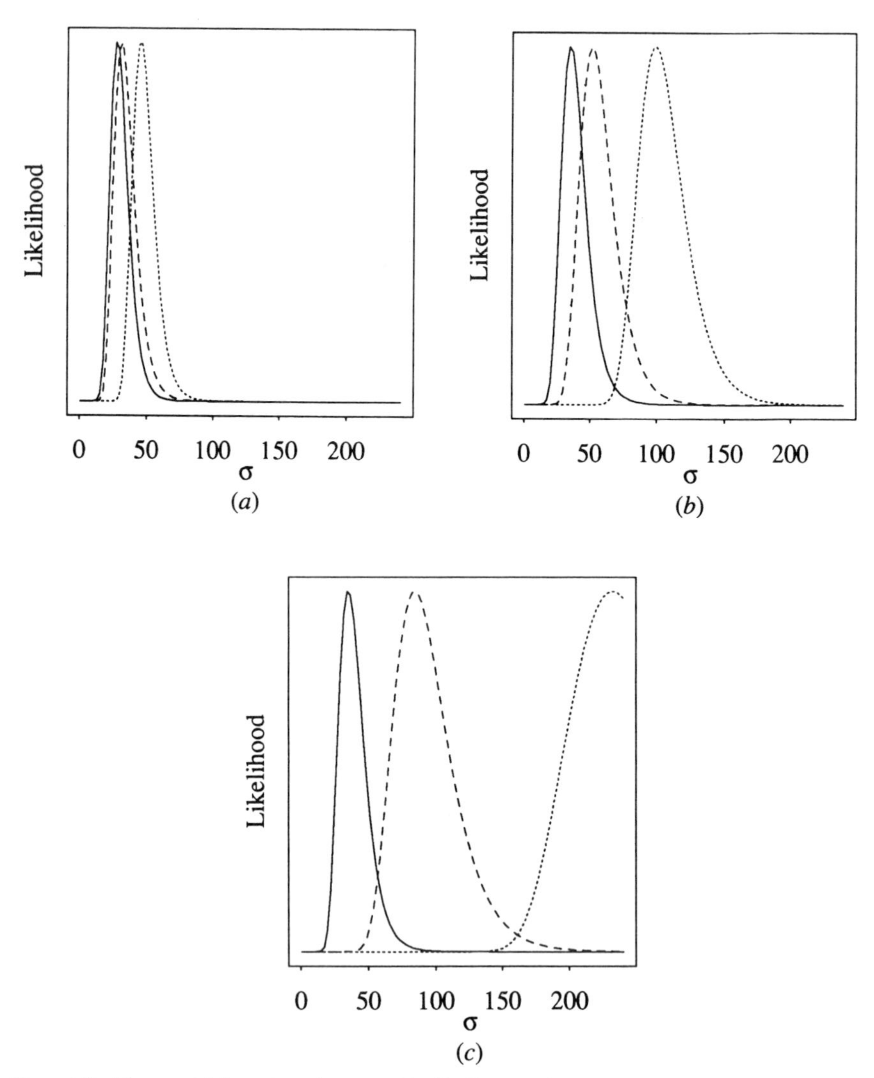

Figure 5.9. The sensitivity of various profile likelihoods for the disk data to an additional observation at x: Gaussian likelihood, dotted line; Laplace likelihood, dashed line; t_3 likelihood, solid line. (*a*) $x = 0$; (*b*) $x = 10\hat{\sigma}$; (*c*) $x = 20\hat{\sigma}$.

likelihood $\hat{\theta}$ be robust when viewed as an estimator but we also require at least the Hessian (the second derivative of ρ) to be robust.

If as an alternative to the Gaussian likelihood, we adopt the likelihood from the Laplace model which is

$$\mathscr{L}(\mu, \sigma) \propto \exp\left\{-\sum_{i=1}^{n} \frac{|z_i - \mu|}{\sigma} - n \log(\sigma)\right\}, \qquad (\mu, \sigma) \in \mathbb{R} \times [0, \infty),$$

the profile likelihood for σ is

$$\mathscr{L}(\text{median}(\mathbf{z}), \sigma) \propto \exp\left\{-\sum_{i=1}^{n} \frac{|z_i\text{-median}(\mathbf{z})|}{\sigma} - n\log(\sigma)\right\}, \qquad \sigma > 0,$$

so the scaled profile likelihood is

$$\exp\left\{-\sum_{i=1}^{n} \frac{|z_i\text{-median}(\mathbf{z})|}{\sigma} + n + n\log\left(n^{-1}\sum_{i=1}^{n} \frac{|z_i\text{-median}(\mathbf{z})|}{\sigma}\right)\right\}, \qquad \sigma > 0. \tag{5.13}$$

This function is less sensitive than (5.12) but, as is shown in Figure 5.9 is not robust either. The reason is that this function is maximized by

$$\hat{\sigma} = n^{-1}\sum_{i=1}^{n} |z_i\text{-median}(\mathbf{z})|$$

which is not robust.

A better choice is to use the likelihood from the Student t model with fixed, small degrees of freedom, say $\nu = 3$. In this case, the robust likelihood is

$$\mathscr{L}(\mu, \sigma) \propto \sigma^{-n}\exp\left\{-\sum_{i=1}^{n} \frac{\nu+1}{2}\log\left(1 + \frac{(z_i - \mu)^2}{\nu\sigma^2}\right)\right\}, \qquad (\mu, \sigma) \in \mathbb{R} \times [0, \infty).$$

To obtain the profile likelihood for σ, we need to fix σ and optimize over μ. That is, for each fixed σ, let $\mu(\sigma)$ satisfy

$$\sum_{i=1}^{n} \frac{Z_i - \mu}{\left\{1 + \frac{(Z_i - \mu)^2}{\nu\sigma^2}\right\}} = 0.$$

Then the profile likelihood for σ is

$$\sigma^{-n}\exp\left\{-\sum_{i=1}^{n} \frac{\nu+1}{2}\log\left(1 + \frac{(z_i - \mu(\sigma))^2}{\nu\sigma^2}\right)\right\}, \qquad \sigma > 0. \tag{5.14}$$

The scaled profile likelihood is plotted in Figure 5.9. Clearly, (5.14) is much less sensitive to the additional observation at x than either (5.12) or (5.13).

As in Section 5.4.9, we can usefully think of the data as arising from a contaminated version $\mathscr{G}$ of the underlying core model $\mathscr{F}$. Although we use a model to generate a robust likelihood, we do not necessarily believe that it actually holds. This is why we can adopt the Student t model with fixed degrees of freedom to generate a likelihood; if we really believed in the model, it would be more natural to treat ν as a parameter and concentrate it out too.

There is more to be said about the choice of robust likelihoods but it is clear that the issues are related to those which arise in robust frequentist inference and therefore that those concepts will be useful in this context.

5.8.3 Sensitivity of Posterior Distributions

In a Bayesian analysis of the stellar velocity data based on the Gaussian model (5.1) and the Jeffreys prior (see Secton 2.3.2)

$$g(\mu, \sigma) \propto \sigma^{-1}, \qquad (\mu, \sigma) \in \mathbb{R} \times [0, \infty),$$

the joint posterior density of (μ, σ) is

$$\begin{aligned} g(\mu, \sigma \mid \mathbf{Z} = \mathbf{z}) &\propto \frac{1}{\sigma^{n+1}} \exp\left\{-\frac{vs^2}{2\sigma^2} - \frac{n(\bar{z} - \mu)^2}{2\sigma^2}\right\} \\ &= \frac{1}{\sigma^{n+1}} \exp\left[-\frac{vs^2}{2\sigma^2}\left\{1 + \frac{n(\bar{z} - \mu)^2}{vs^2}\right\}\right] \end{aligned}$$

so the marginal posterior density for σ is

$$\begin{aligned} g(\sigma \mid \mathbf{Z} = \mathbf{z}) &\propto \int_{-\infty}^{\infty} \frac{1}{\sigma^{n+1}} \exp\left[-\frac{\{vs^2 + n(\bar{z} - \mu)^2\}}{2\sigma^2}\right] d\mu \\ &\propto \frac{1}{\sigma^{v+1}} \exp\left[-\frac{vs^2}{2\sigma^2}\right], \qquad \sigma > 0, \end{aligned}$$

by a standard integration result. That is $\sigma \mid \mathbf{Z} = \mathbf{z} \sim v^{1/2} s \chi_v^{-1}$, where χ_v^{-1} is a random variable with the "inverse χ" distribution with v degrees of freedom. (See 6bv in the Apendix). The marginal posterior density of σ is the same as the profile likelihood for σ so its sensitivity (and consequent lack of robustness) to extreme observations is revealed in Figure 5.9.

5.8.4 Bayesian Robustness

In a Bayesian analysis we need to be concerned about the robustness of both the choice of likelihood and the choice of prior. Procedures which are robust against the choice of prior are said to be *posterior robust* while those that are robust against the choice of model are said to be *inference* (Box and Tiao, 1973, p. 152) or *model* (Berger, 1985, p. 248) *robust*. Actually, there is no fundamental distinction between the two types of robustness since the choice of model can be interpreted as a form of prior specification. This means that the ideas in Section 2.2 on the choice of prior can be viewed as relevant to the choice of model.

The basic idea behind posterior robustness is that it is often difficult to determine a single unique prior distribution for a given problem but that it is fairly easy to determine a set of plausible prior distributions. If the inference based on each of the prior distributions in the plausible class is similar, then the inference is posterior robust. See Section 2.2.7. As applied to the choice of likelihood, these ideas suggest that we consider a model $\mathscr{G}$ which contains our proposed model $\mathscr{F}$ as a special case and then explore the effect on our inferences of using different likelihoods from $\mathscr{G}$. Inference or model robustness holds when the inferences drawn under the assumed model are insensitive to departures from the assumed model.

An important aspect of inference or model robustness is that it is a conditional property in the sense that it depends both on the model $\mathscr{G}$ and the data at hand. The choice of $\mathscr{G}$ is important because if it is too small, robustness is meaningless while if it is too large, robustness will often be unattainable. One way to specify $\mathscr{G}$ is to think about the kinds of departures from the models which ought to be captured. Specifically, if we are concerned about extreme observations, then a class of longer-tailed model distributions should be considered. If $\mathscr{G}$ consists of a class of longer-tailed distributions than in $\mathscr{F}$, then inference or model robustness holds only if there are no extreme observations in the data and, in this case, we can use inferences based on $\mathscr{F}$.

5.8.5 Inference Robustness

Box and Tiao (1973, pp. 156–96) suggested embedding the Gaussian model (5.1) in the *exponential power family* for which

$$\mathscr{G} = \left\{ f(\mathbf{z}; \mu, \sigma, \beta) = \prod_{i=1}^{n} \frac{1}{\sigma} h_\beta\left(\frac{z_i - \mu}{\sigma}\right), \right.$$

$$\left. -\infty \leq z_i \leq \infty; -\infty \leq \mu \leq \infty, \sigma > 0, -1 < \beta \leq 1 \right\}, \quad (5.15)$$

where $h_\beta(x) = c(\beta) \exp\{-[\Gamma\{3(1+\beta)/2\}/\Gamma\{(1+\beta)/2\}]^{1/(1+\beta)}|x|^{2/(1+\beta)}\}$, with $c(\beta) = [\Gamma\{3(1+\beta)/2\}^{1/2}/(1+\beta)\Gamma\{(1+\beta)/2\}^{3/2}]$. The exponential power family is Gaussian for $\beta = 0$ but includes both longer- and shorter-tailed distributions (we obtain the Laplace distribution for $\beta = 1$ and the $U(-\sqrt{3}, \sqrt{3})$ distribution as $\beta \to -1$).

We can also embed the Gaussian model (5.1) in the Student t model

$$\mathscr{G} = \left\{ g(\mathbf{z}; \sigma, \nu) = \prod_{i=1}^{n} \frac{1}{\sigma} g_\nu\left(\frac{z_i - \mu}{\sigma}\right), -\infty \leq z_i \leq \infty, \sigma > 0, \nu > 0 \right\}, \quad (5.16)$$

where $g_\nu(x) = [\Gamma\{(\nu+1)/2\}/(\pi\nu)^{1/2}\Gamma(\nu/2)]\{1 + x^2/\nu\}^{-(\nu+1)/2}$ denotes the density of the Student t distribution with ν degrees of freedom. The Student t distribution has longer tails than the Gaussian distribution (5.1) which we recover as $\nu \to \infty$.

Given the data $\mathbf{z}$, we need to study how inferences (specifically the posterior distributions) change as a function of β or ν. The inferences are inference or model robust if they are insensitive to the value of β or ν. Alternatively, we can treat β or ν as regular parameters and implement a full Bayesian analysis or (analogously to the frequentist approach) use a fixed conservative value of β or ν.

If we use the Student t model with fixed degrees of freedom, we have

$$g(\mu, \sigma \mid \mathbf{Z} = \mathbf{z}) \propto \frac{1}{\sigma^{n+1}} \exp\left\{-\sum_{i=1}^{n} \frac{\nu+1}{2} \log\left(1 + \frac{(z_i - \mu)^2}{\nu\sigma^2}\right)\right\}, \quad (\mu, \sigma) \in \mathbb{R} \times [0, \infty),$$

and hence the marginal posterior density for σ

$$g(\sigma \mid \mathbf{Z} = \mathbf{z}) \propto \int_{-\infty}^{\infty} \frac{1}{\sigma^{n+1}} \exp\left\{-\sum_{i=1}^{n} \frac{\nu+1}{2} \log\left(1 + \frac{(z_i - \mu)^2}{\nu\sigma^2}\right)\right\} d\mu, \qquad \sigma > 0.$$

We can either use numerical integration (Section 2.1.7) or Laplace's method (Sections 4.6.2 and 4.6.4) to approximate the integral to obtain the marginal posterior density for σ. Some other approaches are discussed in Sections 6.8.5–6.8.7.

PROBLEMS

5.8.1. Suppose that we adopt the Weibull model

$$\mathcal{F} = \left\{ f_\kappa(\mathbf{z}; \lambda) = \prod_{i=1}^{n} \kappa\lambda(\lambda z_i)^{\kappa-1} \exp\{-(\lambda z_i)^\kappa\}, z_i > 0 : \lambda > 0 \right\}$$

with fixed κ and the Jeffreys prior for λ for the pressure vessel data. For fixed κ, find the posterior density of λ. Plot the posterior density of λ for selected values of κ and discuss the robustness of inferences based on an exponential model. Repeat the analysis with the data of Problem 1.5.4.

5.8.2. Suppose that we adopt the exponential power family (5.15) and the Jeffreys prior for the change in the volume of urine due to the ingestion of caffeine data presented in Section 1.1.3. Show that the marginal posterior density of μ is

$$g(\mu \mid \mathbf{z}) \propto \left\{ n^{-1} \sum_{i=1}^{n} |z_i - \mu|^{2/(1+\beta)} \right\}^{-n(1+\beta)/2}.$$

Plot the marginal posterior density of μ for selected values of β and discuss the robustness of inferences based on a Gaussian model.

5.8.3. Suppose that we adopt the exponential power family (5.15) and the Jeffreys prior for the velocities of the intermediate stars presented in Section 1.1.4. Adopting the simplifying assumption that $\mu = 0$ is known, plot the posterior density of σ for selected values of β and discuss the robustness of inferences based on a Gaussian model.

5.8.4. Suppose that we adopt the exponential power family of (5.15) and the Jeffreys prior for the velocities of the intermediate and halo stars presented in Section 1.1.4. Assuming that $\mu = 0$ in both cases, show that the posterior density of $\theta = \sigma_1/\sigma_2$ is

$$g(\theta \mid \mathbf{z}) \propto \theta^{n_2}\left\{1 + \theta^{2/(1+\beta)} \frac{\sum_{i=1}^{n_2} |z_{i2}|^{2/(1+\beta)}}{\sum_{i=1}^{n_1} |z_{i1}|^{2/(1+\beta)}}\right\}^{-(n+1)(1+\beta)/2}, \qquad \theta > 0,$$

with $n = n_1 + n_2$. Plot the posterior density of θ for selected values of β and discuss the robustness of inferences based on a Gaussian model.

FURTHER READING

The books by Staudte and Sheather (1990), Rousseeuw and Leroy (1987), Hampel et al. (1986), and Huber (1981) all contain accessible, philosophical material on frequentist robustness but differ in the mathematical level they demand. Box and Tiao (1973, Chapters 3 and 4) and Berger (1985, pp. 248–50) are important references for the Bayesian perspective.

CHAPTER 6

Randomization and Resampling

Randomization is the use of a controlled random process to introduce randomness additional to that inherent in the data into the data collection procedure or the analysis of the data. There are four recognizably different uses of randomization:

1. Randomization can be used after the data have been collected to jitter or smooth the data.
2. Randomization can be used during or after the data collection process to protect the confidentiality of participating units.
3. Randomization can be used in the design of an experiment or sample survey to select units, allocate treatments to units, etc.
4. Randomization can be used as a basis for inference.

Our primary focus on randomization in this chapter is on the use of randomization as a basis for inference (Sections 6.3–6.5) but we will discuss 1–2 briefly below and 3 in Sections 6.1–6.2.

We explicitly used randomization in form 1 in Section 3.7.3 where we overcame the difficulties caused by the discreteness of Z by using the distribution of $Z + U$, with $U \sim \mathrm{U}(0, 1)$ independent of Z, to set exact confidence intervals. We also used randomization in form 1 implicitly in Section 1.5.1 because the kernel density estimate based on $\{Z_1, \ldots, Z_n\}$ can be interpreted as the density of $Z_i + hU_i$, where $\{U_i\}$ are independent of $\{Z_i\}$ and independent and identically distributed with common density function K and $h > 0$. We will discuss this use of randomization further in Section 6.9.4.

Government agencies which collect sensitive data are often required to maintain the confidentiality of the data. This is achieved by, amongst other things, restricting access to the data and, when data is released, providing access only to grouped data or data to which random noise has been added. This use of randomization in form 2 causes serious difficulties for secondary analysis of the data but does contribute to the preservation of confidentiality. A different use for randomization in form 2 to protect the confidentiality of respondents and so encourage truthful response in sample surveys (due to Warner, 1965) is to incorporate randomization into the response to sensitive questions. Suppose we want to estimate the proportion π of cocaine users in a population. Each respondent is asked to conduct a private experiment (which is not observed by the data collector), the outcome of which is either the statement "I am a cocaine user" with probability 3/4 or the statement "I am not a cocaine user" with probability 1/4. The respondent is asked to answer "yes" or "no" according to whether the result of their private experiment is true or not. The data collector does not know the result of the private experiment and hence does not know the true status of the respondent but, under the assumption that P(cocaine user) = π for every member of the target population and that the answers are truthful,

$$P(\text{"Yes"}) = \tfrac{3}{4}\pi + \tfrac{1}{4}(1 - \pi) = \tfrac{1}{2}\pi + \tfrac{1}{4},$$

so cocaine usage in the target population can be estimated from an estimate of the proportion of people in the target population giving "yes" answers. Applications of this methodology to explore drug usage are given by Goodstadt and Gruson (1975) and Brewer (1981).

6.1 EXPERIMENTAL DESIGN

In the caffeine experiment of Bellet et al. (1969), a baseline urine level z_{i0} under a control regime and a urine level z_{i1} after ingesting caffeine were observed for each subject. We decided to base our analysis on the pairwise differences $z_i = z_{i1} - z_{i0}$, $i = 1, \ldots, n$, which represent the change in the volume of urine produced by the ingestion of caffeine and in Section 1.2.3, we modelled $\mathbf{z} = (z_1, \ldots, z_n)$ as a realization of a random variable $\mathbf{Z}$ which is distributed according to the Gaussian model

$$\mathscr{F} = \left\{ f(\mathbf{z}; \mu, \sigma) = \prod_{i=1}^{n} \frac{1}{(2\pi\sigma^2)^{1/2}} \exp\left\{ -\frac{(z_i - \mu)^2}{2\sigma^2} \right\}, \mathbf{z} \in \mathbb{R}^n \colon \mu \in \mathbb{R}, \sigma > 0 \right\}. \quad (6.1)$$

Under this model, the parameter of interest μ represents the mean change in the volume of urine produced and the nuisance parameter σ^2 represents the magnitude of the variability in the data.

The caffeine experiment was described in detail by Bellet et al. (1969) as follows:

> Eighteen normal, young male subjects, aged 18–22 years, were included in this study. None was on drug therapy. Following 10 p.m. of the night prior to the experiment, they were instructed not to smoke, eat or drink; at 7:30 a.m. when arising, after voiding, they drank two glasses of water. All subjects reported to the laboratory at

8:30 a.m. They rested until 9:00 a.m. when they voided and immediately thereafter the experiments were started. In each subject the experiment consisted of the ingestion of 5 Gm. instant coffee (this contains 220 mg. caffeine and is equivalent to 2 cups of coffee) dissolved in 500 ml. water and a control study consisting in the ingestion of 500 ml. water. The sequence of these control and caffeine experiments were [sic] randomized. Between the two studies, there was a free interval of at least five days. During both studies, the subjects rested in an armchair from 9:00 a.m. until 12 noon. Urine voided during this period was collected in bottles which contained 5 ml. HCl as a preservative. The subjects voided at the end of the three-hour period so that the entire amount of urine secreted during the three-hour period of the experiment was collected. During the experiments, the subjects were kept isolated, one in a room, in relaxed surroundings, with no smoking, eating or drinking. The urine collected was analysed for creatinine and catecholamine content (epinephrine and norepinephrine) by the fluorometric method.

Although it is not mentioned explicitly in the above description, the amount of urine voided by each subject under the two regimes was also recorded and is the basis for the modeling in Section 1.2.3 leading to (6.1).

The caffeine experiment is a simple *comparative experiment* in which we observe urine and hormone levels with and without caffeine on $n = 18$ subjects. A properly conducted comparative experiment ensures the possibility of a logically sound conclusion: If the factor of interest (caffeine) is varied while everything else is held constant, changes in the observed response (urine or stress) must be caused by changes in the factor of interest. Obviously, we need to be able to separate changes in the observed response from background variability (the inference problem) and ensure that everything else is held constant (the design problem). The latter objective is achieved in two ways. First, if we can identify subject dependent factors which may affect the response, we may be able to control some of these factors by grouping (or blocking) similar subjects so that comparisons can be made over groups (or blocks) of similar subjects, or explicitly measuring them as covariates so that we can make an appropriate adjustment for their effects in the analysis. Second, those factors which we cannot control, perhaps are not even aware of, are controlled by randomization which ensures that, on average, they have no effect on the response.

6.1.1 Blocking

Suppose we rewrite the model (6.1) as we did in (1.21) so that the effect on subject i of treatment t is

$$Z_{it} = \alpha + \mu t + A_i + U_{it}, \qquad t = 0, 1, \; i = 1, \ldots, n, \tag{6.2}$$

where the A_i are independent with identical Gaussian distributions with mean 0 and variance σ_a^2 and are independent of the U_{ij} which are themselves independent with identical Gaussian distributions with mean 0 and variance σ_u^2. The advantage of writing (6.1) in the form (6.2) is that the variability in the model is explicitly decomposed into a subject effect (A_i) and a measurement effect (U_{it}) and we can use (6.2) to examine the effect of different design strategies on controlling these sources of variability.

One way to carry out the caffeine experiment would be to simply randomly allocate the subjects into one of two groups and give caffeine to one group and the control liquid to the other. In this experiment, inferences about the treatment effect μ assess the sampling variability in terms of

$$\text{Var}\,(Z_{i1} - Z_{i'0}) = 2(\sigma_a^2 + \sigma_u^2), \qquad i \neq i'.$$

Clearly, if the between subject variance σ_a^2 is large and the treatment effect small, the between subject variance will obscure the effect of the treatment.

We can control or reduce the variation in the experiment by *blocking* or grouping together homogeneous experimental units. Blocking could be achieved in the caffeine experiment by allocating subjects to pairs (blocks of two units are conveniently referred to as *pairs*) so that the two subjects in each pair are similar to each other and then randomizing one member from each pair to receive the treatment and the other to receive the control. We can do even better by using each subject as his own control because each subject is obviously perfectly matched to himself. That is, each subject is observed both after ingesting the control liquid and after treatment. In this case, inferences about the treatment effect μ assess the sampling variability in terms of the within subject variability

$$\text{Var}\,(Z_{i1} - Z_{i0}) = 2\sigma_u^2 = \sigma^2.$$

Blocking therefore removes σ_a^2, the between subject variability, and thereby increases the precision of the inferences.

6.1.2 Replication

We need to replicate the urine experiment on n independent subjects in order to estimate the variability (represented in (6.1) by σ^2), increase the precision of our inferences and extend the range of validity of the experiment. We can use the idea of achieving a predetermined precision in our inferences as a basis for specifying the sample size n.

Under (6.1), the sampling variance of the mean $\bar{Z}$ is σ^2/n so if we knew σ, we would set a $100(1-\alpha)\%$ confidence interval for μ as a realization of

$$\bar{Z} \pm n^{-1/2}\sigma\Phi^{-1}(1-\alpha/2).$$

If we want to estimate μ to within a desired level of accuracy Δ, we require

$$2n^{-1/2}\sigma\Phi^{-1}(1-\alpha/2) \leq \Delta$$

or

$$n \geq \frac{4\sigma^2\Phi^{-1}(1-\alpha/2)^2}{\Delta^2}.$$

This calculation requires us to know at least an upper bound for σ^2 and to specify Δ but is otherwise straightforward to use.

Alternative sample size calculations can be formulated in terms of the required power for detecting a particular alternative, in terms of achieving a specified relative standard error (defined as $\sigma/n|\bar{Z}|$) and so on. These alternative calculations suggest slightly different sample sizes.

6.1.3 Randomization

Randomization was used in the caffeine experiment to specify the order in which each subject received the control and the treatment liquids. Formally, a *design* D for the urine experiment is a vector of length n in which the ith entry specifies the order in which the ith subject receives the treatment and control liquids. Notice that the design works within the block structure we have imposed. We can construct the space $\mathscr{D}$ of all possible designs which is the set of all 2^n n-vectors with binary entries and then think of the randomization process as treating all of these designs as equally likely and selecting one of them from $\mathscr{D}$.

In general, the purpose of randomization is threefold:

1. To protect the experimenter against unconscious bias leading to allocations which favour the achievement of desired results.
2. To prevent extraneous factors from having a systematic effect on the results and hence prevent systematic bias and confounding (i.e., the impossibility of separating the effect of the extraneous factors from the effect of the treatment).
3. To ensure that the assumption of independent errors is reasonable and hence that the estimate of experimental error (reflected in the model $\mathscr{F}$ by σ^2) is meaningful.

That randomization achieves these objectives cannot be established under (6.1) but requires the introduction of a different kind of model which considers hypothetical repetitions of the randomization process. This kind of randomization model for the paired design is introduced in Section 6.2 and we establish that these objectives are achieved under the randomization model. The properties of statistics under the randomization model are referred to as *design-based properties* to distinguish them from the *model-based properties* we obtain from models like (6.1).

As we will see in Section 6.2, the design-based properties which justify randomization are properties which hold on average when we consider averaging over the set of possible designs. This means that there is no guarantee that a particular design selected from $\mathscr{D}$ achieves these goals. This is of course characteristic of the frequentist approach (Chapter 3): no particular design or analysis is guaranteed to achieve its objectives, but a large set of designs or analyses do so on average.

6.1.4 The Consequences of a Failure to Randomize

A famous example of an experiment rendered invalid by a failure to randomize treatment allocation is the Lanarkshire milk experiment discussed by "Student" (1931). The experiment involving 20 000 children was intended to evaluate the nutritional value of milk. A total of 5000 children were given $\frac{3}{4}$ pint of raw milk every day for a set period, another 5000 were given $\frac{3}{4}$ pint of pasteurized milk every day for the same period, and 10 000 children were controls who were given no milk. The allocation to treatment groups was carried out in schools by the teachers. Although the initial allocation was random, the teachers were permitted to reallocate the children so that no group in the school over-represented either the poorly nourished or the well-nourished children. The results showed that children in the control group had higher weight gain than the children in the treatment groups. "Student" (1931) argued that the teachers had unconsciously biased the results by ensuring that the smaller children received the milk. He recommended rerunning the experiment with a random treatment allocation but also recommended a smaller paired experiment based on the use of twins.

6.1.5 Practical Considerations

It should be obvious that randomization cannot provide blanket protection against every possible cause of failure of an experiment. In particular, properties which hold on average are often poorly reflected in any particular realization. Nonetheless, it is often possible even before we carry out an experiment to recognize that certain designs are more likely to give misleading results than others. In practice, poor designs are usually rejected and the randomization process is repeated until a sensible design obtains. In formal terms, the design is randomly selected from a subset of the set of all possible designs rather than from the set of all possible designs.

The use of a randomized experiment does not preclude modeling the effects of factors even if randomization was intended to eliminate these factors. Indeed, some conditioning on the results of the particular experiment is essential because it is not generally sensible to ignore a clear trend in the results of an experiment on the basis that on average (over other experiments) the trend would not be there. In the sample survey context, Royall (1976) referred to this as the "*closurization principle*": after using randomization to collect the data, close your eyes to the fact that randomization was used and analyze the data set actually obtained.

6.1.6 The Bayesian Perspective

Bayesian inference is made conditionally on the observed data so properties which hold on average are usually of at best secondary interest to Bayesians. Although it is possible to argue (at least theoretically) that randomization has no role in the design of an experiment which will be subjected to a Bayesian analysis, there are at least two important practical reasons for Bayesians to incorporate randomization into their experiments.

Generally, Bayesians argue that if all the unknowns in a problem have been identified and assigned their true prior distributions, then there is no need to randomize in designed experiments. However, it is essentially impossible to identify all the unknowns including those induced by the experimenter. Moreover, even if all the unknowns could be identified, the prior specification would be immensely complicated. So Rubin (1978) and Berger (1984) have argued that randomization can be used as a basis for simplifying the prior specification.

Royall (1976), Rubin (1978), and others have argued that randomization protects the statistician (whether Bayesian or not) against charges of doctoring the data to achieve a particular outcome. This is actually one of the strongest arguments for incorporating randomization into an experiment and, insofar as it has been adopted by regulatory agencies, can have legal as well as moral force. This is the case in the randomized clinical trials required for the approval of drugs or procedures by the United States Food and Drug Administration (FDA) and ensures that nearly all serious experiments with public interest implications are randomized experiments whether designed by Bayesians or not.

6.2 RANDOMIZATION MODELS

In Section 6.1.3, we presented randomization as a sensible protective device to ensure that, on average, the conclusions of an experiment are valid. It requires some effort to establish this claim but it is sufficiently important to justify the effort and the argument lays the basis for considering the use of randomization as a basis for inference (Sections 6.3–6.4).

Recall that in the caffeine experiment, we based our analysis on the pairwise differences $z_i = z_{i1} - z_{i0}$, $i = 1, \ldots, n$, which represent the change in the volume of urine produced by the ingestion of caffeine. Under the Gaussian model (6.1) for $\mathbf{z} = (z_1, \ldots, z_n)$, frequentist inference about the parameter of interest μ is usually based on $\bar{Z}$ which is a uniformly minimum variance unbiased (UMVU) estimator of μ and S^2/n which is a UMVU estimator of its sampling variance σ^2/n (see Section 3.1.4).

In an independent replication of the experiment, we would ordinarily think of varying all aspects of the experiment including the randomization but we can also imagine a hypothetical replication in which we hold everything fixed except for the randomization which is selected anew from $\mathscr{D}$. The randomization model $\mathscr{R}$ specifies that all possible realizations are equiprobable. (Nonequiprobable randomization is possible – see Section 6.5 – but rare in this kind of experiment.) The randomization model $\mathscr{R}$ can be used to generate the randomization distribution of a statistic (just as the data model $\mathscr{F}$ can be used to generate its sampling distribution) and properties such as unbiasedness can be explored under the randomization distribution. The claims presented in Section 6.1.3 that randomization prevents bias and enables us to obtain a meaningful estimate of the experimental error can be expressed as

$$\mathrm{E}\bar{Z} = \mu \tag{6.3}$$

and

$$\mathrm{E}S^2 = n \operatorname{Var}(\bar{z}), \tag{6.4}$$

where the expectations and variances are calculated with respect to the randomization model $\mathscr{R}$. That is, the sample mean and variance are randomization or design-unbiased for μ and the randomization variance of the normalized sample mean respectively.

6.2.1 The Randomization Model

Let μ_{itv} denote the expected response for the ith subject ($i = 1, \ldots, n$) under treatment t ($t = 0, 1$) when it is applied in the order v ($v = 0, 1$). Let

$$\mu_{...} = (4n)^{-1} \sum_{i,t,v} \mu_{itv}, \qquad \mu_{it.} = (2n)^{-1} \sum_{i,t} \mu_{itv}$$

and so on, so a period subscript means that we have averaged over the subscript in that position. Then we have the additive decomposition

$$\begin{aligned}\mu_{itv} &= \mu_{...} + (\mu_{.t.} - \mu_{...}) + (\mu_{i..} - \mu_{...}) + (\mu_{it.} - \mu_{i..} - \mu_{.t.} + \mu_{...}) + (\mu_{itv} - \mu_{it.})\\ &= \lambda + \alpha_t + \beta_i + \gamma_{it} + \xi_{itv},\end{aligned} \tag{6.5}$$

say, where the terms on the right-hand side are all unknown constants. Their definition and interpretation are given in Table 6.1.

The *unit error* ξ_{itv} is the only term affected by the order in which the treatment is applied to the subject and therefore represents the effect of the order in which the treatment is applied to each subject. From the definitions of the constants in the model (6.5), we have that

$$\alpha_. = \beta_. = \gamma_{i.} = \gamma_{.t} = \xi_{it.} = 0. \tag{6.6}$$

Table 6.1. Definition and Interpretation of the Decomposition of the Expected Response

Constant	Name
$\lambda = \mu_{...}$	Grand mean
$\alpha_t = \mu_{.t.} - \mu_{...}$	Treatment effect
$\beta_i = \mu_{i..} - \mu_{...}$	Subject effect
$\gamma_{it} = \mu_{it.} - \mu_{i..} - \mu_{.t.} + \mu_{...}$	Subject/treatment interaction
$\xi_{itv} = \mu_{itv} - \mu_{it.}$	Unit error

Some authors (Neyman, 1935; Wilk, 1955; Scheffé, 1959, Chapter 9) also include a *technical error* u_{itv} which is a random variable intended to represent the difference between the observed value and the expected value, though others (Welch, 1937a; Pitman, 1937; Kempthorne, 1952, Chapter 7; 1955) do not. When we include a technical error, the response for the ith subject under treatment t when it is applied in the order v is

$$Z_{itv} = \mu_{itv} + u_{itv}. \tag{6.7}$$

Since we defined μ_{itv} to be the expected response, we must have $\mathrm{E}u_{itv} = 0$.

We obviously cannot observe the Z_{itv} in (6.7) over all values of v because the randomization specifies a single value for v. We actually observe

$$Z_{it} = \lambda + \alpha_t + \beta_i + \gamma_{it} + \xi_{it} + u_{it}, \tag{6.8}$$

where ξ_{it} and u_{it} are the ξ_{itv} in (6.5) and u_{itv} in (6.7) for which $v = v(i, t)$ is the realized value of v, and we base inference on the pairwise differences

$$\begin{aligned} Z_i &= Z_{i1} - Z_{i0} \\ &= \alpha_1 - \alpha_0 + \gamma_{i1} - \gamma_{i0} + \xi_{i1} - \xi_{i0} + u_{i1} - u_{i0} \\ &= \mu + \theta_i + \xi_i + u_i, \end{aligned} \tag{6.9}$$

say, where $\mu = \alpha_1 - \alpha_0$ is the caffeine effect, $\theta_i = \gamma_{i1} - \gamma_{i0}$, $\xi_i = \xi_{i1} - \xi_{i0}$, and $u_i = u_{i1} - u_{i0}$, $i = 1, \ldots, n$. Notice that here the pairing enables us to eliminate the subject effect β_i from the model but that in contrast to the analysis in Section 6.1.1 based on (6.1), the subject effect here is a constant.

It is convenient to represent the outcome of the randomization in terms of $4n$ binary random variables $\{d_{itv}\}$ (which are independent of the technical errors $\{u_{itv}\}$) such that $d_{itv} = 1$ if subject i is assigned treatment t in the order v and $d_{itv} = 0$ otherwise. For fixed i, the possible values of d_{itv} can be arranged in a two-way table as

		Order	
		First	Second
	Control	d_{i00}	d_{i01}
Treatment	Treatment	d_{i10}	d_{i11}

The d_{itv} are not independent because a 1 can occur only once in each row and column. There are two ways in which this can happen ($d_{i11} = d_{i00} = 1$ and $d_{i10} = d_{i01} = 1$) and they are treated as equally likely under the randomization. Notice also that $d_{i10} = 1 - d_{i11}$ so that randomization is entirely determined by the sequence of n independent binary random variables $\{d_{i11}\}$.

In terms of $\{d_{itv}\}$, we can write the ξ_{it} and u_{it} of (6.8) as

$$\xi_{it} = \sum_{v=0}^{1} d_{itv}\xi_{itv} \quad \text{and} \quad u_{it} = \sum_{v=0}^{1} d_{itv}u_{itv},$$

and hence ξ_i and u_i of (6.9) as

$$\begin{aligned}
\xi_i &= \xi_{i1} - \xi_{i0} \\
&= \sum_{v=0}^{1} (d_{i1v}\xi_{i1v} - d_{i0v}\xi_{i0v}) \\
&= (d_{i10}\xi_{i10} - d_{i00}\xi_{i00}) + (d_{i11}\xi_{i11} - d_{i01}\xi_{i01}) \\
&= (d_{i10}\xi_{i10} - d_{i11}\xi_{i00}) + (d_{i11}\xi_{i11} - d_{i10}\xi_{i01}) \\
&= (1 - d_{i11})(\xi_{i10} - \xi_{i01}) + d_{i11}(\xi_{i11} - \xi_{i00}) \\
&= (2d_{i11} - 1)(\xi_{i11} - \xi_{i00})
\end{aligned} \tag{6.10}$$

because the constraint $\xi_{it.} = 0$ in (6.6) implies that $\xi_{i01} = -\xi_{i00}$ and $\xi_{i11} = -\xi_{i10}$, and, similarly,

$$u_i = (1 - d_{i11})(u_{i10} - u_{i01}) + d_{i11}(u_{i11} - u_{i00}). \tag{6.11}$$

6.2.2 Design-Unbiasedness of the Mean of the Pairwise Differences

It follows from (6.9) that

$$\bar{Z} = \mu + n^{-1} \sum_{i=1}^{n} (\xi_i + u_i) \tag{6.12}$$

as $\theta_{.} = \gamma_{.1} - \gamma_{.0} = 0$ by (6.6). From (6.10) we have that

$$\mathrm{E}\xi_i = \frac{\xi_{i11} - \xi_{i00}}{2} - \frac{\xi_{i11} - \xi_{i00}}{2} = 0 \tag{6.13}$$

and from (6.11) that

$$\mathrm{E}u_i = \frac{\mathrm{E}(u_{i10} - u_{i01})}{2} + \frac{\mathrm{E}(u_{i11} - u_{i00})}{2} = 0, \tag{6.14}$$

and hence

$$\mathrm{E}\bar{Z} = \mu + n^{-1} \sum_{i=1}^{n} (\mathrm{E}\xi_i + \mathrm{E}u_i) = \mu$$

which establishes (6.3).

6.2.3 Design-Based Variance of the Mean of the Pairwise Differences

To find the variance of $\bar{Z}$, notice that if the $\{u_i\}$ are independent random variables, we have from (6.12) that

$$\begin{aligned}\mathrm{Var}(\bar{Z}) &= \mathrm{E}(\bar{Z} - \mu)^2 \\ &= \mathrm{E}\left\{n^{-1} \sum_{i=1}^{n} (\xi_i + u_i)\right\}^2 \\ &= n^{-2} \sum_{i=1}^{n} \mathrm{E}(\xi_i + u_i)^2 + n^{-2} \sum_{i=1}^{n} \sum_{j \neq i} \mathrm{E}(\xi_i + u_i)(\xi_j + u_j).\end{aligned}$$

Now by (6.10) and (6.11) we have that for all $i, j \in \{1, \ldots, n\}$,

$$\begin{aligned}\mathrm{E}u_i\xi_j &= \mathrm{E}\{(2d_{i11} - 1)(\xi_{i11} - \xi_{i00})\}\{(1 - d_{i11})(u_{i10} - u_{i01}) + d_{i11}(u_{i11} - u_{i00})\} \\ &= (\xi_{i11} - \xi_{i00}) \frac{\{\mathrm{E}(u_{i11} - u_{i00}) - \mathrm{E}(u_{i10} - u_{i01})\}}{2} \\ &= 0.\end{aligned} \tag{6.15}$$

Hence by (6.13), (6.14) and (6.15), we have

$$\begin{aligned}\mathrm{Var}(\bar{Z}) &= n^{-2} \sum_{i=1}^{n} \{\mathrm{Var}(\xi_i) + \mathrm{Var}(u_i)\} \\ &= n^{-1}(\sigma_\xi^2 + \sigma_u^2),\end{aligned} \tag{6.16}$$

where $\sigma_\xi^2 = n^{-1} \sum_{i=1}^{n} \mathrm{Var}(\xi_i)$ and $\sigma_u^2 = n^{-1} \sum_{i=1}^{n} \mathrm{Var}(u_i)$.

6.2.4 Design-Unbiasedness of the Sample Variance of the Pairwise Differences

Applying the usual decomposition to the sum of squares and then (6.9) and (6.16), we obtain

$$\begin{aligned} \mathrm{E} \sum_{i=1}^{n} (Z_i - \bar{Z})^2 &= \sum_{i=1}^{n} \mathrm{E}(Z_i - \mu)^2 - n\mathrm{E}(\bar{Z} - \mu)^2 \\ &= \sum_{i=1}^{n} \mathrm{E}(\theta_i + \xi_i + u)^2 - (\sigma_\xi^2 + \sigma_u^2) \\ &= (n-1)(\sigma_\xi^2 + \sigma_u^2) + \sum_{i=1}^{n} \theta_i^2. \end{aligned}$$

Hence

$$\mathrm{E}S^2 = \mathrm{E}(n-1)^{-1} \sum_{i=1}^{n} (Z_i - \bar{Z})^2 = (\sigma_\xi^2 + \sigma_u^2) + (n-1)^{-1} \sum_{i=1}^{n} \theta_i^2. \quad (6.17)$$

Comparing (6.17) to (6.16), we see that the claim that S^2 is a design-unbiased estimator of the variability is true when, in addition to the assumptions of the randomization model, we have $\theta_i = 0$ for all $i = 1, \ldots, n$. This requirement is basically that the subject/treatment interaction is 0 which means that the effect of caffeine is the same for each subject. This is implicitly assumed in the Gaussian model when we specify a common μ across subjects so (6.4) holds under the usual assumption of constant treatment effect.

6.3 RANDOMIZATION TESTS

We showed in Section 6.2 that the randomization model can be used to establish the design-unbiasedness of the sample mean and variance. In fact, Fisher (1925a) suggested that we go further and actually use the randomization model as the basis for inference by computing p-values under the randomization distribution derived from the randomization model $\mathscr{R}$ (as opposed to the sampling distribution derived from the data model $\mathscr{F}$).

6.3.1 The Randomization Model Under the Null Hypothesis of No Treatment Effect

Suppose we want to test the null hypothesis that there is no treatment effect. Under the randomization model (6.8), this hypothesis entails $\mathrm{H}_0\colon \mu = 0, \gamma_{it} = 0$, $t = 0, 1, i = 1, \ldots, n$. Under H_0, the model for the pairwise differences (6.9) becomes

$$Z_i = \xi_i + u_i, \qquad i = 1, \ldots, n, \quad (6.18)$$

where ξ_i and u_i are defined by (6.10) and (6.11) respectively. We will suppose for simplicity in this and the next section that there are no technical errors so that $u_i = 0$ in (6.18). When technical errors are present, conditional arguments can be used to determine the randomization distribution (see Scheffé, 1959, p. 322). It follows then from (6.18) and (6.10) that, under the randomization model and H_0,

$$Z_i = (2d_{i11} - 1)(\xi_{i11} - \xi_{i00}), \qquad i = 1, \ldots, n. \tag{6.19}$$

From the definition of $\{d_{itv}\}$ in Section 6.2.1, the $\{2d_{i11} - 1\}$ are independent random variables which are independent of u_i such that $2d_{i11} - 1 = 1$ with probability $\frac{1}{2}$ and $2d_{i11} - 1 = -1$ with probability $\frac{1}{2}$. Thus, under the randomization model and H_0, Z_i takes on only two values, $\pm(\xi_{i11} - \xi_{i00})$.

Heuristically, under H_0, the treatment effect is the same as the control effect so Z_{i1} and Z_{i0} measure the same volume of urine and a nonzero difference $Z_i = Z_{i1} - Z_{i0}$ can be attributed to uncontrolled nuisance factors. Thus a different randomization in which the order of receiving the control or treatment is reversed would result only in a change of labels and hence our observing not Z_i but $-Z_i$.

Equation (6.19) implies that under H_0 and replications in which everything except the randomization is held fixed, the sample space of differences is the set of 2^n possible points

$$\mathscr{Z}_R = \{(\pm|\xi_{i11} - \xi_{i01}|, \pm|\xi_{i11} - \xi_{i01}|, \ldots, \pm|\xi_{i11} - \xi_{i01}|)\} \tag{6.20}$$

and these are all equally likely. We obtain the randomization distribution of a statistic by calculating the value of the statistic at all 2^n points in $\mathscr{Z}_R$ and noting that each of these values of the statistic occurs with probability $1/2^n$. Thus, the randomization p-value for a test of H_0 based on a test statistic $T(\mathbf{Z})$ (chosen so large values represent evidence against H_0) is

$$\begin{aligned} p\text{-value} &= P\{T(\mathbf{Z}) \geq T(\mathbf{z}); H_0\} \\ &= \frac{\#\{\text{points in } \mathscr{Z}_R \text{ leading to a } T \text{ as or more extreme than that actually observed}\}}{2^n}. \end{aligned} \tag{6.21}$$

The p-value can always be obtained numerically but sometimes exact or approximate analytic results can be used to simplify the computations.

6.3.2 The Sign Test

As in any frequentist analysis, we need to choose the test statistic on which to base the test. A venerable test considered by Arbuthnot (1710) is the *sign test* based on the sum of the signs of the paired differences. (For some other possibilities, see Sections 6.4 and Section 6.9.1.) Formally, let

$$\Delta_i = \begin{cases} 1 & \text{if } Z_i \geq 0 \\ -1 & \text{otherwise.} \end{cases} \tag{6.22}$$

Then the sign test is based on the statistic

$$S = \sum_{i=1}^{n} \Delta_i. \tag{6.23}$$

Intuitively, under H_0, ignoring variability, we expect the same number of positive as negative observations so $S = 0$ while under any departures from H_0 we expect $|S| > 0$. Of course, we need to take the variability into account by computing the distribution of S under H_0.

The randomization distribution of S under H_0 is derived by evaluating S at each of the 2^n equally likely points in $\mathscr{Z}_R$. This is a tedious calculation which, in the case of the sign test, we can avoid by obtaining the randomization distribution of S under H_0 analytically. Comparing the definition (6.22) of Δ_i to (6.19), we see that

$$\Delta_i = (2d_{i11} - 1), \qquad i = 1, \ldots, n.$$

Since $\{d_{i11}\}$ are independent binomial$(1, \frac{1}{2})$ random variables, S has the same distribution under H_0 as $2B - n$, where $B \sim$ binomial$(n, \frac{1}{2})$.

The exact p-value for testing H_0 with the sign test is readily computed. We have

$$\begin{aligned} p &= \mathrm{P}\{|S| \geq |s|; \mathrm{H}_0\} \\ &= \mathrm{P}\left\{\left|B - \frac{n}{2}\right| \geq \left|\frac{s}{2}\right|; \mathrm{H}_0\right\} \\ &= \mathrm{P}\left\{B \leq \frac{n}{2} - \left|\frac{s}{2}\right|; \mathrm{H}_0\right\} + \mathrm{P}\left\{B \geq \frac{n}{2} + \left|\frac{s}{2}\right|; \mathrm{H}_0\right\} \\ &= \sum_{i=0}^{n/2-|s/2|} \binom{n}{i}\left(\frac{1}{2}\right)^n + \sum_{i=n/2+|s/2|}^{n} \binom{n}{i}\left(\frac{1}{2}\right)^n. \end{aligned}$$

Alternatively, using the Gaussian approximation to the binomial distribution (see Section 3.7.2), we obtain

$$p \approx 2\{1 - \Phi(n^{-1/2}|s|)\}.$$

For the change in the volume of urine data, 13 out of 18 of the pairwise differences are positive so $s = 8$. The p-value is

$$\mathrm{P}\{B \geq 13; \mathrm{H}_0\} + \mathrm{P}\{B \leq 5; \mathrm{H}_0\} = 0.059$$

which is borderline at the usual levels. This finding is much weaker than those based on the model-based analysis of the sample mean presented in Section 3.2.

Notice that if all the pairwise differences are positive ($Z_i > 0$), $S = n$ is as extreme as it can be and $p = 1/2^n$. As we increase n, the p-value decreases and, in general, the greater the cardinality of $\mathscr{Z}_R$, the more extreme an extreme result will be. At the other extreme, if we have n matched pairs of units and we randomly assign one group of n to treatment and the other to control, then $\mathscr{Z}_R = \{(z_1, \ldots, z_n), -(z_1, \ldots, z_n)\}$ and the only possible p-values are 1 and $\frac{1}{2}$ whatever the value of n so it is important to know which randomizations were considered. If two people carry out the same matched pairs experiment and get the same results but selected the design from a different randomization set, they will obtain different significance levels. The choice of test statistic also affects the p-value: if instead of S, we base a test on median(Z_i), we obtain a different p-value.

One difficulty with randomization tests is that it is difficult to formulate meaningful alternative hypotheses in the randomization model. This difficulty is often avoided by treating randomization tests as permutation tests (Section 6.6).

PROBLEMS

6.3.1. A sample of five observations from the urine data of Section 1.1.3 is -90, 410, -5, 87, 40. Compute the randomization distribution (under the hypothesis of no caffeine effect) of the sample mean exactly. How does this compare to the Gaussian approximation? Repeat the computations after replacing observation 40 by 540. Compare the results for the perturbed and the unperturbed data. What does this exercise show about the robustness of the randomization approach? What happens if we use the sample median instead of the sample mean?

6.4 THE RANDOMIZATION BASIS FOR GAUSSIAN MODEL-BASED TESTS

Model-based tests of sharp hypotheses about μ in the Gaussian model (6.1) are constructed from the fact that under (6.1), $n^{1/2}(\bar{Z} - \mu)/S$ has the Student t distribution with $\nu = n - 1$ degrees of freedom (see Section 3.2) or equivalently $n(\bar{Z} - \mu)^2/S^2$ has an F distribution with 1 and ν degrees of freedom. If the sample size is large, a test of the hypothesis H_0 of no caffeine effect on the volume or urine produced can be based on the fact that under $H_0: \mu = 0$

$$n^{1/2}\bar{Z}/S \sim t_\nu \to N(0, 1) \qquad \text{as } n \to \infty,$$

or, equivalently,

$$F = \frac{n\bar{Z}^2}{S^2} = \frac{\nu n\bar{Z}^2}{\sum_{i=1}^{n} Z_i^2 - n\bar{Z}^2} \sim F(1, \nu) \to \chi_1^2 \qquad \text{as } n \to \infty, \tag{6.24}$$

where $\nu = n - 1$.

Suppose that instead of using the sign test (Section 6.3.2) we decide to base a randomization test of the null hypothesis H_0 of no caffeine effect on the volume of urine produced on the usual Gaussian theory F-statistic

$$F = n\bar{Z}^2/S^2 = \nu n\bar{Z}^2 \bigg/ \left\{ \sum_{i=1}^{n} Z_i^2 - n\bar{Z}^2 \right\}.$$

To implement the test, we require the randomization distribution of F or a good approximation to it. It is convenient to approximate the randomization distribution of $U = n\bar{Z}^2/\sum_{i=1}^{n} Z_i^2$ and then use the fact that

$$F = \frac{\nu U}{1 - U} \tag{6.25}$$

to obtain the randomization distribution of F from that of U.

6.4.1 The Approximate Randomization Distribution of U

Under the randomization model (6.9) and H_0, we can apply (6.19) to show that U is a realization of

$$U_R = n^{-1} \frac{\{\sum_{i=1}^{n} (2d_{i11} - 1)|\xi_{i11} - \xi_{i00}|\}^2}{\sum_{i=1}^{n} (\xi_{i11} - \xi_{i00})^2}.$$

From (6.19), $|Z_i| = |\xi_{i11} - \xi_{i00}|$ which is nonstochastic, a fact we emphasize by writing $|Z_i| = |z_i|$, where z_i is the realized value of Z_i. In this notation, we write

$$U_R = n^{-1} \frac{\{\sum_{i=1}^{n} (2d_{i11} - 1)|z_i|\}^2}{\sum_{i=1}^{n} z_i^2}.$$

The randomization distribution of U_R is obtained by calculating U_R over all 2^n points in $\mathscr{X}_R$ and noting that each of these values of U_R occurs with probability $1/2^n$. In the caffeine experiment with $n = 18$, we have to compute $2^{18} = 262\,144$ values of U_R to compute the randomization distribution and hence the randomization p-value. It makes sense to try to approximate the randomization distribution.

Since $\{2d_{i11} - 1\}$ are independent and identically distributed random variables with mean $\mathrm{E}(2d_{i11} - 1) = 0$ and $\mathrm{Var}\,(2d_{i11} - 1) = 1$, the statistic U_{R} is a weighted mean of independent and identically distributed random variables such that

$$\mathrm{E}U_{\mathrm{R}} = n^{-1}\mathrm{E}\,\frac{\{\sum_{i=1}^{n} (2d_{i11} - 1)|z_i|\}^2}{\sum_{i=1}^{n} z_i^2} = n^{-1} \tag{6.26}$$

and

$$\mathrm{Var}\,(U_{\mathrm{R}}) = n^{-2}\mathrm{E}\left[\frac{\{\sum_{i=1}^{n} (2d_{i11} - 1)|z_i|\}^2}{\sum_{i=1}^{n} z_i^2} - 1\right]^2$$

$$= n^{-2}\mathrm{E}\,\frac{[\{\sum_{i=1}^{n} (2d_{i11} - 1)|z_i|\}^2 - \sum_{i=1}^{n} z_i^2]^2}{\{\sum_{i=1}^{n} z_i^2\}^2}$$

$$= n^{-2}\mathrm{E}\,\frac{\{\sum_{i=1}^{n} (2d_{i11} - 1)|z_i|\}^4 - \{\sum_{i=1}^{n} z_i^2\}^2}{\{\sum_{i=1}^{n} z_i^2\}^2}.$$

Since

$$\mathrm{E}\left\{\sum_{i=1}^{n} (2d_{i11} - 1)|z_i|\right\}^4 = \sum_{i=1}^{n} z_i^4 + 3\sum_{i \neq j}^{n} z_i^2 z_j^2 = 3\left\{\sum_{i=1}^{n} z_i^2\right\}^2 - 2\sum_{i=1}^{n} z_i^4,$$

we can write

$$\mathrm{Var}\,(U_{\mathrm{R}}) = 2n^{-2}\,\frac{\{\sum_{i=1}^{n} z_i^2\}^2 - \sum_{i=1}^{n} z_i^4}{\{\sum_{i=1}^{n} z_i^2\}^2}$$

$$= 2n^{-2}\left[1 - \frac{\sum_{i=1}^{n} z_i^4}{\{\sum_{i=1}^{n} z_i^2\}^2}\right]. \tag{6.27}$$

Results (6.26) and (6.27) were obtained by Welch (1937a) and Pitman (1937).

Since the distribution of U_{R} has support on $(0, 1)$, we can approximate it by a (continuous) beta (r, s) distribution. If we want the first two moments to match $r/(r + s)$ and $2rs/(r + s)^2(r + s + 2)$ we find

$$r = \lambda, \qquad s = (n - 1)\lambda,$$

where $\lambda = (1 - n^{-1}V)^{-1} - n^{-1}2$ with $V = (n - 1)^{-1}n[n\sum_{i=1}^{n} z_i^4/\{\sum_{i=1}^{n} z_i^2\}^2 - 1]$. Thus we have the approximation

$$U_{\mathrm{R}} \sim \mathrm{beta}\,(\lambda, (n - 1)\lambda). \tag{6.28}$$

Pitman (1937) showed that the third and fourth moments of U_R also match those of a beta $(\lambda, (n-1)\lambda)$ distribution reasonably well provided $n^{-1}V$ is not too close to 1.

6.4.2 The Approximate Randomization Distribution of the F-Ratio

If $X \sim$ beta (r, s), it follows that $sX/r(1-X) \sim F(r, s)$ so that from (6.25) and (6.28) we have the approximation

$$F_R = \frac{vU_R}{1-U_R} \sim F(\lambda, v\lambda),$$

where $v = n-1$. Notice that when $\lambda \approx 1$ or, equivalently, when $n^{-1}V \approx 0$, we have

$$F_R \sim F(1, v) \tag{6.29}$$

which can be compared to the model-based approximation (6.24).

6.4.3 Large Sample Approximations

We can show that asymptotically, $n^{-1}V \to 0$ and hence $\lambda \to 1$. Notice that

$$\mathrm{E}n^{-1} \sum_{i=1}^{n} (2d_{i11}-1)|z_i| = 0$$

and

$$\mathrm{Var}\left\{n^{-1} \sum_{i=1}^{n} (2d_{i11}-1)|z_i|\right\} = n^{-2} \sum_{i=1}^{n} \mathrm{Var}\,(2d_{i11}-1)z_i^2 = n^{-2} \sum_{i=1}^{n} z_i^2$$

so we can apply the Lyapounov central limit theorem (see the remarks following Theorem 4.1 in Section 4.1.7) to show that

$$\left\{\sum_{i=1}^{n} z_i^2\right\}^{-1/2} \sum_{i=1}^{n} (2d_{i11}-1)z_i \xrightarrow{\mathscr{D}} \mathrm{N}(0, 1),$$

provided $\{\sum_{i=1}^{n} z_i^2\}^{-(1+\gamma/2)} \sum_{i=1}^{n} |z_i|^{2+\gamma} \to 0$, for some $\gamma > 0$. It then follows from Theorem 4.7 that

$$nU_R = \left\{\sum_{i=1}^{n} z_i^2\right\}^{-1} \left\{\sum_{i=1}^{n} (2d_{i11}-1)|z_i|\right\}^2 \xrightarrow{\mathscr{D}} \chi_1^2.$$

Using (6.25) and the fact that $U_{\mathrm{R}} = o_p(1)$, it follows from Theorem 4.2 that

$$F_{\mathrm{R}} \xrightarrow{\mathscr{D}} \chi_1^2, \qquad \text{as } n \to \infty, \tag{6.30}$$

which is the result of Wald and Wolfowitz (1944).

6.4.4 The Quality of the Approximations to the Randomization Distribution

The quality of the approximations (6.29) and (6.30) is illustrated in Figure 6.1. With $n = 18$, there are already too many permutations to compute to get the randomization distribution exactly so we adopt a statistical approach and estimate it from a random sample of permutations. The randomization distribution in Figure 6.1 is obtained by taking a random sample of 10 000 permutations of the data, computing F_{R} in each case and then estimating the quantile function of the distribution of F_{R}. The quantile function of the distribution of F_{R} is plotted against the quantile functions of the $F(\lambda, (n-1)\lambda)$ and χ_1^2 distributions for comparison. Since $\lambda = 1.067$, we expect and see that the two approximations are very similar. The chi-squared approximation is the

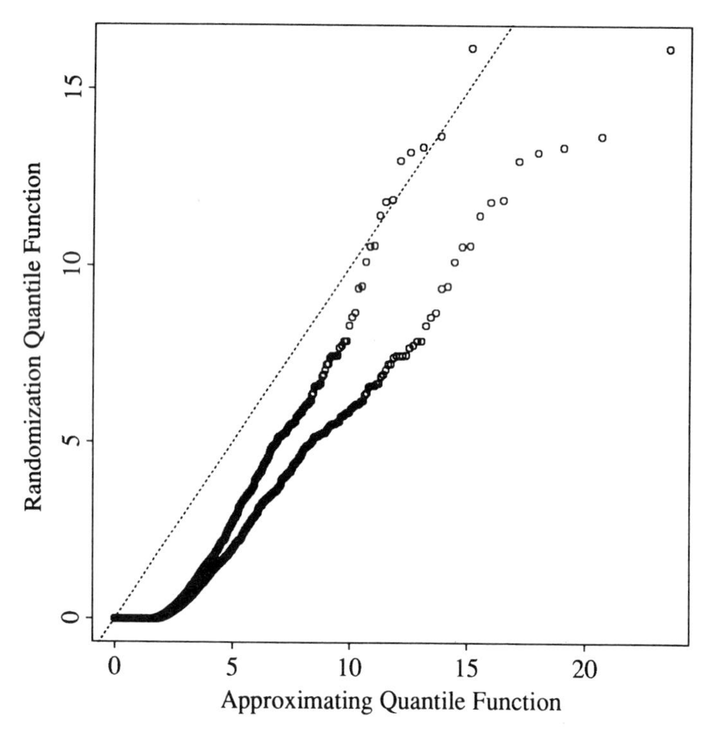

Figure 6.1. A comparison of the randomization distribution of F_{R} for the urine data with the F and χ^2 approximations.

better of the two because the plotted curve is closer to the 45° line than the curve for the F approximation. Both approximations put less of a spike near the origin and more weight over the right tail of the distribution. The observed F-ratio is $3.50^2 = 12.25$ so the p-value is 0.0005, 0.002 and 0.0005 for the randomization distribution, the F approximation and the chi-squared approximation respectively.

6.4.5 The Justification for the Gaussian Theory Results

The similarity between the approximations (6.29) or (6.30) and (6.24) can be interpreted as showing that the design-based approach provides a justification for the usual Gaussian model-based analysis even when this model does not hold. However, the sample space is different in the two analyses: The model-based analysis fixes the design and considers hypothetical independent replications of the data generating process while the design-based analysis fixes the data and considers hypothetical replications of the design.

Although the randomization or design-based approach was suggested by Fisher (1925a), it was not pursued by him and his later views on the subject are unclear (see Basu, 1980). However, unless the probability of selecting a particular design is made to depend on the parameters of interest, the design actually used in an experiment is an ancillary statistic (Sections 3.9.2 and 7.4) which contains no information about the parameter of interest. Thus, according to Fisher's conditionality ideas (which were discussed in Section 3.9 and are discussed further in Section 7.4), we should make inference conditional on the design, and design-based inference conflicts with this advice. In particular, in Fisher's terms, in the randomization test we condition on what is informative and base inference on what is uninformative. The importance of this point is obscured here by the numerical agreement between the two approaches and it becomes more important when the approaches differ numerically as we will see in the next section.

PROBLEMS

6.4.1. We presented data in Section 1.1.3 on the amount of epinephrine voided in urine after ingesting caffeine and after ingesting a control liquid. Carry out randomization tests based on the sample mean and the sample median to see whether the amount of epinephrine in voided urine is affected by the ingestion of caffeine. Compare your inferences.

6.4.2. Suppose we have $n = n_1 + n_2$ homogeneous units, n_1 of which are assigned at random to a control group and the remaining n_2 of which are assigned to a treatment. Describe how to carry out a randomization test of the null hypothesis of no treatment effect.

6.5 INFERENCE FOR FINITE POPULATIONS

The randomization-based or design-based approach to inference provides support for the classical model-based approach to analyzing data from designed experiments. Nonetheless, in practice, a model-based approach is usually adopted for the analysis of data from designed experiments. In contrast, the design-based approach advocated by Neyman (1934) has been and remains the dominant paradigm for inference in finite population sampling theory. In this context, the design-based and model-based approaches often lead to different results.

6.5.1 A Finite Population Problem

To formalize the discussion, suppose that we have a finite population $\mathscr{P}$ consisting of N units which we label $1, \ldots, N$, each of which has associated with it the value Y_i of a *survey variable* Y of interest. We sometimes also have available information about the units in the population which we use to select the sample from the population. We represent this information as a *design variable* X, the values X_i of which are known for all units in the population. The act of drawing a sample of size $n < N$ from $\mathscr{P}$ involves specifying a set $\mathscr{S} = \{i: i \in \{1, \ldots, N\}\}$ of labels of units included in the sample. Equivalently, the sample selection process produces N binary *sample inclusion indicators* I_i which indicate whether unit i is included in the sample or not. Schematically, we can represent $\mathscr{P}$ as in Table 6.2.

Once we have selected the sample, we observe the value Y_i of Y for each of the n units included in the sample. Thus we know all N values of the design variable and the sample indicator but only n values of the survey variable. A common problem is to use the known X_i, $i = 1, \ldots, N$ and the observed Y_i, $i \in \mathscr{S}$ to estimate *finite population parameters* such as the finite population total $T = \sum_{i=1}^{N} Y_i$.

As a simple illustrative example, take $\mathscr{P}$ to be a set of companies in a particular region. Let Y be a measure of economic activity and X a measure of economic size. Then we may be interested in estimating $T = \sum_{i=1}^{N} Y_i$, the total economic activity in the region.

Table 6.2. A Typical Finite Population Structure

Units	Design Variable	Sample Indicator	Survey Variable
1	X_1	I_1	Y_1
2	X_2	I_2	Y_2
⋮			
N	X_N	I_N	Y_N

6.5.2 Randomization in Sample Selection

Many of the issues which arise in designing an experiment (Section 6.1) have analogues in selecting a sample. For example, *stratification*, the partitioning of a heterogeneous population into homogeneous subpopulations or strata, is analogous to blocking (Section 6.1.1) and the use of random sampling methods to select samples is analogous to the random allocation of treatments to units (Section 6.1.3).

Samples selected from a finite population by a randomization procedure are called *random samples*. Formally, a random sample is a sample selected in such a way that we know the probability that any possible sample will be drawn. This means that for a random sampling scheme, we can set up the space of all possible samples of size n from $\mathscr{P}$ and work out the probability of selecting each one.

It is convenient to express the probability of selecting a sample in terms of the *inclusion probabilities* $\pi_i = \mathrm{E}I_i$ which are the probabilities that unit i is included in a sample of size n. To keep the present discussion simple, we allow the inclusion probabilities to depend on X but not on Y. (Technically, this is known as *uninformative sampling* and it ensures that the sample design contains no information about T so is ancillary. See Sections 3.9.2 and 7.4.)

A random sample is a *probability sample* if $\pi_i > 0$ for all i and a *simple random sample* if all possible samples of size n are equally likely. In particular, when this occurs $\pi_i = n/N$. In *stratified random sampling*, we classify the units into strata (on the basis of the design variable) and draw a simple random sample from each stratum. In this case, if we let $h(i)$ denote the stratum to which unit i belongs, $\pi_i = n_{h(i)}/N_{h(i)}$, where n_h is number of units in the sample from stratum h and N_h is the number of units in stratum h. Obviously, these probabilities can be expressed as a function of the design variable X_i since this determines stratum membership. An even more explicit dependence on X_i occurs when X_i is a non-negative measure of size and we use *probability proportional to size sampling* (PPS). In this case, $\pi_i = nX_i/\sum_{j=1}^{N} X_j$.

6.5.3 The Design-Based Approach

The design-based approach to analyzing samples from finite populations treats the Y_i as fixed numbers (so T is an unknown constant) and randomness arises only from the sampling process which is represented by the sample inclusion indicator I_i. Clearly, for statistical methods to apply to this approach, the sample must be a probability sample.

The canonical estimator of T in the design-based approach is the *Horvitz–Thompson* (1952) *estimator*

$$\hat{T}_{\mathrm{H}} = \sum_{i \in \mathscr{S}} \frac{Y_i}{\pi_i} = \sum_{i=1}^{N} \frac{I_i Y_i}{\pi_i}. \tag{6.31}$$

This estimator does not make explicit use of the design variable although of course, π_i may depend on the design variable. Under the design-based approach

$$\mathrm{E}\hat{T}_{\mathrm{H}} = \sum_{i=1}^{N} \frac{\mathrm{E}I_i Y_i}{\pi_i} = \sum_{i=1}^{N} Y_i = T$$

so the estimator is design-unbiased. Also

$$\begin{aligned}\mathrm{Var}(\hat{T}_{\mathrm{H}}) &= \mathrm{E}\left\{\sum_{i=1}^{N}\left(\frac{I_i}{\pi_i} - 1\right)Y_i\right\}^2 \\ &= \sum_{i=1}^{N} \mathrm{E}\left(\frac{I_i}{\pi_i} - 1\right)^2 Y_i^2 + \sum_{i=1}^{N} \sum_{j=1, j\neq i}^{N} \mathrm{E}\left(\frac{I_i}{\pi_i} - 1\right)\left(\frac{I_j}{\pi_j} - 1\right)Y_i Y_j \\ &= \sum_{i=1}^{N}\left(\frac{1}{\pi_i} - 1\right)Y_i^2 + \sum_{i=1}^{N} \sum_{j=1, j\neq i}^{N}\left(\frac{\pi_{ij}}{\pi_i \pi_j} - 1\right)Y_i Y_j,\end{aligned} \tag{6.32}$$

where $\pi_{ij} = \mathrm{P}(i \in \mathscr{S} \text{ and } j \in \mathscr{S})$ is the *joint sample inclusion probability.* The design-based variance (6.32) can be estimated by the design-unbiased estimator

$$\hat{V}_{\mathrm{H}} = \sum_{i\in\mathscr{S}} \frac{(1-\pi_i)Y_i^2}{\pi_i^2} + \sum_{i\in\mathscr{S}} \sum_{j\in\mathscr{S}, j\neq i}\left(1 - \frac{\pi_i\pi_j}{\pi_{ij}}\right)\left(\frac{Y_i}{\pi_i}\right)\left(\frac{Y_j}{\pi_j}\right) \tag{6.33}$$

proposed by Horvitz and Thompson (1952).

If n is fixed, we can re-express (6.32) in a more convenient form. Since $\sum_{i=1}^{N} I_i = n$, we can take expectations of both sides to obtain

$$\sum_{i=1}^{N} \pi_i = n.$$

Also

$$\sum_{j=1, j\neq i}^{N} \pi_{ij} = \sum_{j=1, j\neq i}^{N} \mathrm{E}I_i I_j = \mathrm{E}\{(n-1)I_i\} = (n-1)\pi_i,$$

so we have

$$\sum_{j=1, j\neq i}^{N} (\pi_i\pi_j - \pi_{ij}) = \pi_i(n - \pi_i) - (n-1)\pi_i = \pi_i(1-\pi_i)$$

and we can re-express (6.32) as

$$\begin{aligned}\operatorname{Var}(\hat{T}_{\mathrm{H}}) &= \sum_{i=1}^{N} \pi_i(1-\pi_i)\left(\frac{Y_i}{\pi_i}\right)^2 + \sum_{i=1}^{N} \sum_{j=1, j\neq i}^{N} (\pi_{ij}-\pi_i\pi_j)\left(\frac{Y_i}{\pi_i}\right)\left(\frac{Y_j}{\pi_j}\right) \\ &= \sum_{i=1}^{N} \sum_{j=1, j\neq i}^{N} (\pi_i\pi_j-\pi_{ij})\left(\frac{Y_i}{\pi_i}\right)^2 - \sum_{i=1}^{N} \sum_{j=1, j\neq i}^{N} (\pi_{ij}-\pi_i\pi_j)\left(\frac{Y_i}{\pi_i}\right)\left(\frac{Y_j}{\pi_j}\right) \\ &= \frac{1}{2}\sum_{i=1}^{N} \sum_{j=1, j\neq i}^{N} (\pi_i\pi_j-\pi_{ij})\left\{\left(\frac{Y_i}{\pi_i}\right)^2 + \left(\frac{Y_j}{\pi_j}\right)^2 - 2\left(\frac{Y_i}{\pi_i}\right)\left(\frac{Y_j}{\pi_j}\right)\right\} \\ &= \frac{1}{2}\sum_{i=1}^{N} \sum_{j=1, j\neq i}^{N} (\pi_i\pi_j-\pi_{ij})\left\{\frac{Y_i}{\pi_i} - \frac{Y_j}{\pi_j}\right\}^2.\end{aligned}$$

This suggests the design-unbiased estimator

$$\hat{V}_{\mathrm{Y}} = \frac{1}{2}\sum_{i\in\mathscr{S}} \sum_{j\in\mathscr{S}, j\neq i} \left(\frac{\pi_i\pi_j}{\pi_{ij}} - 1\right)\left\{\frac{Y_i}{\pi_i} - \frac{Y_j}{\pi_j}\right\}^2 \tag{6.34}$$

which was proposed by Yates and Grundy (1953) and Sen (1953) for (6.32) when n is fixed. Somewhat disturbingly, both (6.33) and (6.34) can be negative (Cochran, 1977, Section 9A.7).

6.5.4 The Horvitz–Thompson Estimator Under Simple Random Sampling Without Replacement

In the special case of a *simple random sample selected without replacement*, we have $\pi_i = n/N$, $\pi_{ij} = n(n-1)/N(N-1)$, and the Horvitz–Thompson estimator (6.31) is the so-called *expansion estimator*

$$\hat{T}_{\mathrm{H}} = \frac{N}{n}\sum_{i\in\mathscr{S}} Y_i = N\bar{Y}_s,$$

where $\bar{Y}_s = n^{-1}\sum_{i\in\mathscr{S}} Y_i$. In this case, (6.32) reduces to

$$\begin{aligned}\operatorname{Var}(\hat{T}_{\mathrm{H}}) &= \left(1-\frac{n}{N}\right)\frac{N}{n}\sum_{i=1}^{N} Y_i^2 + \frac{1}{(N-1)}\left(1-\frac{n}{N}\right)\frac{N}{n}\sum_{i=1, i\neq j}^{N}\sum_{j=1}^{N} Y_iY_j \\ &= \left(1-\frac{n}{N}\right)\frac{N}{n}\frac{1}{(N-1)}\left\{(N-1)\sum_{i=1}^{N} Y_i^2 - \sum_{i=1, i\neq j}^{N}\sum_{j=1}^{N} Y_iY_j\right\} \\ &= \left(1-\frac{n}{N}\right)\frac{N}{n}\frac{1}{(N-1)}\left\{N\sum_{i=1}^{N} Y_i^2 - \left(\sum_{i=1}^{N} Y_i\right)^2\right\} \\ &= \left(1-\frac{n}{N}\right)\frac{N^2}{n} S_Y^2,\end{aligned} \tag{6.35}$$

where $S_Y^2 = (N-1)^{-1} \sum_{i=1}^{N} (Y_i - \bar{Y})^2$ is the population variance. The usual estimator of Var $(\hat{T}_H)$ is the design-unbiased estimator

$$\hat{V}_H = \left(1 - \frac{N}{n}\right) \frac{N^2}{n} s_Y^2,$$

where $s_Y^2 = (n-1)^{-1} \sum_{i \in \mathcal{S}} (Y_i - \bar{Y}_s)^2$ is the sample variance.

6.5.5 The Ratio Estimator

One of the difficulties of the design-based approach to analyzing sample surveys is that although design-unbiasedness is the fundamental concept in the design-based approach, design-biased estimators often perform better than the Horvitz–Thompson estimator. For example, an estimator which tries to use some of the information in the design variable is the *ratio estimator*

$$\hat{T}_R = N \frac{\bar{X} \bar{Y}_s}{\bar{X}_s}, \tag{6.36}$$

where $\bar{X}_s = n^{-1} \sum_{i \in \mathcal{S}} X_i$ and $\bar{X} = N^{-1} \sum_{i=1}^{N} X_i$. Using a simple Taylor series estimation (see 2 in the Appendix), we can write (6.36) as

$$\begin{aligned}
\hat{T}_R &= T + N \frac{\bar{X} \bar{Y}_s}{\bar{X}_s} - T \\
&= T + \frac{(N \bar{X} \bar{Y}_s - T \bar{X}_s)}{\bar{X}_s} \\
&= T + \left(N \bar{Y}_s - T \frac{\bar{X}_s}{\bar{X}}\right) \left\{1 - \frac{(\bar{X}_s - \bar{X})}{\bar{X}} + O((\bar{X}_s - \bar{X})^2)\right\} \\
&= T + N\left\{\bar{Y}_s - \bar{X}_s\left(\frac{\bar{Y}}{\bar{X}}\right)\right\}\left\{1 - \frac{(\bar{X}_s - \bar{X})}{\bar{X}} + O((\bar{X}_s - \bar{X})^2)\right\}
\end{aligned} \tag{6.37}$$

Under simple random sampling without replacement

$$\mathrm{E}\bar{X}_s = \bar{X}$$

$$\mathrm{E}(\bar{Y}_s - \bar{Y})(\bar{X}_s - \bar{X}) = \frac{1}{n}\left(1 - \frac{n}{N}\right) S_{XY}$$

and

$$\mathrm{E}(\bar{X}_s - \bar{X})^2 = \frac{1}{n}\left(1 - \frac{n}{N}\right) S_X^2,$$

where $S_{XY} = (N-1)^{-1} \sum_{i=1}^{N} (X_i - \bar{X})(Y_i - \bar{Y})$, and $S_X^2 = (N-1)^{-1} \sum_{i=1}^{N} \times (X_i - \bar{X})^2$, so

$$\mathrm{E}\hat{T}_{\mathrm{R}} \approx T - N\mathrm{E}\left\{\bar{Y}_s - \bar{X}_s \frac{\bar{Y}}{\bar{X}}\right\} \frac{(\bar{X}_s - \bar{X})}{\bar{X}}$$

$$= T + \frac{N}{n}\left(1 - \frac{N}{n}\right)\left\{\frac{S_X^2 \dfrac{\bar{Y}}{\bar{X}} - S_{XY}}{\bar{X}}\right\}.$$

Using (6.37) again, the mean squared error of the ratio estimator is

$$\begin{aligned}
\mathrm{MSE}\,(\hat{T}_{\mathrm{R}}) &= \mathrm{E}(\hat{T}_{\mathrm{R}} - T)^2 \\
&\approx N^2\mathrm{E}\left\{\bar{Y}_s - \bar{X}_s \frac{\bar{Y}}{\bar{X}}\right\}^2 \\
&= N^2\mathrm{E}\left\{(\bar{Y}_s - \bar{Y}) - (\bar{X}_s - \bar{X})\frac{\bar{Y}}{\bar{X}}\right\}^2 \\
&= N^2\left\{\mathrm{E}(\bar{Y}_s - \bar{Y})^2 - 2\mathrm{E}(\bar{Y}_s - \bar{Y})(\bar{X}_s - \bar{X})\frac{\bar{Y}}{\bar{X}} + \mathrm{E}(\bar{X}_s - \bar{X})^2\left(\frac{\bar{Y}}{\bar{X}}\right)^2\right\} \\
&= \left(1 - \frac{n}{N}\right)\frac{N^2}{n}\left\{S_Y^2 - 2S_{XY}\frac{\bar{Y}}{\bar{X}} + S_X^2\left(\frac{\bar{Y}}{\bar{X}}\right)^2\right\},
\end{aligned}$$

where $S_Y^2 = (N-1)^{-1} \sum_{i=1}^{N} (Y_i - \bar{Y})^2$. The mean squared error is conventionally estimated by

$$\hat{M}_{\mathrm{R}} = \left(1 - \frac{n}{N}\right)\frac{N^2}{n}\left\{s_Y^2 - 2s_{XY}\frac{\bar{Y}_s}{\bar{X}_s} + s_X^2\left(\frac{\bar{Y}_s}{\bar{X}_s}\right)^2\right\}, \tag{6.38}$$

where

$$s_X^2 = (n-1)^{-1} \sum_{i \in \mathscr{S}} (X_i - \bar{X}_s)^2 \quad \text{and} \quad s_{XY} = (n-1)^{-1} \sum_{i \in \mathscr{S}} (X_i - \bar{X}_s)(Y_i - \bar{Y}_s).$$

Straightforward algebra can be used to show that

$$\hat{M}_{\mathrm{R}} = \left(1 - \frac{n}{N}\right)\frac{N^2}{n}(n-1)^{-1} \sum_{i \in \mathscr{S}} \left\{Y_i - \frac{\bar{Y}_s}{\bar{X}_s} X_i\right\}^2. \tag{6.39}$$

The finite population central limit theorem (Hajek, 1960) justifies the use of a normal approximation to the design-based sampling distribution (see for example Scott and Wu, 1981) and this can be used to obtain approximate design-based confidence intervals.

Comparing (6.35) to MSE $(\hat{T}_R)$, we see that under simple random sampling without replacement, MSE $(\hat{T}_R) < \text{Var}\,(\hat{T}_H)$ when

$$\frac{S_{XY}}{S_X^2} > \frac{1}{2}\left(\frac{\bar{Y}}{\bar{X}}\right)$$

for n large.

6.5.6 Difficulties with the Design-Based Approach

The Horvitz–Thompson estimator (6.31) is the design-based UMVU estimator (see Section 3.1.4) in the class of linear estimators of the population total given by

$$\mathscr{T}_1 = \left\{ \sum_{i \in \mathscr{S}} w_i Y_i \colon w_i \text{ are fixed weights} \right\}. \tag{6.40}$$

However, the ratio estimator (6.36) which is design-biased under simple random sampling without replacement often performs better than the Horvitz–Thompson estimator.

Part of the explanation lies in the nature and meaning of design-unbiasedness. Suppose we take a simple random sample of companies. The Horvitz–Thompson estimator (6.31) is design-unbiased regardless of the data we actually observe. Now by chance, we may observe a sample which only contains small companies. In this case, the Horvitz–Thompson estimator is design-unbiased but the estimate of T is too small so has a real negative bias.

The situation is even worse with unequal probability sampling as is dramatically illustrated in Basu's (1971) colourful elephant example.

> The circus owner is planning to ship his 50 adult elephants and so he needs a rough estimate of the total weight of his elephants. As weighing an elephant is a cumbersome process, the owner wants to estimate the total weight by weighing just one elephant. Which elephant should he weigh? So the owner looks back on his records and discovers a list of the elephants' weights taken 3 years ago. He finds that 3 years ago Sambo the middle-sized elephant was the average (in weight) elephant in his herd. He checks with the elephant trainer who reassures him (the owner) that Sambo may still be considered to be the average elephant in the herd. Therefore, the owner plans to weigh Sambo and take $50y$ (where y is the present weight of Sambo) as an estimate of the total weight $Y = Y_1 + \cdots + Y_{50}$ of the 50 elephants. But the circus statistician is horrified when he learns of the owner's purposive sampling plan. "How can you get an unbiased estimate of Y this way?" protests the statistician. So, together they work out a compromise plan. With the help of a table of random numbers they devise a plan that allots a selection probability of 99/100 to Sambo and equal selection

> probabilities of 1/4900 to each of the other 49 elephants. Naturally, Sambo is selected and the owner is happy. "How are you going to estimate Y?" asks the statistician. "Why? The estimate ought to be $50y$ of course," says the owner. "Oh! No! That cannot possibly be right," says the statistician, "I recently read an article in the *Annals of Mathematical Statistics* where it is proved that the Horvitz–Thompson estimator is the unique hyperadmissible estimator in the class of all generalized polynomial unbiased estimators." "What is the Horvitz–Thompson estimate in this case?" asks the owner, duly impressed. "Since the selection probability for Sambo in our plan was 99/100," says the statistician, "the proper estimate of Y is $100y/99$ and not $50y$." "And, how would you have estimated Y," inquires the incredulous owner, "if our sampling plan made us select, say, the big elephant Jumbo?" "According to what I understand of the Horvitz–Thompson estimation method," says the unhappy statistician, "the proper estimate of Y would then have been $4900y$, where y is Jumbo's weight." (Reproduced with the permission of the author.)

Clearly, the estimate based on Sambo is a gross underestimate (negative bias) while that based on Jumbo is a gross overestimate (positive bias).

Both of the above examples and the fact that unbiased estimators of variance can be negative (Section 6.5.3) show that properties like design-unbiasedness which hold on average over all possible samples tell us nothing about the properties of an estimator in any particular sample. Thus an estimator can perform well on average but very poorly in any particular sample. See also Problem 6.5.1.

To overcome the failings of design-unbiasedness, we need to relate the units in the sample to those not in the sample. One way to do this is to replace the weights w_i in (6.40) by weights $w_i(\mathscr{S})$ which depend on the sample. Unfortunately, Godambe (1955) showed that there is no design-based UMVU estimator in this modified class so there is no optimal design-based analysis. Godambe (1966) showed further that the design-likelihood (which is a function of $\mathbf{Y} \in \mathbb{R}^N$) is constant for all $\mathbf{Y} \in \mathbb{R}^N$ consistent with the sample and 0 elsewhere. Thus the design-likelihood is uninformative about the nonsample portion of the population and does not allow us to relate the units in the sample to the nonsample units. A different approach to the problem is to try to make the ratio estimator conform more closely to the design-based paradigm by reducing its design bias through conditioning to restrict the sample space. Smith (1984) showed that this does not work either because conditioning often restricts the sample space to the unique sample actually observed, making design-based inference impossible.

In summary, strict design-based inference for finite population parameters seems unable to incorporate information from a particular sample into the inference. However, this can be achieved through the use of models for the variables.

6.5.7 The Model-Based Approach

Models have a long history in the sampling literature, see Cochran (1939; 1946), Deming and Stephan (1941), Madow and Madow (1944) and there are a number of different ways of introducing them. For example, Basu (1969), Kalbfleisch and Sprott (1969), and Ericson (1969) incorporated the conventional design-based structure whereas Brewer (1963), Scott and Smith (1969), and Royall (1970; 1971) formulated a model which relates the values of the survey variable for the observed units to the unobserved units by specifying the relationship between the survey variable Y and the design variable X. For a review of the possibilities, see Cassel et al. (1977).

A commonly used model for the kind of economic survey we are considering (at least for textbook discussions of it) is to suppose that the conditional distribution of Y given X is given by

$$\mathscr{F} = \left\{ f(\mathbf{y} \mid \mathbf{x}; \beta, \sigma) = \prod_{i=1}^{N} \frac{1}{(2\pi\sigma^2 v(x_i))^{1/2}} \exp\left\{ -\frac{(y_i - \beta x_i)^2}{2\sigma^2 v(x_i)} \right\}, \right.$$

$$\left. y_i \in \mathbb{R}: \beta \in \mathbb{R}, \sigma > 0 \right\}. \quad (6.41)$$

Here $v(x)$ is treated as known; it is usually of the form $v(x) = |x|^{1+\gamma}$, where $0 \leq \gamma \leq 1$, corresponding to increasing fan-shaped variation as x increases.

The N population values (Y_i, X_i) are now regarded as realizations of a random variable (Y, X). This kind of model can be interpreted as implying that the population $\mathscr{P}$ is a "sample" from a superpopulation, so is sometimes called a *superpopulation model*. Since, the survey variable Y is regarded as a random variable, T is also a random variable and the problem of estimating T can be viewed as a prediction problem. Note that if we observe all N units, we know the population total T exactly but we do not know the parameters (β, σ) of the underlying model exactly. Thus we have the usual analytic problem of making inferences about the parameters of the superpopulation model and the enumerative problem of predicting parameters like the finite-population total. (See Section 1.4.3.)

If we observe the entire population $\mathscr{P}$, the log-likelihood under $\mathscr{F}$ can be obtained directly from (6.41). However, we observe Y_i only if $i \in \mathscr{S}$ so the sample is incomplete. We can obtain the marginal density of the observed data by integrating the unobserved Y_i out of the model density in $\mathscr{F}$. Since the sampling is uninformative, the marginal log-likelihood for the sample is

$$\log f_S(\mathbf{y} \mid \mathbf{x}; \beta, \sigma) = -n \log(\sigma^2) - \sum_{i \in \mathscr{S}} \frac{(Y_i - \beta X_i)^2}{2\sigma^2 v(X_i)}, \qquad \beta \in \mathbb{R}, \quad \sigma > 0.$$

The estimating equations are therefore

$$0 = \sum_{i \in \mathscr{S}} \frac{X_i(Y_i - \beta X_i)}{\sigma^2 v(X_i)}$$

and

$$0 = -\frac{n}{\sigma^2} + \sum_{i \in \mathscr{S}} \frac{(Y_i - \beta X_i)^2}{\sigma^4 v(X_i)}$$

and the likelihood is maximized at

$$\hat{\beta} = \left\{ \sum_{i \in \mathcal{S}} \frac{X_i^2}{v(X_i)} \right\}^{-1} \sum_{i \in \mathcal{S}} \frac{X_i Y_i}{v(X_i)} \tag{6.42}$$

and

$$\hat{\sigma}^2 = n^{-1} \sum_{i \in \mathcal{S}} \frac{(Y_i - \hat{\beta} X_i)^2}{v(X_i)}.$$

The mean squared prediction error of a predictor $\hat{T}$ of T is $\mathrm{E}\{(\hat{T} - T)^2 \mid \text{Sample}\}$ which, on differentiating with respect to $\hat{T}$ and equating the result to 0, can be shown to be minimized by

$$\hat{T} = \mathrm{E}\left\{ \sum_{i=1}^{N} Y_i \mid \text{Sample} \right\} = \sum_{i \in \mathcal{S}} Y_i + \beta \sum_{i \notin \mathcal{S}} X_i.$$

We estimate the *optimal predictor* by the regression estimator

$$\hat{T} = \sum_{i \in \mathcal{S}} Y_i + \hat{\beta} \sum_{i \notin \mathcal{S}} X_i \tag{6.43}$$

where $\hat{\beta}$ is given by (6.42). Under the model (6.41), the *mean squared prediction error* of $\hat{T}$ (which is in fact the *prediction variance furhter since* $\mathrm{E}(\hat{T} - T) = 0$) is

$$\begin{aligned}
\mathrm{E}\left\{(\hat{T} - T)^2 \,\middle|\, \text{Sample}\right\} &= \mathrm{E}\left\{\left(\hat{\beta} \sum_{i \notin \mathcal{S}} X_i - \sum_{i \in \mathcal{S}} Y_i\right)^2 \,\middle|\, \text{Sample}\right\} \\
&= \mathrm{E}\left[\left\{(\hat{\beta} - \beta) \sum_{i \notin \mathcal{S}} X_i - \sum_{i \notin \mathcal{S}} (Y_i - X_i \beta)\right\}^2 \,\middle|\, \text{Sample}\right] \\
&= \mathrm{E}\{(\hat{\beta} - \beta)^2 \mid \text{Sample}\} \left\{ \sum_{i \notin \mathcal{S}} X_i \right\}^2 \\
&\quad + \sum_{i \notin \mathcal{S}} \mathrm{E}\{(Y_i - X_i \beta)^2 \mid \text{Sample}\} \\
&= \frac{\sigma^2 \left\{ \sum_{i \notin \mathcal{S}} X_i \right\}^2}{\sum_{i \in \mathcal{S}} \frac{X_i^2}{v(X_i)}} + \sum_{i \notin \mathcal{S}} \sigma^2 v(X_i) \\
&= \sigma^2 \frac{\left[\left\{ \sum_{i \notin \mathcal{S}} X_i \right\}^2 + \sum_{j \notin \mathcal{S}} v(X_j) \sum_{i \in \mathcal{S}} \frac{X_i^2}{v(X_i)} \right]}{\sum_{i \in \mathcal{S}} \frac{X_i^2}{v(X_i)}}.
\end{aligned} \tag{6.44}$$

The right-hand side of (6.44) can be estimated by $\hat{V}_{\mathrm{T}}$ which is obtained by replacing σ^2 by $\hat{\sigma}^2$. A large sample approximation to the sampling distribution of the regression estimator under (6.41) has been obtained by Fuller (1975) and yields the approximate $100(1-\alpha)\%$ confidence interval

$$[\hat{T} - \hat{V}_{\mathrm{T}}^{1/2}\Phi^{-1}(1-\alpha/2), \hat{T} - \hat{V}_{\mathrm{T}}^{1/2}\Phi^{-1}(1-\alpha/2)].$$

6.5.8 The Model-Based Ratio Estimator

In the special case that $v(x) = x$, the regression estimator (6.43) reduces to the ratio estimator (6.36):

$$\hat{T}_{\mathrm{R}} = n\bar{Y}_s + (N-n)\frac{\bar{X}_r \bar{Y}_s}{\bar{X}_s} = \frac{N\bar{X}\bar{Y}_s}{\bar{X}_s},$$

where $\bar{X}_r = (N-n)^{-1}\sum_{i\notin\mathcal{S}} X_i$. Under the model (6.41) with $v(x) = x$, (6.44) reduces to

$$\begin{aligned}
\mathrm{E}\{(\hat{T}_{\mathrm{R}} - T)^2 \mid \text{Sample}\} &= \{n\bar{X}_s\}^{-1}\sigma^2\{(N-n)\bar{X}_r\}^2 + (N-n)\sigma^2\bar{X}_r \\
&= \frac{\sigma^2[\{(N-n)\bar{X}_r\}^2 + (N-n)n\bar{X}_s\bar{X}_r]}{n\bar{X}_s} \\
&= \frac{\bar{X}_r\sigma^2(N-n)[(N-n)\bar{X}_r + n\bar{X}_s]}{n\bar{X}_s} \\
&= \sigma^2(N-n)\frac{N\bar{X}_r\bar{X}}{n\bar{X}_s} \\
&= \sigma^2\left(1-\frac{n}{N}\right)\frac{N^2}{n}\frac{\bar{X}_r\bar{X}}{\bar{X}_s}
\end{aligned} \tag{6.45}$$

which again can be estimated from the fact that

$$\hat{\sigma}_{\mathrm{R}}^2 = (n-1)^{-1}\sum_{i\in\mathcal{S}}\frac{(Y_i - (\bar{Y}_s/\bar{X}_s)X_i)^2}{X_i} \tag{6.46}$$

is model-unbiased for σ^2. Estimators which perform better when the model is misspecified have been proposed by Royall and Cumberland (1981).

6.5.9 A Comparison of the Ratio Estimator in the Two Approaches

It is instructive to compare the ratio estimator in the two frameworks. The ratio estimator is design-biased and by (6.38) has design-based mean squared error approximately equal to $(1-n/N)n^{-1}N^2S^2$. Under the model (6.41) with

$v(x) = x$, the ratio estimator is model-unbiased and by (6.45) has model-based mean squared prediction error equal to $\sigma^2(1 - n/N)n^{-1}N^2\bar{X}_r\bar{X}/\bar{X}_s$. Whereas the design-based mean squared error is averaged over all possible samples, the model-based mean squared prediction error depends on the characteristics of the sample reflected through the design variable means $\bar{X}_s$ and $\bar{X}_r$. This means that inferences from the two approaches can be numerically incompatible.

If the model is incorrect, the model-based inference is usually model biased. For example, if there should actually be an intercept in the model (6.41), the ratio estimator is model biased. Design based and model-based properties are both mathematically correct but the design-based properties are valid regardless of what model holds and the model-based properties are valid only if a particular model holds. Thus the design-based approach requires fewer assumptions and therefore seems the more attractive. However, as we noted in Section 6.4.5, the design is ancillary (see Sections 3.9.2 and 7.4) and contains no information about the parameter of interest, so it is strange to base inference exclusively upon the choice of design. In this sense, the model-based approach is preferable and we simply have to confront the issue of model specification.

The two inference frameworks also suggest different sampling strategies. The design-based theory for the ratio estimator requires a simple random sample from the population. However, in the model-based theory, some samples are clearly better than others in the sense that they lead to estimates with smaller variability. For the ratio estimator, the purposive sample consisting of the n largest companies minimizes the prediction variance of the estimator. This seems to eliminate the need for probability or randomization in sample selection. However, as Royall (1976) has argued, random sampling still has a useful role. As indicated in Section 6.1.3, randomization can protect an investigator against unconscious bias leading to allocations which favor the achievement of desired results and can prevent extraneous factors from having a systematic effect on the results and hence prevent systematic bias. Royall (1976) argues (as we noted in Section 6.1.5) that if we have evidence that the sample differs systematically from the whole population, we should make an appropriate adjustment based on this evidence. This is in conflict with the design-based analysis which requires us to ignore the data and any other evidence presented about it and base inference solely on the randomization implemented before the data were collected. In any case, the optimal design for the ratio estimator is not necessarily the best to use in practice since, as with most optimal designs, the design can prevent us from exploring departures from the model effectively. Thus adoption of the optimal design reflects too great a commitment to the model before we see the data.

6.5.10 Missing Data and Outliers in Surveys

In practice, in analyzing sample survey data, we need to deal with the problems of missing data due to nonresponse and outliers in the data. Whatever the basis for the final inference, the model-based perspective is very helpful in formulating and suggesting approaches to dealing with these problems. Little and Rubin (1987) give a general presentation of the analysis of data when some observations are missing. The subtle and challenging problems of dealing with outliers when trying to estimate finite-population parameters are discussed by Chambers (1986) and Welsh and Ronchetti (1996).

PROBLEMS

6.5.1. (Lahiri, 1968) Consider the sampling scheme for a population of N units which allows us to select either a single unit with probability $\frac{1}{2N}$ or the whole population with probability $\frac{1}{2}$. Show that the sample mean $\bar{Y}_s$ is design-unbiased for the population mean $\bar{Y}$ and find its design-based variance. Show that the design-based variance does not describe the precision of $\bar{Y}_s$ effectively.

6.5.2. Show that the Horvitz–Thompson variance estimator (6.33) and the Yates, Grundy, and Sen variance estimator (6.34) are design-unbiased for the variance (6.32) of the Horvitz–Thompson estimator.

6.5.3. Show that under simple random sampling without replacement, the observations in a sample of size n from a finite population of size N are exchangeable but not independent.

6.5.4. Under which model is the expansion estimator $\hat{T}_{\mathrm{H}} = N \sum_{i \in s} Y_i/n = N\bar{Y}_s$ the optimal estimator of the finite population total? Evaluate the prediction variance of $\hat{T}_{\mathrm{H}}$ under this model. Is this model likely to be widely applicable in practice? Discuss how you might use a sample to evaluate the plausibility of the model.

6.5.5. Show that the estimator

$$\hat{V}_{\mathrm{H}} = \left(1 - \frac{n}{N}\right)\frac{N^2}{n}(n-1)^{-1}\sum_{i \in s}(Y_i - \bar{Y}_s)^2,$$

where $\bar{Y}_s = \sum_{i \in s} Y_i$, is design-unbiased for the design-based variance of the expansion estimator and find its model expectation when the model

$$\mathscr{F} = \left\{ f(\mathbf{y} \mid \mathbf{x}; \beta, \sigma) = \prod_{i=1}^{N} \frac{1}{(2\pi\sigma^2 x_i)^{1/2}} \exp\left\{ -\frac{(y_i - \beta x_i)^2}{2\sigma^2 x_i} \right\}, \right.$$
$$\left. -\infty < y_i < \infty : \beta \in \mathbb{R}, \sigma > 0 \right\}$$

holds. Interpret your results.

6.5.6. Suppose that for a finite population such as that described in Section 6.5.1, we decide to base inference on the expansion estimator $N\bar{Y}_s$. To explore the effects of misspecification of the model, suppose that the data are actually generated by the model

$$\mathscr{F} = \left\{ f(\mathbf{y} \mid \mathbf{x}; \beta, \sigma) = \prod_{i=1}^{N} \frac{1}{(2\pi\sigma^2 x_i)^{1/2}} \exp\left\{ -\frac{(y_i - \beta x_i)^2}{2\sigma^2 x_i} \right\}, \right.$$
$$\left. -\infty < y_i < \infty : \beta \in \mathbb{R}, \sigma > 0 \right\}$$

for which the ratio estimator is optimal. Obtain the model-based expectation and prediction variance of the expansion estimator under $\mathscr{F}$. Under what circumstances is the expansion estimator model-biased? Discuss the implications of this result under simple random sampling.

6.5.7. Suppose that for a finite population such as that described in Section 6.5.1, we decide to base inference on the ratio estimator $N\bar{X}\bar{Y}_s/\bar{X}_s$. To explore the effects of misspecification of the model, suppose that the data are actually generated by the model

$$\mathscr{F} = \left\{ f(\mathbf{y} \mid \mathbf{x}; \beta, \sigma) = \prod_{i=1}^{N} \frac{1}{(2\pi\sigma^2 v x_i))^{1/2}} \exp\left\{ -\frac{(y_i - \alpha - \beta x_i)^2}{2\sigma^2 v(x_i)} \right\}, \right.$$
$$\left. -\infty < y_i < \infty : \alpha, \beta \in \mathbb{R}, \sigma > 0 \right\},$$

where $v(x)$ is of the form $v(x) = |x|^{1+\gamma}$, where $0 \le \gamma \le 1$. The ratio estimator is optimal for this model when $\alpha = 0$ and $\gamma = 0$. Explore the model expectation and prediction variance of the ratio estimator when $\alpha \ne 0$ and $\gamma > 0$.

6.6 PERMUTATION TESTS

Permutation tests are numerically identical to randomization tests but are based on sampling distributions derived from models like the Gaussian model (6.1) rather than randomization distributions derived from randomization models like (6.9).

6.6.1 The Permutation Distribution

Consider again the volume of urine experiment. Suppose that the data are realizations of independent and identically distributed random variables $(Z_{01}, Z_{11}), \ldots, (Z_{0n}, Z_{1n})$ distributed like (Z_0, Z_1). If we let $F(z_0, z_1)$ denote the joint distribution of (Z_0, Z_1), the null hypothesis H_0 of no difference between treatment and control implies that $F(z_0, z_1) = F(z_1, z_0)$. Under this null hypothesis, we have that the pairwise differences satisfy

$$P(Z \le z) = P(Z_1 - Z_0 \le z) = P(Z_0 - Z_1 \le z) = P(-Z \le z)$$

so Z and $-Z$ have the same distribution. Hence, under H_0, Z is symmetrically distributed about the origin and we have the nonparametric model (see Section 1.3.3)

$$\mathscr{F}_0 = \left\{ F(\mathbf{z}) = \prod_{i=1}^{n} F(z_i) \colon F \text{ is continuous and } F(x) = 1 - F(x) \right\}. \tag{6.47}$$

The subscript 0 in $\mathscr{F}_0$ is intended to show that this model holds under H_0.

The conditional distribution of $\mathbf{Z}$ given $\{|Z_1|, \ldots, |Z_n|\}$ under (6.47) is discrete with support on the set

$$\mathscr{Z}_R = \{(\pm|z_1|, \pm|z_2|, \ldots, \pm|z_n|)\}. \tag{6.48}$$

Under (6.47), each of these values is equally likely (because Z and $-Z$ have the same distribution) so $\mathbf{Z}$ takes on any one of them with probability $1/2^n$. Formally, the joint density of the data $\mathbf{Z}$ is

$$f(\mathbf{z}) = \prod_{i=1}^{n} f(z_i)$$

and the joint density of the absolute data $(|Z_1|, \ldots, |Z_n|)$ under (6.47) is

$$f(\mathbf{z}) = \prod_{i=1}^{n} \{f(z_i) + f(-z_i)\} = 2^n \prod_{i=1}^{n} f(z_i), \qquad z_i \ge 0.$$

It follows that, under (6.47),

$$f(\mathbf{z} \mid |Z_1| = |z_1|, \ldots, |Z_n| = |z_n|) = 1/2^n, \qquad \mathbf{z} \in \mathscr{Z}_R. \tag{6.49}$$

Thus from (6.49) and the equivalence of (6.48) and (6.20), the conditional sampling distribution of $\mathbf{Z}$ given the absolute observations $(|Z_1|, \ldots, |Z_n|)$ is the same as the randomization distribution of $\mathbf{Z}$. The conditional sampling distribution of $\mathbf{Z}$ given the absolute observations $(|Z_1|, \ldots, |Z_n|)$ is called the *permutation distribution* of $\mathbf{Z}$ because the points in $\mathscr{Z}_R$ can be obtained (as in fact they are in the randomization framework) by considering all possible permutations of the control and treatment labels. Even though we have assumed the model (6.47), the permutation distribution does not depend on F under H_0. This is often referred to as the *distribution free property* of permutation tests.

6.6.2 Permutation Tests

A permutation test of H_0 based on a statistic T (chosen so that large values of T represent departures from H_0) is the test provided by computing the p-value

$$p(n) = P\{T(\mathbf{Z}) \geq T(\mathbf{z}) \mid |Z_1| = |z_1|, \ldots, |Z_n| = |z_n|; H_0\}$$
$$= \frac{\#\{\text{points in } \mathscr{Z}_R \text{ leading to a } T \text{ as or more extreme than that actually observed}\}}{2^n}.$$

From (6.49), $p(n)$ depends on n but not on the data $\mathbf{Z}$. The unconditional p-value is therefore

$$P\{T(\mathbf{Z}) \geq T(z)\ H_0\} = EP\{T(\mathbf{Z}) \geq T(z) \mid |Z_1| = |z_1|, \ldots, |Z_n| = |Z_n|; H_0\} = p(n)$$

which is the same as the conditional p-value. Thus the permutation test can be interpreted unconditionally.

Permutation tests are numerically the same as randomization tests (based on the same test statistic) and we can approximate the permutation distribution using methods like those used in Section 6.4 for randomization tests or saddlepoint approximations (Robinson, 1982).

6.6.3 The Sign Test as a Permutation Test

As in Section 6.3.2, the exact permutation distribution of the sign test (6.22) under H_0 is the same as $2B - n$, where $B \sim \text{binomial}(n, \frac{1}{2})$. An advantage of the permutation test as opposed to the randomization test formulation is that the underlying model makes it easier to construct plausible alternative hypotheses and thereby to compare the power of tests. For example, the null hypothesis that Z is symmetrically distributed about the origin implies that the median of F is 0 or, equivalently, $F(0) = \frac{1}{2}$. One alternative hypothesis of interest is that the data are generated by a shifted version of F, say $P\{Z \leq x\} = F(x - \theta)$, or equivalently, $P\{Z - \theta \leq x\} = F(x)$, for some $\theta \neq 0$ rather than $P\{Z \leq x\} = F(x)$. Under this alternative, $EI(Z > 0) = P(Z > 0) = P(Z - \theta \geq -\theta) = 1 - F(-\theta)$ so that S has the same distribution as $2B(\theta) - n$, where $B(\theta) \sim \text{binomial}(n, 1 - F(-\theta))$, which depends on the underlying distribution. (Obviously in a randomization test this cannot happen.) The asymptotic mean of $n^{-1}B(\theta)$ is $\mu(\theta) = 1 - F(-\theta)$ and the asymptotic variance is $\tau(\theta)^2 = \{1 - F(-\theta)\}F(-\theta)$. Since $\mu'(0) = f(0)$ and $\tau(0)^2 = \frac{1}{4}$, it follows from Section 4.5 and Problem 4.5.2 that the efficacy of the sign test is $2f(0)$. For comparison, under H_1, the sample mean has asymptotic mean θ and variance σ^2 so its efficacy is $1/\sigma$. Hence the Pitman efficiency of the sign test relative to the mean test is

$$4f(0)^2\sigma^2.$$

This can be computed for different distributions; the sign test is more efficient than the mean test when the underlying distribution F has sufficiently long tails.

In contrast to the mean test, the sign test is not obviously based on a location estimator. Nonetheless, it is equivalent to a test based on the sample median. The sample median can be defined to be the minimum of $\sum_{i=1}^{n} |Z_i - \theta|$ so an estimating equation for the sample median is

$$\sum_{i=1}^{n} \{I(Z_i - \theta > 0) - I(Z_i - \theta < 0)\} = 0. \tag{6.50}$$

Since the left-hand side of (6.50) equals S at $\theta = 0$ (see (6.22)), the sign test is simply a score test of $H_0: \theta = 0$ based on the sample median. This means that we can interpret the Pitman efficiency of the two tests as the asymptotic relative efficiencies of the sample mean to the sample median. In addition, by inverting the sign test (as discussed in Section 4.5.14), we can obtain confidence intervals for the population median. The problem of obtaining confidence intervals for the median is pursued in the next section.

6.6.4 Randomization and Permutation Inference

In practical terms, the difference between a randomization test and a permutation test is that in the former we take actions (randomization) which ensure that the conditions for the test hold whereas in the latter, we simply assume that they hold (by adopting a model). Nonetheless, the sample space is quite different in the two cases. In a randomization test, permutation is justified by what would happen under replication of the design conditional on the observed data while in a permutation test, the permutation is justified by a model for the data. Thus permutation inference does not require the change of perspective that randomization inference does.

PROBLEMS

6.6.1. For the effect of caffeine on the volume of urine data (Section 1.1.3), describe how to carry out a permutation test based on Huber's location M-estimator (Section 5.4) to test the null hypothesis that the ingestion of caffeine has no effect on the volume of urine voided.

6.6.2. Compute the Pitman efficiency of the sign test to the t test at the contaminated Gaussian distribution which has density

$$f(z; \varepsilon, \sigma) = (1 - \varepsilon)\phi(z) + \varepsilon\sigma^{-1}\phi\left(\frac{z}{\sigma}\right), \qquad -\infty \le z \le \infty, \quad \sigma > 0, \quad \varepsilon > 0.$$

Try a range of parameter values including $\varepsilon = 0, 0.05, 0.10, 0.20$, and $\sigma = 1, 3$.

6.6.3. Permutation tests always have the stated level by definition and this can be taken to justify the use of Gaussian theory tests even when the model is non-Gaussian. Use the results of your computations in Problem 6.6.2 to comment on the usefulness of this conclusion.

6.7 THE BOOTSTRAP

Consider again the caffeine problem for which we have observations $\{Z_1, \ldots, Z_n\}$ which represent the effect of the ingestion of caffeine on the volume of urine produced. Instead of adopting the usual Gaussian model, we can adopt a model such as

$$\mathscr{F} = \left\{ f(\mathbf{z}) = \prod_{i=1}^{n} f(z_i), \mathbf{z} \in \mathbb{R}^n: f \text{ absolutely continuous} \right\}, \tag{6.51}$$

and let the median $F^{-1}(1/2)$ rather than the mean represent the center of the distribution. We can estimate the median by the sample median which we define in terms of the order statistics $Z_{n1} \le \cdots \le Z_{nn}$ by $\text{med}(\mathbf{Z}) = Z_{nm}$, where $m = n/2$ if n is even and $m = (n+1)/2$ if n is odd. As noted in Section 1.5.3, this definition of the sample median is convenient because it corresponds to $F_n^{-1}(\frac{1}{2}) = \inf\{x: F_n(x) \ge \frac{1}{2}\}$, where

$$F_n(x) = n^{-1} \sum_{i=1}^{n} I(Z_i \le x), \qquad x \in \mathbb{R}, \tag{6.52}$$

is the empirical distribution function defined in Section 1.5.2. The usual definition of the sample median is the same as ours when n is odd but uses $(Z_{nm} + Z_{n,m+1})/2$ when n is even. We avoid this definition because its exact distribution is more complicated than that of Z_{nm}.

6.7.1 Inference for the Median

To base frequentist inference on the sample median, we need to find the sampling distribution of the sample median. Under the model (6.51), we have

$$\begin{aligned} \mathrm{P}\{Z_{nm} \le x\} &= \mathrm{P}\{\tfrac{1}{2} \le F_n(x)\} \\ &= \mathrm{P}\left\{B(n, F(x)) \ge \frac{n}{2}\right\} \\ &= \sum_{k=m}^{n} \binom{n}{k} F(x)^k \{1 - F(x)\}^{n-k}, \end{aligned} \tag{6.53}$$

where $B(n, \frac{1}{2}) \sim \text{binomial}(n, \frac{1}{2})$ and $m = n/2$ if n is even $m = [n/2] + 1 = (n + 1)/2$ if n is odd. Thus if a and b are integers such that $P\{B(n, \frac{1}{2}) \geq b\} \geq \alpha/2$ and $P\{B(n, \frac{1}{2}) \leq a\} \geq \alpha/2$, we have

$$\begin{aligned} P\{Z_{na} \leq F^{-1}(\tfrac{1}{2}) \leq Z_{nb}\} &= P\{a \leq F_n(F^{-1}(\tfrac{1}{2})) \leq b\} \\ &= P\{a \leq B(n, \tfrac{1}{2}) \leq b\} \\ &\geq 1 - \alpha, \end{aligned}$$

so a $100(1 - \alpha)\%$ confidence interval for the median is given by

$$[Z_{na}, Z_{nb}]. \tag{6.54}$$

For $\alpha = 0.05$ and $n = 18$, we find $a = 5$ (as $P\{B(18, \frac{1}{2}) \leq 4\} = 0.015 < 0.025 < P\{B(18, \frac{1}{2}) \leq 5\} = 0.048$) and $b = 13$ so a 95% confidence interval for the median volume of urine voided is

$$[-5, 235].$$

This interval is wider and closer to the origin than the 95% mean based interval [67, 274] obtained under a Gaussian model in Section 3.5.

The calculations for the exact confidence interval for the median become tedious as n increases but we can approximate (6.53) by showing that as $n \to \infty$,

$$n^{1/2}(Z_{nm} - F^{-1}(\tfrac{1}{2})) \xrightarrow{\mathscr{D}} N(0, 1/4f(F^{-1}(\tfrac{1}{2}))^2). \tag{6.55}$$

This large sample approximation to the sampling distribution of the sample median enables us to construct an approximate confidence interval. Typically, in this case, we need to estimate the sparsity function $1/f(F^{-1}(\frac{1}{2}))$; see Sheather (1987).

6.7.2 The Bootstrap Distribution

An alternative approximation to the sampling distribution of the sample median can be obtained by bootstrapping. The bootstrap distribution of Z_{nm} is obtained by treating the empirical distribution (6.52) as the underlying distribution. That is, we replace the underlying distribution F by F_n. We then approximate the sampling distribution by repeated resampling from F_n which is achieved by drawing samples $\mathbf{Z}^{(r)}$ of size n by sampling independently with replacement from $\{Z_1, \ldots, Z_n\}$. If all the observations are distinct, there are $\binom{2n-1}{n}$ distinct samples in $\mathscr{B} = \left\{\mathbf{Z}^{(r)}, r = 1, \ldots, \binom{2n-1}{n}\right\}$ which are equally likely. The *bootstrap distribution* of the sample median Z_{nm} is derived by calculating the realization $Z_{nm}^{(r)}$ for each of the resamples and assigning each one equal probability.

As $n \to \infty$, the empirical distribution F_n converges to the underlying distribution F, so it is intuitively plausible that the bootstrap distribution should converge to the sampling distribution. When this holds, the bootstrap distribution is an asymptotically valid approximation to the sampling distribution of a statistic. Mammen (1993) showed that the bootstrap distribution is an asymptotically valid approximation to the sampling distribution when the Gaussian approximation is asymptotically valid. However, in small samples, the two approximations can give rather different results.

6.7.3 The Exact Bootstrap Distribution of the Sample Median

In the particular case of the median, the bootstrap distribution can be found exactly as the distribution of the median under sampling from a multinomial$(n, n^{-1}, \ldots, n^{-1})$ model supported on $\mathbf{Z}$.

Since the median must equal one of the observations, its bootstrap distribution has support on the data $\{Z_1, \ldots, Z_n\}$. Define $M_j^{(r)}$ to be the number of observations in the rth bootstrap sample which equal the jth order statistic Z_{nj} so

$$M_j^{(r)} = \#\{Z_i^{(r)} = Z_{nj}\}.$$

Then, the event that the median of the rth bootstrap sample is greater than the kth order statistic, namely $\{Z_{nm}^{(r)} > Z_{nk}\}$, implies and is implied by the event that the number of observations in the rth bootstrap sample less than or equal to $Z_{nk} = \sum_{j=1}^{k} M_j^{(r)}$ is less than m which we write as $\{\sum_{j=1}^{k} M_j^{(r)} \le m - 1\}$. Hence

$$\mathrm{P}\{Z_{nm}^{(r)} > Z_{nk}\} = P\left\{\sum_{j=1}^{k} M_j^{(r)} \le m - 1\right\} = \mathrm{P}\left\{B\left(n, \frac{k}{n}\right) \le m - 1\right\},$$

because the fact that $Z_i^{(r)}$ has a multinomial distribution implies that $\sum_{j=1}^{k} M_j^{(r)}$ which is the number of observations less than or equal to Z_{nk} in a sample of size n has a binomial$(n, k/n)$ distribution. It follows that

$$\begin{aligned} \mathrm{P}\{Z_{nm}^{(r)} = Z_{nk}\} &= \mathrm{P}\{Z_{nm}^{(r)} > Z_{n,k-1}\} - \mathrm{P}\{Z_{nm}^{(r)} > Z_{nk}\} \\ &= \mathrm{P}\left\{\mathrm{B}\left(n, \frac{k-1}{n}\right) \le m - 1\right\} - \mathrm{P}\left\{\mathrm{B}\left(n, \frac{k}{n}\right) \le m - 1\right\} \\ &= p_{nk}, \end{aligned} \tag{6.56}$$

say. That is, the bootstrap distribution of the median is itself a multinomial distribution. The bootstrap distribution of the median for the urine data is tabulated in Table 6.3. The bootstrap density function is plotted in Figure 6.2.

Table 6.3. The Bootstrap Distribution of the Sample Median Applied to the Urine Data

Order statistics	−90	−45	−20	−5	−5	40
p_{nk}	0	0.00005	0.00108	0.00771	0.02876	0.0700
Distribution fuction	0	0.00005	0.00113	0.00884	0.03760	0.1076
Order statistics	47	75	87	117	228	230
p_{nk}	0.12437	0.17143	0.18933	0.16971	0.12316	0.07104
Distribution function	0.23197	0.40341	0.59274	0.76245	0.88561	0.95665
Order statistics	235	254	325	410	555	635
p_{nk}	0.03134	0.00986	0.00195	0.00019	0.00001	0
Distribution function	0.98799	0.99785	0.99980	0.99999	1	1

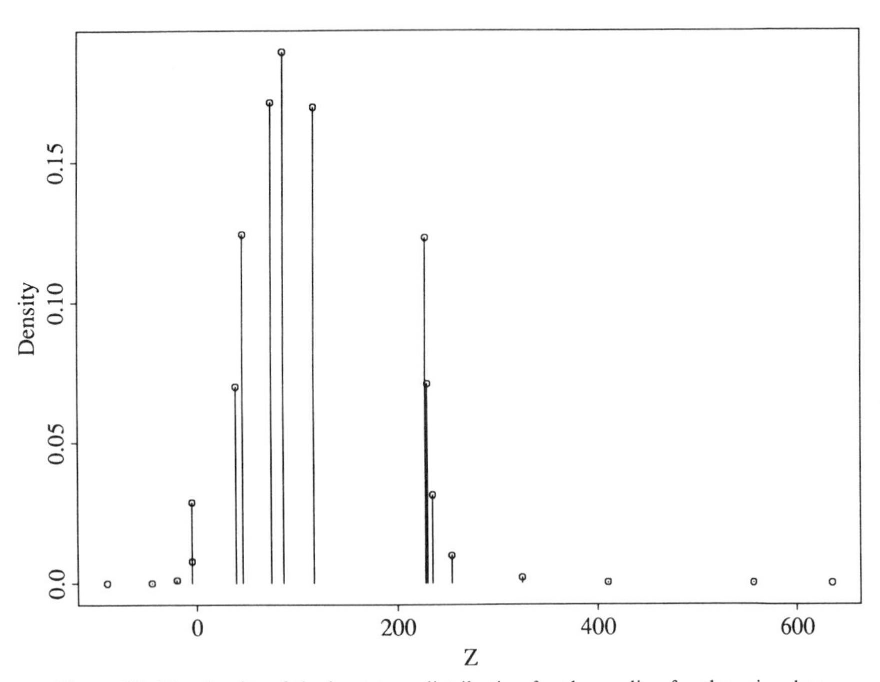

Figure 6.2. The density of the bootstrap distribution for the median for the urine data.

6.7.4 Using the Bootstrap Distribution

The bootstrap distribution can be used to estimate the bias and standard error of the sample median. We estimate the bias by the difference between the mean of the bootstrap distribution (6.56) and the sample median, namely

$$\sum_{k=1}^{n} p_{nk} z_{nk} - z_{nm} = 112.63 - 87 = 25.63.$$

We estimate the standard error of the sample median from the bootstrap distribution by calculating its standard deviation. Since the bootstrap distribution (6.56) is multinomial, the standard error of the median is estimated by

$$\left\{\sum_{k=1}^{n} p_{nk} z_{nk}^2 - \left\{\sum_{k=1}^{n} p_{nk} z_{nk}\right\}^2\right\}^{1/2} = 72.04.$$

This estimator of the standard error of the median was proposed by Maritz and Jarrett (1978) and Efron (1979).

The quantile function of the bootstrap distribution (6.56) is plotted against the quantile function of the approximate sampling distribution (6.55) of the sample median, namely $N(112.63, (72.04)^2)$, in Figure 6.3. The bootstrap distribution is asymmetric with a shorter left tail and a longer upper tail than the approximating Gaussian distribution. If we correct the bias and use the bootstrap standard error in conjunction with the Gaussian approximation, we

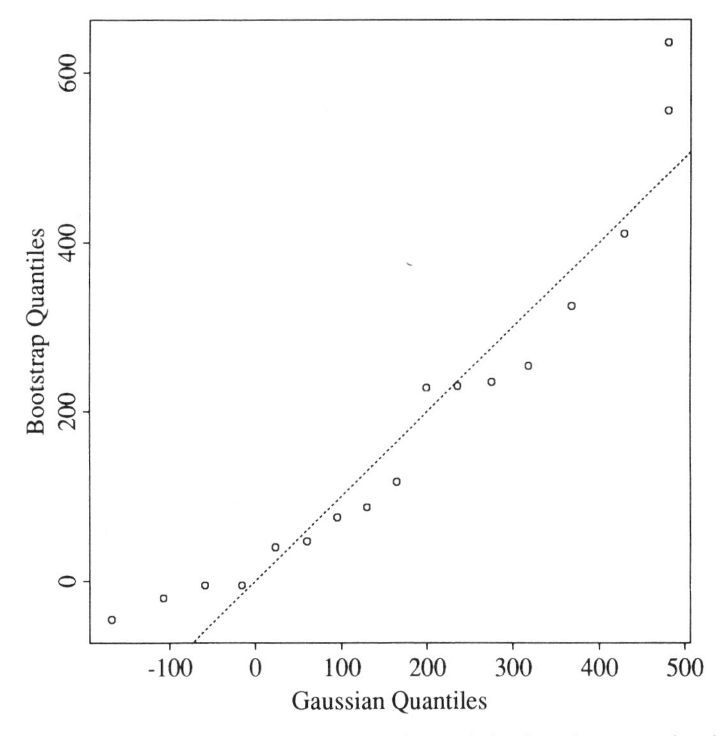

Figure 6.3. A comparison of the bootstrap distribution and the Gaussian approximation to the sampling distribution of the sample median for the urine data.

obtain an approximate 95% confidence interval for the population median

$$[-28.57, 253.83].$$

This interval is slightly wider than the exact interval.

The bootstrap distribution can also be used to construct approximate confidence intervals directly. The distribution function of the bootstrap distribution is $F_{\mathrm{B}}(z) = n^{-1} \sum_{k=1}^{n} p_{nk} I(Z_{nk} \leq z)$ so a $100(1-\alpha)\%$ bootstrap confidence interval for the population median is

$$[F_{\mathrm{B}}^{-1}(\alpha), F_{\mathrm{B}}^{-1}(1-\alpha)], \tag{6.57}$$

where we define $F_{\mathrm{B}}^{-1}(u) = \inf\{x: F_{\mathrm{B}}(x) \geq u\}$ to take into account the fact that F_{B} is a step function. For the change in the volume of urine example, we obtain the exact binomial interval (6.54).

6.7.5 Simulating the Bootstrap Distribution

Even if we cannot obtain an analytic expression for the bootstrap distribution of a statistic, we can in principle compute all $\binom{2n-1}{n}$ possible values of the statistic and so obtain the multinomial probabilities. However, the computation is infeasible for even moderate n so we need alternative approaches. The simplest and most widely used alternative is to simulate the bootstrap distribution by drawing a number B of samples and using them to estimate the bootstrap distribution. For example, for the sample median, the estimated bootstrap distribution is multinomial with

$$\mathrm{P}_{\mathrm{B}}\{Z_{nm} = Z_{nk}\} = \frac{\#\{\text{Samples } \mathbf{Z}^{(r)} \text{ for which } Z_{nm}^{(r)} = Z_{nk}\}}{B}.$$

We use this distribution exactly as we use the actual bootstrap distribution (and indeed the estimated bootstrap distribution is often referred to simply as the bootstrap distribution). The bootstrap quantile function estimate from $B = 2000$ resamples is plotted against the actual bootstrap quantile function in Figure 6.4. The approximation is good except in the extreme right tail where the simulated tail is too short.

We can estimate the bias by the difference between the sample mean of the bootstrap estimates and the sample estimate, and we can estimate the standard error from the sample standard deviation of the bootstrap estimates. An approximate $100(1-\alpha)\%$ bootstrap confidence interval for the population median is particularly easy to obtain as

$$[Z_{nm}^{\{B\alpha\}}, Z_{nm}^{\{B-B\alpha\}}], \tag{6.58}$$

where $Z_{nm}^{\{1\}} \leq Z_{nm}^{\{2\}} \leq \cdots \leq Z_{nm}^{\{B\}}$. The interval (6.58) can be compared to (6.57).

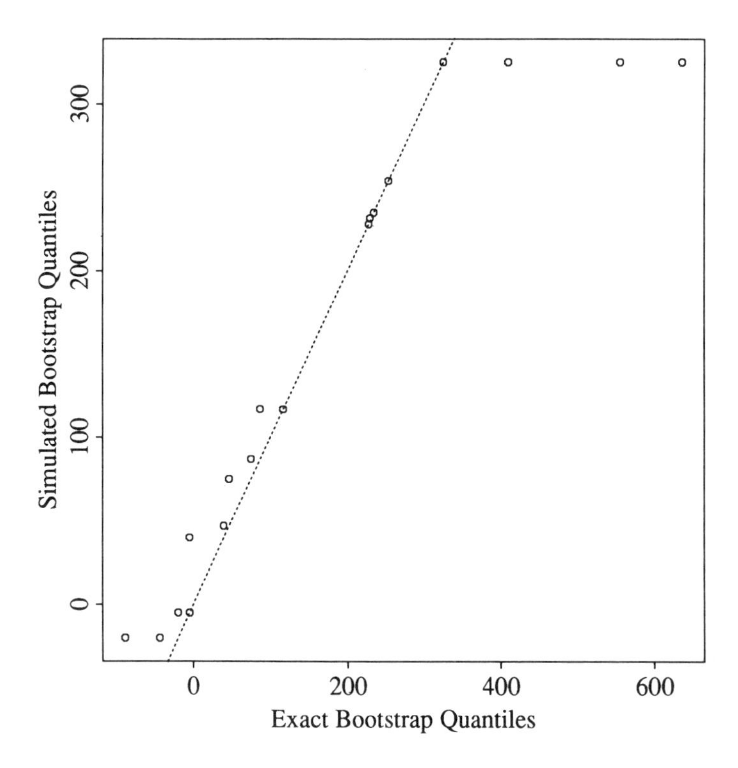

Figure 6.4. A comparison of the bootstrap distribution and the simulated bootstrap distribution of the sample median for the urine data.

The power of the simulated bootstrap distribution is that it can be obtained for almost any statistic. For example, we can redefine the median for n even as $(Z_{nm} + Z_{n,m+1})/2$ and simulate the bootstrap distribution very simply.

6.7.6 The Saddlepoint Approximation to the Bootstrap Distribution

An alternative to simulating the bootstrap (which avoids the need for explicit resampling) is to use a saddlepoint approximation to the bootstrap distribution. Essentially, the formulae derived in Section 4.4.3 are applied to the empirical distribution F_n rather than the true underlying distribution F_0 (Problem 6.7.9). The resulting approximation approximates the bootstrap and hence the actual sampling distribution of the statistic. For more details see Davison and Hinkley (1988), Wang (1992), and Ronchetti and Welsh (1994).

6.7.7 Bootstrap and Permutation Inference

Bootstrap and permutation inferences are based on underlying models for the data and so differ from randomization inference which is based on randomization

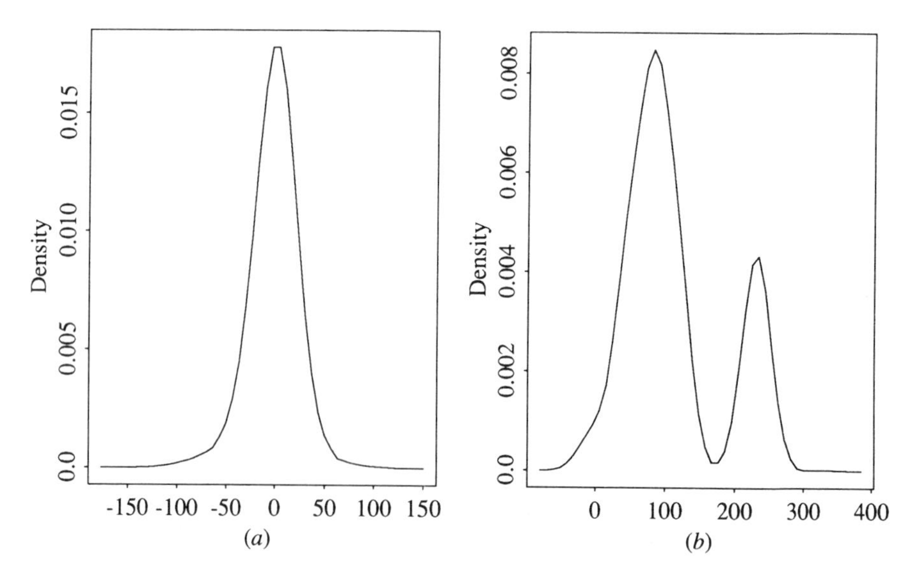

Figure 6.5. A comparison of the randomization/permutation distribution and the simulated bootstrap distribution of the median for the urine data. The densities are obtained using the Gaussian kernel and $h = 80$: (*a*) randomization/permutation density; (*b*) simulated bootstrap density.

models. Bootstrap and permutation inferences are computed conditionally but interpreted unconditionally. In the case of the bootstrap, the bootstrap distribution is simply an approximation to the sampling distribution. The permutation distribution for the urine problem is obtained by permuting the treatment and control observations which is equivalent to considering the sample space $\{\pm Z_1, \ldots, \pm Z_n\}$. Under the null hypothesis of no difference between treatment and control, the pairwise differences are symmetrically distributed about the origin so the permutation distribution is centered about the origin. On the other hand, the bootstrap distribution uses the sample space $\mathscr{B} = \left\{ \mathbf{Z}^{(r)}, r = 1, \ldots, \binom{2n-1}{n} \right\}$, where $\mathbf{Z}^{(r)}$ are independent samples of size n drawn independently with replacement from $\{Z_1, \ldots, Z_n\}$. The permutation and (simulated) bootstrap distributions for the urine data are shown in Figure 6.5. The bootstrap distribution is centered about the location estimate used to estimate the difference between the effect of the treatment and the control while the permutation distribution is centered (as is appropriate since it is computed under the null hypothesis of no treatment effect) at the origin.

6.7.8 Other Bootstrap Procedures

The version of the bootstrap presented above is the simplest available. It is convenient to give a slightly different but equivalent formulation. Suppose that we want to make inferences about a parameter $\theta = \theta(F_0)$ based on the statistic $\hat{\theta} = \theta(F_n)$, where F_0 is the underlying distribution and F_n is the empirical distribution. Let $\mathbf{Z}^{(r)}$ denote a sample of size n obtained by sampling independently with replacement from $\{Z_1, \ldots, Z_n\}$ and let $F_n^{(r)}$ denote the empirical distribution of $\mathbf{Z}^{(r)}$. We can approximate the sampling distribution of $n^{1/2}(\hat{\theta} - \theta)$ by the bootstrap distribution F_{P} which is the distribution of $n^{1/2}(\hat{\theta}^{(r)} - \hat{\theta}) = n^{1/2}(\theta(F_n^{(r)}) - \theta(F_n))$ and thereby obtain a $100(1-\alpha)\%$ bootstrap confidence interval for θ as

$$[\hat{\theta} - n^{-1/2}F_{\mathrm{P}}^{-1}(\alpha), \hat{\theta} - n^{-1/2}F_{\mathrm{P}}^{-1}(1-\alpha)].$$

Note that this interval is equivalent to (6.57) and (6.58) as it follows from the definition of F_{B} as the bootstrap definition of $\hat{\theta}$ that $F_{\mathrm{B}}^{-1}(q) = \hat{\theta} - n^{-1/2}F_{\mathrm{P}}^{-1}(q)$. This bootstrap is often called the *percentile method.*

If we can calculate the standard error $\hat{\tau}/n^{1/2}$ of $\hat{\theta}$, we can apply the bootstrap procedure to the studentized quantity $n^{1/2}(\hat{\theta} - \theta)/\hat{\tau}$ to obtain the bootstrap distribution F_{S} of $n^{1/2}(\hat{\theta}^{(r)} - \hat{\theta})/\hat{\tau}^{(r)}$. A $100(1-\alpha)\%$ bootstrap confidence interval for θ based on F_{S} is

$$[\hat{\theta} - n^{-1/2}\hat{\tau}F_{\mathrm{S}}^{-1}(\alpha), \hat{\theta} - n^{-1/2}\hat{\tau}F_{\mathrm{S}}^{-1}(1-\alpha)].$$

This is known as the *bootstrap-t method.*

To compare the percentile and the bootstrap-t methods, Hall (1988) showed that, under some regularity conditions,

$$F_P(x) = \mathrm{P}\{n^{1/2}(\hat{\theta}^{(r)} - \hat{\theta}) \leq x \mid F_n\} = \mathrm{P}\{n^{1/2}(\hat{\theta} - \theta) \leq x\} + O_p(n^{-1/2})$$

and

$$F_{\mathrm{S}}(x) = \mathrm{P}\left\{n^{1/2}\frac{(\hat{\theta}^{(r)} - \hat{\theta})}{\hat{\tau}^{(r)}} \leq x \mid F_n\right\} = \mathrm{P}\left\{n^{1/2}\frac{(\hat{\theta} - \theta)}{\hat{\tau}} \leq x\right\} + O_p(n^{-1}).$$

That is, the bootstrap distribution obtained by bootstrapping the studentized quantity $n^{1/2}(\hat{\theta} - \theta)/\hat{\tau}$ provides a closer approximation to the sampling distribution of $n^{1/2}(\hat{\theta} - \theta)/\hat{\tau}$ than the bootstrap distribution obtained by bootstrapping $n^{1/2}(\hat{\theta} - \theta)$ provides to the sampling distribution of $n^{1/2}(\hat{\theta} - \theta)$. Intuitively, the bootstrap variance of $n^{1/2}(\hat{\theta}^{(r)} - \hat{\theta})$ is $\hat{\tau}^2$, the sampling variance of $n^{1/2}(\hat{\theta} - \theta)$ is τ^2 and Edgeworth expansion arguments (Section 4.1.4) show that the difference between the bootstrap distribution and the sampling distribution is determined by $\hat{\tau}^2 - \tau^2 = O_p(n^{-1/2})$. The bootstrap-$t$ procedure provides a better approximation because the bootstrap variance of $n^{1/2}(\hat{\theta}^{(r)} - \hat{\theta})/\hat{\tau}^{(r)}$ is $1 + O_p(n^{-1})$ and the sampling variance of $n^{1/2}(\hat{\theta} - \theta)/\hat{\tau}$ is $1 + O(n^{-1})$ and these differ by $O_p(n^{-1})$. See also Martin (1989). In practice, to use the bootstrap-t procedure, we need to be able to compute the standard error readily and stably. See Efron and Tibshirani (1993, pp. 162–6).

Although we do not pursue them here, a number of alternative bootstrap methods involving various adjustments (such as smoothing F_n, including parametric information, using more informative resampling schemes, iterating the bootstrap, etc.) have also been investigated. For a recent review, see Efron and Tibshirani (1993).

PROBLEMS

6.7.1. For the epinephrine data (Section 1.1.3) where $n = 18$, compare the bootstrap distribution of $F_n^{-1}(\frac{1}{2}) = Z_{nm}$ with that of the usual sample median $(Z_{nm} + Z_{n,m+1})/2$, where $m = n/2 = 9$.

6.7.2. Use the sample median to estimate the median survival time from the pressure vessel data given in Section 1.1.2 and Problem 1.5.4. Compare your inferences to those obtained in Problem 5.4.5.

6.7.3. Consider the stellar velocity data described in Section 1.1.4. Construct bootstrap confidence intervals for the spread of the stellar velocity distributions when these are estimated using the standard deviation and Huber's M-estimator (Section 5.4). Compare the results to the large sample results obtained in Chapter 5.

6.7.4. Consider the negative binomial model

$$\mathscr{F} = \left\{ f(\mathbf{z}; \lambda, \delta) = \prod_{i=1}^{n} \left\{ z_i! \Gamma\left(\frac{\lambda}{\delta}\right)\right\}^{-1} \Gamma\left(z_i + \frac{\lambda}{\delta}\right) \delta^{z_i} (1+\delta)^{-(z_i + \lambda/\delta)}, \right.$$
$$\left. z_i = 0, 1, 2, \ldots, \lambda > 0, \delta > 0 \right\},$$

for the Mosteller and Wallace (1964/1984, p. 33) data on the number of occurrences of the word "may" in 262 blocks of text written by James Madison. Suppose we want to carry out a test of $H_0\colon \delta = 0$ using the method of moments estimator $\hat{\delta}$ derived in Problem 3.1.1. Simulate the data from the Poisson model to approximate the sampling distribution of $\hat{\delta}$. What is the simulation-based p-value? How would you obtain a bootstrap p-value for this test?

6.7.5. Suppose we adopt the gamma model

$$\mathscr{F} = \left\{ f(\mathbf{y}, \lambda, \kappa) = \prod_{i=1}^{n} \frac{1}{\Gamma(\kappa)} \lambda(\lambda y_i)^{\kappa-1} \exp(-\lambda y_i), y_i > 0 \colon \lambda > 0 \right\}$$

for the pressure vessel data of Section 1.1.2. Obtain and explore the bootstrap distribution of the method of moments estimator of κ which was derived in Section 4.3.1. Set 95% bootstrap confidence intervals for κ and λ.

6.7.6. Suppose we adopt the Gaussian regression model

$$\mathcal{F} = \left\{ f(\mathbf{y}; \alpha, \beta, \sigma) = \prod_{i=1}^{n} \frac{1}{(2\pi\sigma^2)^{1/2}} \exp\left\{ -\frac{(y_i - \alpha - x_i\beta)^2}{2\sigma^2} \right\}, \right.$$

$$\left. -\infty \leq y_i \leq \infty : \alpha, \beta \in \mathbb{R}, \sigma > 0 \right\}$$

to relate catecholamine excretion to the volume of urine produced during the ingestion of caffeine (Section 1.1.3). To apply the bootstrap here, we need estimators $\hat{\alpha}$ and $\hat{\beta}$ of α and β so that we can compute the residuals

$$r_i = y_i - \hat{\alpha} - \hat{\beta}x_i, \qquad i = 1, \ldots, n.$$

We then resample independently, with replacement from the residuals to obtain a bootstrap sample $r_1^*, \ldots, r_n^*$ of residuals and construct the bootstrap observations

$$Y_i^* = \hat{\alpha} + \hat{\beta}x_i + r_i^*, \qquad i = 1, \ldots, n.$$

We use $(Y_1^*, x_1), \ldots, (Y_n^*, x_n)$ to re-estimate α and β. We construct the bootstrap distribution of these parameters by repeating the resampling process B times. Construct the bootstrap distribution of the least squares estimator of β and use it to set an approximate 95% confidence interval for β. Compare the result to that obtained in Problem 3.6.7.

6.7.7. Construct the bootstrap distributions of the least squares estimator and the Huber M-estimator of the temperature coefficient in a linear regression model for the enamel covered steel plate data presented in Section 1.1.5. Construct approximate 95% confidence intervals for this parameter based on the two estimators. Compare the bootstrap distributions and intervals. To explore the effects of misspecification, repeat the above analysis omitting the time variable from the model. How does the bootstrap perform under misspecification?

6.7.8. Suppose that we have observations $Z_1, \ldots, Z_n$ on the model (6.51) and we want to use the sample mean $\bar{z}$ to make inferences about the mean of f. Show that the saddlepoint approximation (Section 4.4.3) to the bootstrap density of $\bar{z}$ is given by

$$f_{\hat{\theta}}(x; \theta_0) \sim \frac{n^{1/2}}{(2\pi)^{1/2}|A_{F_x}(x)|^{1/2}} \exp\{nK_n^*(\alpha(x)) - n\alpha(x)x\},$$

where $\alpha(x)$ satisfies

$$0 = \sum_{i=1}^{n} (z_i - x) \exp\{\alpha(z_i - x)\}, \; K_n^*(\alpha) = \log\left\{n^{-1} \sum_{i=1}^{n} \exp(z_i \alpha)\right\},$$

and

$$A_{F_x}(x) = n^{-1} \sum_{i=1}^{n} (z_i - x)^2 \exp\{\alpha(x) z_i - K_n^*(\alpha(x))\}.$$

6.7.9. Consider an estimator $\hat{\theta}_n$ of θ based on $Z_1, \ldots, Z_n$. Let $\hat{\theta}_{n,-i}$ denote the estimator calculated from the $n - 1$ observations after Z_i has been excluded. The *jackknife estimator* of θ is defined to be

$$J(\hat{\theta}_n) = n^{-1} \sum_{i=1}^{n} J_i(\hat{\theta}_n),$$

where $J_i(\hat{\theta}_n) = \hat{\theta}_n - (n - 1)(\hat{\theta}_{n,-i} - \hat{\theta}_n)$, $i = 1, \ldots, n$. Show that if

$$\mathrm{E}\hat{\theta}_n = \theta + \frac{a}{n} + \frac{b}{n^2} + O\left(\frac{1}{n^3}\right),$$

then

$$\mathrm{E}J(\hat{\theta}_n) = \theta - \left(\frac{n}{n-1}\right)\frac{b}{n^2} + O\left(\frac{1}{n^3}\right).$$

Interpret this result. Find the bias of $\hat{\sigma}^2 = (1/n) \sum_{i=1}^{n} (z_i - \bar{z})^2$ and show that jackknifing $\hat{\sigma}^2$ produces an unbiased estimator of the underlying variance σ^2.

6.7.10. The variance of a jackknife estimator $J(\hat{\theta}_n)$ can be estimated by

$$V = \frac{1}{n-1} \sum_{i=1}^{n} \{J_i(\hat{\theta}_n) - J(\hat{\theta}_n)\}^2.$$

What is the jackknife estimate of the variance of $J(\hat{\sigma}^2)$, where $\hat{\sigma}^2 = (1/n) \sum_{i=1}^{n} (z_i - \bar{z})^2$? Compare this to the actual variance of $\hat{\sigma}^2$ obtained in Section 5.1.

6.7.11. For the stellar velocity data for the disk stars (Section 1.1.4), compute and then plot the quantities $J_i(\hat{\sigma}^2) - J(\hat{\sigma}^2)$ against Z_i, where $\hat{\sigma}^2 = (1/n)\sum_{i=1}^{n}(z_i - \bar{z})^2$. How does this compare to the sensitivity curve introduced in Section 5.1? How does the curve change if we jackknife $\hat{\sigma}$ instead of $\hat{\sigma}^2$? Use Tukey's (1958) suggestion that $n^{1/2}\{J(\hat{\theta}_n) - \theta\}/V^{1/2}$ has approximately a Student t distribution with $n - 1$ degrees of freedom to construct a 95% jackknife confidence interval for σ. Compare this interval to the bootstrap interval from Problem 6.7.3 and compare these intervals to the other intervals obtained in Section 5.1.

6.8 OTHER RESAMPLING METHODS

The bootstrap approximates the sampling distribution of a statistic by sampling from the empirical distribution F_n or, equivalently, by resampling from the observed data. In this section, we describe some methods which use resampling (from distributions other than F_n) to produce simulated data from a specified distribution P.

6.8.1 The Inverse Probability Transformation

As noted in Section 3.10, the general problem of simulating realizations from a specified distribution is discussed in Devroye (1986) and Ripley (1987). The simplest method of generating a (univariate) random variable from a distribution P is by generating a random variable from the uniform U(0, 1) distribution and applying the inverse probability or quantile transformation $P^{-1}(u)$, where P is the distribution function of the distribution. The difficulty with this method is that we require the quantile function P^{-1}. The resampling methods we present below provide a way to generate random variables from the distribution P given only its density function p.

These methods are particularly useful for approximating posterior distributions, posterior moments, and so on in Bayesian analysis, but because they have much wider applicability we will present a general description of them before discussing specific Bayesian implementations. For the purpose of general discussion, we suppose that we want to generate a sample of random vectors $\theta_1, \ldots, \theta_m$ from the distribution with density function proportional to $p(\theta)$. It is important that we do not need to know the normalizing constant because in many Bayesian applications it is unknown.

6.8.2 The Rejection Method

If we can find a distribution with density function q such that we can generate θ from q and there is an $M > 0$ satisfying

$$\frac{p(\theta)}{q(\theta)} \leq M,$$

then an exact sampling procedure known as the *rejection method* is defined by:

1. Generate θ from q and u from U(0, 1).
2. If $u \le \dfrac{p(\theta)}{q(\theta)M}$, accept θ into the sample; otherwise repeat 1–2.
3. Repeat 1–2 independently until we have accepted m realizations of θ into the sample.

This is a resampling method in the sense that we generate a large set of realizations of θ from q (in fact a random number of them) and then resample from this set according to the acceptance procedure. To see that the rejection method works, notice that

$$\begin{aligned}
\mathrm{P}(\theta_i \le t \mid \theta_i \text{ accepted}) &= \frac{\mathrm{P}(\theta_i \le t, \theta_i \text{ accepted})}{\mathrm{P}(\theta_i \text{ accepted})} \\
&= \frac{\displaystyle\iint I\left\{\theta \le t, u \le \frac{p(\theta)}{q(\theta)M}\right\} q(\theta)\, du\, d\theta}{\displaystyle\iint I\left\{u \le \frac{p(\theta)}{q(\theta)M}\right\} q(\theta)\, du\, d\theta} \\
&= \frac{\displaystyle\iint I\{\theta \le t\}\, \frac{p(\theta)}{q(\theta)M}\, q(\theta)\, d\theta}{\displaystyle\iint \frac{p(\theta)}{q(\theta)M}\, q(\theta)\, d\theta} \\
&= \frac{\displaystyle\int I\{\theta \le t\} p(\theta)\, d\theta}{\displaystyle\int p(\theta)\, d\theta}.
\end{aligned}$$

That is, the conditional distribution of θ given that θ is accepted is the distribution with density proportional to p as required.

6.8.3 The Sampling-Importance-Resampling Method

An alternative method which does not require the bound M in the rejection method is the method which Rubin (1988) called the *sampling-importance-resampling* (SIR) algorithm:

1. Generate a sample $\theta_1, \ldots, \theta_m$ from $q(\theta)$ and compute the sample weights $w(\theta_i) = \dfrac{p(\theta_i)}{q(\theta_i)}$.

2. Draw $\theta_1^*, \ldots, \theta_m^*$ with replacement from the multinomial distribution with atoms $\theta_1, \ldots, \theta_m$ and probability

$$\left\{\frac{w(\theta_1)}{\sum_{i=1}^m w(\theta_i)}, \ldots, \frac{w(\theta_m)}{\sum_{i=1}^m w(\theta_i)}\right\}.$$

The set $\{\theta_1^*, \ldots, \theta_m^*\}$ is approximately distributed as an independent sample with density proportional to $p(\theta)$. To see this, notice that

$$\begin{aligned} \mathrm{P}(\theta^* \leq t) &= \frac{\sum_{i=1}^m w(\theta_i) I(\theta_i \leq t)}{\sum_{i=1}^m w(\theta_i)} \\ &\to \frac{\int w(\theta) \mathrm{I}(\theta \leq t) q(\theta)\, d\theta}{\int w(\theta) q(\theta)\, d\theta} \\ &= \frac{\int I(\theta \leq t) p(\theta)\, d\theta}{\int p(\theta)\, d\theta}. \end{aligned}$$

Observe that

$$n^{-1} \sum_{i=1}^n w(\theta_i) = n^{-1} \sum_{i=1}^n \frac{p(\theta_i)}{q(\theta_i)} \to \int p(\theta)\, d\theta$$

is an estimate of the normalizing constant. The SIR algorithm is similar to the bootstrap except that we resample from the sample from q rather than from the data. We also resample with unequal probability which is possible but not a requirement with the bootstrap.

Both the rejection and the SIR method generally work better when q is close to p. We also need to ensure that simulating realizations from q is straightforward, so the choice of q usually represents a tradeoff between closeness to p and the simplicity of simulating realizations from q.

6.8.4 Markov Chain Methods

A different class of approaches to generating a random variable θ from $p(\theta)$ is by constructing a Markov chain $\{\theta^{(k)}\}$ whose state space is Ω and whose equilibrium distribution is $p(\theta)$. If we run the chain for a long time, until $k = K$ say, then $\theta^{(K)}$ has approximately the distribution $p(\theta)$ in the sense that

$$\theta^{(k)} \xrightarrow{\mathscr{D}} \theta \sim p(\theta), \qquad \text{as } k \to \infty.$$

Thus, if we make parallel independent runs of the chain, we can simulate a sample of independent observations from $p(\theta)$. A single run of the chain suffices from some purposes because the ergodic theorem ensures that

$$\frac{1}{k}\sum_{m=1}^{k} g(\theta^{(m)}) = \int g(\theta)p(\theta)\,d\theta + o_p(1), \qquad \text{as } k \to \infty.$$

That is, we can estimate moments from a single run of the chain. Markov chain methods are not usually regarded as resampling methods but they do involve iteratively resampling from conditional distributions to construct the chains.

The *Gibbs sampling algorithm* (Gelfand and Smith, 1990; Gelfand et al., 1990) uses the so-called *full conditional densities* of the individual components of θ, namely

$$p(\theta_i \mid \theta_j, j \neq i),$$

to generate a chain. Given arbitrary initial values $(\theta_2^{(0)}, \ldots, \theta_p^{(0)})$, we simulate the chain by repeatedly making the following sequence of steps: For $k = 0, 1, 2, \ldots, K$

$$\begin{aligned}
&\text{sample } \theta_1^{(k+1)} \text{ from } p(\theta_1 \mid \theta_2^{(k)}, \ldots, \theta_p^{(k)})\\
&\text{sample } \theta_2^{(k+1)} \text{ from } p(\theta_2 \mid \theta_1^{(k)}, \theta_3^{(k)}, \ldots, \theta_p^{(k)})\\
&\text{sample } \theta_3^{(k+1)} \text{ from } p(\theta_3 \mid \theta_1^{(k)}, \theta_2^{(k+1)}, \theta_4^{(k)}, \ldots, \theta_p^{(k)})\\
&\text{sample } \theta_4^{(k+1)} \text{ from } p(\theta_4 \mid \theta_1^{(k)}, \theta_2^{(k+1)}, \theta_3^{(k+1)}, \theta_5^{(k)}, \ldots, \theta_p^{(k)})\\
&\vdots\\
&\text{sample } \theta_p^{(k+1)} \text{ from } p(\theta_p \mid \theta_1^{(k)}, \theta_2^{(k+1)}, \ldots, \theta_{p-1}^{(k+1)}).
\end{aligned}$$

The sequence

$$\theta^{(1)}, \theta^{(2)}, \theta^{(3)}, \ldots$$

is called a *Gibbs sequence* and has the property that for large k, $\theta^{(k)} \sim p$. Obviously, the components of $\theta^{(k)}$ are realizations from the marginal distributions of p so we can use them estimate marginal properties. Notice that we can estimate the marginal densities of the components by computing

$$\hat{p}_i(\theta_i) = m^{-1}\sum_{k=1}^{m} p(\theta_i \mid \theta_j^{(k)}, j \neq i).$$

6.8.5 Applications to Bayesian Inference

Just as samples from the empirical distribution can be useful in frequentist inference, samples from the posterior distribution

$$g(\theta \mid \mathbf{z}) \propto f(\mathbf{z} \mid \theta)g(\theta)$$

can be useful in Bayesian inference, particularly if we can generate the samples without having to integrate to find the normalizing constant and the quantile function. We can use the samples to estimate the normalizing constant, marginal posterior densities, marginal tail probabilities, and moments of the posterior distribution.

The rejection method can be applied directly using an arbitrary distribution with density function q or using the prior distribution for q. In the latter case step 1 entails generating θ from the prior $g(\theta)$ and u from the U(0, 1) distribution. Then since

$$\frac{g(\theta \mid \mathbf{z})}{g(\theta)} = f(\mathbf{z} \mid \theta),$$

we can take $M = f(\mathbf{z} \mid \hat{\theta})$, where $\hat{\theta}$ maximizes $f(\mathbf{z} \mid \theta)$. That is, in step 2, we accept θ if

$$u \leq \frac{f(\mathbf{z} \mid \theta)}{f(\mathbf{z} \mid \hat{\theta})}.$$

The resulting sample is a sample of m independent observations from the posterior distribution.

If we use the prior distribution for q in the SIR algorithm, step 1 requires generating a sample $\theta_1, \ldots, \theta_m$ from the prior distribution $g(\theta)$ and computing the sample weights

$$w(\theta_i) = \frac{g(\theta_i \mid \mathbf{z})}{g(\theta_i)} = f(\mathbf{z} \mid \theta_i).$$

In step 2, we sample $\theta_1^*, \ldots, \theta_m^*$ with replacement from the multinomial distribution with atoms $\theta_1, \ldots, \theta_m$ and probability

$$\left\{f(\mathbf{z} \mid \theta_1)\Big/\sum_{i=1}^{m} f(\mathbf{z} \mid \theta_i), \ldots, f(\mathbf{z} \mid \theta_m)\Big/\sum_{i=1}^{m} f(\mathbf{z} \mid \theta_i)\right\}.$$

The set $\{\theta_1^*, \ldots, \theta_m^*\}$ is then approximately distributed as an independent sample from the posterior distribution $g(\theta \mid \mathbf{z})$. In this case, the normalizing constant for the posterior distribution can be estimated by

$$m^{-1} \sum_{i=1}^{m} f(\mathbf{z} \mid \theta_i) \to \int f(\mathbf{z} \mid \theta) g(\theta)\, d\theta.$$

6.8.6 Sensitivity Analysis for the Effect of Caffeine Problem

Suppose as part of a sensitivity analysis, we adopt the Student t model with degrees of freedom fixed at $v = 3$ for the effect of caffeine on the volume of urine data. In this case, the likelihood is

$$\mathscr{L}(\mu, \sigma) \propto \sigma^{-n} \exp\left\{-\sum_{i=1}^{n} \frac{\nu+1}{2} \log\left(1 + \frac{(z_i - \mu)^2}{\nu\sigma^2}\right)\right\}, \qquad (\mu, \sigma) \in \mathbb{R} \times [0, \infty).$$

Suppose that as a vague prior we assume that $\mu \sim \mathrm{N}(0, 10\,000)$ independent of $\sigma^2 \sim \Gamma(1, 10\,000)$. If we generate a set of realizations from this prior and transform them using either the rejection method or the SIR algorithm, we find that there are very few distinct points in the posterior distribution. This is not surprising because the likelihood is negligible over most of the support of the prior distribution or, in other words, the prior density playing the role of q is very different from the posterior density playing the role of p. However, the distinct points in the posterior distribution suggest the region where the likelihood is non-negligible (in this case $-50 \leq \mu \leq 250$ and $0 \leq \sigma^2 \leq 450$) so we can apply the principle of precise measurement (Section 2.2.9), approximate the prior by a uniform distribution over this region, and then repeat the simulation to obtain a useful sample from the posterior distribution. A set of 2000 realizations from the approximating uniform prior distribution with density

$$g(\mu, \sigma^2) \propto \frac{1}{300 \times 450} I(-50 \leq \mu \leq 250, 0 \leq \sigma^2 \leq 450),$$

is shown in the first panel of Figure 6.6. The same set of observations after transformation into a sample of (μ, σ) from the posterior distribution using the SIR algorithm is shown in the second panel of Figure 6.6 and kernel density estimates (Section 1.5.1) of the marginal posterior distributions of μ and σ (obtained from the first and second components of the realizations of (μ, σ) respectively) are shown in Figure 6.7. The Gaussian qq-plot (Section 1.5.3) in Figure 6.8 shows that the marginal posterior distribution of μ is shorter tailed than the Gaussian distribution so we can use a Gaussian approximation to it. The posterior mean of μ is estimated by the sample mean of the realizations of μ from the posterior distribution, namely 133.40, and the posterior standard deviation by the sample standard deviation of the realizations of μ from the posterior distribution, namely 47.40. An approximate 95% credibility interval for μ is then given by $133.40 \pm 1.96 \times 47.40$ or

$$[40.50, 226.30],$$

which is shifted to the left and slightly narrower than the interval [67, 274] which we obtained in Section 2.3.4 under the Gaussian model. Alternatively, we can construct a 95% Bayesian credibility interval for μ from the 0.025 and 0.975 quantiles of the sample from the marginal posterior distribution of μ.

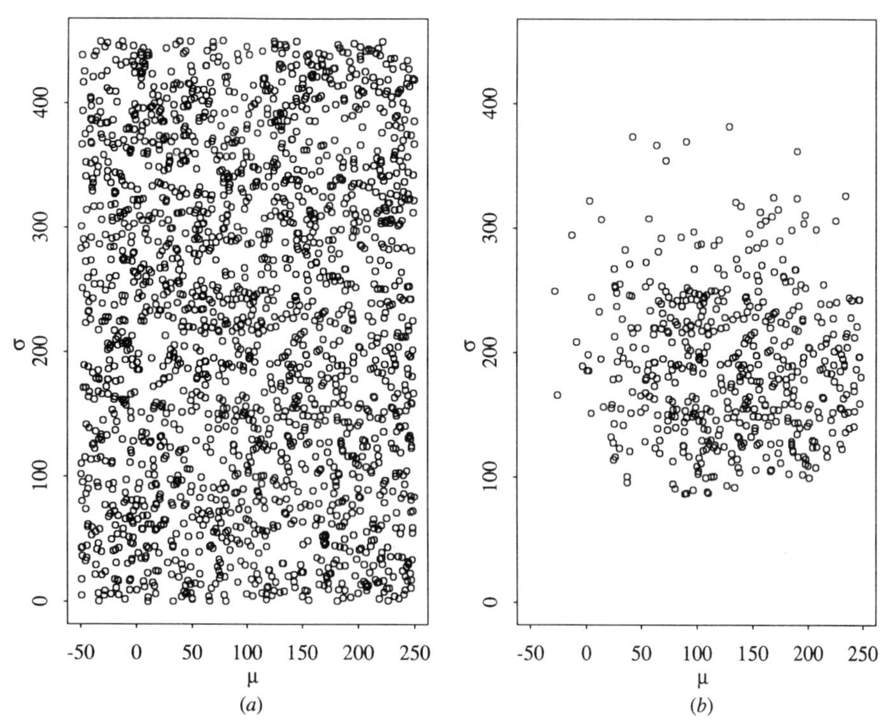

Figure 6.6. Simulated observations from a uniform prior distribution and from the posterior distribution of (μ, σ) for the urine data obtained via the SIR algorithm: (*a*) prior distribution; (*b*) posterior distribution.

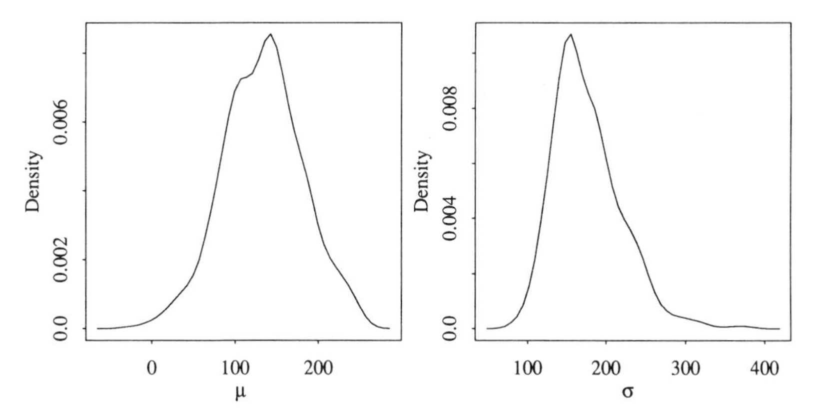

Figure 6.7. Simulated marginal posterior densities of μ and σ for the urine data. The densities are obtained using the Gaussian kernel and $h = 50$: (*a*) marginal posterior density of μ; (*b*) marginal posterior density of σ.

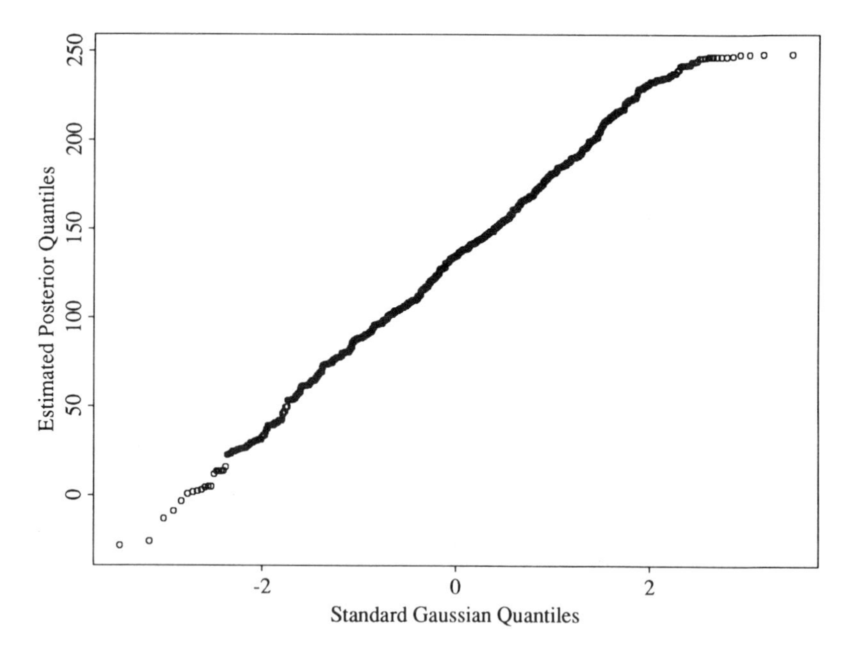

Figure 6.8. Gaussian qq-plot of the simulated quantiles from the marginal posterior distribution for μ.

We obtain

$$[37.18, 231.40],$$

which is very similar to the interval obtained using the Gaussian approximation.

The use of the prior distribution for q in these resampling algorithms provides insight into the nature of Bayesian analysis because the initial sample from the prior is transformed into the final sample from the posterior distribution. As pointed out by Smith and Gelfand (1992) and Albert (1993), this is analogous to the transformation of the prior density into the posterior density by means of Bayes' theorem.

Markov chain methods do not require us to sample from the prior distribution which can therefore be improper. If we adopt the Jeffreys prior with our Student t model, the posterior density is

$$g(\mu, \sigma \mid \mathbf{z}) \propto \sigma^{-n-1} \exp\left\{-\sum_{i=1}^{n} \frac{\nu+1}{2} \log\left(1 + \frac{(z_i + \mu)^2}{\nu\sigma^2}\right)\right\}.$$

However, the conditional posterior densities of μ given σ and of σ given μ do not have recognizable forms so it is not straightforward to generate data from them. An alternative Markov chain approach to computing posterior distributions is provided by the data augmentation algorithm of Tanner and Wong (1987).

6.8.7 The Role of Simulation in Bayesian Analysis

It is noticeable that the way we use the simulated realizations from the posterior distribution to estimate the normalizing constant, marginal posterior densities, marginal tail probabilities, and moments of the posterior distribution is essentially frequentist rather than Bayesian. This is a concern to some Bayesians (see for example O'Hagan, 1987) who argue that non-Bayesian simulation methods should not be used in a Bayesian analysis.

PROBLEMS

6.8.1. Suppose we adopt the Student t_3 model of Section 6.8.6 for the stellar velocity data for the halo stars (Section 1.1.4). Use the rejection method or the SIR algorithm to generate a sample from the posterior distribution of (μ, σ). Begin with a vague prior to determine the region over which the posterior distribution has non-negligible support and then use uniform priors over the region with non-negligible support. Set an approximate 95% Bayesian credibility interval for σ. Construct a 95% credibility interval for σ using a Gaussian model and compare the two intervals.

6.8.2. Suppose we adopt the gamma model of Problem 6.7.5 for the pressure vessel failure data of Section 1.1.2. Use the rejection method or the SIR algorithm to generate a sample from the posterior distribution of (λ, κ). Use a vague prior for λ and an independent U(0, 3) prior for κ. Obtain the marginal distributions of λ and δ and then set approximate 95% Bayesian credibility intervals for these parameters. What do you conclude about the model?

6.8.3. Suppose that we adopt the negative binomial model of Problem 6.7.4 for Rutherford and Geiger's (1910) alpha particle emissions data presented in Problem 1.5.1. Use the rejection method or the SIR algorithm to generate a sample from the posterior distribution of (λ, δ). Use a vague prior for λ and an independent exponential prior with mean 1 for δ. Obtain the marginal distributions of λ and δ and then set approximate 95% Bayesian credibility intervals for these parameters. What do you conclude about the model?

6.8.4. Suppose that we have observations $\mathbf{Z}$ on the Gaussian model (6.1) and we adopt the Jeffreys prior for (μ, σ). Find the conditional posterior distribution of μ given σ and $\mathbf{z}$ and the conditional posterior distribution of σ given μ and $\mathbf{z}$. For the effect of caffeine on the volume of urine data of Section 1.1.3, apply the Gibbs sampling scheme to generate a random sample from the posterior distribution of (μ, σ). (Generate 1000 realizations from 50 steps of the chain each time.) Compare the samples from the marginal posterior distributions generated by Gibbs sampling to the known marginal posterior distributions.

6.8.5. Consider the variance component model presented in Section 1.3.6

$$\mathscr{F} = \left\{ f(\mathbf{z}; \mu, \sigma_a, \sigma_u) = \frac{1}{(2\pi)^{mg/2}|\Sigma|^{1/2}} \exp\{-(\mathbf{z} - \mu)^{\mathrm{T}}\Sigma^{-1}(\mathbf{z} - \mu)/2\}, \right.$$
$$\left. -\infty < z_{ij} < \infty \colon \mu \in \mathbb{R}, \sigma_a > 0, \sigma_u > 0 \right\},$$

where Σ is the block diagonal matrix with blocks $\sigma_a^2\mathbf{J} + \sigma_u^2 I$, where J is the $m \times m$ matrix with all elements equal to 1 and I is the $m \times m$ identity matrix. Suppose we adopt the prior of Problem 2.7.4 for $(\mu, \tau_a = m\sigma_a^2 + \sigma_u^2, \tau_u = \sigma_u^2)$. Find the required conditional posterior distributions and show how to use Gibbs sampling to estimate the marginal posterior distributions of the parameters.

6.8.6. Discuss the use of Gibbs sampling in the context of Problem 6.8.2. How would you generate the realizations of κ?

6.9 NONPARAMETRIC METHODS

Design-based inference described in Sections 6.3–6.5 does not require the assumption of a parametric model for the data. Such inference is often described as *distribution free* or *nonparametric*. Neither description is entirely satisfactory; the randomization distribution and the derived sampling distribution are used in the inference and the quantities of interest are still parameters. Nonetheless, it is clear that the procedures are rather different in nature from those considered in earlier chapters in that they have a weaker reliance on simple parametric models for the data. We will follow convention and describe such procedures as nonparametric, but it is important to keep in mind the limitations of this description.

We have implicitly encountered the main types of nonparametric methods already but it is useful to make the scope of nonparametric methods explicit by describing the procedures from a different perspective. Many of these procedures are more properly semiparametric procedures (Section 1.3.3) but, in conformity with current usage, we will ignore this distinction in this section.

6.9.1 Rank-Based Methods

An important class of nonparametric methods are those based on the ranks of the data. The rank of Z_i among $Z_1, \ldots, Z_n$ is the position of Z_i in $Z_{n1} \le Z_{n2} \le \cdots \le Z_{nn}$ or, equivalently, the number of observations less than Z_i. For a paired data problem like the caffeine experiment, let R_i be the rank of the ith absolute pairwise difference $|Z_i|$ among $|Z_1|, \ldots, |Z_n|$, and consider the class of rank tests based on rank statistics of the form

$$\sum_{i=1}^{n} \Delta_i a(R_i),$$

where $0 = a(0) \le a(1) \le \cdots \le a(n)$ is a nonconstant sequence of scores and $\{\Delta_i\}$ is the set of signs of the pairwise differences defined in (6.22). The sign test (Sections 6.3.2 and 6.6.2) is a trivial rank test with $a(x) = 1$, $x > 0$. A nontrivial rank test is the *Wilcoxon* (1945) *signed-rank test* which has $a(x) = x$.

The Wilcoxon signed-rank statistic can also be written as

$$W = 2 \sum_{k=1}^{n} kI(\Delta_k > 0) - \frac{n(n+1)}{2}$$
$$= 2W^* - \frac{n(n+1)}{2}, \tag{6.59}$$

where $W^* = \sum_{k=1}^{n} kI(\Delta_k > 0)$. Note that the test based on W is equivalent to that based on W^*. Since under H_0, $I(\Delta_k > 0) \sim \text{binomial}(1, 1/2)$, we have that

$$\text{E}W^* = \sum_{k=1}^{n} k\text{E}I(\Delta_k > 0) = \frac{n(n+1)}{4}$$

and

$$\text{Var}\,(W^*) = \sum_{k=1}^{n} k^2\, \text{EVar}\, I(\Delta_k > 0) = \sum_{k=1}^{n} k^2 = \frac{n(n+1)(2n+1)}{24}.$$

As W^* is a weighted sum of independent binary random variables, we can apply a central limit theorem to obtain a (distribution free) Gaussian approximation to its sampling distribution. The distribution theory under a general alternative is quite complicated (see for example Section 2.5 of Hettmansperger, 1984) but it can be shown that, under shift alternatives, the efficacy of the Wilcoxon test is $(12)^{1/2} \int f_0(x)^2\, dx$. It can also be shown that the Wilcoxon signed rank statistic is equivalent to a score test based on the median of the Walsh averages med $\{(Z_i + Z_j)/2\}$ which is known as the *Hodges–Lehmann estimator*. (The connection between Walsh averages and the Wilcoxon signed rank statistic is shown in Section 6.9.3.)

6.9.2 Empirical Distribution Function Methods

Another important class of nonparametric methods are those based on the empirical distribution function (6.52). The basic idea is that we specify a parameter $T(F_0)$ as a function of the underlying distribution F_0, whatever F_0 may be, and then base inference on the Fisher consistent estimator $T(F_n)$ of $T(F_0)$. The systematic study of methods of this type along the lines indicated in Section 5.1.6 was initiated by von Mises (1947); recent references include Serfling (1980, Chapter 6) and Fernholz (1983). The bootstrap (Section 6.7) is an obvious example of a method of this type as are many of the methods described in Chapters 4 and 5. Estimators defined by estimating equations under a general model (Sections 4.2.2–4.2.5, 4.4.2–4.4.3, 4.5.6–4.5.11, 4.5.14, 5.1.1, and 5.4–5.6) and the methods used in nonparametric robustness (described in Section 5.7.9) are clearly of this type.

Although the class of estimators defined by estimating equations is very broad, the class of empirical distribution function procedures includes estimators which are not defined by estimating equations. A class of procedures in this category are procedures based on *linear combinations of order statistics* which are of the form

$$L_n = n^{-1} \sum_{i=1}^{n} J\left(\frac{i}{n}\right) Z_{ni} \tag{6.60}$$

where J is an integrable weight function on $[0, 1]$. With

$$J(u) = \frac{1}{1 - 2\alpha} I(\alpha \le u \le 1 - \alpha), \qquad 0 \le \alpha < 1/2,$$

in (6.60) we obtain the *α-trimmed mean* which is usually written in the asymptotically equivalent form

$$T_n = \frac{1}{n - 2[n\alpha]} \sum_{i=[n\alpha]+1}^{n-[n\alpha]} Z_{ni}.$$

As $\alpha \to 0$, T_n tends to the sample mean and, as $\alpha \to 1/2$, the sample median which was considered in Section 6.7.

If we use the use the fact that the order statistics $Z_{n1} \le \cdots \le Z_{nn}$ can be expressed in terms of the empirical quantile function (Section 1.5.3) as $Z_{ni} = F_n^{-1}(i/n)$, we can show that linear functions of the order statistics can be written as

$$L_n \sim \sum_{i=1}^{n} \left[\int_{(i-1)/n}^{i/n} J(u)\, du \right] Z_{ni} = \int_0^1 F_n^{-1}(u) J(u)\, du$$

so that linear functions of the order statistics are estimators of

$$L(F_0) = \int_0^1 F_0^{-1}(u)J(u)\,du.$$

The α-trimmed mean is an estimator of

$$T(F_0) = \frac{1}{1-2\alpha}\int_\alpha^{1-\alpha} F_0^{-1}(u)\,du = \frac{1}{1-2\alpha}\int_{F_0^{-1}(\alpha)}^{F_0^{-1}(1-\alpha)} y\,dF_0(y),$$

and, provided the α and $1-\alpha$ quantiles of F_0 are unique, inference about $T(F_0)$ can be based on the fact that

$$n^{1/2}\frac{(T_n - T(F_0))}{\hat{\sigma}(\gamma)} \xrightarrow{\mathscr{D}} \mathrm{N}(0, 1),$$

where

$$\hat{\sigma}^2(\alpha) = \frac{1}{(1-2\alpha)^2}\Bigg[n^{-1}\sum_{i=[n\alpha]+1}^{n-[n\alpha]}(Z_{ni} - T_n)^2 + \alpha\{F_n^{-1}(\alpha) - T_n\}^2$$
$$+ \alpha\{F_n^{-1}(1-\alpha) - T_n\}^2 - \alpha^2\{F_n^{-1}(1-\alpha) + F_n^{-1}(\alpha) - 2T_n\}^2\Bigg].$$

See for example Serfling (1980, Chapter 8) who gives a review of mcthods for establishing the theoretical properties of linear combinations of order statistics. As we noted in Problem 5.1.5, Tukey and McLaughlin (1963) showed that the Student t distribution with $n - 2[n\alpha]$ degrees of freedom often provides a good small sample approximation to the sampling distribution of $n^{1/2}(T_n - T(F))/\hat{\sigma}(\alpha)$.

6.9.3 *U*-Statistics

A useful class of estimators for a set of independent and identically distributed observations $\{Z_i\}$ with common distribution function F_0, is the class of *U-statistics* (Hoeffding, 1948) which are estimators of the form

$$U_n = \frac{1}{\binom{n}{k}}\sum_{1\le i_1<\cdots<i_k\le n} h(Z_{i_1}, \ldots, Z_{i_k}), \tag{6.61}$$

where $h(x_1, \ldots, x_k)$ is a kernel function of order k which is symmetric in its arguments. The sample mean and the empirical distribution function (6.52) are trivial U-statistics with kernels of degree 1. The sample variance (which we considered in Sections 5.1–5.2) can be written as

$$s^2 = \frac{1}{(n-1)} \sum_{i=1}^{n} (Z_i - \bar{Z})^2 = \frac{2}{n(n-1)} \sum_{1 \le i < j \le n} (Z_i - Z_j)^2/2$$

which is a U-statistic (6.61) with kernel $h(x_1, x_2) = (x_1 - x_2)^2/2$ of degree 2. The rank R_i of a positive Z_i among $|Z_1|, \ldots, |Z_n|$ is the number of Z_j satisfying $|Z_j| \le Z_i$ or equivalently $Z_j + Z_i > 0$ so as Tukey (1949) showed, the Wilcoxon signed-rank statistic (6.59) is equivalent to

$$W^* = \sum_{k=1}^{n} kI(\Delta_k > 0) = \sum_{1 \le i < j \le n} I(Z_i + Z_j > 0)$$

which is $n(n-1)/2$ times a U-statistic (6.61) with kernel $h(x_1, x_2) = I(x_1 + x_2 > 0)$ of degree 2.

U-statistics are unbiased estimators of

$$U(F_0) = \int \cdots \int h(x_1, \ldots, x_k)\, dF_0(x_1) \cdots dF_0(x_k).$$

If we adopt a model $\mathscr{F}$ which specifies that the distribution F_0 is in a rich enough class of distributions, then U-statistics are UMVU estimators (Section 3.1.4) of their expectations $U(F_0)$. This holds for example when $\mathscr{F}$ is the class of all absolutely continuous distributions. See for example Lehmann (1983, p. 101).

U-statistics are related to empirical distribution function procedures through the fact that

$$U_n = \frac{n^k}{\binom{n}{k}} U(F_n)$$

but they are usually treated separately. Suppose that the kernel function satisfies $\mathrm{E}_{F_0} h(Z_1, Z_2, \ldots, Z_k)^2 < \infty$ and $\mathrm{E}_{F_0}\{\tilde{h}_1(Z) - U(F_0)\}^2 > 0$, where

$$\tilde{h}_1(x) = \mathrm{E}_{F_0} h(x, Z_2, \ldots, Z_k).$$

Then the large sample approximation to the sampling distribution of U_n is

$$n^{1/2}(U_n - U(F_0)) \xrightarrow{\mathscr{D}} \mathrm{N}(0, k^2 \mathrm{E}_{F_0}\{\tilde{h}_1(Z_i) - U(F_0)\}^2).$$

See for example Serfling (1980, Chapter 5). The asymptotic variance of U_n can be estimated by Sen's (1960) estimator

$$k^2(n-1)^{-1}\sum_{i=1}^{n}(U_{n(i)}-U_n)^2,$$

where

$$U_{n(i)}=\frac{1}{\binom{n-1}{k-1}}\sum_{\substack{1\le i_2<\cdots<i_k\le n\\ i_j\ne i}}h(Z_i,Z_{i_2},\ldots,Z_{i_k}).$$

6.9.4 Smoothing

The final class of nonparametric procedures we consider is the class of estimators of smooth functions. An important example of a procedure of this type is the kernel density estimator discussed in Section 1.5.1. Kernel smoothing methods extend to estimating distribution functions, quantile functions and derivatives of these in a straightforward way.

A different but still related application of smoothing methods is to curve estimation. Suppose we have the nonparametric regression model

$$Y_i=m(X_i)+e_i,$$

where $\mathrm{E}(Y_i \mid X_i)=m(X_i)$ is a smooth but otherwise unspecified function and $\{e_i\}$ are independent and identically distributed with mean 0 and variance σ^2. If m is parametric (as is the case when $m(x)=\alpha+\beta x$), we can estimate it using the least squares procedure

$$\min_m\sum_{i=1}^{n}(Y_i-m(X_i))^2.$$

This does not work when m is nonparametric (the solution interpolates the data points) but we can use the same idea at a point x by making the least squares criterion local (to x) by introducing a weight function $K\{(X_i-x)/h\}$, $h>0$, which gives weight to X_i close to x and no weight to X_i far from x, and then making a (local) Taylor expansion (see 2 in the Appendix) for $m(X_i)$ about $m(x)$. If we use a first order Taylor expansion, the resulting local least squares criterion is to estimate $m(x)$ and $m'(x)$ by $\hat{m}(x)$ and $\hat{m}'(x)$ which satisfy

$$\min_{m(x),\,m'(x)}\sum_{i=1}^{n}(Y_i-m(x)-m'(x)(X_i-x))^2K\left(\frac{X_i-x}{h}\right). \tag{6.62}$$

Notice that $\hat{m}(x)$ is the intercept or equivalently the fitted value at x of the weighted linear regression of Y_i on (X_i-x) with weights $K\{(X_i-x)/h\}$. If we repeat the local regression fit on a grid of x values, we obtain an estimate of the function $\hat{m}(x)$. This *local regression method* is due to Stone (1977) and Cleveland (1979).

If we let A_x denote the $n \times 2$ matrix with ith row $(1, X_i - x)$, $K_x(h) = \text{diag}\,(K\{(X_1 - x)/h\}, \ldots, K\{(X_n - x)/h\})$, $y = (Y_1, \ldots, Y_n)^T$ and $a = (1, 0)^T$, we have that

$$\hat{m}(x) = a^T\{A_x^T K_x(h) A_x\}^{-1} A_x^T K_x(h) y.$$

It follows that with $m = (m(X_1), \ldots, m(X_n))^T$,

$$E\{\hat{m}(x) \mid X_1, \ldots, X_n\} = a^T\{A_x^T K_x(h) A_x\}^{-1} A_x^T K_x(h) m,$$

so the estimator is biased, and

$$\text{Var}\,\{\hat{m}(x) \mid X_1, \ldots, X_n\} = \sigma^2 a^T\{A_x^T K_x(h) A_x\}^{-1} A_x^T K_x(h)^2 A_x \{A_x^T K_x(h) A_x\}^{-1} a. \tag{6.63}$$

Both of these expressions are complicated functions of the kernel K and h. If $h \to 0$ such that $nh \to \infty$ as $n \to \infty$, K satisfies $|K(x)| < \infty$, $K(x) = 0$ for $|x| > c$, $\int K(z)\,dz = 1$, $\int zK(z)\,dz = 0$, $\int z^2K(z)\,dz < \infty$, and $\int K(z)^2\,dz < \infty$, and x is an interior point of the support of the density f_X of X, we can show that

$$E\{\hat{m}(x) \mid X_1, \ldots, X_n\} - m(x) \sim \frac{h^2}{2} m''(x) \int z^2 K(z)\,dz$$

and

$$\text{Var}\,\{\hat{m}(x) \mid X_1, \ldots, X_n\} \sim \frac{\sigma^2}{nhf_X(x)} \int K(z)^2\,dz,$$

as $n \to \infty$ (Ruppert and Wand, 1994). These results show that the choice of h has the same effect as in density estimation (Section 1.5.1): Small h leads to low bias but increased variance while large h leads to large bias but decreased variance. We can choose h informally (as we did in Section 1.5.1) or more formally by minimizing an estimate of the asymptotic mean squared error.

A local regression estimate of the regression function is shown superimposed on the scatterplot of the change in catecholamine against the change in the volume of urine voided due to the ingestion of caffeine in Figure 6.9. The fit over a range of h supports the assumption that the relationship is linear. If we assume that the bias is small, we can set a standard error band using (6.63) with σ^2 estimated from the residual sum of squares. Of more value to the present problem is the fact that we can carry out a formal test of linearity using the method of Azzalini and Bowman (1993).

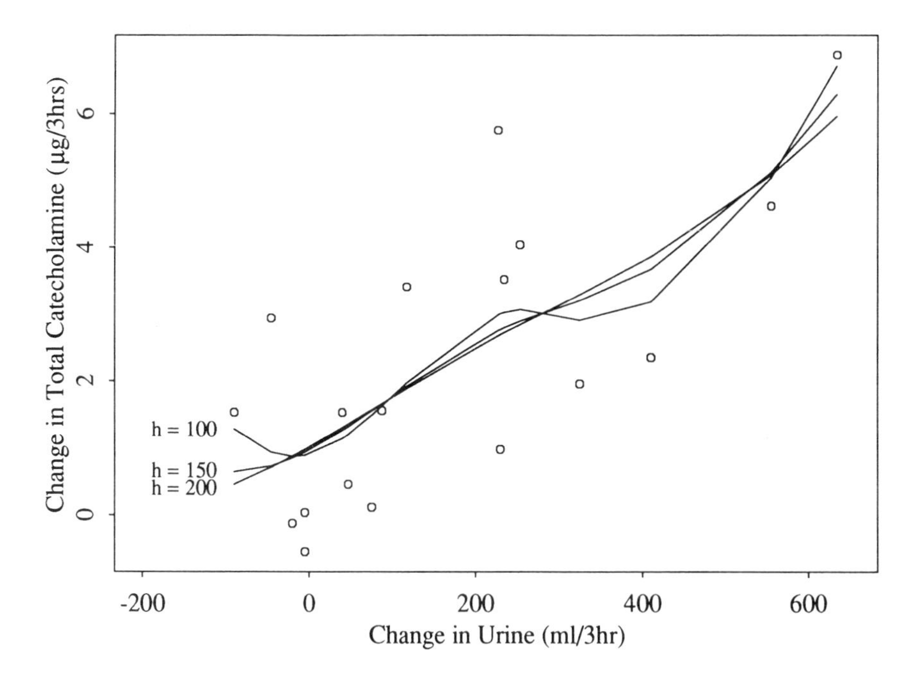

Figure 6.9. A scatterplot of the change in total catecholamine against the change in the volume of urine produced with the estimated regression function superimposed.

If we make only the zero order Taylor expansion $m(X_i) \approx m(x)$, (which is like setting $m'(x) = 0$ in (6.62), we obtain the *Nadaraya–Watson* (Nadaraya, 1964; Watson, 1964) *estimator*

$$\hat{m}(x) = \frac{\sum_{i=1}^{n} Y_i K\left(\frac{X_i - x}{h}\right)}{\sum_{i=1}^{n} K\left(\frac{X_i - x}{h}\right)}. \tag{6.64}$$

Higher order expansions can also be used to estimate the derivatives of $m(x)$ and the approach extends to multiple explanatory variables (see Ruppert and Wand, 1994). The estimators obtained by minimizing (6.62) are not robust against outliers but we can apply the same approach to robust regression criteria (see Section 5.5) to obtain robust estimators. See Fan et al. (1994), and Welsh (1996).

As we noted at the start of this chapter, kernel density estimation uses a form of randomization to smooth the data. The same is true of curve estimation. This is most easily seen for the Nadaraya–Watson estimator (6.64) which can be interpreted as the conditional expectation of $Y_i + gv_i$ given $X_i + hu_i$ where $\{v_i\}$ are independent and identically distributed with common density function

K_1 independent of $\{u_i\}$ which are independent and identically distributed with common density function K, provided $\int K_1(z)\,dz = 1$ and $\int zK_1(z)\,dz = 0$ (Problem 6.9.6). This brings us back to the theme of randomization with which we started this chapter.

6.9.5 Limitations of Nonparametric Methods

The use of nonparametric models is sometimes advocated as a way of accommodating the approximate nature of statistical models. Nonparametric models can be very useful but they are not a panacea for inference: They still rest on assumptions (such as independence and symmetry which are often ignored), they often apply only in relatively simple, unstructured problems, they often require reasonably large samples to be fitted effectively, they are often not parsimonious, and they often fail to answer the substantive question when the data contains outliers.

PROBLEMS

6.9.1. Apply the Wilcoxon signed rank test to the Cushny and Peebles (1905) data (Problem 5.1.5) to test the hypothesis of no difference in the effect of the two optimal isomers. Compare your results to those obtained in Problem 5.1.5.

6.9.2. Suppose that $\{Z_i\}$ are independent and identically distributed random variables with mean μ and variance σ^2. Show that the sample variance is a U-statistic with kernel $h(x_1, x_2) = (x_1 - x_2)^2/2$. Then show that $\tilde{h}_1(x) = \{(x - \mu(F_0))^2 - \sigma^2(F_0)\}/2$ and confirm that the large sample approximation to the sampling distribution of s^2 given in Section 6.9.3 agrees with that obtained in Section 5.1.1.

6.9.3. Suppose that $\{Z_i\}$ are independent and identically distributed random variables with common distribution function F. Obtain a large sample approximation to the sampling distribution of W^* in (6.59).

6.9.4. Use local regression smoothing to explore the relationship between the change in epinephrine and the change in the volume of urine voided following the ingestion of caffeine and then between the change in norepinephrine and the change in the volume of urine voided following the ingestion of caffeine.

6.9.5. Suppose that $\{Z_i\}$ are independent and identically distributed random variables with common density function f. Consider estimating $f(x)$ with the kernel density estimator

$$\hat{f}(x) = (nh)^{-1} \sum_{i=1}^{n} K\left(\frac{y - Z_i}{h}\right),$$

where K satisfies $|K(x)| < \infty$, $K(x) = 0$ for $|x| > c$, $\int K(z)\,dz = 1$, $\int zK(z)\,dz = 0$, $\int z^2K(z)\,dz < \infty$, and $\int K(z)^2\,dz < \infty$. If $h \to 0$ such that $nh \to \infty$ as $n \to \infty$, show that

$$\mathrm{E}\hat{f}(x) - f(x) \sim \frac{h^2}{2} f''(x) \int z^2K(z)\,dz.$$

(Hint: Make a change of variables and then expand f in a Taylor series.) Similarly, show that

$$\mathrm{Var}\,\{\hat{f}(x)\} \sim \frac{f(x)}{nh} \int K(z)^2\,dz$$

and find the value of h which minimizes the asymptotic mean square error.

6.9.6. Suppose that $\{(Y_i, X_i)\}$ are independent and identically distributed random variables with common joint density function f. Show that provided $\int K_1(z)\,dz = 1$ and $\int zK_1(z)\,dz = 0$, the Nadaraya–Watson estimator (6.64) is the estimator of

$$\mathrm{E}(Y \mid X = x) = \frac{\int yf(y, x)dy}{\int f(y, x)\,dy}$$

obtained by taking the ratio of the estimates

$$\int y\hat{f}(y, x)\,dy \quad \text{and} \quad \int \hat{f}(y, x)\,dy,$$

where $\hat{f}(y, x) = (nhg)^{-1} \sum_{i=1}^{n} K_1\{(y - Y_i)/g\}K\{(x - X_i)/h\}$.

6.9.7. Suppose that $\{(Y_i, X_i)\}$ are independent and identically distributed random variables with common joint density function f. If $h \to 0$ such that $nh \to \infty$ as $n \to \infty$, K satisfies $|K(x)| < \infty$, $K(x) = 0$ for $|x| > c$, $\int K(z)\,dz = 1$, $\int zK(z)\,dz = 0$, $\int z^2K(z)\,dz < \infty$, and $\int K(z)^2\,dz < \infty$, and x is an interior point of the support of the density f_X of X, show that the Nadaraya–Watson estimator (6.64) satisfies

$$\mathrm{E}\{\hat{m}(x) \mid X_1, \ldots, X_n\} - m(x) \sim \frac{h^2}{2}\left\{m''(x) + \frac{f'_X(x)}{f_X(x)} m'(x)\right\} \int z^2K(z)\,dz$$

and

$$\text{Var}\{\hat{m}(x) \mid X_1, \ldots, X_n\} \sim \frac{\sigma^2}{nhf_X(x)} \int K(z)^2 \, dz,$$

as $n \to \infty$. (Hint: Find the bias and variance of the numerator and combine these with the results obtained in Problem 6.9.6.)

FURTHER READING

An accessible but more general presentation of the randomization model and inference based on it is given by Scheffé (1959). Finite population inference is discussed by Royall (1976; 1992), and Särndal et al. (1992). Rank methods are accessibly presented by Lehmann (1975) and Hettmansperger (1984). Sheather (1987) discusses inference for the sample median in considerable detail. Efron (1982) and Efron and Tibshirani (1993) are useful references on bootstrapping. Resampling methods in Bayesian analysis are discussed by Smith and Gelfand (1992), Albert (1993), and Bernado and Smith (1994, pp. 350–6). Casella and George (1992) give a simple introduction to Gibbs sampling. There is a huge literature on smoothing. Cleveland (1993) shows the power of scatterplot smoothing techniques in applied data analysis.

CHAPTER 7

Principles of Inference

The axiomatic method is so fundamental to mathematics that it is not surprising there have been a number of attempts to apply it to develop foundational systems for statistical inference. That is, to develop a set of axioms or principles from which we can derive rules for selecting inference procedures in a variety of contexts.

The key issues in an axiomatic development are the reasonableness and acceptability of the underlying principles (which were traditionally viewed as self-evident), the correctness of the logic used in deducing the consequences of the principles, and the interpretation of the consequences of the principles. Once the logic has been established, the system rests on the underlying principles and the interpretation of their consequences. Their discussion can be presented in the language of formal logic or by means of colourful examples which illustrate the issues. We will adopt the second approach because it is more accessible.

The issues arising from attempts to clarify the foundations of statistical inference are interesting and important but need to be kept in perspective. A broadly acceptable foundational system which is flexible enough to deal with the range of possible problems may not exist. Moreover there is no essential requirement for there to be a single, normative approach to inference. The choice between approaches can be a matter of personal preference which can depend on the nature of the problem at hand. Ultimately, the various approaches to statistical inference must be justified by their usefulness in solving substantive problems rather than by their foundational basis and they can be compared on the basis of an overall appreciation of their nature and consequences for inference. This perspective is reflected in the structure of this book which begins with the presentation of data and substantive problems and relegates discussion of foundational issues to this, the final chapter.

7.1 THE COHERENCY PRINCIPLE

Ramsey in 1926 (see Ramsey, 1931), de Finetti (1937), Savage (1954; 1961), Pratt et al. (1964), Heath and Sudderth (1978), Lane and Sudderth (1983), and others have argued that inference is like a betting game in which we are required to make bets about an unknown parameter and that it is sensible to avoid placing bets which we expect to lose.

The following development is due to Cornfield (1969). Suppose that a parameter θ can take on J possible values θ_j and that there are I possible samples $\mathbf{z}_i$. That is, both the parameter space $\Omega = \{\theta_1, \ldots, \theta_J\}$ and the sample space $\mathscr{Z} = \{\mathbf{z}_1, \ldots, \mathbf{z}_I\}$ are finite. We will use j to index parameter values and i sample values throughout this section. Let the probability of realizing the ith sample point $\mathbf{z}_i$ given θ_j be denoted by $f(\mathbf{z}_i \mid \theta_j)$, where $\sum_{i=1}^{I} f(\mathbf{z}_i \mid \theta_j) = 1$ for all j. Let C_k, $k = 1, \ldots, 2^J$, denote one of the 2^J possible subsets of $\Omega = \{\theta_1, \ldots, \theta_J\}$. Let $\delta_{jk} = 1$ if $\theta_j \in C_k$ and 0 otherwise.

A statistician B is required to make probability assignments $P_{ik} = g(C_k \mid \mathbf{z}_i)$ to each C_k given the probabilities $f(\mathbf{z}_i \mid \theta_j)$ and the fact that $\mathbf{z}_i$ has been realized. Given the same information as B, an antagonist A (representing nature) is permitted to bet any amount for or against any combination of the 2^J intervals. Thus B assigns probabilies P_{ik} and A assigns the stake S_{ik}. For the interval C_k, A therefore pays $P_{ik}S_{ik}$ to B, and receives back either S_{ik} if C_k is correct or nothing if it is not. Thus A is risking $P_{ik}S_{ik}$ to win $(1 - P_{ik})S_{ik}$. If the true parameter value is θ_j, B's gain on this bet when sample point $\mathbf{z}_i$ has been realized is given by

$$G_{ijk} = (P_{ik} - \delta_{jk})S_{ik}.$$

Summing over all possible subsets,

$$G_{ij.} = \sum_{k=1}^{2^J} (P_{ik} - \delta_{jk})S_{ik}$$

and B's expected gain when θ_j obtains is

$$G_{.j.} = \sum_{i=1}^{I} f(\mathbf{z}_i \mid \theta_j) \sum_{k=1}^{2^J} (P_{ik} - \delta_{jk})S_{ik}. \tag{7.1}$$

We can now give a formal definition of coherent assignment.

We say that B's assignments are incoherent if for any stake S_{ik}, B's expected gain $G_{.j.} \leq 0$ for all j and $G_{.j.} < 0$ for some j. That is, B never expects to win and sometimes expects to lose. If B's assignments are not incoherent, we say that they are coherent.

That betting behavior should be coherent is arguably a minimal requirement for sensible betting and therefore can be set down as a basic principle. If we accept that the problem of making inferences about an unknown parameter is analogous to placing bets on statements about the parameter, then it follows that a basic principle of inference is that it should correspond to coherent betting behavior. This is the coherency principle.

Coherency principle: Inferences about unknown parameters should follow the rules for coherent betting behavior for placing bets on statements about the parameters.

7.1.1 Bayesian Assignment is Coherent

It is simple to show that Bayesian probability assignments are coherent. Notice that if there exists a set of finite q_j satisfying $q_j > 0$ for all $j = 1, \ldots, J$ such that

$$0 = \sum_{j=1}^{J} q_j G_{.j.}, \tag{7.2}$$

then either $G_{.j.} = 0$ for all j or if $G_{.j.} < 0$ for some values of j, there must be at least one value of j for which $G_{.j.} > 0$ so the assignment is coherent. To find such an assignment, we substitute (7.1) into the right-hand side of (7.2) and then change the order of summation to obtain

$$\begin{aligned} 0 &= \sum_{j=1}^{J} q_j \sum_{i=1}^{I} f(\mathbf{z}_i \mid \theta_j) \sum_{k=1}^{2^J} (P_{ik} - \delta_{jk}) S_{ik} \\ &= \sum_{i=1}^{I} \sum_{k=1}^{2^J} S_{ik} \sum_{j=1}^{J} q_j f(\mathbf{z}_i \mid \theta_j)(P_{ik} - \delta_{jk}). \end{aligned} \tag{7.3}$$

The right-hand side of (7.3) can only be identically 0 for all S_{ik} if

$$\sum_{j=1}^{J} q_j f(\mathbf{z}_i \mid \theta_j)(P_{ik} - \delta_{jk}) = 0$$

or

$$\begin{aligned} P_{ik} &= \frac{\sum_{j=1}^{J} q_j f(\mathbf{z}_i \mid \theta_j)\delta_{jk}}{\sum_{j=1}^{J} q_j f(\mathbf{z}_i \mid \theta_j)} \\ &= \frac{\sum_{\theta_j \in C_k} q_j f(\mathbf{z}_i \mid \theta_j)}{\sum_{j=1}^{J} q_j f(\mathbf{z}_i \mid \theta_j)}. \end{aligned}$$

If we identify $q_j / \sum_{j=1}^{J} q_j$ with the prior probability that θ_j is the true parameter, then P_{ik} is the posterior probability that $\theta_j \in C_k$. It follows that if each θ_j is assigned probability $P_{ij} = q_j f(\mathbf{z}_i \mid \theta_j) / \sum_{j=1}^{J} q_j f(\mathbf{z}_i \mid \theta_j)$, where $q_j > 0, j = 1, \ldots, J$, that if the probability of any set of θ_j is assigned the sum of the probabilities of each θ_j in that set, and that if a set of no θ_j is assigned probability 0, the assignment is coherent.

7.1.2 Coherent Assignments Are Bayesian

In fact, coherence also implies that the probability assignment must be Bayesian. The first step is to prove that coherence implies that the assigned probabilities P_{ij} satisfy the usual probability axioms, namely:

1. $0 \leq P_{ij} \leq 1$ for all i and j.
2. $\sum_{j=1}^{J} P_{ij} = 1$ for all i.
3. If $C_k = C_u \cup C_v$, where C_u and C_v are disjoint sets, then $P_{ik} = P_{iu} + P_{iv}$.

To prove 1, note that for a stake S_{ij}, B's gain is $P_{ij}S_{ij} - S_{ij} = (P_{ij} - 1)S_{ij}$ when θ_j obtains and $P_{ij}S_{ij}$ when it does not. Thus if $P_{ij} < 0$, B has a certain loss if $S_{ij} > 0$, while if $P_{ij} > 1$, B has a certain loss if $S_{ij} < 0$. To prove 2, let A select S_i for each sample point $\mathbf{z}_i$. A then risks $S_i \sum_{j=1}^{J} P_{ij}$, and, since one of the states must obtain, is paid S_i. B's gain is therefore $S_i(\sum_{j=1}^{J} P_{ij} - 1)$ which can be made negative by an appropriate choice of S_i unless $\sum_{j=1}^{J} P_{ij} = 1$. Finally, to prove 3), suppose A selects S_{ik}, S_{iu} and S_{iv}. Then B's gains for $\mathbf{z}_i$ when C_u obtains, when C_v obtains, and when neither obtains are given by

$$G_u = (P_{iu} - 1)S_{iu} + P_{iv}S_{iv} + (P_{ik} - 1)S_{ik}$$

$$G_v = P_{iu}S_{iu} + (P_{iv} - 1)S_{iv} + (P_{ik} - 1)S_{ik}$$

and

$$G_k = P_{iu}S_{iu} + P_{iv}S_{iv} + P_{ik}S_{ik},$$

respectively. This system of equations will have solutions (S_{iu}, S_{iv}, S_{ik}) which make G_{iu}, G_{iv} and G_{ik} negative, unless we choose the P_{ij} such that

$$\begin{aligned} 0 &= \begin{vmatrix} P_{iu} - 1 & P_{iv} & P_{ik} - 1 \\ P_{iu} & P_{iv} - 1 & P_{ik} - 1 \\ P_{iu} & P_{iv} & P_{ik} \end{vmatrix} \\ &= (P_{iu} - 1)(P_{iv} - 1)P_{ik} - (P_{iu} - 1)P_{iv}(P_{ik} - 1) - P_{iv}P_{iu}P_{ik} \\ &\quad + P_{iv}(P_{ik} - 1)P_{iu} + (P_{ik} - 1)P_{iu}P_{iv} - (P_{ik} - 1)(P_{iv} - 1)P_{iu} \\ &= P_{ik} - P_{iu} - P_{iv}. \end{aligned}$$

That is, 3 holds.

Now consider any two sample points $\mathbf{z}_1$ and $\mathbf{z}_2$ and any two-set partition of $\Omega = C_1 \cup C_2$. If B assigns P_{ij} to each θ_j then by 2 above B must assign $\sum_{m=1}^{J} P_{im}\delta_{m1}$ to C_1. (Recall that $\delta_{m1} = 1$ if $\theta_m \in C_1$ and zero otherwise.) If A selects the stakes,

Sample point 1	$S_{11} = kS$	$S_{12} = 0$
Sample point 2	$S_{21} = -k$	$S_{22} = 0$
Sample point $3, \ldots, I$	$S_{i1} = 0$	$S_{i2} = 0,$

then B's expected gain if the true parameter is in C_1 is

$$\begin{aligned} G_1 &= f(\mathbf{z}_1 \mid \theta_u)\left(kS \sum_{m=1}^{J} P_{1m}\delta_{m1} - kS\right) + f(\mathbf{z}_2 \mid \theta_u)\left(-k \sum_{m=1}^{J} P_{2m}\delta_{m1} + k\right) \\ &= -kSf(\mathbf{z}_1 \mid \theta_u)\left(1 - \sum_{m=1}^{J} P_{1m}\delta_{m1}\right) + kf(\mathbf{z}_2 \mid \theta_u)\left(1 - \sum_{m=1}^{J} P_{2m}\delta_{m1}\right), \\ & \qquad \text{for all } \theta_u \in C_1, \end{aligned}$$

while if the true parameter is in C_2, it is

$$\begin{aligned} G_2 &= f(\mathbf{z}_1 \mid \theta_v)\left(kS \sum_{m=1}^{J} P_{1m}\delta_{m1}\right) + f(\mathbf{z}_2 \mid \theta_v)\left(-k \sum_{m=1}^{J} P_{2m}\delta_{m1}\right) \\ &= kSf(\mathbf{z}_1 \mid \theta_v) \sum_{m=1}^{J} P_{1m}\delta_{1m} - kf(\mathbf{z}_2 \mid \theta_v) \sum_{m=1}^{J} P_{2m}\delta_{m1}, \qquad \text{for all } \theta_v \in C_2. \end{aligned}$$

The expected gains G_1 and G_2 will both be negative whenever

$$\frac{f(\mathbf{z}_2 \mid \theta_u)}{f(\mathbf{z}_1 \mid \theta_u)} \frac{1 - \sum_{m=1}^{J} P_{2m}\delta_{m1}}{1 - \sum_{m=1}^{J} P_{1m}\delta_{m1}} < S < \frac{f(\mathbf{z}_2 \mid \theta_v)}{f(\mathbf{z}_1 \mid \theta_v)} \frac{\sum_{m=1}^{J} P_{2m}\delta_{m1}}{\sum_{m=1}^{J} P_{1m}\delta_{m1}}$$

for all $\theta_u \in C_1$, $\theta_v \in C_2$, and $k > 0$, or (7.4)

$$\frac{f(\mathbf{z}_2 \mid \theta_u)}{f(\mathbf{z}_1 \mid \theta_u)} \frac{1 - \sum_{m=1}^{J} P_{2m}\delta_{m1}}{1 - \sum_{m=1}^{J} P_{1m}\delta_{m1}} > S > \frac{f(\mathbf{z}_2 \mid \theta_v)}{f(\mathbf{z}_1 \mid \theta_v)} \frac{\sum_{m=1}^{J} P_{2m}\delta_{m1}}{\sum_{m=1}^{J} P_{1m}\delta_{m1}}$$

for all $\theta_u \in C_1$, $\theta_v \in C_2$ and $k < 0$. To prevent losses, B must therefore select P_{ij} so that for any two-set partition of Ω,

$$\frac{\sum_{m=1}^{J} P_{2m}\delta_{m2}}{\sum_{m=1}^{J} P_{1m}\delta_{m2}} \bigg/ \frac{\sum_{m=1}^{J} P_{2m}\delta_{m1}}{\sum_{m=1}^{J} P_{1m}\delta_{m1}} = \frac{f(\mathbf{z}_2 \mid \theta_v)}{f(\mathbf{z}_1 \mid \theta_v)} \bigg/ \frac{f(\mathbf{z}_2 \mid \theta_u)}{f(\mathbf{z}_1 \mid \theta_u)} \qquad \text{for all } \theta_u \in C_1, \quad \theta_v \in C_2.$$

This in turn requires that

$$\frac{P_{2v}}{P_{1v}} \bigg/ \frac{P_{2u}}{P_{1u}} = \frac{f(\mathbf{z}_2 \mid \theta_v)}{f(\mathbf{z}_1 \mid \theta_v)} \bigg/ \frac{f(\mathbf{z}_2 \mid \theta_u)}{f(\mathbf{z}_1 \mid \theta_u)} \qquad \text{for all } v, u. \tag{7.5}$$

because otherwise, A can always find a set C_1 such that the inequalities in (7.4) obtain. Equation (7.5) implies that

$$P_{ij} = k_i f(\mathbf{z}_i \mid \theta_j) q_j \qquad \text{for } i = 1, 2 \text{ and all } j,$$

where, by 3 above, $k_i = 1/\sum_{j=1}^{J} f(\mathbf{z}_i \mid \theta_j) q_j$. Since the argument applies to any pair of sample points $\mathbf{z}_1$ and $\mathbf{z}_2$, this relationship must hold for all i. Finally, the q_j must be of the same sign to ensure that 1 holds and no q_j can be 0 since if q_h is 0, A bets $S_{ij} = 0$ for all $j \neq h$ and wins S_{ih} when θ_j obtains by risking $P_{ih} S_{ih} = 0$.

7.1.3 Discussion

The arguments of Sections 7.1.1–7.1.2 establish the following theorem.

Theorem 7.1 *Suppose that $f(\mathbf{z}_i \mid \theta_j)$ is the probability of realizing $\mathbf{z}_i$ given θ_j, and that $\mathbf{z}_i$ has been realized. Probability assignments are coherent if and only if we assign probabilities to the element θ_j of the parameter space Ω according to*

$$P_{ij} = \frac{f(\mathbf{z}_i \mid \theta_j) q_j}{\sum_{j=1}^{J} f(\mathbf{z}_i \mid \theta_j) q_j} \qquad \text{for all } i \text{ and } j,$$

where $0 < q_j < \infty$. That is, as we would do in a Bayesian analysis with proper prior probabilities q_j.

This result implies that we must make Bayesian inferences if we want to satisfy the coherency principle.

It is quite legitimate to reject the idea that inference is analogous to gambling. This viewpoint vitiates the entire development but may well carry implications for other approaches to inference too.

Perhaps a more serious objection is that coherency is a form of self-consistency which ensures internal (to an individual) rather than external validity. This is reflected in the fact that Theorem 7.1 holds for any proper prior including priors in conflict with the true prior. The introduction of a prior prevents us from making incoherent assignments but provides no guarantee that we satisfy any other requirements such as that the analysis recovers the true parameter value.

Implicitly in our development and explicitly in Savage's (1954) development of the coherency principle, we require the statistician to have an attitude to every uncertain event which can be measured by a probability (i.e. we have to assign a probability to every possible subset C_k of Ω) and we require these probabilities to be comparable. Wolfowitz (1962) argued that the requirement that we have an attitude to every uncertain event is unreasonably strong. He argued that it ought to be enough to make one reasonable choice of set to contain Θ without having to order all other possibilities. Kempthorne (1969) and LeCam (1977) have also argued that inference need not require the allocation of probabilities to all possible subsets of the parameter space Ω and coherency may not be compelling if we only assign probabilities to selected subsets of Ω.

7.2 THE LIKELIHOOD PRINCIPLE

As we noted in Section 2.6.2, Fisher argued that the likelihood contains all the information in the data. If we accept this argument, then logically, if two likelihoods for a parameter θ are proportional, we should make the same inferences for θ regardless of which likelihood we use. Formally, this is known as the likelihood principle.

Likelihood principle: Inference for given models $\mathscr{F} = \{f(\cdot;\theta): \theta \in \Omega\}$ *and* $\mathscr{G} = \{g(\cdot;\theta): \theta \in \Omega\}$ *should be identical for any two data sets* $\mathbf{y}$ *on* $\mathscr{F}$ *and* $\mathbf{z}$ *on* $\mathscr{G}$ *for which*

$$f(\mathbf{y};\theta) = h(\mathbf{y}, \mathbf{z})g(\mathbf{z};\theta).$$

The likelihood principle was developed from Fisher's ideas and advocated by Barnard (1947a,b; 1949), Barnard et al. (1962), and Birnbaum (1962) who used it to try to justify the likelihood approach to inference. Berger and Wolpert (1984) provide a valuable review.

The statement of the likelihood principle depends on the models $\mathscr{F}$ and $\mathscr{G}$. While these do not in principle have to be simple parametric models, satisfying the principle when they are not seems prohibitively difficult and the principle is usually viewed as applying only to parametric models.

The likelihood principle has important implications for the role of the sample space in inference. Two examples illustrate these implications.

7.2.1 Sequential Sampling

Suppose that we observe the number of successes Z in n independent Bernoulli trials with success probability θ. Then the model is

$$\mathscr{F} = \left\{ f(z;\theta) = \binom{n}{z}\theta^z(1-\theta)^{n-z}, z = 0, 1, \ldots, n: 0 \le \theta \le 1 \right\}$$

and the data is a realization of Z. If instead of holding a fixed number of Bernoulli trials, we decide to sample until we observe z successes and then observed the realization of N, the number of trials to the zth success, N has a negative binomial distribution so the model is

$$\mathscr{G} = \left\{ g(n; \theta) = \binom{n-1}{z-1} \theta^z (1-\theta)^{n-z}, n = z, z+1, \ldots : 0 \le \theta \le 1 \right\}.$$

We have that

$$g(n; \theta) = \frac{\binom{n-1}{z-1} f(z; \theta)}{\binom{n}{z}}.$$

The sampling scheme (and hence the sample space) is different in the two models but the likelihood principle states that we should ignore this and make the same inferences in both cases.

7.2.2 Possible Censoring

Suppose that we observe a realization $\mathbf{z}$ of $\mathbf{Z}$ generated by the exponential model

$$\mathscr{F} = \left\{ f(\mathbf{z}; \theta) = \prod_{i=1}^{n} \theta \exp(-\theta z_i), z_i > 0 : \theta > 0 \right\} \tag{7.6}$$

so that the likelihood is

$$f(\mathbf{z}; \theta) = \theta^n \exp\left\{ -\theta \sum_{j=1}^{n} z_j \right\}, \qquad \theta > 0,$$

Now suppose that the observations are censored at $c > 0$ so that instead of $\mathbf{z}$ we actually observe $\mathbf{y}$, where

$$y_j = z_j I(z_j \le c) + c I(z_j > c), \qquad 1 \le j \le n.$$

The y_j are realizations of independently distributed random variables Y_j which have density $\theta \exp(-\theta x)$ if $x < c$ and equal c with probability

$$\mathrm{E}I(Z_j > c) = \mathrm{P}(Z_j > c) = \int_c^{\infty} \theta\, e^{-\theta x}\, dx = e^{-\theta c}.$$

Thus in the censored case, the data are generated by the model

$$\mathscr{G} = \left\{ g(\mathbf{y}; \theta) = \prod_{j=1}^{n} \{\theta \exp(-\theta z_j)\}^{I(z_j \le c)} \{e^{-\theta c}\}^{I(z_j > c)}, 0 < z_j < c : \theta > 0 \right\}. \tag{7.7}$$

The likelihood under (7.7) is

$$g(\mathbf{y}; \theta) = \theta^r \exp\left[-\theta \sum_{j=1}^{n} \{z_j I(z_j \le c) + cI(z_j > c)\}\right]$$
$$= \theta^r \exp\left(-\theta \sum_{j=1}^{n} y_j\right) \qquad \theta > 0,$$

where $r = \sum_{j=1}^{n} I(z_j \le c)$ is the (random) number of uncensored observations.

If we draw a sample from (7.7) in which none of the observations is actually greater than c, no censoring occurs and we have $z_j = y_j$, $r = n$, and $\bar{z} = \bar{y}$ so that

$$g(\mathbf{y}; \theta) = f(\mathbf{z}; \theta).$$

The likelihood principle asserts that we should make the same inferences in both cases. That is, if censoring is possible but does not occur, inference should be the same as when censoring is impossible.

7.2.3 Analyses in Conflict With the Likelihood Principle

Suppose that we are in the situation described in Section 7.2.2 and we want to do a Bayesian analysis using the Jeffreys prior (Section 2.2.11). Under (7.6), the Fisher information is $1/\theta^2$ so the Jeffreys prior is

$$j(\theta) \propto 1/\theta, \qquad \theta > 0.$$

However, under (7.7), the Fisher information is

$$I(\theta) = \mathrm{E}\,\frac{r}{n\theta^2} = n^{-1} \sum_{j=1}^{n} \frac{\mathrm{E}I(Z_j \le c)}{\theta^2} = \frac{1 - e^{-\theta c}}{\theta^2}$$

so the Jeffreys prior is

$$j_c(\theta) \propto (1 - e^{-\theta c})^{1/2}/\theta \qquad \theta > 0.$$

Even if no censoring occurs in the sample, these priors lead to different posterior distributions and hence different inferences. Thus using the Jeffreys prior violates the likelihood principle. If however, the prior is chosen without reference to the sample space, the posterior distributions will be the same and the likelihood principle will be satisfied.

Now consider a frequentist analysis. The likelihood under (7.6) is maximized at $\hat{\theta} = 1/\bar{z}$. The expected and the observed Fisher information are equal and are given by $I(\hat{\theta}) = \mathscr{I} = 1/\hat{\theta}^2 = \bar{z}^2$ (Section 4.2.9) so an approximate $100(1 - \alpha)\%$

confidence interval for θ based on (4.24) is given by

$$\frac{1}{\bar{z}} \pm \frac{1}{n^{1/2}\bar{z}} \Phi^{-1}(1 - \alpha/2). \tag{7.8}$$

The likelihood under (7.7) is maximized at $\hat{\theta} = r/n\bar{y}$. The observed Fisher information (Section 4.2.9) is $\mathscr{I} = r/n\hat{\theta}^2 = r\bar{y}^2/n$ and the expected Fisher information (Section 4.2.9) is

$$I(\hat{\theta}) = \frac{1 - e^{-\hat{\theta}c}}{\hat{\theta}^2} = \frac{n^2\bar{y}^2}{r^2}\left\{1 - \exp\left(-\frac{cr}{n\bar{y}}\right)\right\}.$$

Under (7.7), an approximate $100(1 - \alpha)\%$ confidence interval for θ based on the expected Fisher information is

$$\frac{1}{\bar{y}} \pm \frac{1}{n^{1/2}(n\bar{y}/r)\{1 - \exp(-cr/n\bar{y})\}^{1/2}} \Phi^{-1}(1 - \alpha/2). \tag{7.9}$$

When no censoring occurs, (7.9) reduces to

$$\frac{1}{\bar{z}} \pm \frac{1}{n^{1/2}\bar{z}\{1 - \exp(-c/\bar{z})\}^{1/2}} \Phi^{-1}(1 - \alpha/2) \tag{7.10}$$

which is wider than (7.8) so the use of (7.10) conflicts with the likelihood principle.

The difference between (7.8) and (7.10) is that the asymptotic variances (based on the expected Fisher information) reflect the dependence of the sampling distribution on the sampling scheme. If we use the observed information which is $\mathscr{I} = r/n\hat{\theta}^2 = r\bar{y}^2/n$ in the censored case, we find that an approximate $100(1 - \alpha)\%$ confidence interval for θ is

$$\frac{1}{\bar{y}} \pm \frac{1}{r^{1/2}\bar{y}} \Phi^{-1}(1 - \alpha/2) \tag{7.11}$$

and (7.11) reduces to (7.8) when censoring does not actually occur. That is, using the observed information in a confidence interval obeys the likelihood principle because the maximum likelihood estimate and the observed information are identical for any two models with proportional likelihoods. This fortuitous occurrence is due to the fact that (7.11) is numerically the same as a large sample approximation to a conditional confidence interval, a credibility interval or a likelihood interval. See Sections 4.6.2–4.6.3.

The issues raised by the possibility of censoring in the frequentist approach have been colourfully illustrated by Pratt (1962).

> An engineer draws a random sample of electron tubes and measures the plate voltages under certain conditions with a very accurate volt-meter, accurate enough so that measurement error is negligible compared with the variability of the tubes. A statistician examines the measurements, which look normally distributed and vary from 75 to 99 volts with a mean of 87 and a standard deviation of 4. He makes the ordinary normal analysis, giving a confidence interval for the true mean. Later he visits the engineer's laboratory, and notices that the volt-meter used reads only as far as 100, so the population appeared to be "censored." This necessitates a new analysis, if the statistician is orthodox. However, the engineer says he has another meter, equally accurate and reading to 1000 volts, which he would have used if any voltage had been over 100. This is a relief to the orthodox statistician, because it means the population was effectively uncensored after all. But the next day the engineer telephones and says, "I just discovered my high-range volt-meter was not working the day I did the experiment you analyzed for me." The statistician ascertains that the engineer would not have held up the experiment until the meter was fixed, and informs him that a new analysis will be required. The engineer is astounded. He says, "But the experiment turned out just the same as if the high-range meter had been working. I obtained the precise voltages of my sample anyway, so I learned exactly what I would have learned if the high-range meter had been available. Next you'll be asking about my oscilloscope." Reprinted with permission from the *Journal of the American Statistical Association*. Copyright © (1962) The American Statistical Association.

Notice that the actual model used is not particularly important since the effect of changes to the model rather than the model itself is at issue. For this reason, the validity of the initial (uncensored model) is not at issue either. The quotation used the Gaussian model but could just as well have used the (simpler) exponential or any other model. The initial analysis is to compute (7.8) (or an appropriate analog). The orthodox statistician here is a strict unconditional frequentist. On discovering the possibility that censoring might have occurred, a strict unconditional frequentist would compute (7.10) (or an appropriate analog) and obtain a different (more conservative) inference. Since anything that might have occurred affects strict unconditional frequentist inference, a strict unconditional frequentist has to be very careful to take everything that might possibly have happened into account. The rest of the quotation points out that this can be difficult to do and that one person's view of what might have happened is not necessarily the same as another's.

The general point is that a frequentist has to specify the relevant sample space (the set of possible outcomes) of the experiment and these choices impact on the inference.

7.2.4 The Bartlett–Armitage Example

Bartlett and Armitage (described by Armitage, 1961) produced an interesting example to show that it is not always sensible to obey the likelihood principle.

Suppose that we fix a value $k > 0$ and then draw realizations z_i of Z_i from

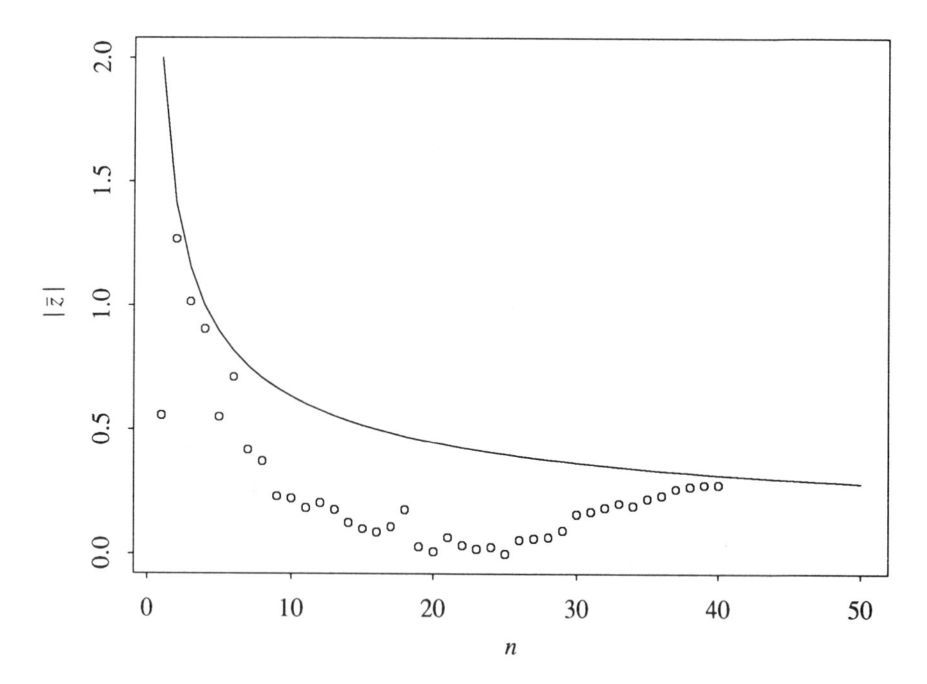

Figure 7.1. A typical sample path in the Bartlett–Armitage problem.

a N(θ, 1) distribution until $|\bar{z}| \geq n^{-1/2}k$, where $\bar{z} = n^{-1}\sum_{i=1}^{n} z_i$. To simplify the presentation, suppose that when we stop sampling, $\bar{z}$ is on the boundary so that $\bar{z} = n^{-1/2}k$. A plot of a typical sample path is given in Figure 7.1. The problem is to test H_0: $\theta = 0$ against H_1: $\theta \neq 0$.

Given the data ($\mathbf{z}$, n), the likelihood is just

$$f(\mathbf{z}, n; \theta) = \prod_{i=1}^{n} (2\pi)^{-1/2} \exp\left\{-\frac{(z_i - \theta)^2}{2}\right\}, \qquad \theta \in \Omega$$

$$= (2\pi)^{-n/2} \exp\left\{-\sum_{i=1}^{n} \frac{(z_i - \bar{z})^2}{2} - \frac{n(\bar{z} - \theta)^2}{2}\right\}, \qquad \theta \in \Omega.$$

The maximum likelihood estimate of θ is $\hat{\theta} = \bar{z}$ so the likelihood ratio is

$$\frac{f(\mathbf{z}, n; \bar{z})}{f(\mathbf{z}, n; 0)} = \exp\left(\frac{n\bar{z}^2}{2}\right)$$

$$= \exp\left(\frac{k^2}{2}\right)$$

$$= \begin{cases} \text{small (no evidence against H)} & \text{if } k \text{ is small} \\ \text{large (evidence against H)} & \text{if } k \text{ is large.} \end{cases}$$

The conclusion here is based not on the data but on the value of k which we chose before we began to collect the data. Most statisticians would regard this as unacceptable.

Barnard (see Basu, 1973) has argued that in the above analysis, we have been finessed into testing $H_0: \theta = 0$ against $H_2: \theta = \bar{z}$ rather than testing $H_0: \theta = 0$ against $H_1: \theta \neq 0$. Since the data always supports $\theta = \bar{z}$, the conclusion that there is evidence against H_0 in favor of H_2 is correct and indeed to be expected. He then argued that we should not allow the alternative hypothesis to be determined by the observed data but should restrict attention to tests of $H_0: \theta = 0$ against $H_3: \theta = \theta_k$, where θ_k is some predetermined real number. The likelihood ratio in this case is

$$\begin{aligned}\frac{f(\mathbf{z}, n; \theta_k)}{f(\mathbf{z}, n; \theta)} &= \exp\left(n\theta_k \bar{z} - \frac{n\theta_k^2}{2}\right) \\ &= \exp\left(kn^{1/2}\theta_k - \frac{n\theta_k^2}{2}\right) \\ &= \exp\left\{-n\left(\frac{\theta_k^2}{2} - n^{-1/2}k\theta_k\right)\right\} \\ &\to 0, \qquad \text{as } n \to \infty.\end{aligned}$$

Thus if n is very large, we have no evidence against H_0 which is intuitively reasonable. However, in practice, it is difficult to specify θ_k.

If we adopt the prior

$$g(\theta) = \frac{1}{2} I(\theta = 0) + \frac{1}{2c} I(\theta \neq 0, |\theta| \leq c), \qquad \theta = \mathbb{R},$$

and if we stop on the boundary so $\bar{z} = n^{-1/2}k$, we obtain the posterior odds ratio for testing $H_0: \theta = 0$ against $H_1: \theta \neq 0$ from (2.31) as

$$\begin{aligned}\frac{2c(n/2\pi)^{1/2} \exp(-n\bar{z}^2/2)}{\Phi\{n^{1/2}(c - \bar{z}) - \Phi\{-n^{1/2}(c + \bar{z})\}} &= \frac{2c(n/2\pi)^{1/2} \exp(-k^2/2)}{\Phi\{n^{1/2}c - k\} - \Phi\{-n^{1/2}c - k\}}. \\ &= \begin{cases}\text{small (evidence against } H_0) & \text{if } n \text{ is small} \\ \text{large (no evidence against } H_0) & \text{if } n \text{ is large,}\end{cases}\end{aligned}$$

for fixed c. Basu (1973) has argued that we should use a bounded stopping rule of the form

"stop when $|\bar{Z}| \geq n^{-1/2}k$ or $n = M$, whichever occurs soonest"

and that we should test $H_0': |\theta| \leq \delta$ against $H_1': |\theta| > \delta$. If we stop on the boundary with $n < M$, the posterior odds ratio using the Jeffreys prior is

$$\frac{\Phi\{n^{1/2}(\delta - \bar{z})\} - \Phi\{n^{1/2}(-\delta - \bar{z})\}}{1 - \Phi\{n^{1/2}(\delta - \bar{z})\} + \Phi\{n^{1/2}(-\delta - \bar{z})\}} = \frac{\Phi(n^{1/2}\delta - k) - \Phi(-n^{1/2}\delta - k)}{1 - \Phi(n^{1/2}\delta - k) + \Phi(-n^{1/2}\delta - k)}$$

$$= \begin{cases} \text{small (evidence against } H_0) & \text{if } n \text{ is small} \\ \text{large (no evidence against } H_0) & \text{if } n \text{ is large.} \end{cases}$$

This shows that while both Bayesian and likelihood methods ignore the sampling scheme, they do so in different ways and can reach different conclusions.

It is clear that the important information in the observed sample is the sample size n. A frequentist test can be based on the random sample size N (using the fact that small values of N are evidence against H_0) provided we can find the sampling distribution of N.

7.2.5 Implications of the Likelihood Principle

The likelihood principle does not conflict with the coherency principle but we saw in Section 7.2.2 that whether Bayesian inference obeys the likelihood principle or not depends on how we choose the prior. The likelihood approach (Section 2.7) should obey the likelihood principle but some of the operations on high dimensional likelihoods that are used in the likelihood approach conflict with it. Fiducial inference (Section 2.6) is usually held to conflict with the likelihood principle because the argument leading to the fiducial distribution depends on information which is not contained in the likelihood. Frequentist inference depends on the sampling distribution and therefore conflicts with the likelihood principle.

The likelihood principle has far-reaching consequences but, to many statisticians, is difficult to accept. For this reason, Barnard et al. (1962) and Birnbaum (1962) presented developments of the likelihood principle from two principles which they felt might be more widely accepted. These are the sufficiency and conditionality principles. We will present and discuss these two principles in Sections 7.3 and 7.4 respectively and then discuss their relationship to the likelihood principle in Section 7.5.

PROBLEMS

7.2.1. Compare the Jeffreys priors for θ under binomial and negative binomial sampling.

7.2.2. Suppose that in the binomial/negative binomial problem of Section 7.2.1 we want to test $H_0\colon \theta = \frac{1}{2}$. If we observe $n = 10$ and $z = 8$, show that the

p-value under binomial sampling is

$$P\{|B(10, \tfrac{1}{2}) - 5| \geq 3\} = 0.1094$$

and the p-value under negative binomial sampling is

$$P\{|NB(8, \tfrac{1}{2}) - 16| \geq 6\} = 0.1493.$$

7.2.3. In a truncated exponential problem we observe realizations of X independent from the model with density $\theta \exp(-\theta x)/\{1 - e^{-\theta c}\}$ if $x < c < \infty$ and 0 elsewhere. Show that the likelihood principal has nothing to say about the relationship between inference under the truncated exponential model and the exponential model (7.6).

7.3 THE SUFFICIENCY PRINCIPLE

We introduced Fisher's (1922) concept of sufficiency in Section 2.5 to justify his claim that the likelihood contains all the information in the data. If we accept this idea, sufficiency is clearly relevant to the likelihood principle.

7.3.1 Sufficient Statistics

In general, the information in the data about a model $\mathscr{F}$ is contained in the class of sufficient statistics for $\mathscr{F}$.

Suppose that $\mathbf{Z}$ *is an observation on a model* $\mathscr{F}$. *A statistic* $t(\mathbf{Z})$ *is a sufficient statistic for* $\mathscr{F}$ *if the conditional distribution of* $\mathbf{Z}$ *given* $t(\mathbf{Z}) = t$ *is the same for all the distributions in* $\mathscr{F}$.

This definition does not actually require the existence of a likelihood. However, in the case that a likelihood exists, the factorization theorem establishes that the likelihood satisfies this definition.

Theorem 7.2 (Halmos and Savage, 1949) *Suppose that* $\mathbf{Z}$ *is an observation on a model* $\mathscr{F}$. *A statistic* $t(\mathbf{Z})$ *is sufficient for the model* $\mathscr{F} = \{f(\cdot; \theta): \theta \in \Omega\}$ *if and only if*

$$f(\mathbf{z}; \theta) = g(t(\mathbf{z}); \theta)h(\mathbf{z}),$$

where h *does not depend on* θ.

This result means that statements about sufficiency are essentially statements about the likelihood function, that to find a sufficient statistic we need only examine the likelihood function and that the likelihood function is itself obviously a sufficient statistic.

7.3.2 Minimal Sufficient Statistics

It is clear from the definition of a sufficient statistic that the data $\mathbf{Z}$ is always a sufficient statistic (the distribution of $\mathbf{Z}$ given $\mathbf{Z} = \mathbf{z}$ is degenerate for all distributions in $\mathscr{F}$) so that restricting attention to a sufficient statistic need not result in much simplification. It is desirable to search for sufficient statistics which have dimension less than the sample size n and indeed to search for sufficient statistics with minimal dimension. Such sufficient statistics correspond to a maximal reduction of the data without incurring any loss of information about the model.

Suppose that $\mathbf{Z}$ is an observation on a model $\mathscr{F}$. A statistic $t(\mathbf{Z})$ is a minimal sufficient statistic for $\mathscr{F}$ if $t(\mathbf{Z})$ can be written as a function of any other sufficient statistic for $\mathscr{F}$.

Notice that any 1–1 function of a minimal sufficient statistic is also a minimal sufficient statistic. We say that two minimal sufficient statistics which can be written as functions of each other are *equivalent*. To find a minimal sufficient statistic we use the following result.

Theorem 7.3 (Lehmann and Scheffé, 1950) *Suppose that $\mathscr{F} = \{f(\cdot; \theta): \theta \in \Omega\}$. If we can find a statistic $t(\mathbf{Z})$ such that $t(\mathbf{y}) = t(\mathbf{z})$ for $\mathbf{y}, \mathbf{z} \in \mathscr{Z}$ if and only if $f(\mathbf{y}; \theta)/f(\mathbf{z}; \theta)$ is the same for all distributions in $\mathscr{F}$, then $t(\mathbf{Z})$ is a minimal sufficient statistic for $\mathscr{F}$.*

For the k-parameter exponential family model (Section 1.3.1) which is given by

$$\mathscr{F} = \left\{ f(\mathbf{z}; \theta) = \prod_{i=1}^{n} \exp\left[\psi_1 a_1(z_i) + \cdots + \psi_k a_k(z_i) + \phi + b(z_i)\right], \mathbf{z} \in A : \theta = (\psi_1, \ldots, \psi_k, \phi) \in \Omega \right\},$$

we find that

$$\frac{f(\mathbf{y}; \theta)}{f(\mathbf{z}; \theta)} = \exp\left[\psi_1 \sum_{i=1}^{n} \{a_1(y_i) - a_1(z_i)\} + \cdots + \psi_k \sum_{i=1}^{n} \{a_k(y_i) - a_k(z_i)\} + \sum_{i=1}^{n} \{b(y_i) - b(z_i)\}\right].$$

This ratio does not depend on θ if and only if $\sum_{i=1}^{n} a_j(y_i) = \sum_{i=1}^{n} a_j(z_i)$, $j = 1, \ldots, k$, so $\{\sum_{i=1}^{n} a_j(Z_i), j = 1, \ldots, k\}$ is a minimal sufficient statistic for $\mathscr{F}$. In particular, for the binomial model ($k = 1$ and $a_1(z) = z$), $\sum_{i=1}^{n} Z_i$ is a minimal sufficient statistic and for the Gaussian model ($k = 2$, $a_1(z) = z^2$ and $a_2(z) = z$), $\{\sum_{i=1}^{n} Z_i^2, \sum_{i=1}^{n} Z_i\}$ is a minimal sufficient statistic.

Models for which minimal sufficient statistics are of the same dimension as the parameter space are the exception rather than the rule. In general, minimal sufficient statistics are of the same dimension as the sample space. For example, for the Weibull model

$$\mathscr{F} = \left\{ f(\mathbf{z}; \kappa, \lambda) = \prod_{i=1}^{n} \kappa\lambda(\lambda z_i)^{\kappa-1} \exp\{-(\lambda z_i)^{\kappa}\}, z_i > 0: \lambda > 0 \right\},$$

the ratio

$$\frac{f(\mathbf{y}; \kappa, \lambda)}{f(\mathbf{z}; \kappa, \lambda)} = \left(\prod_{i=1}^{n} \frac{y_i}{z_i} \right)^{\kappa-1} \exp\left\{ \lambda^{\kappa} \sum_{i=1}^{n} (z_i^{\kappa} - y_i^{\kappa}) \right\}$$

is free of (λ, κ) if and only if $\prod_{i=1}^{n} y_i = \prod_{i=1}^{n} z_i$ and $\sum_{i=1}^{n} y_i^{\kappa} = \sum_{i=1}^{n} z_i^{\kappa}$. However, these relationships involve κ so do not identify statistics. We find that the ratio is free of (λ, κ) if and only if the z_i's are a permutation of the y_i's. Thus $\{Z_{n1}, \ldots, Z_{nn}\}$, where $Z_{n1} \le Z_{n2} \le \cdots \le Z_{nn}$ are the order statistics, is a minimal sufficient statistic for $\mathscr{F}$. This is not a substantial reduction from the data $\mathbf{Z}$ itself but the assumption of independence is being exploited to ensure that the order in which the data were collected does not matter. This argument works quite generally even for semiparametric and nonparametric models provided the observations are independent.

7.3.3 The Sufficiency Principle

Recall that a minimal sufficient statistic corresponds to the greatest reduction of the data that can be achieved without losing information about the model. It is reasonable to base inference on the reduced data set represented by the minimal sufficient statistic rather than on the original data set. The sufficiency principle goes further and asserts that we should use the minimal sufficient statistic only.

Sufficiency principle: Inference for a given model $\mathscr{F}$ *which admits a minimal sufficient statistic* $t(\cdot)$ *should be identical for any two data sets* $\mathbf{z}_1$ *and* $\mathbf{z}_2$ *for which* $t(\mathbf{z}_1) = t(\mathbf{z}_2)$.

It follows from Theorem 7.3 that for models based on distributions which have density functions, the sufficient principle is equivalent to the following principle:

Weak likelihood principle: Inference for a given model $\mathscr{F} = \{f(\cdot; \theta): \theta \in \Omega\}$ *should be identical for any two data sets* $\mathbf{z}_1$ *and* $\mathbf{z}_2$ *for which* $f(\mathbf{z}_1; \theta) = h(\mathbf{z}_1, \mathbf{z}_2) f(\mathbf{z}_2; \theta)$.

This principle requires that two data sets on a model $\mathscr{F}$ which result in proportional likelihoods should result in the same inferences whereas the

Table 7.1. The Sample Space for 3 Bernoulli Observations Partitioned by the Number of Successes

$z = 0$	$z = 1$	$z = 2$	$z = 3$
(0, 0, 0)	(1, 0, 0)	(1, 1, 0)	(1, 1, 1)
	(0, 1, 0)	(1, 0, 1)	
	(0, 0, 1)	(0, 1, 1)	
1	3	3	1

likelihood principle allows for more complicated situations in which we have observations on different models.

The sufficiency principle insists that inference should be identical when we have a single observation from a binomial(n, θ) distribution and when we have n independent Bernoulli(θ) (=binomial(1, θ)) observations. In other words, given the total number of successes Z (which is a minimal sufficient statistic), there is no additional information in knowing the exact pattern of outcomes. If for the moment we take $n = 3$, it is clear that we can organize the sample space into sets of the different observations which lead to a given number z of successes. The sample space is shown in Table 7.1.

Thus the minimal sufficient statistic Z has induced an exhaustive partition of the sample space. For general n, the minimal sufficient statistic induces the partition $\mathscr{Z} = \{\mathscr{S}_1, \ldots, \mathscr{S}_n\}$ where card $(\mathscr{S}_z) = \binom{n}{z}$. The sufficiency principle says that data sets in the same partition set should yield the same inference for the model.

Any statistic $s(\cdot)$ (and hence in particular any minimal sufficient statistic $s(\cdot)$) induces a partition $\mathscr{S} = \{\mathscr{S}_r : r \in \text{range}\,(s(\cdot))\}$ where $\mathscr{S}_r = \{\mathbf{z} \in \mathscr{Z} : s(\mathbf{z}) = r\}$ on the sample space $\mathscr{Z}$. A schematic representation of this partition is given in Figure 7.2. The sufficiency principle says that data sets in the same partition set of a partition induced by a minimal sufficient statistic should yield the same inference for the model.

7.3.4 Discussion

Sufficiency, at least when it provides a substantial reduction of the data, depends strongly on the assumed model $\mathscr{F}$. For example, for the Gaussian model, $\{\sum_{i=1}^n Z_i^2, \sum_{i=1}^n Z_i\}$ is a minimal sufficient statistic, but in a contamination neighborhood of this model, the order statistics are a minimal sufficient statistic. This means that sufficient statistics need not be robust to changes in $\mathscr{F}$ and, in general, if we require robustness, the sufficiency principle implies only that we should base inferences on the observed data. While this seems to be disappointingly weak, it is still strong enough to preclude the possibility of using postdata

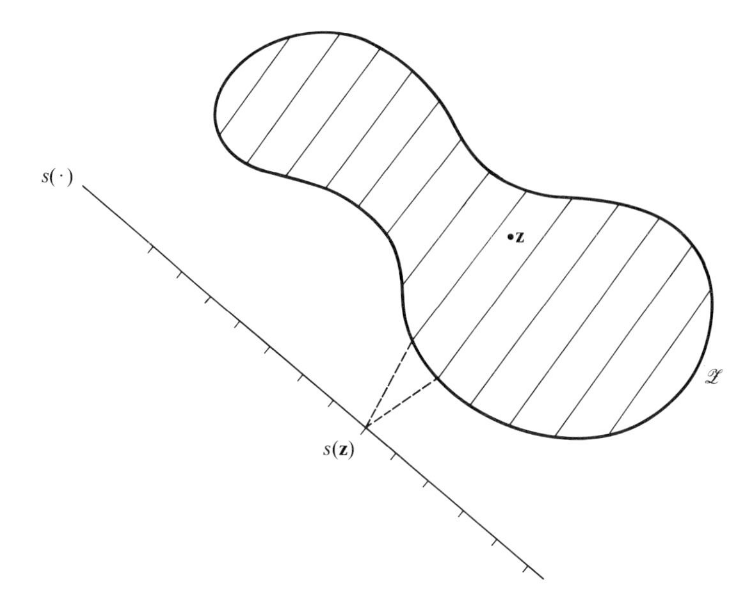

Figure 7.2. A schematic representation of the partition of a sample space $\mathscr{Z}$ induced by a real valued statistic $s: \mathscr{Z} \to \mathbb{R}$.

randomization to construct inferences. That is, it precludes the use of randomized confidence intervals (or tests) for discrete distributions such as that presented for the binomial model in Section 3.7.3, the use of randomized confidence intervals for the Behrens–Fisher problem (Section 3.8.1), and a variety of smoothing procedures (Section 6.9.4).

The sufficiency principle is by no means universally accepted. The main objection is that it is naive and simplistic to assert that all the information is contained in a minimal sufficient statistic and that no other information is relevant to inference. Objections of this type are also raised against the Bayesian paradigm by LeCam (1977), Kiefer (1977a), Efron (1986), and others.

PROBLEMS

7.3.1. Suppose we have observations $\mathbf{Z}$ on a model $\mathscr{F}$. Find minimal sufficient statistics in each of the following cases:

1. $\mathscr{F} = \left\{ f(\mathbf{y}; \theta) = \prod_{i=1}^{n} \frac{1}{\theta} I(0 \le y_i \le \theta); \theta \ge 0 \right\}$

2. $\mathscr{F} = \left\{ f(\mathbf{y}; \theta) = \prod_{i=1}^{n} I(\theta - \frac{1}{2} \le y_i \le \theta + \frac{1}{2}): \theta \in \mathbb{R} \right\}$

3. $\mathscr{F} = \left\{ f(\mathbf{y}; \lambda, \kappa) = \prod_{i=1}^{n} \frac{1}{\Gamma(\kappa)} \lambda(\lambda y_i)^{\kappa - 1} \exp(-\lambda y_i), y_i > 0 : \lambda > 0 \right\}$

4. $\mathscr{F} = \left\{ f(\mathbf{y}; \lambda) = \prod_{i=1}^{n} \frac{\kappa \lambda^{\kappa}}{y_i^{\kappa+1}}, y_i > \lambda : \kappa, \lambda > 0 \right\}$

5. $\mathscr{F} = \left\{ f(\mathbf{y}; \mu, \sigma) = \frac{\Gamma\left(\frac{\nu + n}{2}\right)}{(\nu\pi)^{n/2}\sigma^n \Gamma\left(\frac{\nu}{2}\right)} \frac{1}{\left\{1 + \sum_{i=1}^{n} \frac{(y_i - \mu)^2}{\nu\sigma^2}\right\}^{(\nu+n)/2}}, \right.$

$$\left. -\infty \leq y_i \leq \infty; \mu \in \mathbb{R}, \sigma > 0 \right\}$$

6. $\mathscr{F} = \left\{ f(\mathbf{y}; \mu, \sigma) = \prod_{i=1}^{n} \frac{\Gamma\left(\frac{\nu + 1}{2}\right)}{(\nu\pi)^{1/2}\sigma \Gamma\left(\frac{\nu}{2}\right)} \frac{1}{\left\{1 + \frac{(y_i - \mu)^2}{\nu\sigma^2}\right\}^{(\nu+1)/2}}, \right.$

$$\left. -\infty \leq y_i \leq \infty : \mu \in \mathbb{R}, \sigma > 0 \right\}$$

7. $\mathscr{F} = \left\{ f(\mathbf{z}; \lambda, \delta) = \prod_{i=1}^{n} \left\{ z_i! \Gamma\left(\frac{\lambda}{\delta}\right) \right\}^{-1} \Gamma\left(z_i + \frac{\lambda}{\delta}\right) \delta^{z_i} (1 + \delta)^{-(z_i + \lambda/\delta)}, \right.$

$$\left. z_i = 0, 1, 2, \ldots, \lambda > 0, \delta > 0 \right\}$$

8. $\mathscr{F} = \left\{ f(\mathbf{y}; \beta, \sigma) = \prod_{i=1}^{n} \frac{1}{(2\pi\sigma^2 v(x_i))^{1/2}} \exp\left\{ -\frac{(y_i - x_i\beta)^2}{2\sigma^2 v(x_i)} \right\}, \right.$

$$\left. y_i \in \mathbb{R}; \beta \in \mathbb{R}, \sigma > 0 \right\}$$

9. $\mathscr{F} = \left\{ f(\mathbf{z}; \theta) = \prod_{i=1}^{n} \theta z_i^{\theta - 1} \exp(-z_i^{\theta}), z_i > 0 : \theta > 0 \right\}$

10. $\mathscr{F} = \left\{ f(\mathbf{z}; \mu, \sigma_a, \sigma_u) = \frac{1}{(2\pi)^{mg/2} |\Sigma|^{1/2}} \exp\left\{ -\frac{(\mathbf{z} - \mu)^{\mathrm{T}} \Sigma^{-1} (\mathbf{z} - \mu)}{2} \right\}, \right.$

$$\left. -\infty < z_{ij} < \infty : \mu \in \mathbb{R}, \sigma_a > 0, \sigma_u > 0 \right\},$$

where Σ is the block diagonal matrix with blocks $\sigma_a^2 J + \sigma_u^2 I$, where J is the $m \times m$ matrix with all elements equal to 1 and I is the $m \times m$ identity matrix.

7.3.2. Suppose that we have observations $\mathbf{Z}$ on the nonparametric model

$$\mathscr{F} = \left\{ f(\mathbf{z}) = \prod_{i=1}^{n} f(z_i) \right\}.$$

Let $Z_{n1} \leq Z_{n2} \leq \cdots \leq Z_{nn}$ denote the order statistics from the sample, Show that the order statistics are sufficient for $\mathscr{F}$.

7.4 THE CONDITIONALITY PRINCIPLE

It is important in making inferences to ensure that the data on which the inference is based are in some sense relevant to the problem. That valid frequentist inferences can be based on irrelevant data was shown by Basu (1981). Suppose that we are interested in a parameter $\theta \in (0, 1)$ and we generate a realization of a random variable $Z \sim \mathrm{U}(0, 1)$. It is intuitively clear that a realization of Z tells us nothing about θ and should not be used as a basis for inference. However, for any fixed set $B \subset (0, 1)$, the set

$$C(Z) = \begin{cases} B & \text{if } 0 < Z \leq 0.05 \\ (0, 1) & \text{if } 0.05 < Z < 0.95 \\ B^{c} & \text{if } 0.95 \leq Z < 1 \end{cases}$$

is a 95% confidence set for θ because

$$\begin{aligned} \mathrm{P}\{\theta \in C(Z)\} &= \mathrm{P}\{0.05 < Z < 0.95\} + \mathrm{P}\{0 < Z < 0.05\}\mathrm{I}(\theta \in B) \\ &\quad + \mathrm{P}\{0.95 < Z < 1\}\mathrm{I}(\theta \in B^{c}) \\ &= 0.90 + 0.05(\mathrm{I}(\theta \in B) + \mathrm{I}(\theta \in B^{c})) \\ &= 0.95. \end{aligned}$$

Since observing Z tells us nothing about θ, we need some principle to preclude the use of this kind of inference.

7.4.1 Ancillary Statistics

If we study Basu's example carefully, we see that Z is uninformative about θ because its distribution is free of θ. Fisher (1934) called any such statistic an *ancillary statistic*. There are a number of different definitions of an ancillary statistic but we will adopt the definition of Cox and Hinkley (1974, pp. 31–2) to avoid paradoxical results (see Problem 7.4.1).

A statistic $c(\mathbf{Z})$ is an ancillary statistic for a model $\mathscr{F} = \{F(\cdot; \theta): \theta \in \Omega\}$ if $c(\mathbf{Z})$ is a component of a minimal sufficient statistic for $\mathscr{F}$ and the marginal distribution of $c(\mathbf{Z})$ does not depend on θ.

We see that a statistic $c(\mathbf{z})$ is ancillary if we can factorize the likelihood as

$$f(\mathbf{z}; \theta) = g_1(s(\mathbf{z}) \mid c(\mathbf{z}); \theta) h(c(\mathbf{z})).$$

Ancillarity is in a sense the dual of sufficiency; sufficiency describes where all the information in the data is contained whereas ancillarity describes where there is no information.

In practice, ancillary statistics may not exist, may exist but be hard to find, and may not be unique. For example, if (n_1, n_2, n_3, n_4) have a multinomial$(n, (1-\theta)/6, (1+\theta)/6, (2-\theta)/6, (2+\theta)/6)$ distribution then

$$\{n_1 + n_2, n_3 + n_4\} \sim \text{multinomial}(n, \tfrac{1}{3}, \tfrac{2}{3})$$

and

$$\{n_1 + n_4, n_2 + n_3\} \sim \text{multinomial}(n, \tfrac{1}{2}, \tfrac{1}{2})$$

are both ancillary statistics (Basu, 1964; Cox, 1971). These difficulties hinder the formulation of a simple general theory based on ancillary statistics.

7.4.2 The Conditionality Principle

In Basu's example, Z is an ancillary statistic so, according to Fisher, we should condition on it. This ensures that inference about θ cannot be based on Z. The conditionality principle states that we should always condition on ancillary statistics.

Conditionality principle: If an ancillary statistic exists for a model $\mathscr{F}$, then inference for the model should be made conditional on the ancillary statistic.

The conditionality principle enables us to simplify certain analyses. If the sample size n is a realization of a random variable N which has a distribution which does not depend on the unknown model parameters, then N is an ancillary statistic and, according to the conditionality principle, we should make inferences conditionally on N, i.e., treating $N = n$ as fixed. Of course, if the distribution of N depends on the unknown model parameters of interest, then N is informative about the model and we should not condition on its value.

We can obtain a simple visual representation of the effect of conditioning by considering the problem in which $Z \mid N = n \sim \text{binomial}(n, \theta)$ and the marginal distribution of N does not depend on θ. We first plot z against n in Figure 7.3. Recall from the discussion of minimal sufficiency in Section 7.2 that each point on the plot represents a partition set in the partition of the sample space $\mathscr{Z}$ which is generated by the minimal sufficient statistic $\{Z, N\}$. We can write $\mathscr{Z} = \bigcup_n \mathscr{Z}_n$, where $\mathscr{Z}_n$ is the sample space of Z given that $N = n$. Conditioning on $N = n$ means that we restrict attention to $\mathscr{Z}_n$ rather than $\mathscr{Z}$. This corresponds to fixing a vertical line through n and has the effect of reducing the two dimensional minimal sufficient statistic to a one dimensional statistic and restricting the sample space.

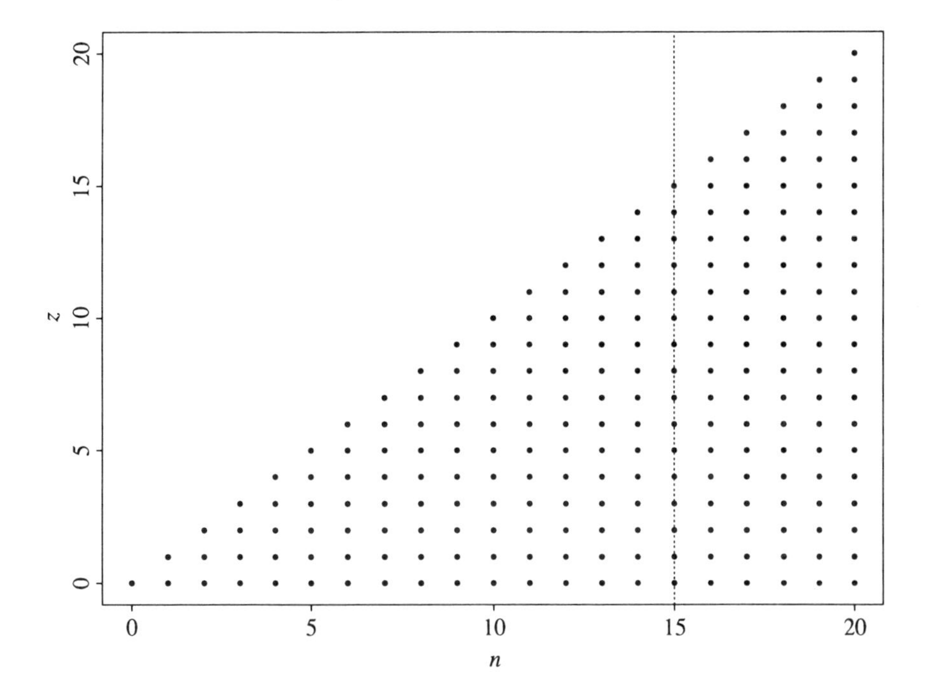

Figure 7.3. The sample space for a binomial experiment with a random number of trials. Points on the vertical line represent the conditional sample space given that $n = 15$.

We consider only the value n of N actually observed but we allow Z to vary over the data sets which are possible for the choice of n.

Formally, the minimally sufficient statistic induces a doubly subscripted partition of $\mathscr{Z}$ which we write as $\mathscr{S} = \{\mathscr{S}_{rn}: r \in \text{range}\,(s(\cdot)), n \in \mathbb{Z}\}$ where $\mathscr{S}_{rn} = \{\mathbf{z} \in \mathscr{Z}_n: s(\mathbf{z}) = r\}$ and the conditioning specifies a subpartition $\mathscr{S}_n = \{\mathscr{S}_{rn}: r \in \text{range}\,(s(\cdot))\}$ to which we restrict attention. We sometimes call the sample space $\mathscr{Z}$ the *frame of reference* and $\mathscr{S}_n$ the *conditional frame of reference*. Similar results holds for models other than the Bernoulli model. A schematic representation is given in Figure 7.4.

7.4.3 Discussion

Whether a statistic is ancillary or not depends on the model so, in this sense, the concept is not robust. However, there are aspects of problems which do not depend on the data distribution and so are ancillary irrespective of the model for the data. For example, in a randomized experiment (such as the caffeine experiment), we use randomization to make choices (including selecting units, allocating units to treatments, selecting measuring instruments, and so on) in the design stage of data collection. As we noted in Section 5.1, we can think of a random variable D which takes values in $\mathscr{D}$, the space of all possible

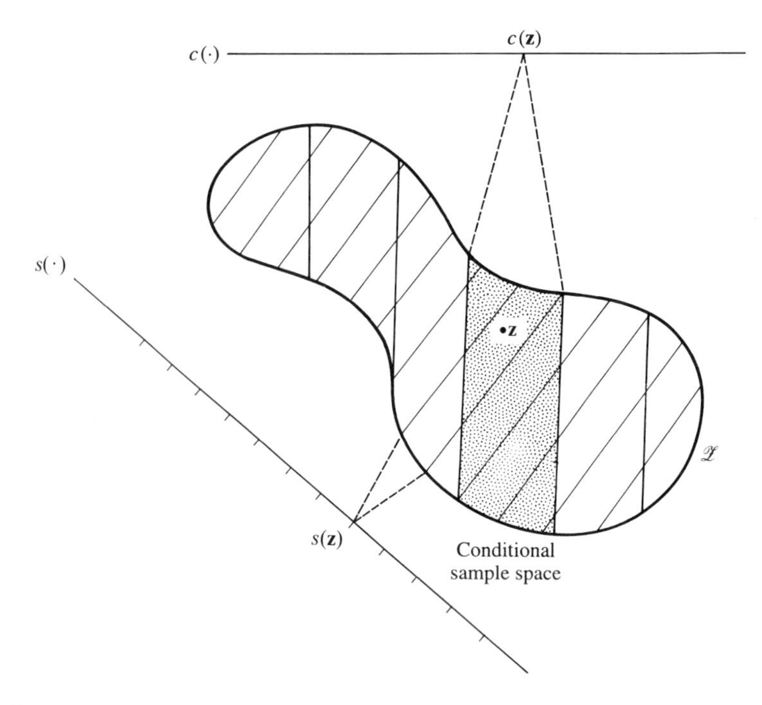

Figure 7.4. A schematic representation of the partition of a sample space $\mathscr{Z}$ induced by real valued statistics $s: \mathscr{Z} \rightarrow \mathbb{R}$ and $c: \mathscr{Z} \rightarrow \mathbb{R}$ showing the conditional sample space for s given c.

designs, and which specifies which design we should use. Randomization generates a realization of the random variable D by drawing D from some known (often uniform) distribution over $\mathscr{D}$. Given the realization D, we observe $\mathbf{Z}$. The observed data from the experiment consists of $(\mathbf{Z}, D)$. Since the distribution of D is known, it cannot depend on any unknown parameters and it is ancillary. The conditionality principle requires inferences to be made conditionally given D and therefore rules out the possibility of design-based inference. See Sections 6.2–6.5.

7.4.4 Partial Ancillarity

Finally, in some problems, only a subset of the model parameters are of interest. In this case, it seems reasonable to weaken the definition of an ancillary statistic so that its distribution may depend on the nuisance parameters. To this end we adopt the following definition from Cox and Hinkley (1974, p. 35)

A statistic $c(\mathbf{Y})$ is a partial ancillary statistic for a model $\mathscr{F} = \{F(\cdot;\theta)$: $\theta = (\psi, \lambda) \in \Omega_\psi \times \Omega_\lambda\}$ if $c(\mathbf{Z})$ is a component of a minimal sufficient statistic for $\mathscr{F}$, the conditional distribution of $\mathbf{Z} \mid c(\mathbf{Z}) = c$ depends on ψ but not λ for all c and the marginal distribution of $c(\mathbf{Z})$ depends on λ but not ψ.

Although this definition is complicated to express in words, we see that a statistic $c(\mathbf{Z})$ is partially ancillary if we can factorize the likelihood as

$$f(\mathbf{z}; \theta) = g(\mathbf{z} \mid c(\mathbf{z}); \psi) h(c(\mathbf{z}); \lambda).$$

We will take the conditionality principle to apply to partial ancillary statistics as well as to ancillary statistics.

PROBLEMS

7.4.1. Suppose that we have observations $\mathbf{Z}$ on the model

$$\mathscr{F} = \left\{ f(\mathbf{z}) = \prod_{i=1}^{n} \frac{1}{\theta} I(0 \le z_i \le \theta); 0 \le \theta \right\},$$

where $I(\cdot)$ is the indicator function. Show that $\{Z_{n1}, Z_{nn}/Z_{n1}\}$, where $Z_{n1} \le Z_{n2} \le \cdots \le Z_{nn}$, is sufficient for $\mathscr{F}$. Then show that the distribution of Z_{nn}/Z_{n1} does not depend on θ. It is tempting to argue that we should base inference on Z_{n1} given Z_{nn}/Z_{n1}. Show that this conflicts with the sufficiency principle.

7.4.2. Suppose that an experiment is conducted to measure a physical constant θ. During the experiment, n measurements are made using two instruments labelled 0 and 1. Before carrying out the experiment n independent Bernoulli trials are carried out to determine which instrument will be used for each measurement. Thus, the instrument for the ith trial is selected by a random mechanism which results in a random variable A_i such that for some known p, $0 < p < 1$,

$$\mathrm{P}(A_i = 0) = 1 - p \quad \text{and} \quad \mathrm{P}(A_i = 1) = p, \qquad 1 \le i \le n.$$

Given that $A_i = a_i$, the ith measurement is a random variable with a

$$Y_i \mid A_i = a_i \sim \mathrm{N}(\theta, \sigma_{a_i}^2)$$

distribution, where σ_0 and σ_1 are known and $\sigma_0 \ne \sigma_1$. At the conclusion of the experiment, the data consists of the pairs of observations $(a_1, y_1), \ldots, (a_n, y_n)$. Let $a = \sum_{i=1}^{n} a_i$ be the number of times instrument 1 is used. Relabel the data if necessary so that the a measurements made on instrument 1 are observations $y_1, \ldots, y_a$ and the $n - a$ observations on instrument 0 are $y_{a+1}, \ldots, y_n$. Show that the maximum likelihood estimator of θ is given by

$$\hat{\theta} = \frac{\sum_{i=a+1}^{n} y_i/\sigma_0^2 + \sum_{i=1}^{a} y_i/\sigma_1^2}{(n-a)/\sigma_0^2 + a/\sigma_1^2}$$

and that $E\hat{\theta} = \theta$. Show that the observed Fisher information is

$$\mathscr{I}_a = \frac{n-a}{n\sigma_0^2} + \frac{a}{n\sigma_1^2},$$

and that $\text{Var}(\hat{\theta} \mid a) = n^{-1}\mathscr{I}_a^{-1}$. Prove that $\mathscr{I}_a$ converges in probability to $(1-p)/\sigma_0^2 + p/\sigma_1^2$ as $n \to \infty$. Write down the conditional p-value for testing $H: \theta = \theta_0$. Is this very different from the unconditional p-value in large samples? Explain.

7.4.3. The large sample confidence intervals for $1/\theta$ obtained in Problem 4.1.6 under binomial and negative binomial sample are identical. Explain this result.

7.4.4. Suppose we have n independent observations (y_i, x_i), $i = 1, 2, \ldots, n$, which we believe are approximately linearly related. The slope parameter in the linear relationship can be estimated by

$$\hat{\beta} = \left\{\sum_{i=1}^{n} (x_i - \bar{x})^2\right\}^{-1} \sum_{i=1}^{n} (x_i - \bar{x}) y_i.$$

Consider carrying out a simulation to evaluate the properties of this estimator. Discuss the relative merits of using the same set of x_i values throughout the simulation and choosing a new set of (y_i, x_i) each time. Describe how you would carry out the simulation in each case.

7.4.5. Suppose there are two groups of individuals of size r_1 and r_2 and that we observe the number of individuals z_1 and z_2 in each group who fall into category A and the remainder $r_1 - z_1$ and $r_2 - z_2$ who fall into category B. The data is often presented in a 2×2 table, as in Table 7.2.

We assume that individuals respond independently of each other and that the probability of success in the two groups is θ_1 and θ_2 respectively. A useful model is that $Z_1 \sim B(r_1, \theta_1)$ and independently $Z_2 \sim B(r_2, \theta_2)$. It is then convenient to set

$$\log\left(\frac{\theta_1}{1-\theta_1}\right) = \alpha \quad \text{and} \quad \log\left(\frac{\theta_2}{1-\theta_2}\right) = \alpha + \Delta.$$

Table 7.2. A 2 × 2 Table

	Group 1	Group 2	Total
Category A	z_1	z_2	$z_1 + z_2$
Category B	$r_1 - z_1$	$r_2 - z_2$	$r_1 + r_2 - z_1 - z_2$
Total	r_1	r_2	$r_1 + r_2$

Suppose we are interested in Δ. It is often recommended (Lehmann, 1959/1991, pp. 154–5 and Fisher, 1956a, p. 89) that inference about Δ should be made conditionally on Z_2 given the marginal total $Z_1 + Z_2$.

Fisher's argument is that the value of $Z_1 + Z_2$ does not enable us to draw inferences about Δ. The value of $Z_1 + Z_2$ determines the precision with which inferences should be drawn and it is therefore appropriate to condition on $Z_1 + Z_2$. Find the distribution of $Z_1 + Z_2$. Discuss Fisher's argument in terms of the principles of inference. Does anything change if we are testing the null hypothesis $H_0: \Delta = 0$ of no difference between the groups?

Show that the distribution of Z_2 given $Z_1 + Z_2$ does not depend on α. This means that conditioning on $Z_1 + Z_2$ eliminates the nuisance parameter α which is the standard Neyman–Pearson approach to eliminating nuisance parameters.

7.4.6. When exploring the relationship between two variables Y and X as expressed through the conditional distribution of Y given X say, there is often an advantageous choice of scale on which to explore the relationship. One approach to choosing the scale is to estimate it from the data. If we estimate the appropriate transformation, we then have to decide whether to condition on the transformation or whether we need to take the variability in the estimated transformation into account in our inferences. Read the papers by Bickel and Doksum (1981) and Hinkley and Runger (1984) and prepare a brief summary of the two opposing viewpoints.

7.5 THE DEVELOPMENT OF THE LIKELIHOOD PRINCIPLE

The likelihood principle obviously implies the weak likelihood principle and hence the sufficiency principle. It also implies the conditionality principle; if we have an ancillary statistic $c(\mathbf{z})$, we can factorize the likelihood as

$$f(\mathbf{z}; \theta) = g_1(s(\mathbf{z}) \mid c(\mathbf{z}); \theta) h(c(\mathbf{z}))$$

and the likelihood principle implies that inferences based on $f(\mathbf{z}; \theta)$ and $g_1(s(\mathbf{z}) \mid c(\mathbf{z}); \theta)$ should be the same. That is, we should condition on the ancillary statistic $c(\mathbf{z})$. However, the likelihood principle has more far reaching consequences than either of these principles. For this reason, Birnbaum's (1962) arguments that the sufficiency and conditionality principles imply the likelihood principle have attracted considerable discussion; see Berger and Wolpert (1984, p. 42). Our development follows that of Berger and Wolpert (1984, pp. 24–8) for the discrete case.

Suppose that we have an observation $\mathbf{z}$ on $\mathscr{F} = \{f(\cdot\,; \theta): \theta \in \Omega\}$ and an observation $\mathbf{y}$ on $\mathscr{G} = \{g(\cdot\,; \theta): \theta \in \Omega\}$ such that

$$f(\mathbf{z}; \theta) = h(\mathbf{y}, \mathbf{z})g(\mathbf{y}; \theta). \tag{7.12}$$

Consider a mixture model $\mathscr{H}$ which involves a Bernoulli($\frac{1}{2}$) random variable B such that if $B = 0$ we make an observation from $\mathscr{F}$ while if $B = 1$ we make an observation from $\mathscr{G}$. (Most disquiet about this argument centers on this combination of the two problems into a single mixture problem.) Then the data for inference about θ is $(B, \mathbf{x}_B)$, where $\mathbf{x}_0$ is an observation from $\mathscr{F}$ and $\mathbf{x}_1$ is an observation from $\mathscr{G}$. Since B is an ancillary statistic, the conditionality principle states that we should condition on B and, having done so, we have added nothing by expanding our data collection process to the mixture experiment $\mathscr{H}$. Now consider the statistic

$$T(B, \mathbf{x}_B) = \begin{cases} (0, \mathbf{z}) & \text{if } (b, \mathbf{x}_b) = (0, \mathbf{y}) \\ (b, \mathbf{x}_b) & \text{otherwise.} \end{cases}$$

This statistic maps both the outcomes $(0, \mathbf{z})$ and $(1, \mathbf{y})$ to the same value. If we can show that T is a sufficient statistic for the mixed model, the sufficiency principle will imply that we should reach the same conclusion whether we observe $(0, \mathbf{z})$ or $(1, \mathbf{y})$ which implies the likelihood principle.

It remains to show that T is a sufficient statistic. This follows from the definition since, using (7.12),

$$\begin{aligned} \mathrm{P}\{(B, \mathbf{x}_B) = (0, \mathbf{z}) \mid T = (0, \mathbf{z}); \theta\} &= \frac{0.5f(\mathbf{z}; \theta)}{0.5f(\mathbf{z}; \theta) + 0.5g(\mathbf{y}; \theta)} \\ &= \frac{h(\mathbf{y}, \mathbf{z})}{h(\mathbf{y}, \mathbf{z}) + 1} \end{aligned}$$

which is free of θ. Hence.

$$\mathrm{P}\{(B, \mathbf{x}_B) = (1, \mathbf{y}) \mid T = (0, \mathbf{z}); \theta\} = 1 - \mathrm{P}\{(B, \mathbf{x}_B) = (1, \mathbf{z}) \mid T = (1, \mathbf{z}); \theta\}$$

is free of θ. Finally

$$\mathrm{P}\{(B, \mathbf{x}_B) = (b, \mathbf{x}_b) \mid T = t \neq (0, \mathbf{z}); \theta\} = \begin{cases} 1 & \text{if } (b, \mathbf{x}_b) = t \\ 0 & \text{otherwise} \end{cases}$$

is also free of θ and the result follows.

7.6 THE REPEATED SAMPLING PRINCIPLE

Frequentist inferences (see Chapter 3) involve a statistic and its sampling distribution. The sampling distribution describes the distribution of the statistic under hypothetical repetitions of the sampling scheme which we believe generated the data. Insofar as there is a principle underlying frequentist inferences, it is that they must use the sampling distribution to interpret and evaluate inferences. The formal statement of this idea is the repeated sampling principle.

Repeated sampling principle: Inference procedures should be interpreted and evaluated in terms of their behavior in hypothetical repetitions under the same conditions.

Confidence intervals (Section 3.6) and hypothesis tests (Section 3.3) are explicitly constructed to satisfy the repeated sampling principle. They are interpreted in terms of their level and evaluated in terms of their length or power, all of which are derived from sampling distributions.

The repeated sampling principle is in conflict with the coherency principle (Section 7.1) because Bayesian inferences do not necessarily have good repeated sampling properties. On the other hand, frequentist inferences with good repeated sampling properties are incoherent. The repeated sampling principle is also in conflict with the likelihood principle (Section 7.2.3).

Coherency is a form of self-consistency which ensures internal (to an individual) rather than external validity. Internal validity may be all that is available, but the repeated sampling principle is an attempt to ensure external validity. Basu's example (Section 7.4) shows that the repeated sampling principle does not on its own achieve this elusive goal. Nonetheless, the choice between the coherency and repeated sampling principles depends to a great extent upon where we draw the balance between trying to achieve internal or external validity.

7.7 OTHER PRINCIPLES

Although the principles we have enunciated in the preceding sections all seek to impose constraints on the way in which we make inferences, they still leave a number of aspects of the inference problem unspecified. For example, the principles all assume that we have data $\mathbf{Z}$ and a model $\mathscr{F}$ and assert how we should make inferences given $\mathbf{Z}$ and $\mathscr{F}$; they are, however, silent on how we should obtain either $\mathbf{Z}$ or $\mathscr{F}$ and, in particular, what the relationship between $\mathbf{Z}$ and $\mathscr{F}$ should be. This is in accordance with Fisher's (1922) view that we should separate the choice of $\mathscr{F}$ from the inference problem but it can complicate thinking about the application of the principles in practice. One consequence of being vague about $\mathbf{Z}$ and $\mathscr{F}$ is that in principle there is no conflict between the principles of inference and concern for robustness – the model $\mathscr{F}$ does not have to be a "true" data generating mechanism in any sense. We can even take the lack of specification of a relationship to an extreme and, at the risk of making foolish inferences, apply the principles in situations where there is no relationship between $\mathbf{Z}$ and $\mathscr{F}$. However, in robust frequentist inference, we may use one model to produce the inference and another to evaluate its repeated sampling properties. In this case, there is an issue as to which model to apply the principles of inference because, at least implicitly, they seem to imply that the models should be the same.

It makes sense to try to introduce principles which guide data collection and model choice since these are germane to statistical inference. Fisher's three principles of experimental design – randomization, replication, and control (see Section 6.1) – set out at least a framework for data collection but there are no comparable principles for model choice. Even if we allow for the fact that principles have to be broken at times, it is difficult to specify principles for model choice because there are so many aspects of context and judgement involved; see Lehmann (1990) and Cox (1990). Obviously, robustness is an important concern but we need appropriate robustness which typically means stability over the nuisance aspects of the model. We do not want to use a procedure that is robust against misspecification of the functional form if we are trying to explore functional form. Also, we do not want robustness to imply simply that we should always use large nonparametric models. (Even if we think of robustness in terms of choosing an inference procedure, we saw in Section 5.4.9 that there is often an implied model underlying the choice and we can reformulate the choice in terms of this model.) Thus, the recognition that there is a robustness principle does not make it easy to state. This is also true of the other aspects of model choice.

Reactions to the different approaches to statistical inference and *a fortiori* to the principles on which they are based tend to reflect personal attitudes to uncertainty and learning. Given the complexity of the inference problem, this is not altogether surprising. It is almost inevitable and arguably healthy that there will be different approaches to inference and therefore ongoing arguments about the relative merits of the different approaches. In particular, it is unrealistic to expect a single approach to always work and always be better than any other approach. Arguably, which approach to inference we use is less important than being clear about how we reach our conclusions, the basis for the assumptions we have made and how sensitive our conclusions are to these assumptions. It is valuable to keep the substantive problem in mind, to adopt a cautious, systematic approach and to include the unquantifiable uncertainties inherent in the analysis in its interpretation. It is important to recognize that alternative approaches are available and to facilitate reanalysis from either the same or different perspectives by, whenever possible, publishing and making the raw data widely available.

Although recognition of the difficulties of making inferences can lead to dissatisfaction and a feeling that statistics is a subject without firm foundations and limited applicability to idealized problems, it is important to recognize that the inference problem is a very difficult problem. Judgement, so favoured by Fisher as a means of overcoming theoretical shortcomings, cannot be eliminated from the process of making inferences. On the other hand, mitigating against the rejection of statistical inference as mere "ad hocery" is the fact that the methodology of statistical inference is useful in making sense of complicated substantive problems. We do need to maintain a perspective about what can reasonably be expected to be achieved.

FURTHER READING

The principles of inference are presented and discussed in the books by Cox and Hinkley (1974), Barnett (1982), and Berger and Wolpert (1984).

Appendix: Some Useful Facts

1. Jensen's inequality

If g is convex and the expectations exist, $g(\mathrm{E}X) \leq \mathrm{E}g(X)$

If the expectations exist, $\mathrm{E} \log (X) \leq \log (EX)$

2. Expansions

$$\log (1 + x) = x - \frac{x^2}{2} + \frac{x^3}{3} - \frac{x^4}{4} \cdots$$

$$\exp (x) = 1 + x + \frac{x^2}{2!} + \frac{x^3}{3!} + \frac{x^4}{4!} \cdots$$

$$f(x + h) = f(x) + f'(x)h + f''(x)\frac{h^2}{2!} + \cdots \qquad (x \in \mathbb{R})$$

$$f(x + h) = f(x) + f'(x)^{\mathrm{T}}h + h^{\mathrm{T}}f''(x)\frac{h}{2!} + \cdots \qquad (x \in \mathbb{R}^p).$$

3. Inverses and determinants of matrices

(a) If A is an $m \times m$ matrix of the form $A = uI + aJ$, where $u \neq 0$, I is the $m \times m$ identity matrix and J is the $m \times m$ matrix of 1s, we have

$$A^{-1} = cI + dJ \quad \text{with} \quad c = \frac{1}{u} \qquad d = -\frac{a}{u(u + ma)} = \frac{1}{m(u + ma)} - \frac{1}{mu}$$

and

$$|A| = u^{m-1}(u + ma).$$

(b) If A is a nonsingular $m \times m$ matrix and x is an m-vector,

$$(A + xx^{\mathrm{T}})^{-1} = A^{-1} - \frac{A^{-1}xx^{\mathrm{T}}A^{-1}}{1 + x^{\mathrm{T}}A^{-1}x}$$

and

$$x^{\mathrm{T}}(A + xx^{\mathrm{T}})^{-1}x = \frac{x^{\mathrm{T}}A^{-1}x}{1 + x^{\mathrm{T}}A^{-1}x}.$$

If we partition A so $A = \begin{pmatrix} A_{11} & A_{12} \\ A_{21} & A_{22} \end{pmatrix}$, we write $A^{-1} = \begin{pmatrix} A^{11} & A^{12} \\ A^{21} & A^{22} \end{pmatrix}$.

(c) If A_{11} and A_{22} are nonsingular, let

$$A_{11.2} = A_{11} - A_{12}A_{22}^{-1}A_{21} \quad \text{and} \quad A_{22.1} = A_{22} - A_{21}A_{11}^{-1}A_{12}.$$

Then

$$\begin{aligned} A^{11} &= A_{11.2}^{-1} \\ A^{22} &= A_{22.1}^{-1} \\ A^{12} &= -A_{11}^{-1}A_{12}A^{22} \\ A^{21} &= -A_{22}^{-1}A_{21}A^{11}. \end{aligned}$$

(d) If A_{22} is singular but A_{11} and $A_{22.1}$ are nonsingular, and A is symmetric

$$\begin{aligned} A^{11} &= A_{11}^{-1} + A_{11}^{-1}A_{12}A_{22.1}^{-1}A_{21}A_{11}^{-1} \\ A^{22} &= A_{22.1}^{-1} \\ A^{12} &= (A^{21})^{\mathrm{T}} = -A_{11}^{-1}A_{12}A^{22}. \end{aligned}$$

In the special case that $A_{22} = 0$, $A_{22.1} = -A_{21}A_{11}^{-1}A_{12}$.

(e) The determinant of A is

$$|A| = \begin{cases} |A_{22}|\,|A_{11.2}| & \text{if } A_{22} \text{ is nonsingular} \\ |A_{11}|\,|A_{22.1}| & \text{if } A_{11} \text{ is nonsingular.} \end{cases}$$

4. Integrals

(a) Gaussian integrals

$$\int \exp\left\{-\frac{(ax^2 - 2bx)}{2}\right\} dx = \left(\frac{2\pi}{a}\right)^{1/2} \exp\left(\frac{b^2}{2a}\right) \qquad (x \in \mathbb{R})$$

$$\int \exp\left\{-\frac{a(x-b)^2+c(x-d)^2}{2}\right\} dx$$

$$= \left\{\frac{2\pi}{a+c}\right\}^{1/2} \exp\left\{\left[\frac{(ab+cd)^2}{a+c} - ab^2 - cd^2\right]\Big/2\right\} \qquad (x \in \mathbb{R})$$

$$\int \exp\left\{-\frac{x^{\mathrm{T}}Ax - 2b^{\mathrm{T}}x]}{2}\right\} dx = \left(\frac{(2\pi)^p}{|A|}\right)^{1/2} \exp\left(\frac{b^{\mathrm{T}}A^{-1}b}{2}\right) \qquad (x \in \mathbb{R}^p)$$

$$\int \exp\left\{-\frac{[(x-b)^{\mathrm{T}}A(x-b)+(x-d)^{\mathrm{T}}C(x-d)]}{2}\right\} dx$$

$$= \left(\frac{(2\pi)^p}{|A+C|}\right)^{1/2} \exp\left\{\frac{(Ab+Cd)^{\mathrm{T}}(A+C)^{-1}(Ab+Cd) - b^{\mathrm{T}}Ab - d^{\mathrm{T}}Cd}{2}\right\}$$

$$(x \in \mathbb{R}^p).$$

(b) Truncated Gaussian integrals

$$\int_{-c}^{c} x^2\phi(x)\, dx = 2\Phi(c) - 1 - 2c\phi(c)$$

$$\int_{-c}^{c} x^{2k}\phi(x)\, dx = (2k-1)\int_{-c}^{c} x^{2k-2}\phi(x)\, dx - 2c^{2k-1}\phi(c).$$

(c) Gamma integrals

$$\int_0^\infty x^{\kappa-1}\exp(-\lambda x^\alpha)\, dx = \int_0^\infty \frac{1}{x^{\kappa+1}}\exp(-\lambda x^{-\alpha})\, dx = \frac{1}{\alpha\lambda^{\kappa/\alpha}}\Gamma\left(\frac{\kappa}{\alpha}\right).$$

5. Discrete distributions

(a) Hypergeometric distribution

$$\mathrm{P}\{Z = z; \pi\} = \frac{\binom{N\pi}{z}\binom{N(1-\pi)}{n-z}}{\binom{N}{n}},$$

$$\max(0, n - N(1-\pi)) \le z \le \min(n, N\pi), \qquad 0 \le \pi \le 1.$$

Moments: $\mathrm{E}Z = n\pi$ and $\mathrm{Var}(Z) = n\pi(1-\pi)(N-n)/(N-1)$.

(b) Binomial distribution binomial(r, π)

$$P(Z = z; \pi) = \binom{r}{z}\pi^z(1-\pi)^{r-z}, \qquad z = 0, 1, \ldots, r, \quad 0 \le \pi \le 1.$$

Moments: $EZ = r\pi$, $\text{Var}(Z) = r\pi(1-\pi)$ and $Ee^{tZ} = [\pi e^t + (1-\pi)]^r$.

Convolution of independent distributions: binomial(r_1, π) + binomial(r_2, π) = binomial$(r_1 + r_2, \pi)$.

When $r = 1$, the binomial distribution is often called the *Bernoulli distribution.*

(c) Multinomial distribution mn $(r, \pi_1, \ldots, \pi_p)$

$$P(\mathbf{Z} = \mathbf{z}; \pi) = \frac{r!}{z_1! \cdots z_p!}\pi_1^{z_1}\cdots\pi_p^{z_p}, \qquad z_i = 0, 1, \ldots, r,$$

$$\sum_{i=1}^{p} z_i = r, \qquad 0 \le \pi_i \le 1, \quad \sum_{i=1}^{p} \pi_i = 1.$$

Moments: $EZ_i = r\pi_i$, $\text{Var}(Z_i) = r\pi_i(1-\pi_i)$ and $\text{Cov}(Z_i, Z_j) = -r\pi_i\pi_j$, $i \ne j$.

Provided $S_1, \ldots, S_q$ is an exhaustive partition of disjoint subsets of $\{1, \ldots, p\}$,

$$\left(\sum_{i\in S_1} Z_i, \ldots, \sum_{i\in S_q} Z_i\right) \sim \text{mn}\left(r, \sum_{i\in S_1}\pi_i, \ldots, \sum_{i\in S_q}\pi_i\right).$$

The two-category multinomial distribution mn (r, π_1, π_2) is the *binomial distribution* binomial(r, π_1).

(d) Poisson distribution Poisson(λ)

$$P(Y = y; \lambda) = \frac{\lambda^y \exp(-\lambda)}{y!}, \qquad y = 0, 1, 2, \ldots, \lambda > 0$$

Moments: $Ey = \lambda$, $\text{Var}(Y) = \lambda$ and $Ee^{tY} = \exp(\lambda(e^t - 1))$.

Convolution of independent distributions: Poisson(λ_1) + Poisson(λ_2) = Poisson$(\lambda_1 + \lambda_2)$.

(e) Negative binomial distribution

(i) The number Y of independent binomial$(1, \pi)$ trials required to observe r 1s has distribution

$$P(Y=y;\pi)=\binom{y-1}{r-1}\pi^r(1-\pi)^{y-r},\qquad y=r,r+1,\ldots,\quad 0\le\pi\le 1.$$

Moments: $\mathrm{E}Y=\frac{r}{\pi}$ and $\mathrm{Var}\,(Y)=r\frac{(1-\pi)}{\pi^2}$.

The negative binomial distribution with $r=1$ is often called the *geometric distribution.*

(ii) The number $Z=Y-r$ of 0s obtained in independent binomial$(1,\pi)$ trials before we observe r 1s has distribution

$$P(Z=z;\pi)=\binom{z+r-1}{z}\pi^r(1-\pi)^z,\qquad z=0,1,\ldots:\quad 0\le\pi\le 1.$$

Moments: $\mathrm{E}Z=r\frac{(1-\pi)}{\pi}$ and $\mathrm{Var}\,(Z)=r\frac{(1-\pi)}{\pi^2}$.

(iii) For modeling departures from the Poisson distribution, take $Z\mid\mu\sim$ Poisson(μ) and $\mu\sim\Gamma(\lambda/\delta,\delta^{-1})$ to obtain

$$P(Z=z;\lambda,\delta)=\left\{z!\,\Gamma\left(\frac{\lambda}{\delta}\right)\right\}^{-1}\Gamma\left(z+\frac{\lambda}{\delta}\right)\delta^z(1+\delta)^{-(z+\lambda/\delta)},$$

$$z=0,1,2,\ldots,\lambda>0,\quad \delta>0.$$

Moments: $\mathrm{E}Z=\lambda$, $\mathrm{Var}\,(Z)=\lambda(1+\delta)$, $\mathrm{E}(Z-\lambda)^3=\lambda(1+\delta)(1+2\delta)$, $\mathrm{E}(Z-\lambda)^4=\lambda(1+\delta)(1+6\delta+6\delta^2)+3\lambda^2(1+\delta)^2$ and

$$Ee^{tZ}=\exp\{-\lambda/\delta\log(1+\delta(1-e^t))\}.$$

Convolution of independent distributions: $\mathrm{NB}(\lambda_1,\delta)+\mathrm{NB}(\lambda_2,\delta)=\mathrm{NB}(\lambda_1+\lambda_2,\delta)$.

This distribution can be written in the form (ii) with $r=\frac{\lambda}{\delta}$ and

$$\pi=\frac{\delta}{1+\delta}.$$

6. Continuous univariate distributions

(a) Gaussian (or normal) distribution $\mathrm{N}(\mu,\sigma^2)$

$$f(z;\mu,\sigma)=\frac{1}{\sqrt{2\pi\sigma^2}}\exp\left\{-\frac{(z-\mu)^2}{2\sigma^2}\right\},$$

$$-\infty\le z\le\infty,\quad -\infty\le\mu\le\infty,\quad \sigma>0.$$

Moments, $EZ = \mu$, $\text{Var}(Z) = \sigma^2$, $E(Z-\mu)^3 = 0$, $E(Z-\mu)^4 = 3\sigma^4$ and $Ee^{tZ} = \exp(\mu t + \sigma^2 t^2/2)$.

Convolution of independent distributions: $N(\mu_1, \sigma_1^2) + N(\mu_2, \sigma_2^2) = N(\mu_1 + \mu_2, \sigma_1^2 + \sigma_2^2)$.

$(Z-\mu)/\sigma \sim N(0, 1)$ if and only if $Z \sim N(\mu, \sigma^2)$.

(b) Gamma distribution $\Gamma(\kappa, \lambda)$

$$f(z; \kappa, \lambda) = \frac{1}{\Gamma(\kappa)} \lambda(\lambda z)^{\kappa-1} \exp(-\lambda z), \qquad z > 0, \quad \kappa > 0, \quad \lambda > 0.$$

Moments: $EZ = \frac{\kappa}{\lambda}$, $\text{Var}(Z) = \frac{\kappa}{\lambda^2}$, $EZ^r = \Gamma(\kappa + r)/\lambda^r\Gamma(\kappa)$ and $Ee^{tZ} = \left(\frac{\lambda}{\lambda - t}\right)^{\kappa}$, $t < \lambda$.

Convolution of independent distributions: $\Gamma(\kappa_1, \lambda) + \Gamma(\kappa_2, \lambda) = \Gamma(\kappa_1 + \kappa_2, \lambda)$.

$$Z \sim \Gamma(\kappa, \lambda) \text{ if and only if } \lambda Z \sim \Gamma(\kappa, 1).$$

The gamma distribution is called the *exponential distribution* when $\kappa = 1$.

(i) Z has an inverse gamma distribution $\Gamma(\kappa, \lambda)^{-1}$ if $1/Z \sim \Gamma(\kappa, \lambda)$ or, equivalently, $\lambda/Z \sim \Gamma(\kappa, 1)$. The inverse gamma distribution has density function

$$f(z; \kappa, \lambda) = \frac{1}{\Gamma(\kappa)} \lambda^{\kappa} \left(\frac{1}{z}\right)^{\kappa+1} \exp\left(-\frac{\lambda}{z}\right), \qquad z > 0, \quad \kappa > 0, \quad \lambda > 0.$$

(ii) Z has a χ_ν^2 distribution if $Z \sim \Gamma(\nu/2, 1/2)$. Conversely, if $Z \sim \Gamma(\kappa, \lambda)$ then $2\lambda Z \sim \Gamma((2\kappa)/2, 1/2) = \chi_{2\kappa}^2$ or $Z \sim (1/2\lambda)\chi_{2\kappa}^2$. The χ_ν^2 distribution has density function

$$f(Z; \nu) = \frac{1}{2^{\nu/2}\Gamma\left(\frac{\nu}{2}\right)} z^{\nu/2-1} \exp\left(-\frac{z}{2}\right), \qquad z > 0, \quad \nu > 0.$$

For independent $Z_i \sim N(0, 1)$, $\sum_{i=1}^n Z_i^2 \sim \chi_n^2$. If $Z_i \sim N(\mu_i, 1)$, then the distribution of $\sum_{i=1}^n Z_i^2$ is a noncentral $\chi_n^2(\psi)$ distribution with n degrees of freedom and noncentrality parameter $\psi = \sum_{i=1}^n \mu_i^2$. Note that $\chi_n^2(\psi) \approx N(n + \psi, 2n + 4\psi)$ for n large.

(iii) Z has a χ_ν distribution if $Z^2 \sim \chi_\nu^2$. The χ_ν distribution has density function

$$f(z; \nu) = \frac{1}{2^{\nu/2}\Gamma\left(\frac{\nu}{2}\right)} z^{\nu-1} \exp\left(-\frac{z^2}{2}\right), \qquad z > 0, \quad \nu > 0.$$

(iv) Z has an inverse χ_ν^2 distribution χ_ν^{-2} if $1/Z$ has a χ_ν^2 distribution. The χ_ν^{-2} distribution has density function

$$f(z;\nu) = \frac{1}{2^{\nu/2}\Gamma\left(\frac{\nu}{2}\right)}\left(\frac{1}{z}\right)^{\nu/2+1} \exp\left(-\frac{1}{2z}\right), \qquad z > 0, \quad \nu > 0.$$

(v) Z has an inverse χ_ν distribution χ_ν^{-1} if $1/Z$ has a χ_ν distribution. The χ_ν^{-1} distribution has density function

$$f(z;\nu) = \frac{1}{2^{\nu/2-1}\Gamma\left(\frac{\nu}{2}\right)}\left(\frac{1}{z}\right)^{\nu+1} \exp\left(-\frac{1}{2z^2}\right), \qquad z > 0, \quad \nu > 0.$$

(c) The Student t_ν distribution

$$f(z;\nu) = \frac{\Gamma\left(\frac{\nu+1}{2}\right)}{\sqrt{\pi\nu}\,\Gamma\left(\frac{\nu}{2}\right)} \frac{1}{\left(1+\frac{z^2}{\nu}\right)^{(\nu+1)/2}}, \qquad -\infty \leq z \leq \infty, \quad \nu > 0.$$

Moments: $\mathrm{E}Z = 0$, $\nu \geq 2$, $\mathrm{Var}(Z) = \frac{\nu}{\nu-2}$, $\nu \geq 3$, $\mathrm{E}Z^3 = 0$, $\nu \geq 4$, and $\mathrm{E}Z^4 = \frac{3\nu^2}{(\nu-2)(\nu-4)}$, $\nu \geq 5$.

When $\nu = 1$, the Student t_1 distribution is called the *Cauchy distribution.*

For independent numerator and denominator $\nu^{1/2}\mathrm{N}(0,1)/\chi_\nu \sim t_\nu$. The distribution of $\nu^{1/2}\mathrm{N}(\mu,1)/\chi_\nu$ is called the noncentral t distribution with noncentrality parameter μ.

(d) F distribution $F(\lambda,\nu)$

$$f(z;\lambda,\nu) = \frac{\Gamma\left(\frac{\nu+\lambda}{2}\right)}{\Gamma\left(\frac{\lambda}{2}\right)\Gamma\left(\frac{\nu}{2}\right)}\left(\frac{\lambda}{\nu}\right)^{\lambda/2} z^{(\lambda-2)/2}\left(1+\frac{\lambda z}{\nu}\right)^{-(\nu+\lambda)/2}, \qquad z > 0, \quad \lambda, \nu > 0.$$

Moments: $\mathrm{E}Z = \frac{\nu}{\nu-2}$, $\nu \geq 3$, and $\mathrm{Var}(Z) = \frac{2\nu^2(\lambda+\nu-2)}{\lambda(\nu-2)^2(\nu-4)}$, $\nu \geq 5$.

For independent numerator and denominator $\dfrac{\chi_\lambda^2/\lambda}{\chi_\nu^2/\nu} \sim F(\lambda, \nu)$. If any of the χ^2 distributions are noncentral, we obtain a noncentral F distribution. The most important of these occurs when the numerator has a noncentral χ^2 distribution.

(e) Beta distribution beta (r, s)

$$f(z; r, s) = \frac{\Gamma(r+s)}{\Gamma(r)\Gamma(s)} x^{r-1}(1-x)^{s-1}, \quad 0 \le x \le 1, \quad r, s > 0.$$

Moments: $\mathrm{E}Z = \dfrac{r}{r+s}$, $\mathrm{Var}\,(Z) = \dfrac{rs}{(r+s)^2(r+s+1)}$.

$sZ/r(1-Z) \sim F(2r, 2s)$ and the case $r = s = 1$ is the *uniform distribution* $\mathrm{U}(0, 1)$.

(f) Uniform distribution $\mathrm{U}(a, b)$

$$f(z; b, a) = \frac{1}{b-a} I(a \le z \le b), \quad a < b.$$

Moments: $\mathrm{E}Z = \dfrac{(b+a)}{2}$, $\mathrm{Var}\,(Z) = \dfrac{(b-a)^2}{12}$, and $\mathrm{E}e^{tZ} = \dfrac{\{e^{bt} - e^{at}\}}{(b-a)t}$.

(g) Weibull distribution

$$f(z; \kappa, \lambda) = \kappa\lambda(\lambda z)^{\kappa-1} \exp\{-(\lambda z)^\kappa\}, \qquad z > 0, \quad \kappa, \lambda > 0.$$

Moments: $\mathrm{E}Z = \Gamma\left(\dfrac{1}{\kappa} + 1\right)$, $\mathrm{Var}\,(Z) = \Gamma\left(\dfrac{2}{\kappa} + 1\right) - \Gamma\left(\dfrac{1}{\kappa} + 1\right)^2$, and $\mathrm{E}Z^r = \Gamma\left(\dfrac{r}{\kappa} + 1\right)$.

The Weibull distribution with $\kappa = 1$ is the *exponential distribution.*

(h) Laplace (or double exponential) distribution

$$f(z; \mu, \sigma) = \frac{1}{2\sigma} \exp\left\{-\frac{|z-\mu|}{\sigma}\right\},$$

$$-\infty \le z \le \infty, \quad -\infty \le \mu \le \infty, \quad \sigma > 0.$$

Moments: $\mathrm{E}Z = \mu$, $\mathrm{Var}\,(Z) = 2\sigma^2$, and $\mathrm{E}(Z-\mu)^r = 0$ for r odd and $r!\sigma^r$ for r even.

(i) Pareto distribution

$$f(z; \kappa, \lambda) = \frac{\kappa\lambda^{\kappa}}{z^{\kappa+1}}, \qquad z > \lambda, \quad \kappa, \lambda > 0.$$

Moments: $\mathrm{E}Z = \dfrac{\lambda\kappa}{\kappa - 1}$, $\kappa > 1$, $\mathrm{Var}\,(Z) = \dfrac{\lambda^2\kappa}{(\kappa-1)^2(\kappa-2)}$, $\kappa > 2$, and $EZ^r = \dfrac{\lambda^r\kappa}{\kappa - r}$, $\lambda > r$.

(j) The exponential power family of distributions

$$f(z; \mu, \sigma, \beta) = \frac{1}{\sigma} h_{\beta}\left(\frac{z - \mu}{\sigma}\right),$$

$$-\infty \leq z \leq \infty; \quad -\infty \leq \mu \leq \infty, \quad \sigma > 0, \quad -1 < \beta \leq 1,$$

where

$$h_{\beta}(x) = \frac{\Gamma\left(\dfrac{3(1+\beta)}{2}\right)^{1/2}}{(1+\beta)\Gamma\left(\dfrac{1+\beta}{2}\right)^{3/2}} \exp\left\{ -\left[\frac{\Gamma\left(\dfrac{3(1+\beta)}{2}\right)}{\Gamma\left(\dfrac{1+\beta}{2}\right)}\right]^{1/(1+\beta)} |x|^{2/(1+\beta)} \right\}.$$

Moments: $\mathrm{E}Z = \mu$ and $\mathrm{Var}\,(Z) = \sigma^2 2^{1+\beta} \dfrac{\Gamma\left(\dfrac{3(1+\beta)}{2}\right)}{\Gamma\left(\dfrac{1+\beta}{2}\right)}$.

The exponential power family of distributions is *Gaussian* when $\beta = 0$, *Laplace* when $\beta = 1$, and $(Z - \mu)/\sigma \to \mathrm{U}(-\sqrt{3}, \sqrt{3})$ when $\beta \to -1$.

7. Continuous multivariate distributions

(a) Multivariate Gaussian (or normal) distribution $\mathrm{N}_p(\mu, \Sigma)$

$$f(y; \mu, \Sigma) = \frac{1}{(2\pi)^{p/2}|\Sigma|^{1/2}} \exp\left\{-\tfrac{1}{2}(y - \mu)^{\mathrm{T}}\Sigma^{-1}(y - X\beta)\right\},$$

$$y \in \mathbb{R}^p, \quad \mu \in \mathbb{R}^p, \quad \Sigma \text{ nonsingular}.$$

Moments, $\mathrm{E}Y = \mu$, $\mathrm{Var}\,(Y) = \Sigma$.

Linear transformation: If $Y \sim \mathrm{N}_p(\mu, \Sigma)$ and A is a $p \times q$ matrix $A^T Y + b \sim \mathrm{N}_q(A^T\mu + b, A^{\mathrm{T}}\Sigma A)$.

Marginal distributions: If $Y \sim \mathrm{N}_p(\mu, \Sigma)$, and Y_1 denotes the first q components of Y, then $Y_1 \sim \mathrm{N}_q(\mu_1, \Sigma_{11})$.

Conditional distributions: If $Y \sim \mathrm{N}_p(\mu, \Sigma)$, and Y_1 denotes the first q components of Y, then $Y_1 \mid Y_2 \sim \mathrm{N}_q(\mu_1 + \Sigma_{12}\Sigma_{22}^{-1}(Y_2 - \mu_2), \Sigma_{11.2})$, where $\Sigma_{11.2}$ is defined in 3(c).

(b) Multivariate Student t_v distribution $\mathrm{St}_p(\mu, \Sigma, v)$

$$f_{\mathrm{St}}(y; \mu, \Sigma, v) = \frac{\Gamma\left(\frac{v+p}{2}\right)}{(v\pi)^{p/2}\Gamma\left(\frac{v}{2}\right)|\Sigma|^{1/2}} \frac{1}{\left\{1 + \frac{(y-\mu)^{\mathrm{T}}\Sigma^{-1}(y-\mu)}{v}\right\}^{(v+p)/2}},$$

$$y \in \mathbb{R}^p,\ u \in \mathbb{R}^p,\ \Sigma \text{ nonsingular}.$$

Moments: $\mathrm{E}(Y) = \mu$, $v \geq 2$, $\mathrm{Var}(y) = \dfrac{v}{v-2}\Sigma$, $v \geq 3$, and

$$E(Y_i - \mu_i)^4 = \frac{3v^2\Sigma_{ii}^2}{(v-2)(v-4)}, \qquad v \geq 5.$$

Linear transformation: If $Y \sim \mathrm{St}_p(\mu, \Sigma, v)$ and A is a $p \times q$ matrix $A^{\mathrm{T}}Y + b \sim \mathrm{St}_q(A^{\mathrm{T}}\mu + b, A^{\mathrm{T}}\Sigma A, v)$.

Marginal distributions: If $Y \sim \mathrm{St}_p(\mu, \Sigma, v)$ and Y_1 denotes the first q components of Y, then $Y_1 \sim \mathrm{St}_q(\mu_1, \Sigma_{11}, v)$.

Conditional distributions: If $Y \sim \mathrm{St}_p(\mu, \Sigma, v)$, and Y_1 denotes the first q components of Y, then $Y_1 \mid Y_2 \sim \mathrm{St}_q(\mu_1 + \Sigma_{12}\Sigma_{22}^{-1}(Y_2 - \mu_2),$ $\{v + (Y_2 - \mu_2)^{\mathrm{T}}\Sigma_{22}^{-1}(Y_2 - \mu_2)\}\Sigma_{11.2}/(v + p - q), v + p - q)$ where $\Sigma_{11.2}$ is defined in 3(c).

If $Y \sim \mathrm{St}_p(\mu, \Sigma, v)$, then Y can be represented in distribution as

$$Y = \mu + h^{-1/2}\Sigma^{1/2}Z,$$

where $Z \sim \mathrm{N}_p(0, I)$, $vh \sim \chi_v^2$ and h and Z are independent.

8. Transformations of random variables

Suppose that Θ has density g. For a one-to-one transformation $\theta \to \lambda(\theta)$, the determinant of the partial derivatives of the inverse transformation $J(\lambda) = |\partial\theta(\lambda)/\partial\lambda|$ is called the *Jacobian* of the transformation. The density of $\Lambda = \lambda(\Theta)$ is given by

$$h(\lambda) = g(\theta(\lambda))|J(\lambda)|.$$

9. Order statistics

Suppose that $Z_1, \ldots, Z_n$ are independent and identically distributed random variables with common continuous distribution $F(;\theta)$. The order statistics are denoted $Z_{n1} \le Z_{n2} \le \cdots \le Z_{nn}$. The joint density of the order statistics is

$$f_{1,\ldots,n}(z_{n1}, \ldots, z_{nn}; \theta) = n! \prod_{i=1}^{n} f(z_{ni}; \theta), \qquad z_{n1} \le \cdots \le z_{nn}.$$

(a) The joint density of (Z_{n1}, Z_{nn}) is

$$f_{1,n}(z_{n1}, z_{nn}; \theta) = n(n-1)(F(z_{nn}; \theta) - F(z_{n1}; \theta))^{n-2} f(z_{nn}; \theta) f(z_{n1}; \theta), \qquad z_{n1} \le z_{nn}.$$

The marginal distribution of Z_{nn} is

$$\mathrm{P}(Z_{nn} \le z_{nn}; \theta) = F(z_{nn}; \theta)^n$$

so

$$f_n(z_{nn}; \theta) = nF(z_{nn}; \theta)^{n-1} f(z_{nn}; \theta).$$

The marginal distribution of Z_{n1} is

$$\mathrm{P}(Z_{n1} \le z_{n1}; \theta) = 1 - \{1 - F(z_{n1}; \theta)\}^n$$

so

$$f_1(z_{n1}; \theta) = n\{1 - F(z_{n1}; \theta)\}^{n-1} f(z_{n1}; \theta).$$

The marginal distribution of the jth order statistic Z_{nj} is

$$\mathrm{P}(Z_{nj} \le z_{nj}; \theta) = \sum_{k=j}^{n} \binom{n}{k} F(z_{nj}; \theta)^k \{1 - F(z_{nj}; \theta)\}^{n-k}$$

so

$$f_j(z_{nn}; \theta) = \frac{n!}{(j-1)!(n-j)!} F(z_{nj}; \theta)^{j-1} \{1 - F(z_{nj}; \theta)\}^{n-j} f(z_{nj}; \theta).$$

(b) When $f(x; \mu) = I(-\frac{1}{2} \le x - \mu \le \frac{1}{2})$, the joint density of (Z_{n1}, Z_{nn}) is

$$f_{1,n}(z_{n1}, z_{nn}) = n(n-1)(z_{nn} - z_{n1})^{n-2}, \; \mu - \tfrac{1}{2} \le z_{n1} \le z_{nn} \le \mu + \tfrac{1}{2}.$$

Making the transformation

$$z_{n1} = m - w/2, \qquad z_{nn} = m + w/2$$

which has Jacobian 1, we obtain the joint density of the midrange and range

$$f(m, w) = n(n-1)w^{n-2}, \qquad \mu - \tfrac{1}{2} \le m - \frac{w}{2} \le m + \frac{w}{2} \le \mu + \tfrac{1}{2}.$$

The region of nonzero density can be expressed as either

$$0 < w < 2(\mu + \tfrac{1}{2} - m), \qquad \mu \le m \le \mu + \tfrac{1}{2}$$
$$0 < w < 2(m - \mu + \tfrac{1}{2}), \qquad \mu - \tfrac{1}{2} \le m \le \mu$$

or

$$\mu - \tfrac{1}{2} + \frac{w}{2} \le m \le \mu + \tfrac{1}{2} - \frac{w}{2}, \qquad 0 \le w \le 1.$$

This region is shown in Figure A.1. Integrating the joint density over w, we find that the marginal density of the midrange is

$$f(m) = \begin{cases} n2^{n-1}(m - \mu + \frac{1}{2})^{n-1} & \mu - \frac{1}{2} \le m \le \mu \\ n2^{n-1}(\mu + \frac{1}{2} - m)^{n-1} & \mu \le m \le \mu + \frac{1}{2} \end{cases}$$

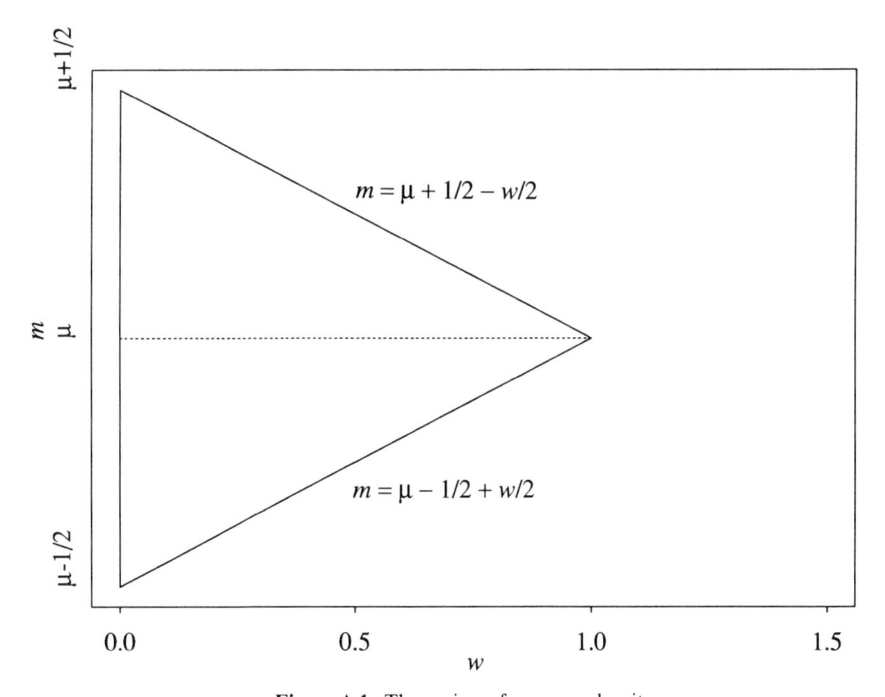

Figure A.1. The region of nonzero density.

which has distribution function

$$F(x) = \begin{cases} 2^{n-1}(x + \frac{1}{2} - \mu)^n & \mu - \frac{1}{2} \le x \le \mu \\ 1 - 2^{n-1}(\mu + \frac{1}{2} - x)^n & \mu \le x \le \mu + \frac{1}{2}. \end{cases}$$

Integrating the joint density over m, we find that the marginal distribution of the range is

$$f(w) = n(n-1)w^{n-2}(1-w), \qquad 0 \le w \le 1.$$

The conditional distribution of the midrange given the range is then

$$f(m \mid w) = (1-w)^{-1}, \qquad \mu - \tfrac{1}{2} + \frac{w}{2} \le m \le \mu + \tfrac{1}{2} - \frac{w}{2},$$

which has distribution function

$$F(m \mid w) = \left\{m - \mu + \tfrac{1}{2} - \frac{w}{2}\right\} \Big/ (1-w), \qquad \mu - \tfrac{1}{2} + \frac{w}{2} \le m \le \mu + \tfrac{1}{2} - \frac{w}{2}.$$

10. Sampling from a Gaussian model

The density of a sample **Z** from the Gaussian model

$$\mathcal{F} = \left\{ f(\mathbf{y}; \mu, \sigma) = \prod_{i=1}^{n} \frac{1}{(2\pi\sigma^2)^{1/2}} \exp\left\{-\frac{(y_i - \mu)^2}{2\sigma^2}\right\}, \right.$$
$$\left. -\infty \le y_i \le \infty \colon \mu \in \mathbb{R}, \sigma > 0 \right\}$$

is

$$\begin{aligned} f(\mathbf{y}; \mu, \sigma) &\propto \exp\left\{-\sum_{j=1}^{n} \frac{(y_j - \mu)^2}{2\sigma^2}\right\}, \\ &= \exp\left\{-\sum_{j=1}^{n} \frac{(y_j - \bar{y})^2}{2\sigma^2} - n\frac{(\bar{y} - \mu)^2}{2\sigma^2}\right\}, \qquad y \in \mathbb{R}^n. \end{aligned}$$

(The proportionality constant is as always determined by the requirement that the density integrate to 1.) The first step is to eliminate $\bar{y}$ from $\sum_{j=1}^{n} (y_j - \bar{y})^2$. This is achieved by making Helmert's orthogonal transformation $u = Hy$ where

$$H = \begin{bmatrix} 1/\sqrt{2} & -1/\sqrt{2} & 0 & 0 & \dots & 0 \\ 1/\sqrt{6} & 1/\sqrt{6} & -2/\sqrt{6} & 0 & \dots & 0 \\ \cdot & \cdot & \cdot & \cdot & \cdot & \cdot \\ 1/\sqrt{n} & 1/\sqrt{n} & 1/\sqrt{n} & 1/\sqrt{n} & \dots & 1/\sqrt{n} \end{bmatrix}.$$

Since $y = H^{\mathrm{T}}u$, we have $\sum_{j=1}^{n} y_j^2 = y^{\mathrm{T}}y = u^{\mathrm{T}}HH^{\mathrm{T}}u = u^{\mathrm{T}}u = \sum_{j=1}^{n} u_j^2$ and hence

$$\begin{aligned}\sum_{j=1}^{n} (y_j - \bar{y})^2 + n(\bar{y} - \mu)^2 &= \sum_{j=1}^{n} y_j^2 - n\bar{y}^2 + n(\bar{y} - \mu)^2 \\ &= \sum_{j=1}^{n-1} u_j^2 + n(n^{-1/2}u_n - \mu)^2.\end{aligned}$$

The Jacobian of an orthogonal transformation is 1 so

$$f(\mathbf{u}; \mu, \sigma) \propto \exp\left\{-\sum_{j=1}^{n-1} \frac{u_i^2}{2\sigma^2} - n\frac{(n^{-1/2}u_n - \mu)^2}{2\sigma^2}\right\}, \qquad u \in \mathbb{R}^n.$$

Next, make the transformation to polar co-ordinates

$$\begin{aligned}u_1 &= (vw)^{1/2} \sin(\theta_1) \\ u_2 &= (vw)^{1/2} \cos(\theta_1) \sin(\theta_2) \\ &\ldots \\ u_{n-1} &= (vw)^{1/2} \cos(\theta_1) \cos(\theta_2) \cdots \cos(\theta_{n-2}) \\ u_n &= n^{1/2}u.\end{aligned}$$

Since $w = v^{-1} \sum_{j=1}^{n-1} u_j^2$ and the Jacobian is proportional to

$$w^{(n-3)/2} \cos(\theta_1)^{n-2} \cos(\theta_2)^{n-3} \cdots \cos(\theta_{n-2}),$$

we obtain

$$\begin{aligned}&f(u, w, \theta_1, \ldots, \theta_{n-2}; \mu, \sigma) \\ &\quad \propto w^{(n-3)/2} \exp\left\{-\frac{vw}{2\sigma^2} - n\frac{(u - \mu)^2}{2\sigma^2}\right\} \cos(\theta_1)^{n-2} \cos(\theta_2)^{n-3} \cdots \cos(\theta_{n-2}), \\ &\qquad u \in \mathbb{R}, \quad w > 0, \quad -\pi/2 \le \theta_1, \ldots, \theta_{n-3} \le \pi/2, \quad 0 \le \theta_{n-2} \le 2\pi.\end{aligned}$$

Integrating over $\theta_1, \ldots, \theta_{n-2}$, we find that the joint density of $\bar{Z}$ and s^2 is

$$\begin{aligned}f(u, w; \mu, \sigma) &\propto w^{(n-3)/2} \exp\left\{-\frac{vw}{2\sigma^2} - n\frac{(u - \mu)^2}{2\sigma^2}\right\} \\ &= w^{(n-3)/2} \exp\left[-\frac{vw}{2\sigma^2}\left\{1 + n\frac{(u - \mu)^2}{vw}\right\}\right], \qquad u \in \mathbb{R}, w > 0.\end{aligned}$$

That is, s^2 and $\bar{Z}$ are independent with $\sigma^2\chi^2_{n-1}/(n-1)$ and $N(\mu, \sigma^2/n)$ distributions respectively. Making the final transformation

$$u = \mu + n^{-1/2}v^{1/2}t, \qquad w = v$$

which has Jacobian proportional to $v^{1/2}$, and integrating over v yields

$$f_{\mu,\sigma}(t) \propto \int_0^\infty v^{n/2-1} \exp\left[-\frac{\nu v}{2\sigma^2}\left\{1 + \frac{t^2}{\nu}\right\}\right] du \propto \frac{1}{\left\{1 + \frac{t^2}{\nu}\right\}^{n/2}}, \; t \in \mathbb{R},$$

which is "Student's" (1908) result. With appropriate modifications, these results extend to more complicated Gaussian models such as the regression model.

References

The original sources for the references have been used throughout the reference list. However, some of the older references are also conveniently available in volumes of collected works including:

Statistical Information and Likelihood: A Collection of Critical Essays by D. Basu. Ghosh, J.K., Ed., New York: Springer-Verlag, 1988.

The Collected Works of George E.P. Box (Two Volumes). Tiao, G., Ed., Belmont CA: Wadsworth, 1984.

Collected Papers of R.A. Fisher (Five Volumes). Bennett, J.H., Ed., The University of Adelaide, 1964.

The Writings of Leonard Jimmie Savage: A Memorial Selection. The American Statistical Association and The Institute for Mathematical Statistics, 1982.

"*Student's*" *Collected Papers.* Pearson, E.S. and Wishart, J., Eds., London: Biometrika Office, 1947.

The Collected Works of John W. Tukey (Six Volumes). Brillinger, D.R., Ed., Belmont CA: Wadsworth, 1984.

Selected Papers in Statistics and Probability by Abraham Wald. Anderson, T.W., Ed., New York: McGraw-Hill, 1955.

S.S. Wilks: Collected Papers, Contributions to Mathematical Statistics. Anderson, T.W., Ed., New York: Wiley, 1967.

Jacob Wolfowitz Selected Papers. Kiefer, J. Ed., New York: Springer Verlag, 1980.

Note: Following each reference, the section or problem (P) to which the entry refers is given. The suffix ".0" refers to the beginning of a chapter or section and a single digit refers to the Further Reading sections at the end of the chapter.

Abramowitz, M. and Stegun, I.A., Eds, (1970). *Handbook of Mathematical Functions.* New York: Dover. [2.1.7, 4.4.5]

Aitken, M. (1991). Posterior Bayes factors (with discussion). *J.R. Statist. Soc. B* **53** 11–142. [2.5.3]

Akaike, H. (1973). Information theory and an extension of the entropy maximisation principle. *Intl. Symposium Information Theory* **2** 267–281. [5.6.4]

Akaike, H. (1978). A new look at the Bayes procedure. *Biometrika* **65** 53–59. [2.4.2]

Albert, J.H. (1993). Teaching Bayesian statistics using sampling methods and MINITAB. *Amer. Statist.* **47** 182–191. [6.8.6, 6]

Arbuthnot, J. (1710). An argument for Divine Providence, taken from the constant regularity observ'd in the birth of both sexes. *Philos. Trans.* **27** 186–190. Reprinted in Kendall, M.G. and Plackett, R.L., Eds., *Studies in the History of Statistics and Probability, Vol. II.* London: Charles Griffin, 1977, 30–34. [3.2.0, 6.3.2]

Armitage, P. (1961). Discussion of Smith, C., Consistency in statistical inference and decision (with discussion). *J.R. Statist. Soc. B* **23** 269–326. [7.2.4]

Atkinson, A.C. (1985). *Plots, Transformations and Regression: An Introduction to Diagnostic Methods of Regression Analysis.* Oxford: Oxford University Press. [1.5.4]

Azzalini, A. and Bowman, A. (1993). On the use of nonparametric regression for checking linear relationships. *J.R. Statist. Soc. B* **55** 549–557. [6.9.4]

Bahadur, R.R. (1960). Stochastic comparison of tests. *Ann. Math. Statist.* **31** 276–295. [4.5.5]

Bahadur, R.R. (1964). On Fisher's bound for asymptotic variances. *Ann. Math. Statist.* **35** 1545–1552. [4.3.4]

Bahadur, R.R. (1971). *Some Limit Theorems in Statistics.* Philadelphia: SIAM. [4]

Barlow, R.E., Toland, R.H., and Freeman, T. (1984). A Bayesian analysis of stress rupture life of Kevlar/epoxy spherical pressure vessels. In Dwivedi, T.D., Ed., *Proceedings of the Canadian Conference in Applied Statistics 1981.* New York: Marcel Dekker, 179–197. [P1.5.4, P2.2.10, P3.6.9].

Barnard, G.A. (1947a). A review of "Sequential Analysis" by Abraham Wald. *J. Amer. Statist. Assoc.* **42** 658–664. [7.2.0]

Barnard, G.A. (1947b). The meaning of significance level. *Biometrika* **34** 179–182. [7.2.0]

Barnard, G.A. (1949). Statistical inference (with discussion). *J.R. Statist. Soc. B* **11** 115–149. [2.7.0, 7.2.0]

Barnard, G.A. (1973). Conditionality, pivotals and robust estimation. In Barndorff-Nielsen, O., Blaesild, P. and Schou, G., Eds., *Proceedings of the Conference on Foundational Questions in Statistical Inference.* Denmark: Aarhus University, 61–80. [2.6.6.]

Barnard, G.A. (1990). Must clinical trials be large? The interpretation of p-values and the combination of test results. *Statist. Med.* **9** 601–604. [3.2.4, 3.4.5]

Barnard, G.A., Jenkins, G.M., and Winsten, C.B. (1962). Likelihood inference and time series (with discussion). *J.R. Statist. Soc. A* **125** 321–372. [2.7.0, 7.2.0, 7.2.5]

Barndorff–Nielsen, O. (1983). On a formula for the distribution of a maximum likelihood estimator. *Biometrika* **70** 343–365. [4.5.15]

Barndorff-Nielsen, O. and Cox, D.R. (1989). *Asymptotic Techniques for Use in Statistics.* London: Chapman & Hall. [4.4.3, 4]

Barndorff-Nielsen, O. and Hall, P. (1988). On the level error after Bartlett adjustment of the likehood ratio statistic. *Biometrika* **75** 374–378. [4.5.16]

Barnett, V. (1982). *Comparative Statistical Inference* (Second edition). New York: Wiley. [2, 3, 7]

Bartlett, M.S. (1937). Properties of sufficiency and statistical tests. *Proc.. R. Soc. A* **160** 268–282. [4.5.16, P4.5.9]

Bartlett, M.S. (1957). A comment on D.V. Lindley's statistical paradox. *Biometrika* **44** 533–534. [2.5.3]

Basu, D. (1964). Recovery of ancillary information. *Sankhyā A* **26** 3–16. [3.9.6, 7.4.1]

Basu, D. (1969). Role of the sufficiency and likelihood principles in sample survey theory. *Sankhyā A* **31** 441–454. [6.5.7]

Basu, D. (1971). An essay on the logical foundations of survey sampling, part 1. In Godambe, V.P. and Sprott, D.A., Eds., *Foundations of Statistical Inference.* Toronto: Holt, Rinehart and Winston, 203–243. [6.5.6.]

Basu, D. (1973). Statistical information and likelihood. In Barndorff-Nielsen, O., Blaesild, P. and Schou, G., Eds., *Proceedings of the Conference on Foundational Questions in Statistical Inference.* Denmark: Aarhus University, 139–236. Republished in *Sankhyā A* (1975) **37** 1–71. [2.6.3, 2, 7.2.4]

Basu, D. (1980). Randomization analysis of experimental data: The Fisher randomization test (with discussion). *J. Amer. Statist. Assoc.* **75** 575–595. [6.4.5]

Basu, D. (1981). On ancillary statistics, pivotal quantities, and confidence statements. In Chaubey, Y.-P. and Dwivedi, T., Eds., *Topics in Applied Statistics.* Montreal: Concordia University, 1–29. [7.4.0].

Bayes, T. (1763). An essay towards solving a problem in the doctrine of chances. *Phil. Trans. R. Soc. Lond.* **53** 370–418. Reprinted in Pearson, E.S. and Kendall, M.G., Eds., *Studies in the History of Statistics and Probability.* London: Charles Griffin, 1970, 131–153. [2.0, 2.2.10]

Bednarski, T. and Clarke, B.R. (1993). Trimmed likelihood estimation of location and scale of the normal distribution. *Austral. J. Statist.* **35** 141–154. [5.4.10]

Behrens, W.-U. (1929). Ein Beitrag zur Fehlen-Berechnung bei wenigen Beobachtungen. *Landw. Jb.* **68** 807–837. [3.8.1]

Bellet, S., Roman, L., DeCastro, O., Kim, K.E., and Kershbaum, A. (1969). Effect of coffee ingestion on catecholamine release. *Metabolism* **18** 288–291. [1.1.3, 6.1.0]

Beran, J. (1991). *M*-estimators of location for Gaussian and related processes with slowly decaying serial correlations. *J. Amer. Statist. Assoc.* **86** 704–708. [5.2.3]

Berger, J. and Delampady, M. (1987). Testing precise hypotheses (with discussion). *Statist. Sci.* **3** 317–352. [3.4.5]

Berger, J. and Selke, T. (1987). Testing a point null hypothesis: The irreconcilability of p-values and evidence (with discussion). *J. Amer. Statist. Assoc.* **82** 112–139. [3.4.5]

Berger, J.O. (1984). The robust Bayesian viewpoint (with discussion). In Kadane, J.B., Ed., *Robustness of Bayesian Analyses.* Amsterdam: Elsevier, 64–144. [2.2.7, 6.1.6]

Berger, J.O. (1985). *Statistical Decision Theory and Bayesian Analysis.* New York: Springer-Verlag. [2.2.4, 2.2.7, 5.8.4, 5]

Berger, J.O. and Bernardo, J.M. (1989). Estimating a product of means: Bayesian analysis with reference priors. *J. Amer. Statist. Assoc.* **84** 200–207. [2.4.2]

Berger, J.O. and Wolpert, R.L. (1984). *The Likelihood Principle.* IMS Lecture Notes – Monograph Series. [7.2.0, 7.5, 7]

Bernado, J.M. (1979). Reference posterior distributions for Bayesian inference (with discussion). *J.R. Statist. Soc. B* **41** 113–147. [2.4.2]

Bernado, J.M. and Smith, A.F.M. (1994). *Bayesian Theory.* New York: Wiley, [1, 2.4.3, 2, 4, 6]

Bernoulli, D. (1777). The most probable choice between several discrepant observations and the formation therefrom of the most likely induction. *Acta Acad. Petrop.* 3–33. Reprinted in Pearson, E.S. and Kendall, M.G., Eds., *Studies in the History of*

Statistics and Probability. English translation by Allen, C.G. London: Charles Griffin, 1970, 157–167. [3.1.2]

Bickel, P.J. (1978). Some recent developments in robust statistics. Presented at the 4th Australian Statistical Conference. [5.7.6]

Bickel, P.J. and Doksum, K.A. (1977). *Mathematical Statistics: Basic Ideas and Selected Topics*. San Francisco: Holden-Day. [3, 4].

Bickel, P.J. and Doksum, K.A. (1981). An analysis of transformations revisited (with discussion). *J. Amer. Statist. Assoc.* **76** 296–311. [P7.4.6]

Bickel, P.J. and Herzberg, A.M. (1979). Robustness of design against autocorrelation in time I: Asymptotic theory, optimality for location and linear regression. *Ann. Statist.* **7** 77–95. [5.2.2]

Bickel, P.J. and Lehmann, E.L. (1975a). Descriptive statistics for nonparametric models I: Introduction. *Ann. Statist.* **3** 1038–1044. [5.7.9]

Bickel, P.J. and Lehmann, E.L. (1975b). Descriptive statistics for nonparametric models II: Location. *Ann. Statist.* **3** 1045–1069. [5.7.9]

Bickel, P.J. and Lehmann, E.L. (1976). Descriptive statistics for nonparametric models III. *Ann. Statist.* **4** 1139–1158. [5.7.9]

Birnbaum, A. (1962). On the foundation of statistical inference (with discussion). *J. Amer. Statist. Assoc.* **57** 269–326. [2.7.0, 7.2.0, 7.2.5, 7.5]

Boole, G. (1854). *The Laws of Thought*. London: Macmillan. Reprinted in 1951 by Dover, New York. [2.4.3]

Bowley, A.L. (1934). Discussion of Neyman, J., On the two different aspects of the representative method: the method of stratified sampling and the method of purposive selection (with discussion). *J.R. Statist. Soc.* **97** 558–606. [3.6.7]

Box, G.E.P. (1953). Non-normality and tests on variances. *Biometrika* **40** 318–335. [5.1.2]

Box, G.E.P. and Andersen, S.L. (1955). Permutation theory in the derivation of robust criteria and the study of departures from assumptions (with discussion). *J.R. Statist. Soc. B* **17** 1–34. [5.1.2]

Box, G.E.P., Hunter, W.G., and Hunter, J.S. (1978). *Statistics for Experimenters: An Introduction to Design, Data Analysis and Model Building*. New York: Wiley. [1.2.1]

Box, G.E.P. and Tiao, G.C. (1973). *Bayesian Inference in Statistical Analysis*. Reading MA: Addison-Wesley. [2.4.2, P2.7.4, 2, 5.8.4, 5.8.5, 5]

Brewer, K.R.W. (1963). Ratio estimation and finite populations: Some results deducible from the assumption of an underlying stochastic process. *Austral. J. Statist.* **5** 93–105. [6.5.7]

Brewer, K.R.W. (1981). Estimating marihuana usage using randomized response – some paradoxical findings. *Austral. J. Statist.* **23** 139–148. [6.0]

Brillinger, D.R. (1962). Examples bearing on the definition of fiducial probability with a bibliography. *Ann. Math. Statist.* **33** 1349–1355. [2.6.5]

Brown, L.D. (1967). The conditional level of Student's t-test. *Ann. Math. Statist.* **38** 1068–1071. [3.9.6]

Bryson, M. (1976). The *Literary Digest* poll: Making of a statistical myth. *Amer. Statist.* **30** 184–185. [1.4.2]

Buehler, R.J. (1959). Some validity criteria for statistical inferences. *Ann. Math. Statist.* **30** 845–863. [3.9.1]

Buehler, R.J. and Federson, A.P. (1963). Note on a conditional property of Student's t. *Ann. Math. Statist.* **34** 1098–1100. [3.9.6]

Carroll, R.J. and Welsh, A.H. (1988). A note on asymmetry and robustness in linear regression. *Amer. Statist.* **42** 285–287. [5.5.2]

Casella, G. (1992). Conditional inference from confidence sets. In Ghosh, M. and Pathak, P.K., Eds., *Current Issues in Statistical Inference: Essays in Honour of D. Basu.* IMS Lecture Notes. [3].

Casella, G. and Berger, R.L. (1987). Reconciling Bayesian and frequentist evidence in the one-sided testing problem. *J. Amer. Statist. Assoc.* **82** 106–111. [3.4.5]

Casella, G. and Berger, R.L. (1990). *Statistical Inference.* Belmont, CA: Wadsworth. [3].

Casella, G. and George, E.I. (1992). Explaining the Gibbs sampler. *Amer. Statist.* **46** 167–174. [6]

Cassel, C.-M., Särndal, C.-E., and Wretman, J.H. (1977). *Foundations of Inference in Survey Sampling.* New York: Wiley [6.5.7]

Chambers, R.L. (1986). Outlier robust finite population estimation. *J. Amer. Statist. Assoc.* **81** 1063–1069. [1.2.4, 6.5.10]

Chernoff, H. (1952). A measure of asymptotic efficiency for tests of an hypothesis based on the sum of observations. *Ann. Math. Statist.* **23** 493–507. [4.5.5]

Chernoff, H. (1954). On the distribution of the likelihood ratio. *Ann. Math. Statist.* **25** 573–578. [4.5.12]

Chernoff, H. (1972). *Sequential Analysis and Optimal Design.* Philadelphia: SIAM. [4]

Cleveland, W.S. (1979). Robust locally weighted regression and smoothing scatterplots. *J. Amer. Statist. Assoc.* **74** 829–836. [6.9.4]

Cleveland, W.S. (1993). *Visualizing Data.* Summit NJ: Hobart Press. [1.2.5, 1, 6]

Cochran, W.G. (1939). The use of analysis of variance in enumeration by sampling. *J. Amer. Statist. Assoc.* **34** 492–510. [6.5.7]

Cochran, W.G. (1946). Relative accuracy of systematic and stratified random samples for a certain class of populations. *Ann. Math. Statist.* **17** 164–177. [6.5.7]

Cochran, W.G. (1964). Approximate significance levels of the Behrens–Fisher test. *Biometrics* **20** 191–195. [3.8.1, P3.10.2]

Cochran, W.G. (1977). *Sampling Techniques* (third edition). New York: Wiley. [6.5.3]

Cook, R.D. and Weisberg, S. (1989). Regression diagnostics with dynamic graphics. *Technometrics* **31** 277–291. [5.5.4]

Cook, R.D. and Weisberg, S. (1994). *An Introduction to Regression Graphics.* New York: Wiley. [1.5.4, 1]

Cornfield, J. (1969). The Bayesian outlook and its application (with discussion). *Biometrics* **25** 617–657. [2, 7.1.0]

Cornish, E.A. and Fisher, R.A. (1937). Moments and cumulants in the specification of distributions. *Rev. Intl. Statist. Inst.* **5** 307–322. [4.1.4]

Cox, D.R. (1958). Some problems connected with statistical inference. *Ann. Math. Statist.* **29** 357–372. [3.3.10, P3.9.1]

Cox, D.R. (1971). The choice between alternative ancillary statistics. *J.R. Statist. Soc. B* **37** 251–255. [7.4.1]

Cox, D.R. (1972). Regression models and life tables (with discussion) *J.R. Statist. Soc. B* **74** 187–220. [2.7.5]

Cox, D.R. (1975). Partial likelihood. *Biometrika* **62** 269–276. [2.7.5]

Cox, D.R. (1990). Role of models in statistical analysis. *Statist. Sci.* **5** 169–174. [1, 7.7]

Cox, D.R. and Hinkley, D.V. (1974). *Theoretical Statistics.* London: Chapman & Hall. [2, 3, 7.4.1, 7.4.4, 7]

Cox, D.R. and Reid, N. (1987). Parameter orthogonality and approximate conditional inference (with discussion). *J.R. Statist. Soc. B* **49** 1–39. [4.4.1, 4.5.15]

Cox, D.R. and Snell, E.J. (1981). *Applied Statistics: Principles and Examples.* London: Chapman & Hall. [1]

Cramer, H. (1946). *Mathematical Methods of Statistics.* Princeton NJ: Princeton University Press. [4.1.8, 4.2.10]

Cramer, H. and Wold, H. (1936). Some theorems on distribution functions. *J. London. Math. Soc.* **11** 290–295. [4.2.3]

Creasy, M.A. (1954). Limits for the ratios of means. *J.R. Statist. Soc. B* **16** 186–194. [3.8.2]

Cushny, A.R. and Peebles, A.R. (1905). The action of optical isomers. II. Hyoscines. *J. Physiol.* **32** 501–510. [P5.1.5, P6.9.1]

Daniels, H.E. (1954). Saddlepoint approximations in statistics. *Ann. Math. Statist.* **25** 631–650. [4.4.3]

Daniels, H.E. (1980). Exact saddlepoint approximations. *Biometrika* **67** 59–63. [4.4.4]

Davison, A.C. and Hinkley, D.V. (1988). Saddlepoint approximations in resampling methods. *Biometrika* **75** 417–431. [6.7.6]

de Finetti, B. (1937). Le prévision: ses lois logiques, ses sources subjectives. *Ann. Inst. Poincare* **7** 1–68. Reprinted as "Foresight: its logical laws, its subjective sources" in *Studies in Subjective Probability*, Kyburg, H.E. and Smokler, H.E., eds., New York: Dover, 93–158. [2.2.0, 2.2.2, 2.2.3, 7.1.0]

de Finetti, B. (1970). *Theory of Probability.* Torino: Einaudi. English translation (1975) by Machi, A. and Smith, A. New York: Wiley. [2.2.2]

Deming, W.E. and Stephan, F. (1941). On the interpretation of censuses as samples. *J. Amer. Statist. Assoc.* **46** 45–49. [6.5.7]

Deming, W.E. (1950). *Some Theory of Sampling.* New York: Wiley. [1.4.2]

Deming, W.E. (1953). On the distinction between enumerative and analytic surveys. *J. Amer. Statist. Assoc.* **48** 244–255. [1.4.2]

Dempster, A.P. (1964). On the difficulties inherent in Fisher's fiducial argument. *J. Amer. Statist. Assoc.* **59** 56–66. [2]

Dempster, A.P. (1973). The direct use of likelihood for significance testing. In Barndorff-Nielsen, O. Blaesild, P. and Schou, G., Eds., *Proceedings of the Conference on Foundational Questions in Statistical Inference.* Denmark: Aarhus University, 335–357. [3.5.2]

Denby, L. and Mallows, C.L. (1977). Two diagnostic displays for robust regression analysis. *Technometrics* **19** 1–13. [5.5.2]

Devroye, L. (1986). *Non-uniform Random Variate Generation.* New York: Springer-Verlag. [3.10.1, 6.8.1]

Diabetic Retinopathy Study Research Group (1976). Preliminary report on effects of photocoagulation therapy. *Amer. J. Ophthalmology* **81** 383–396. [1.1.1, P1.2.3]

DiCiccio, T.J. and Stern, S.E. (1994). Frequentist and Bayesian Bartlett correction of test statistics based on adjusted profile likelihoods. *J.R. Statist. Soc. B* **56** 397–408. [4.5.16]

Donoho, D.L. and Liu, R.C. (1988). The automatic robustness of minimum distance functionals. *Ann. Math. Statist.* **16** 552–586. [5.4.10]

Eddington, A.S. (1914). *Stellar Movements and the Structure of the Universe.* London: Macmillan. [5.1.7]

Edgeworth, F.Y. (1905). The law of error. *Trans. Camb. Phil. Soc.* **20** 36–65 and 113–141. [4.1.4]

Edgeworth, F.Y. (1908–9). On the probable errors of frequency constants. *J.R. Statist. Soc.* **71** 381–397, 499–512, 651–678 and **72** 81–90. [3.1.2, 4.2.2]

Edwards, A.W.F. (1972). *Likelihood.* Cambridge University Press. [2.7.0, 2]

Edwards, W., Lindman, H., and Savage, L.J. (1963). Bayesian statistical inference for psychological research. *Psychological Review* **70** 193–242. [2.2.9]

Efron, B. (1979). Bootstrap methods: Another look at the jackknife. *Ann. Statist.* **7** 1–26. [6.7.4]

Efron, B. (1982). *The Jackknife, the Bootstrap, and other Resampling Plans.* Philadelphia: SIAM. [6]

Efron, B. (1986). Why isn't everyone a Bayesian? (with discussion). *Amer. Statist.* **40** 1–11. [2.4.3, 7.3.4]

Efron, B. and Hinkley, D.V. (1978). Assessing the accuracy of the maximum likelihood estimator: Observed versus expected Fisher information. *Biometrika* **65** 457–487. [4.6.3]

Efron, B. and Tibshirani, R. (1993). *An Introduction to the Bootstrap.* New York: Chapman & Hall. [6.7.8, 6]

Ericson, W.A. (1969). Subjective Bayesian models in sampling finite populations (with discussion). *J.R. Statist. Soc. B* **31** 195–233. [6.5.7]

Esscher, F. (1932). On the probability function in collective risk theory. *Scand. Act. J.* **15** 175–195. [4.4.3]

Fan, J., Hu, T.-C., and Truong, Y.K. (1994). Robust nonparametric function estimation. *Scand. J. Statist.* **21** 433–446. [6.9.4]

Feller, W. (1935). Über den Zentralengrenswertsatz der Wahrscheinlichkeitsrechnung. *Math. Zeit.* **40** 521–559. [4.1.7]

Feinstein, S.B., Keller, M.W., Kerber, R.E., Vandenberg, B., Hoyte, J., Kutruff, C., Bingle, J., Fraker, T.D., Chappell, R., and Welsh, A.H. (1989). Sonicated echocardiographic contrast agents: Reproducibility studies. *J. Amer. Soc. Echocardiography* **2** 125–131. [P1.5.2, P2.2.4, P3.1.2, P3.2.4]

Fernholz, L.T. (1983). *Von Mises Calculus for Statistical Functionals.* Lecture Notes in Statistics No. 19. New York: Springer-Verlag. [5.1.6, 6.9.2]

Field, C.E. (1982). Small sample asymptotic expansions for multivariate M-estimates. *Ann. Statist.* **10** 672–689. [4.4.3]

Field, C.A. and Ronchetti, E. (1990). *Small Sample Asymptotics.* IMS Monograph Series. [4.4.3, 4.4.5, 4]

Fieller, E.C. (1940). The biological standardisation of insulin (with discussion). *Supplement to J.R. Statist. Soc.* **7** 1–64. [3.8.2]

Fisher, R.A. (1920). A mathematical examination of the methods of determining the accuracy of an observation by the mean error and by the mean square error. *Mon. Not. Roy. Astr. Soc.* **80** 758–770. [5.1.7]

Fisher, R.A. (1922). On the mathematical foundations of theoretical statistics. *Philos. Trans. R. Soc. London A* **222** 309–368. [1.4.2, 2.1.1, 2.4.3, 2.6.0, 2.6.2, 2.7.0, 2.7.1, 3.1.2, 4.3.4, 5.1.2, 7.3.0, 7.7]

Fisher, R.A. (1925a) *Statistical Methods for Research Workers.* Edinburgh: Oliver and Boyd. [3.2.0, 6.3.0, 6.4.5]

Fisher, R.A. (1925b). Expansion of "Student's" integral in powers of n^{-1}. *Metron* **5** 109–120. [3.2.3]

Fisher, R.A. (1930). Inverse probability. *Proc. Camb. Phil. Soc.* **26** 528–535. [2.6.0]

Fisher, R.A. (1934). Two new properties of mathematical likelihood. *Proc. R. Soc. A* **144** 285–307. [3.9.3, 7.4.1]

Fisher, R.A. (1935a). *The Design of Experiments.* Edinburgh: Oliver and Boyd. Eighth edition published in 1966. [2.4.3, 2.6.0]

Fisher, R.A. (1935b). The fiducial argument in statistical inference. *Ann. Eug.* **6** 391–398. [3.8.1]

Fisher, R.A. (1936). Has Mendel's work been rediscovered? *Ann. Sci.* **1** 115–137. [3.2.4]

Fisher, R.A. (1956a). *Statistical Methods and Scientific Inference.* Edinburgh: Oliver and Boyd. Third edition published in 1973. [2.4.3, 2.6.0, 2, P7.4.5]

Fisher, R.A. (1956b). On a test of significance in Pearson's *Biometrika* tables (No. 11). *J.R. Statist. Soc. B* **18** 56–60. [3.9.1]

Foutz, R.V. and Srivastava, R.C. (1977). The performance of the likelihood ratio test when the model is incorrect. *Ann. Statist.* **5** 1183–1194. [4.5.11]

Fraser, D.A.S. (1968). *The Structure of Inference.* New York: Wiley. [2.6.6]

Fuller, W.A. (1975). Regression analysis for sample survey. *Sankhyā C* **37** 117–132. [6.5.7]

Fuller, W.A. (1976). *Introduction to Statistical Time Series.* New York: Wiley. [5.2.2]

Fuller, W.A. (1987). *Measurement Error Models.* New York: Wiley. [5.5.6]

Gastwirth, J.L. and Rubin, H. (1975). The behavior of robust estimators on dependent data. *Ann. Statist.* **3** 1070–1100. [5.2.2]

Gayen, A.K. (1950). The distribution of the variance ratio in random samples of any size drawn from non-normal universes. *Biometrika* **37** 236–255. [5.1.2]

Gauss, C.F. (1809). Theoria motus corporum celestium. Hamburg: Perthes et Bessar. Translated 1857 as *Theory of Motion of the Heavenly Bodies Moving about the Sun in Conic Sections* by C.H. Davis. Boston: Little, Brown. Reprinted 1963. [3.1.2]

Geary, R.C. (1936). The distribution of "Student's" ratio for non-normal samples. *J.R. Statist. Soc. Suppl.* **3** 178–184. [5.1.2]

Geisser, S. (1993). *Predictive Inference: An Introduction.* New York: Chapman & Hall. [1.4.3, 1, 2.2.10]

Gelfand, A.E. and Smith, A.F.M. (1990). Sampling based approaches to calculating marginal densities. *J. Amer. Statist. Assoc.* **85** 398–409. [6.8.4]

Gelfand, A.E., Hills, S.E., Racine-Poon, A., and Smith, A.F.M. (1990). Illustration of Bayesian inference in normal data models using Gibbs sampling. *J. Amer. Statist. Assoc.* **85** 972–985. [6.8.4]

Gigerenzer, G., Swijtink, Z., Porter, T., Daston, L., Beatty, J., and Kruger, L. (1989). *The Empire of Chance: How Probability Changed Science and Everyday Life.* Cambridge University Press. [3.4.0, 3]

Godambe, V.P. (1955). A unified theory of sampling from finite populations. *J.R. Statist. Soc. B* **17** 269–278. [6.5.6]

Godambe, V.P. (1960). An optimal property of regular maximum likelihood estimation. *Ann. Math. Statist.* **31** 1208–1212. [4.2.2]

Godambe, V.P. (1966). A new approach to sampling from a finite universe, Parts I and II. *J.R. Statist. Soc. B* **28** 310–328. [6.5.6]

Godambe, V.P. (1991). *Estimating Functions.* Oxford: Oxford University Press. [4.2.2]

Good, I.J. (1950). *Probability and the Weighting of Evidence.* London: Griffin. [2.2.3]

Good, I.J. (1965). *The Estimation of Probabilities: An Essay on Modern Bayesian Methods.* Cambridge MA: MIT Press. [2.2.6]

Good, I.J. (1986). A flexible Bayesian model for comparing two treatments. *J. Statist. Comput. Simulation* **26** 301–305. [3.4.5]

Good, I.J. (1988). The interface between statistics and philosophy of science (with discussion). *Statist. Sci.* **3** 386–412. [3.4.5]

Goodstadt, M.S. and Gruson, V. (1975). The randomized response technique: A test on drug use. *J. Amer. Statist. Assoc.* **70** 814–818. [6.0]

Greenwood, J.A. and Durand, D. (1960). Aids for fitting the gamma distribution by maximum likelihood. *Technometrics* **2** 55–65. [4.2.1]

Hahn, G.J. and Meeker, W.Q. (1993). Assumptions for statistical inference. *Amer. Statist.* **47** 1–11. [1]

Hajek, J. (1960). Limiting distributions in simple random sampling from a finite population. *Periodica Math.* **5** 361–374. [6.5.4]

Hall, P. (1988). Theoretical comparison of bootstrap confidence intervals (with discussion). *Ann. Statist.* **16** 927–985. [6.7.8]

Halmos, P.R. and Savage, L.J. (1949). Applications of the Radon–Nikodym theorem to the theory of sufficient statistics. *Ann. Math. Statist.* **20** 225–241. [7.3.1]

Hampel, F.R. (1968). *Contributions to the Theory of Robust Estimation.* PhD Thesis, University of California, Berkeley. [5.1.1, 5.1.5, 5.7.0, 5.7.1]

Hampel, F.R. (1971). A general qualitative definition of robustness. *Ann. Math. Statist.* **42** 1887–1896. [5.7.0, 5.7.1]

Hampel, F.R. (1973). Some small sample asymptotics. In Hajek, J., ed., *Proceedings of the Prague Symposium on Asymptotic Statistics*, 109–126. Prague: Charles University. [4.4.3]

Hampel, F.R. (1974). The influence curve and its role in robust estimation. *J. Amer. Statist. Assoc.* **69** 383–393. [5.1.1, 5.7.0]

Hampel, F.R., Rousseeuw, P.J., and Ronchetti, E.M. (1981). The change-of-variance curve and optimal redescending M-estimators. *J. Amer. Statist. Assoc.* **76** 643–648. [5.7.3]

Hampel, F.R., Rousseeuw, P.J., Ronchetti, E.M., and Stahel, W.A. (1986). *Robust Statistics: The Approach based on Influence Functions.* New York: Wiley. [5.2.3, 5.5.6, 5.5.7, 5.6.5, 5.7.0, 5.7.1, 5.7.2, 5.7.3, 5]

Harter, H.L. (1964). *New Tables of the Incomplete Gamma Function Ratio and of*

Percentage Points of the Chi-square and Beta Distributions. Washington DC: U.S. Govt. Printing Office. [4.1.0]

Harville, D.A. (1977). Maximum likelihood approaches to variance component estimation and related problems. *J. Amer. Statist. Assoc.* **72** 320–340. [4.5.15]

Heath, D. and Sudderth, W. (1976). De Finetti's theorem on exchangeable variables. *Amer. Statist.* **30** 188–189. [2.2.2]

Heath, D. and Sudderth, W.D. (1978). On finitely additive priors, coherence and extended admissibility. *Ann. Statist.* **6** 333–345. [7.1.0]

Heritier, S. and Ronchetti, E. (1994). Robust bounded-influence tests in general parametric models. *J. Amer. Statist. Assoc.* **89** 897–904. [4.5.11, 5.6.5]

Hettmansperger, T.P. (1984). *Statistical Inference Based on Ranks.* New York: Wiley. [6.9.1, 6]

Hettmansperger, T.P. and Sheather, S. (1992). A cautionary note on the method of least median of squares. *Amer. Statist.* **46** 79–83. Correspondence, *ibid.* (1993) **47** 160–163. [5.7.4]

Hill, R.W. and Holland, P.W. (1977). Two robust alternatives to least squares regression. *J. Amer. Statist. Assoc.* **72** 828–833. [5.5.2]

Hinkley, D.V. (1978). Likelihood inference about location and scale parameters. *Biometrika* **65** 253–261. [4.6.3]

Hinkley, D.V. and Runger, G. (1984). The analysis of transformed data (with discussion). *J. Amer. Statist. Assoc.* **79** 302–320. [P7.4.6]

Hoaglin, D.C. (1980). A Poissonness plot. *Amer. Statist.* **34** 146–149. [P1.5.1]

Hodges, J.L., Jr., and Lehmann, E.L. (1956). The efficiency of some nonparametric competitors of the t-test. *Ann. Math. Statist.* **27** 324–335. [4.5.5]

Hoeffding, W. (1948). A class of statistics with asymptotically normal distribution. *Ann. Math. Statist.* **19** 293–325. [6.9.3]

Hoeffding, W. (1965). Asymptotically optimal tests for multinomial distributions (with discussion). *Ann. Math. Statist.* **36** 369–408. [4.5.5]

Holland, P.W. and Welsch, R.E. (1977). Robust regression using iteratively reweighted least squares. *Commun. Statist. A* **6** 813–827. [5.5.2]

Horvitz, D.G. and Thompson, D.J. (1952). A generalization of sampling without replacement from a finite universe. *J. Amer. Statist. Assoc.* **47** 663–685. [6.5.3]

Huber, P.J. (1964). Robust estimation of a location parameter. *Ann. Math. Statist.* **35** 73–101. [4.2.2, 5.4.1, 5.7.5]

Huber, P.J. (1965). A robust version of the probability ratio test. *Ann. Math. Statist.* **36** 1753–1758. [5.7.8]

Huber, P.J. (1967). The behaviour of maximum likelihood estimates under non-standard conditions. *Proc. Fifth Berkeley Symp. Math. Statist. Prob.* Vol. I 221–233. [4.2.4]

Huber, P.J. (1968). Robust confidence limits. *Z. Wahrsch. verw. Geb.* **10** 269–278. [5.7.8]

Huber, P.J. (1981). *Robust Statistics.* New York: Wiley. [5.4.6, 5.7.1, 5]

Huber, P.J. (1984). Finite sample breakdown of M- and P-estimators. *Ann. Statist.* **12** 119–126. [5.4.10]

Huber-Carol, C. (1970). *Etude asymptotique de tests robustes.* PhD Thesis, ETH Zurich. [5.7.6]

Jahn, R.G., Dunne, B.J., and Nelson, R.D. (1987). Engineering anomalies research. *J. Scien. Exploration* **1** 21–50. [3.4.5, 3.5.3]

James, A.T., Wilkinson, G.N., and Venables, W.N. (1974), Interval estimates for a ratio of means. *Sankhyā A* **36** 177–183. [3.8.2]

Jaynes, E.T. (1968). Prior probabilities. *IEEE Trans. Systems Science and Cybernetics* **4** 227–241. [2.4.2]

Jefferys, W.H. (1990). Bayesian analysis of random event generator data. *J. Scien. Exploration* **4** 153–169. [3.4.5, 3.5.3]

Jefferys, W.H. (1992). Response to Dobyns. *J. Scien. Exploration* **6** 47–57. [3.4.5, 3.5.3]

Jeffreys, H. (1939/1961). *Theory of Probability*. Oxford: Clarendon Press. Third edition (with corrections) published in 1967. [2.2.3, 2.5.1, 2.5.3, 2.5.4, 2, 3.2.5, 3.5.2, 3.5.3]

Jeffreys, H. (1940). Note on the Behrens–Fisher formula. *Ann. Eug.* **10** 48–51. [3.8.1]

Jeffreys, H. (1946). An invariant form for the prior probability in estimation problems. *Proc. Roy. Soc. A* **186** 453–461. [2.2.3, 2.2.11, 2.4.2]

Johns, M.V. (1979). Robust Pitman-like estimators. In Launder, R.L. and Wilkinson, G.N., Eds., *Robustness in Statistics*. New York: Academic Press, 49–60. [5.4.10]

Johnson, N.L. and Leone, F.C. (1964). *Statistics and Experimental Design: In Engineering and the Physical Sciences I*. New York: Wiley. [1.1.5, P1.5.9]

Kahn, W.D. (1987). A cautionary note for Bayesian estimation of the binomial parameter n. *Amer. Statist.* **41** 38–39. [P2.4.4]

Kalbfleisch, J.D. and Sprott, D.A. (1969). Application of likelihood and fiducial probability to sampling finite populations. In Johnson, N.L. and Smith, H. Jr., Eds., *New Developments in Survey Sampling*. New York: Wiley, 358–389. [6.5.7]

Kariya, T. (1980). Locally robust tests for serial correlation in least squares regression. *Ann. Statist.* **8** 1065–1070. [5.2.2]

Keating, J.P., Glaser, R.E., and Ketchum, N.S. (1990). Testing hypotheses about the shape parameter of a gamma distribution. *Technometrics* **32** 67–82. [1.1.2]

Kempthorne, O. (1952). *The Design and Analysis of Experiments*. New York: Wiley. [6.2.1]

Kempthorne, O. (1955). The randomisation theory of statistical inference. *J. Amer. Statist. Assoc.* **50** 946–967. [6.2.1]

Kempthorne, O. (1969). Discussion of Cornfield, J., The Bayesian outlook and its application (with discussion). *Biometrics* **25** 617–657. [7.1.3]

Kendall, M. and Stuart, A. (1979). *The Advanced Theory of Statistics, Vol. II* (fourth edition). London: Charles Griffin. [2]

Kent, J.T. (1982). Robust properties of likelihood ratio tests. *Biometrika* **69** 19–27. [4.5.11]

Keynes, J.M. (1921). *A Treatise on Probability*. London: Macmillan. [2.2.10]

Kiefer, J.C. (1977a). The foundations of statistics – Are there any?. *Synthèse* **36** 161–176. [2.4.3, 3.3.10, 7.3.4]

Kiefer, J.C. (1977b). Conditional confidence statements and confidence estimators (with discussion). *J. Amer. Statist. Assoc.* **72** 789–827. [3.9.5]

Kimball, A.W. (1957). Errors of the third kind in statistical consulting. *J. Amer. Statist. Assoc.* **57** 133–142. [3.3.2]

Knüsel, L.F. (1969). *Über Minimum-Distance-Schätzungen.* PhD Thesis, ETH. [5.4.10]

Lahiri, D.B. (1968). On the unique sample, the surveyed one. Paper presented at the Chapel Hill Symposium. [P6.5.1]

Lane, W.A. and Sudderth, W.D. (1983). Coherent and continuous inference. *Ann. Statist.* **11** 114–120. [7.1.0]

Laplace, P.S. (1774). Mémoire sur la probabilité des causes le évènemens. *Mémoires de l'Académie Royale des Sciences Presentés par Divers Savans* **6** 621–656. (English translation by S. M. Stigler (1986) *Statist. Sci.* **1** 364–378.) [2.0, 2.2.10]

Laplace, P.S. (1812). *Théorie Analytique des Probabilités.* Paris. [2.0]

Lawless, J.F. (1987). Negative binomial and mixed Poisson regression. *Canad. J. Statist.* **15** 209–225. [P4.3.3]

Lawley, D.N. (1956). A general method for approximating to the distribution of likelihood ratio criteria. *Biometrika* **43** 295–303. [4.5.16]

LeCam, L. (1977). A note on metastatistics or "An essay towards stating a problem in the doctrine of chances." *Synthèse* **36** 133–160. [2.4.3, 7.1.3, 7.3.4]

Lehmann, E.L. (1959/1991). *Testing Statistical Hypotheses.* New York: Wiley. [3.3.4, 3.3.5, 4.2.11, P7.4.5]

Lehmann, E.L. (1975). *Nonparametrics: Statistical Methods Based on Ranks.* San Francisco: Holden-Day. [6]

Lehmann, E.L. (1983). *Theory of Point Estimation.* New York: Wiley. [3.1.4, 3.1.5, 3.9.5, 3, 4.2.11, 6.9.3]

Lehmann, E.L. (1990). Model specification: The views of Fisher, Neyman and later developments. *Statist. Sci.* **5** 160–168. [1, 7.7]

Lehmann, E.L. and Scheffé, H. (1950). Completeness, similar regions and unbiased estimation – Part I. *Sankhyā* **10** 305–340. [7.3.2]

Leonard, T. (1982). Comment on Lejeune, M. and Faulkenberry, G.D., A simple predictive density function. *J. Amer. Statist. Assoc.* **77** 657–658. [4.6.5]

Lindeberg, J.W. (1922). Eine neue Herleitung des Exponentialgesetz in der Wahrscheinlichkeitsrechnung. *Math. Zeit.* **15** 211–225. [4.1.7]

Lindley, D.V. (1957). A statistical paradox. *Biometrika* **44** 187–192. [3.5.3]

Lindley, D.V. (1958). Fiducial distributions and Bayes' theorem. *J.R. Statist. Soc. B* **20** 102–107. [2.6.4]

Lindley, D.V. (1972). *Bayesian Statistics, A Review.* Philadelphia: SIAM. [2.4.3, 2, 3.1.5]

Lindley, D.V. and Smith, A.F.M. (1972). Bayes estimates for the linear model (with discussion). *J.R. Statist. Soc. B* **135** 1–41. [2.2.6]

Linnik, Yu.V. (1968). *Statistical Problems with Nuisance Parameters.* Translations of mathematical monographs, no. 20 (from the 1966 Russian edition). New York: American Mathematical Society. [3.8.1]

Little, R.J.A. and Rubin, D.B. (1987). *Statistical Analysis with Missing Data.* New York: Wiley. [6.5.10]

Lugannani, R. and Rice, S. (1980). Saddle point approximation for the distribution of the sum of independent random variables. *Adv. Appl. Probab.* **12** 475–490. [4.4.6]

Madow, W.G. and Madow, L.H. (1944). On the theory of systematic sampling. *Ann. Math. Statist.* **15** 1–24. [6.5.7]

Malik, H.J. (1970). Estimation of the parameters of the Pareto distribution. *Metrika* **16** 126–132. [P4.2.4]

Mallows, C. (1973). Some comments on C_p. *Technometrics* **15** 661–675. [5.6.4]

Mammen, E. (1993). Bootstrap and wild bootstrap for high dimensional linear models. *Ann. Statist.* **21** 255–285. [6.7.2]

Maritz, J.S. and Jarrett, R. (1978). A note on estimating the variance of the sample median. *J. Amer. Statist. Assoc.* **73** 194–196. [6.7.4]

Markatou, M. and Hettmansperger, T.P. (1990). Robust bounded influence tests in linear models. *J. Amer. Statist. Assoc.* **85** 187–190. [5.6.1]

Markatou, M., Stahel, W., and Ronchetti, E. (1991). Robust M-type testing procedures for linear models. In Stahel, W. and Weisberg, S., Eds. *Directions in Robust Statistics and Diagnostics I*. New York: Springer-Verlag, 201–220. [5.6.1]

Martin, M.A. (1989). *The Bootstrap and Confidence Intervals*. PhD Thesis, ANU. [6.7.8]

Martin, R.D. and Zamar, R.H. (1989). Asymptotically min–max robust M-estimates of scale for positive random variables. *J. Amer. Statist. Assoc.* **84** 494–501. [5.7.5]

McCullagh, P.M. and Nelder, J. (1989). *Generalized Linear Models* (second edition). London: Chapman & Hall. [1.3.4, 2.7.5]

McCullagh, P.M. and Tibshirani, R. (1990). A simple method for the adjustment of profile likelihoods. *J.R. Statist. Soc. B* **52** 325–344. [4.5.15]

McKean, J.W., Sheather, S.J., and Hettmansperger, T.P. (1993). The use and interpretation of residuals based on robust estimation. *J. Amer. Statist. Assoc.* **88** 1254–1263. [5.5.4]

Moran, P.A.P. (1971). Maximum likelihood estimation in non-standard conditions. *Proc. Camb. Philos. Soc.* **70** 441–450. [4.2.4, 4.5.12]

Morgenthaler, S. (1986). Robust confidence intervals for a location parameter: The configural approach. *J. Amer. Statist. Assoc.* **81** 518–525. [5.7.7]

Morgenthaler, S. (1987). Confidence intervals for scale. *Austral. J. Statist.* **29** 278–292. [5.7.7]

Morrison, H.L., Flynn, C.M., and Freeman, K.C. (1990). Where does the disk stop and the halo begin? – Kinematics in a rotation field. *Astronomical Journal* **100** 1191–1222. [1.1.4]

Mosteller, F. and Wallace, D.L. (1964/1984). *Inference and Disputed Authorship: The Federalist*. Reading MA: Addison-Wesley, Second edition published in 1984 by Springer-Verlag. [2, P3.2.2, P4.3.3, P6.7.4]

Nadaraya, E.A. (1964). On estimating regression. *Theor. Probab. Applic.* **10** 186–190. [6.9.4]

Neyman, J. (1934). On the two different aspects of the representative method: the method of stratified sampling and the method of purposive selection (with discussion). *J. R. Statist. Soc.* **97** 558–625. [2.6.6, 3.6.0, 6.5.0]

Neyman, J. (1935). Statistical problems in agricultural experimentation (with discussion). *J.R. Statist. Soc. Suppl 2* 107–180, with the co-operation of K. Iwaszkiewicz and S. Kolodziejcayk. [6.2.1]

Neyman, J. (1937). Outline of a theory of statistical estimation based on the classical theory of probability. *Phil. Trans. R. Soc. A* **236** 333–380. [2.6.6, 3.6.0]

Neyman, J. (1977). Frequentist probability and frequentist statistics. *Synthèse* **36** 97–131. [3.0]

Neyman, J. and Pearson, E.S. (1928). On the use and interpretation of certain test criteria for purposes of statistical inference. Part 1. *Biometrika* **20** 175–240. [3.3.0]

Neyman, J. and Pearson, E.S. (1933). On the problem of the most efficient tests of statistical hypotheses. *Phil. Trans. R. Soc. A* **231** 289–337. [3.3.0, 3.3.3]

Novick, M.R. (1969). Multiparameter Bayesian indifference procedures (with discussion). *J.R. Statist. Soc. B* **31** 29–64. [2.4.2]

O'Hagan, A. (1987). Monte Carlo is fundamentally unsound. *The Statistician* **36** 247–249. [6.8.7]

O'Hagan, A. (1995). Fractional Bayes factors for model comparison (with discussion). *J.R. Statist. Soc. B* **57** 99–138. [2.5.3]

Parr, W.C. and Schucany, W.R. (1980). Minimum distance and robust estimation. *J. Amer. Statist. Assoc.* **75** 616–624. [5.4.10]

Patterson, H.D. and Thompson, R. (1971). Recovery of inter-block information when block sizes are unequal. *Biometrika* **58** 545–554. [2.7.5, 4.5.15]

Pearson, K.P. (1894). Contributions to the mathematical theory of evolution. *Phil. Trans. R. Soc. London A* **185** 71–110. [3.1.1]

Pearson, K.P. (1900). On the criterion that a given system of deviations from the probable in the case of a correlated system of variables is such that it can reasonably be supposed to have arisen from random sampling. *Phil. Mag. Series 5* **50** 157–175. [P3.2.1]

Pearson, E.S. (1929). The distribution of frequency constants in small samples from non-normal symmetrical and skew populations. *Biometrika* **21** 259–286. [5.1.2]

Pearson, E.S. (1931). The analysis of variance in cases of non-normal variation. *Biometrika* **23** 114–133. [5.1.2]

Pierce, D.A. (1973). On some difficulties with a frequentist theory of inference. *Ann. Statist.* **1** 241–250. [3.9.6]

Pitman, E.J.G. (1937). Significance tests which can be applied to samples from any population III: The analysis of variance test. *Biometrika* **29** 322–335. [6.2.1, 6.4.1]

Pitman, E.J.G. (1938). The estimation of the location and scale parameters of a continuous population of any given form. *Biometrika* **30** 391–421. [2.6.6, 3.1.5, 3.9.2, 3.9.3]

Pitman, E.J.G. (1949). Notes on non-parametric statistical inference. Unpublished lecture notes. [4.5.3]

Pitman, E.J.G. (1979). *Some Basic Theory for Statistical Inference.* London: Chapman & Hall. [3.3.10]

Popper, K.R. (1935). *The Logic of Scientific Discovery.* London: Hutchinson. [3.2.5]

Portnoy, S.L. (1977). Robust estimation in dependent situations. *Ann. Statist.* **5** 22–43. [5.2.2]

Portnoy, S.L. (1979). Further remarks on robust estimation in dependent situations. *Ann. Statist.* **7** 224–231. [5.2.2]

Pratt, J.W. (1962). Discussion of Birnbaum. A., On the foundations of statistical inference (with discussion). *J. Amer. Statist. Assoc.* **57** 314–316. [7.2.3]

Pratt, J.W. (1976). F.Y. Edgeworth and R.A. Fisher on the efficiency of maximum likelihood estimation. *Ann. Statist.* **4** 501–514. [3]

Pratt, J.W., Raiffa, H., and Schlaifer, R. (1974). The foundations of decision under uncertainty: An elementary exposition. *J. Amer. Statist. Assoc.* **59** 353–375. [7.1.0]

Proschan, F. (1963). Theoretical explanation of observed decreasing failure rate. *Technometrics* **5** 375–383. [P1.5.11, P2.2.6, P3.6.3, P3.9.5]

Ramsey, F.P. (1931). Truth and probability. In *The Foundation of Mathematics and Other Essays*. London: Kegan Paul. [2.2.3, 7.1.0]

Reid, N. (1988). Saddlepoint methods and statistical inference. *Statist. Sci.* **3** 213–238. [4.4.3]

Richardson, A.M. and Welsh, A.H. (1996). Covariate screening in mixed linear models. *J. Mult. Anal.* To appear. [5.6.2]

Rieder, H. (1978). A robust asymptotic testing model. *Ann. Statist.* **6** 1080–1094. [5.7.6]

Rieder, H. (1980). Estimates derived from robust tests. *Ann. Statist.* **8** 106–115. [5.7.6]

Rieder, H. (1991). Robust testing of functionals. In Stahel, W. and Weisberg, S., Eds., *Directions in Robust Statistics and Diagnostics II*. New York: Springer-Verlag. [5.7.9]

Ripley, B.D. (1987). *Stochastic Simulation*. New York: Wiley. [3.10.1, 6.8.1]

Robinson, G.K. (1975). Some counterexamples to the theory of confidence intervals. *Biometrika* **62** 155–161. [3.9.6]

Robinson, G.K. (1979a). Conditional properties of statistical procedures. *Ann. Statist.* **7** 742–755. [3.9.6]

Robinson, G.K. (1979b). Conditional properties of statistical procedures for location and scale parameters. *Ann. Statist.* **7** 756–771. [3.9.6]

Robinson, J. (1982). Saddlepoint approximations for permutation tests and confidence intervals. *J.R. Statist. Soc. B* **44** 91–101. [6.6.2]

Ronchetti, E. (1982). *Robust Testing in Linear Models: The Infinitesimal Approach*. PhD Thesis. ETH Zurich. [4.5.11, 5.6.5]

Ronchetti, E. and Staudte, R.G. (1994). A robust version of Mallow's C_p. *J. Amer. Statist. Assoc.* **89** 550–559. [5.6.4]

Ronchetti, E. and Welsh, A.H. (1994). Empirical saddlepoint approximations for multivariate M-estimators. *J.R. Statist. Soc. B* **56** 313–326. [6.7.6]

Rousseeuw, P.J. (1981). A new infinitesimal approach to robust estimation. *Z. Wahrsch. verw. Geb.* **56** 127–132. [5.7.3]

Rousseeuw, P.J. (1984). Least median of squares. *J. Amer. Statist. Assoc.* **79** 871–880. [5.7.4]

Rousseeuw, P.J. and Leroy, A.M. (1987). *Robust Regression and Outlier Detection*. New York: Wiley. [5]

Rousseeuw, P.J. and van Zomeren, B.C. (1990). Unmarking multivariate outliers and leverage points (with discussion). *J. Amer. Statist. Assoc.* **85** 633–651. [5.7.4]

Rousseeuw, P.J. and Yohai, V. (1984). Robust estimation by means of S-estimators. In Franke, J., Härdle, W., and Martin, R.D., Eds., *Robust and Nonlinear Time Series Analysis*. Lecture Notes in Statistics 26, New York: Springer-Verlag, 256–272. [5.5.7, 5.7.4]

Royall, R.M. (1970). On finite population sampling under certain linear regression models. *Biometrika* **57** 377–387. [6.5.7]

Royall, R.M. (1971). Linear regression models in finite population sampling theory (with discussion). In Godambe, V.P. and Sprott, D.A., Eds., *Foundations of Statistical Inference*, Toronto: Holt, Rinehart and Winston, 259–279. [6.5.7]

Royall, R.M. (1976). Current advances in sampling theory: Implications for human observational studies (with discussion). *Amer. J. Epidem.* **104** 463–477. [6.1.5, 6.1.6, 6.5.9, 6]

Royall, R.M. (1986). The effect of sample size on the meaning of significance tests. *Amer. Statist.* **40** 313–315. [3.4.5]

Royall, R.M. (1992). The model-based (prediction) approach to finite population sampling theory. In Ghosh, M. and Pathak, P.K., Eds., *Current Issues in Statistical Inference: Essays in Honour of D. Basu.* IMS Lecture Notes 225–240. [6]

Royall, R.M. and Cumberland, W.G. (1981). An empirical study of the ratio estimator and estimators of its variance (with discussion). *J. Amer. Statist. Assoc.* **76** 66–88. [6.5.8]

Rubin, D.B. (1978). Bayesian inference for causal effects: The role of randomisation. *Ann. Statist.* **6** 34–58. [6.1.6]

Rubin, D.B. (1984). Bayesianly justifiable and relevant frequency calculations for the applied statistician. *Ann. Statist.* **12** 1151–1172. [3.6.6]

Rubin, D.B. (1988). Using the SIR algorithm to simulate posterior distributions. In Bernado, J.M., DeGroot, M.H., Lindley, D.V., and Smith, A.F.M., Eds. *Bayesian Statistics 3.* New York: Oxford University Press. 395–402 [6.8.3]

Rubin, H. and Sethuraman, J. (1965). Bayes risk efficiency. *Sankhyā* **27A** 347–356. [4.5.5]

Ruppert, D. and Aldershof, B. (1989). Transformations to symmetry and homoscedasticity. *J. Amer. Statist. Assoc.* **84** 437–446. [5.5.2]

Ruppert, D. and Wand, M.P. (1994). Multivariate locally weighted least squares regression. *Ann. Statist.* **22** 1346–1370. [6.9.4]

Rutherford, E. and Geiger, H. (1910). The probability variations in the distribution of α particles. *Philosophical Magazine Sixth Ser.* **20** 698–704. [P1.5.1, P2.1.3, P2.2.5, P3.2.1, P4.1.4, P6.8.3]

Särndal, C.-E., Swensson, B. and Wretman, J. (1992). *Model Assisted Survey Sampling.* New York: Springer-Verlag. [6]

Savage, L.J. (1954). *The Foundations of Statistics.* New York: Wiley. [2.2.3, 7.1.0, 7.1.3]

Savage, L.J. (1961). The foundation of statistics reconsidered. *Proc. 4th Berkeley Symp. Math. Statist. Prob.* **1** 575–586. [2.2.3, 7.1.0]

Savage, L.J. (1962a). Subjective probability and statistical practice. In *The Foundations of Statistical Inference: A discussion opened by Professor L.J. Savage at a meeting of the Joint Statistics Seminar, Birkbeck and Imperial Colleges, in the University of London.* London: Methuen. [2.2.3, 2.2.9]

Savage, L.J. (1962b). Discussion of Birnbaum, A., On the foundations of statistical inference (with discussion). *J. Amer. Statist. Assoc.* **57** 269–306. [2.6.0]

Scheffé, H. (1959). *The Analysis of Variance.* New York: Wiley. [5.1.2, 6.2.1, 6.3.1, 6]

Schmoyer, R.L. (1991). Nonparametric analyses for two-level single-stress accelerated life tests. *Technometrics* **33** 175–186. [P1.5.4]

Schrader, R.M. and Hettmansperger, T.P. (1980). Robust analysis of variance based upon a likelihood ratio criterion. *Biometrika* **67** 93–101. [5.6.1]

Scott, A. and Smith, T.M.F. (1969). Estimation in multi-stage surveys. *J. Amer. Statist. Assoc.* **64** 830–840. [6.5.7]

Scott, A. and Wu, C-F. (1981). On the asymptotic distribution of ratio and regression estimators. *J. Amer. Statist. Assoc.* **76** 98–102. [6.5.5]

Sen, A.R. (1953). On the estimate of variance in sampling with varying probabilities. *J. Indian Soc. Agric. Statist.* **5** 119–127. [6.5.3]

Sen, P.K. (1960). On some convergence properties of U-statistics. *Calc. Statist. Assoc. Bull.* **10** 1–18. [6.9.3]

Sen, P.K. and Singer J.M. (1993). *Large Sample Methods in Statistics.* New York: Chapman & Hall. [4]

Serfling, R.J. (1980). *Approximation Theorems in Mathematical Statistics.* New York: Wiley. [4.5.5, 4, 5.1.6, 6.9.2, 6.9.3]

Sheather, S.J. (1987). Assessing the accuracy of the sample median: Estimated standard errors versus interpolated confidence intervals. In Dodge, Y., Ed., *Statistical Data Analysis Based on the L1 Norm and Related Methods.* Amsterdam: North-Holland, 203–215. [6.7.1, 6]

Silverman, B.W. (1986). *Density Estimation for Statistics and Data Analysis.* London: Chapman & Hall. [1.5.1]

Silvey, S.D. (1959). The Lagrangian multiplier test. *Ann. Math. Statist.* **30** 389–407. [4.5.10]

Silvey, S.D. (1970). *Statistical Inference.* London, Chapman & Hall. [3, 4]

Simpson, D.G., Ruppert, D., and Carroll, R.J. (1992). One-step GM-estimates and stability of inferences in linear regression. *J. Amer. Statist. Assoc.* **87** 439–450. [5.5.6, 5.5.7]

Slutsky, E.E. (1925). Über stochastiche Asymptoter und Grenzwerte. *Math. Analen* **5** 93. [4.1.8]

Smith, A.F.M. and Gelfand, A.E. (1992). Bayesian statistics without tears: A sampling–resampling perspective. *Amer. Statist.* **46** 84–88. [6.8.6, 6]

Smith, A.F.M., Skene, A.M., Shaw, J.E.H., and Naylor, J.C. (1987). Progress with numerical and graphical methods for practical Bayesian statistics. *The Statistician* **36** 75–82. [2.1.7]

Smith, A.F.M., Skene, A.M., Shaw, J.E.H., Naylor, J.C., and Dransfield, M. (1985). The implementation of the Bayesian paradigm. *Commun. Statist. A* **14** 1079–1102. [2.1.7]

Smith, T.M.F. (1984). Sample surveys: Present position and potential developments: Some personal views. *J.R. Statist. Soc. A* **147** 208–221. [6.5.6]

Staudte, R.G. and Sheather, S.J. (1990). *Robust Estimation and Testing.* New York: Wiley. [5.7.1, 5]

Stein, C. (1959). An example of wide discrepancy between fiducial and confidence intervals. *Ann. Math. Statist.* **30** 877–880. [2.4.2, P2.4.5, P3.8.3]

Stigler, S.M. (1980). Studies in the history of probability and statistics XXXVIII: R.H. Smith, a Victorian interested in robustness. *Biometrika* **67** 217–221. [P5.4.2]

Stigler, S.M. (1986). *The History of Statistics: The Measurement of Uncertainty before 1900.* Cambridge MA: Belknap Press. [2.2.10, 3.2.6]

Stone, C.J. (1977). Consistent nonparametric regression. *Ann. Statistic.* **5** 595–620. [6.9.4]

Stone, M. (1976). Strong inconsistency from uniform priors (with discussion). *J. Amer. Statist. Assoc.* **71** 114–125. [2.4.2]

Stone, M. (1977). An asymptotic equivalence of choice of model by cross-validation and Akaike's criterion. *J.R. Statist. Soc. B* **39** 44–47. [5.6.4]

Stone, M. and Dawid, A.P. (1972). Un-Bayesian implications of improper Bayesian inference in routine statistical problems. *Biometrika* **59** 369–375. [2.4.2]

"Student" (1908). The probable error of a mean. *Biometrika* **67** 1–25. [3.2.3, Appendix]

"Student" (1931). The Lanarkshire milk experiment. *Biometrika* **23** 398–406. [6.1.4]

Tanner, M.A. and Wong, W.H. (1987). The calculation of posterior distributions by data augmentation (with discussion). *J. Amer. Statist. Assoc.* **82** 528–550. [6.8.6]

Taylor, R.L., Daffer, P.Z., and Patterson, R.F. (1985). *Limit Theorem for Sums of Exchangeable Random Variables.* New Jersey: Rowman & Allengeld. [2.2.1, 2.2.2]

Thisted, R. and Efron, B. (1987). Did Shakespeare write a newly discovered poem? *Biometrika* **74** 445–455. [P3.2.5]

Tierney, L. and Kadane, J.B. (1986). Accurate approximations for posterior moments and marginal densities. *J. Amer. Statist. Assoc.* **81** 82–86. [4.6.5]

Tingley, M. and Field, C. (1990). Small sample confidence intervals. *J. Amer. Statist. Assoc.* **85** 427–434. [4.4.7]

Tukey, J.W. (1949). The simplest signed rank tests. Princeton University Statist. Res. Group memo, Report No. 17. [6.9.3]

Tukey, J.W. (1957a). Some example with fiducial relevance. *Ann. Math. Statist.* **28** 687–695. [2.6.5]

Tukey, J.W. (1957b) The present state of fiducial probability. In, Mallows, C.L., Ed., *The Collected Works of John W. Tukey VI: More Mathematical.* Belmont CA: Wadsworth 55–118. [2].

Tukey, J.W. (1958). Bias and confidence in not-quite large samples (abstract). *Ann. Math. Statist.* **29** 614. [P6.7.12]

Tukey, J.W. (1960). A survey of sampling from contaminated distributions. In Olkin, I., Ghurye, S.G., Hoeffding, W., Madow, W.G., and Mann, H.B., Eds., *Contributions to Probability and Statistics: Essays in Honour of Harold Hotelling.* Stanford CA: Stanford University Press, 448–485. [1.2.4, 5.1.5]

Tukey, J.W. (1981). Some advanced thoughts on the data analysis involved in configural polysampling directed towards high performance estimates. Tech. Report 189, Series 2. Department of Statistics, Princeton University. [5.7.7]

Tukey, J.W. and McLaughlin, G. (1963). Less vulnerable confidence and significance procedures for location based on a single sample. *Sankhyā A* **35** 331–352. [P5.1.5, 6.9.2]

Von Mises, R. (1947). On the asymptotic distribution of differentiable statistical functions. *Ann. Math. Statist.* **18** 309–348. [5.1.6, 6.9.2]

Wald, A. (1943). Tests of statistical hypotheses concerning several parameters when the number of observations is large. *Trans. Amer. Math. Soc.* **54** 426–482. [4.5.7]

Wald, A. (1949). Note on the consistency of the maximum-likelihood estimate. *Ann. Math. Statist.* **20** 595–601. [4.2.10]

Wald, A. (1950). *Statistical Decision Functions.* New York: Wiley. [3.3.10]

Wald, A. and Wolfowitz, J. (1944). Statistical tests based on permutations of the observations. *Ann. Math. Statist.* **15** 358–372. [6.4.3]

Walker, A.M. (1969). On the asymptotic behaviour of posterior distributions. *J.R. Statist. Soc. B* **31** 80–88. [4.6.2]

Wallace, D.L. (1980). The Behrens–Fisher and Fieller–Creasy problems. In Fienberg, S.E. and Hinkley, D.V., Eds. *R.A. Fisher: An Appreciation.* Berlin: Springer-Verlag, 119–147. [3]

Wang, S. (1992). General saddlepoint approximations in the bootstrap. *Statist. Probab. Letters* **13** 61–66. [6.7.6]

Warner, S.L. (1965). Randomized response: A survey technique for eliminating evasive answer bias. *J. Amer. Statist. Assoc.* **60** 63–69. [6.0]

Watson, G.S. (1964). Smooth regression analysis. *Sankhyā A* **26** 359–372. [6.9.4]

Weisberg, S. (1985). *Applied Linear Regression Analysis.* New York: Wiley. [1.5.4]

Welch, B.L. (1937a). On the z-test in randomised blocks and latin squares. *Biometrika* **29** 21–52. [6.2.1, 6.4.1]

Welch, B.L. (1937b). The significance of the difference between two means when the population variances are unequal. *Biometrika* **29** 350–362. [3.8.1, P3.10.2, P4.5.7]

Welch, B.L. (1939). On confidence limits and sufficiency, with particular reference to parameters of location. *Ann. Math. Statist.* **10** 58–69. [3.9.5]

Welch, B.L. (1947). The generalisation of "Student's" problem when several different population variances are involved. *Biometrika* **34** 28–35. [3.8.1, 3.9.1]

Welsh, A.H. (1996). Robust estimation of smooth regression and spread functions and their derivatives. *Statist. Sinica.* **6** 347–366 [6.9.4]

Welsh, A.H. and Morrison, H.L. (1990). Robust L-estimation of scale with an application in astronomy. *J. Amer. Statist. Assoc.* **85** 729–743. [5.4.7]

Welsh, A.H. and Ronchetti, E. (1996). Bias-calibrated estimation from sample surveys containing outliers. Preprint [6.5.10]

Wilcoxon, F. (1945). Individual comparisons by ranking methods. *Biometrics* **1** 80–83. [6.9.1]

Wilk, M.B. (1955). *Linear Models and Randomised Experiments.* Iowa State College PhD Thesis. [6.2.1]

Wilkinson, G.N. (1977). On resolving the controversy in statistical inference (with discussion). *J.R. Statist. Soc. B* **39** 119–171. [2.6.5]

Wilks, S.S. (1938a). Shortest average confidence intervals from large samples. *Ann. Math. Statist.* **9** 166–175. [3.6.4, 4.3.3]

Wilks, S.S. (1938b). The large-sample distribution of the likelihood ratio for testing composite hypotheses. *Ann. Math. Statist.* **9** 554–560. [4.5.11]

Wilson, E.B. and Hilferty, M.M. (1931). The distribution of chi-square. *Proc. Natl. Acad. Sci. USA* **17** 684–688. [4.1.10]

Wolfowitz, J. (1962). Bayesian inference and the axioms of consistent decision. *Econometrica* **30** 470–479. [7.1.3]

Wolfowitz, J. (1967). Remarks on the theory of testing hypotheses. *New York Statist.* **18** 1–3. [3.3.2]

Yates, F. and Grundy, P.M. (1953). Selection without replacement from within strata probability proportional to size. *J.R. Statist. Soc. B* **15** 253–261. [6.5.3]

Zabell, S.L. (1992). Fisher and the fiducial argument. *Statist. Sci.* **7** 369–387. [2]

Zellner, A. (1977). Maximal data information prior distributions. In Aykac, A. and Brumat, C., Eds., *New Developments in the Application of Bayesian Methods.* Amsterdam: North Holland. [2.4.2]

Author Index[1]

[1]Note: Following each reference, the section or problem (P) to which the entry refers is given. The suffix ".0" refers to the beginning of a chapter or section and a single digit refers to the reading notes at the end of the chapter. An "*" indicates that a section or problem contains references to multiple papers by the author and a "†" indicates that the reference is in the et al. form.

Data and Analysis Index[1]

[1]This index lists the data sets used in the book alphabetically, cross references them to similar data sets and references both data analysis of the data set and theoretical analysis which is immediately relevant to the data set.

Subject Index

WILEY SERIES IN PROBABILITY AND STATISTICS

ESTABLISHED BY WALTER A. SHEWHART AND SAMUEL S. WILKS

Probability and Statistics

ANDERSON · An Introduction to Multivariate Statistical Analysis, *Second Edition*
*ANDERSON · The Statistical Analysis of Time Series
ARNOLD, BALAKRISHNAN, and NAGARAJA · A First Course in Order Statistics
BACCELLI, COHEN, OLSDER, and QUADRAT · Synchronization and Linearity: An Algebra for Discrete Event Systems
BARTOSZYNSKI and NIEWIADOMSKA-BUGAJ · Probability and Statistical Inference
BERNARDO and SMITH · Bayesian Statistical Concepts and Theory
BHATTACHARYYA and JOHNSON · Statistical Concepts and Methods
BILLINGSLEY · Convergence of Probability Measures
BILLINGSLEY · Probability and Measure, *Second Edition*
BOROVKOV · Asymptotic Methods in Queuing Theory
BRANDT, FRANKEN, and LISEK · Stationary Stochastic Models
CAINES · Linear Stochastic Systems
CAIROLI and DALANG · Sequential Stochastic Optimization
CHEN · Recursive Estimation and Control for Stochastic Systems
CONSTANTINE · Combinatorial Theory and Statistical Design
COOK and WEISBERG · An Introduction to Regression Graphics
COVER and THOMAS · Elements of Information Theory
CSÖRGŐ and HORVÁTH · Weighted Approximations in Probability Statistics
*DOOB · Stochastic Processes
DUDEWICZ and MISHRA · Modern Mathematical Statistics
DUPUIS · A Weak Convergence Approach to the Theory of Large Deviations
ETHIER and KURTZ · Markov Processes: Characterization and Convergence
FELLER · An Introduction to Probability Theory and Its Applications, Volume 1, *Third Edition*, Revised; Volume II, *Second Edition*
FREEMAN and SMITH · Aspects of Uncertainty: A Tribute to D. V. Lindley
FULLER · Introduction to Statistical Time Series, *Second Edition*
FULLER · Measurement Error Models
GHOSH · Sequential Estimation
GIFI · Nonlinear Multivariate Analysis
GUTTORP · Statistical Inference for Branching Processes
HALD · A History of Probability and Statistics and Their Applications before 1750
HALL · Introduction to the Theory of Coverage Processes
HANNAN and DEISTLER · The Statistical Theory of Linear Systems
HEDAYAT and SINHA · Design and Inference in Finite Population Sampling
HOEL · Introduction to Mathematical Statistics, *Fifth Edition*
HUBER · Robust Statistics
IMAN and CONOVER · A Modern Approach to Statistics
JUREK and MASON · Operator-Limit Distributions in Probability Theory
KAUFMAN and ROUSSEEUW · Finding Groups in Data: An Introduction to Cluster Analysis
KOTZ · Leading Personalities in Statistical Sciences from the Seventeenth Century to the Present
LAMPERTI · Probability: A Survey of the Mathematical Theory, *Second Edition*

*Now available in a lower priced paperback edition in the Wiley Classics Library.

Probability and Statistics (Continued)

LARSON · Introduction to Probability Theory and Statistical Inference, *Third Edition*
LESSLER and KALSBEEK · Nonsampling Error in Surveys
LINDVALL · Lectures on the Coupling Method
MANTON, WOODBURY, and TOLLEY · Statistical Applications Using Fuzzy Sets
MARDIA · The Art of Statistical Science: A Tribute to G. S. Watson
MORGENTHALER and TUKEY · Configural Polysampling: A Route to Practical Robustness
MUIRHEAD · Aspects of Multivariate Statistical Theory
OLIVER and SMITH · Influence Diagrams, Belief Nets and Decision Analysis
*PARZEN · Modern Probability Theory and Its Applications
PRESS · Bayesian Statistics: Principles, Models, and Applications
PUKELSHEIM · Optimal Experimental Design
PURI and SEN · Nonparametric Methods in General Linear Models
PURI, VILAPLANA, and WERTZ · New Perspectives in Theoretical and Applied Statistics
RAO · Asymptotic Theory of Statistical Inference
RAO · Linear Statistical Inference and Its Applications, *Second Edition*
*RAO and SHANBHAG · Choquet-Deny Type Functional Equations with Applications to Stochastic Models
RENCHER · Methods of Multivariate Analysis
ROBERTSON, WRIGHT, and DYKSTRA · Order Restricted Statistical Inference
ROGERS and WILLIAMS · Diffusions, Markov Processes, and Martingales, Volume I: Foundations, *Second Edition;* Volume II: Îto Calculus
ROHATGI · An Introduction to Probability Theory and Mathematical Statistics
ROSS · Stochastic Processes
RUBINSTEIN · Simulation and the Monte Carlo Method
RUBINSTEIN and SHAPIRO · Discrete Event Systems: Sensitivity Analysis and Stochastic Optimization by the Score Function Method
RUZSA and SZEKELY · Algebraic Probability Theory
SCHEFFE · The Analysis of Variance
SEBER · Linear Regression Analysis
SEBER · Multivariate Observations
SEBER and WILD · Nonlinear Regression
SERFLING · Approximation Theorems of Mathematical Statistics
SHORACK and WELLNER · Empirical Processes with Applications to Statistics
SMALL and McLEISH · Hilbert Space Methods in Probability and Statistical Inference
STAPLETON · Linear Statistical Models
STAUDTE and SHEATHER · Robust Estimation and Testing
STOYANOV · Counterexamples in Probability
STYAN · The Collected Papers of T. W. Anderson: 1943–1985
TANAKA · Time Series Analysis: Nonstationary and Noninvertible Distribution Theory
THOMPSON and SEBER · Adaptive Sampling
WELSH · Aspects of Statistical Inference
WHITTAKER · Graphical Models in Applied Multivariate Statistics
YANG · The Construction Theory of Denumerable Markov Processes

Applied Probability and Statistics

ABRAHAM and LEDOLTER · Statistical Methods for Forecasting
AGRESTI · Analysis of Ordinal Categorical Data
AGRESTI · Categorical Data Analysis
AGRESTI · An Introduction to Categorical Data Analysis
ANDERSON and LOYNES · The Teaching of Practical Statistics

*Now available in a lower priced paperback edition in the Wiley Classics Library.

ANDERSON, AUQUIER, HAUCK, OAKES, VANDAELE, and WEISBERG · Statistical Methods for Comparative Studies
ARMITAGE and DAVID (editors) · Advances in Biometry
*ARTHANARI and DODGE · Mathematical Programming in Statistics
ASMUSSEN · Applied Probability and Queues
*BAILEY · The Elements of Stochastic Processes with Applications to the Natural Sciences
BARNETT and LEWIS · Outliers in Statistical Data, *Second Edition*
BARTHOLOMEW, FORBES, and McLEAN · Statistical Techniques for Manpower Planning, *Second Edition*
BATES and WATTS · Nonlinear Regression Analysis and Its Applications
BECHHOFER, SANTNER, and GOLDSMAN · Design and Analysis of Experiments for Statistical Selection, Screening, and Multiple Comparisons
BELSLEY · Conditioning Diagnostics: Collinearity and Weak Data in Regression
BELSLEY, KUH, and WELSCH · Regression Diagnostics: Identifying Influential Data and Sources of Collinearity
BERRY · Bayesian Analysis in Statistics and Econometrics: Essays in Honor of Arnold Zellner
BERRY, CHALONER, and GEWEKE · Bayesian Analysis in Statistics and Econometrics: Essays in Honor of Arnold Zellner
BHAT · Elements of Applied Stochastic Processes, *Second Edition*
BHATTACHARYA and WAYMIRE · Stochastic Processes with Applications
BIEMER, GROVES, LYBERG, MATHIOWETZ, and SUDMAN · Measurement Errors in Surveys
BIRKES and DODGE · Alternative Methods of Regression
BLOOMFIELD · Fourier Analysis of Time Series: An Introduction
BOLLEN · Structural Equations with Latent Variables
BOULEAU · Numerical Methods for Stochastic Processes
BOX · R. A. Fisher, the Life of a Scientist
BOX and DRAPER · Empirical Model-Building and Response Surfaces
BOX and DRAPER · Evolutionary Operation: A Statistical Method for Process Improvement
BOX, HUNTER, and HUNTER · Statistics for Experimenters: An Introduction to Design, Data Analysis, and Model Building
BROWN and HOLLANDER · Statistics: A Biomedical Introduction
BUCKLEW · Large Deviation Techniques in Decision, Simulation, and Estimation
BUNKE and BUNKE · Nonlinear Regression, Functional Relations and Robust Methods: Statistical Methods of Model Building
CHATTERJEE and HADI · Sensitivity Analysis in Linear Regression
CHATTERJEE and PRICE · Regression Analysis by Example, *Second Edition*
CLARKE and DISNEY · Probability and Random Processes: A First Course with Applications, *Second Edition*
COCHRAN · Sampling Techniques, *Third Edition*
*COCHRAN and COX · Experimental Designs, *Second Edition*
CONOVER · Practical Nonparametric Statistics, *Second Edition*
CONOVER and IMAN · Introduction to Modern Business Statistics
CORNELL · Experiments with Mixtures, Designs, Models, and the Analysis of Mixture Data, *Second Edition*
COX · A Handbook of Introductory Statistical Methods
*COX · Planning of Experiments
COX, BINDER, CHINNAPPA, CHRISTIANSON, COLLEDGE, and KOTT · Business Survey Methods
CRESSIE · Statistics for Spatial Data, *Revised Edition*
DANIEL · Applications of Statistics to Industrial Experimentation

*Now available in a lower priced paperback edition in the Wiley Classics Library.

Applied Probability and Statistics (Continued)

DANIEL · Biostatistics: A Foundation for Analysis in the Health Sciences, *Sixth Edition*
DAVID · Order Statistics, *Second Edition*
*DEGROOT, FIENBERG, and KADANE · Statistics and the Law
*DEMING · Sample Design in Business Research
DILLON and GOLDSTEIN · Multivariate Analysis: Methods and Applications
DODGE and ROMIG · Sampling Inspection Tables, *Second Edition*
DOWDY and WEARDEN · Statistics for Research, *Second Edition*
DRAPER and SMITH · Applied Regression Analysis, *Second Edition*
DUNN · Basic Statistics: A Primer for the Biomedical Sciences, *Second Edition*
DUNN and CLARK · Applied Statistics: Analysis of Variance and Regression, *Second Edition*
ELANDT-JOHNSON and JOHNSON · Survival Models and Data Analysis
EVANS, PEACOCK, and HASTINGS · Statistical Distributions, *Second Edition*
FISHER and VAN BELLE · Biostatistics: A Methodology for the Health Sciences
FLEISS · The Design and Analysis of Clinical Experiments
FLEISS · Statistical Methods for Rates and Proportions, *Second Edition*
FLEMING and HARRINGTON · Counting Processes and Survival Analysis
FLURY · Common Principal Components and Related Multivariate Models
GALLANT · Nonlinear Statistical Models
GLASSERMAN and YAO · Monotone Structure in Discrete-Event Systems
GNANADESIKAN · Methods for Statistical Data Analysis of Multivariate Observations, *Second Edition*
GREENWOOD and NIKULIN · A Guide to Chi-Squared Testing
GROSS and HARRIS · Fundamentals of Queueing Theory, *Second Edition*
GROVES · Survey Errors and Survey Costs
GROVES, BIEMER, LYBERG, MASSEY, NICHOLLS, and WAKSBERG · Telephone Survey Methodology
HAHN and MEEKER · Statistical Intervals: A Guide for Practitioners
HAND · Discrimination and Classification
*HANSEN, HURWITZ, and MADOW · Sample Survey Methods and Theory, Volume I: Methods and Applications
*HANSEN, HURWITZ, and MADOW · Sample Survey Methods and Theory, Volume II: Theory
HEIBERGER · Computation for the Analysis of Designed Experiments
HELLER · MACSYMA for Statisticians
HINKELMAN and KEMPTHORNE: · Design and Analysis of Experiments, Volume 1: Introduction to Experimental Design
HOAGLIN, MOSTELLER, and TUKEY · Exploratory Approach to Analysis of Variance
HOAGLIN, MOSTELLER, and TUKEY · Exploring Data Tables, Trends and Shapes
HOAGLIN, MOSTELLER, and TUKEY · Understanding Robust and Exploratory Data Analysis
HOCHBERG and TAMHANE · Multiple Comparison Procedures
HOCKING · Methods and Applications of Linear Models: Regression and the Analysis of Variables
HOEL · Elementary Statistics, *Fifth Edition*
HOGG and KLUGMAN · Loss Distributions
HOLLANDER and WOLFE · Nonparametric Statistical Methods
HOSMER and LEMESHOW · Applied Logistic Regression
HØYLAND and RAUSAND · System Reliability Theory: Models and Statistical Methods
HUBERTY · Applied Discriminant Analysis
IMAN and CONOVER · Modern Business Statistics
JACKSON · A User's Guide to Principle Components
JOHN · Statistical Methods in Engineering and Quality Assurance

*Now available in a lower priced paperback edition in the Wiley Classics Library.

Applied Probability and Statistics (Continued)

JOHNSON · Multivariate Statistical Simulation
JOHNSON and KOTZ · Distributions in Statistics
Continuous Univariate Distributions—2
Continuous Multivariate Distributions
JOHNSON, KOTZ, and BALAKRISHNAN · Continuous Univariate Distributions, Volume 1, *Second Edition*
JOHNSON, KOTZ, and BALAKRISHNAN · Discrete Multivariate Distributions
JOHNSON, KOTZ, and KEMP · Univariate Discrete Distributions, *Second Edition*
JUDGE, GRIFFITHS, HILL, LÜTKEPOHL, and LEE · The Theory and Practice of Econometrics, *Second Edition*
JUDGE, HILL, GRIFFITHS, LÜTKEPOHL, and LEE · Introduction to the Theory and Practice of Econometrics, *Second Edition*
JUREČKOVÁ and SEN · Robust Statistical Procedures: Aymptotics and Interrelations
KADANE · Bayesian Methods and Ethics in a Clinical Trial Design
KADANE AND SCHUM · A Probabilistic Analysis of the Sacco and Vanzetti Evidence
KALBFLEISCH and PRENTICE · The Statistical Analysis of Failure Time Data
KASPRZYK, DUNCAN, KALTON, and SINGH · Panel Surveys
KISH · Statistical Design for Research
*KISH · Survey Sampling
LAD · Operational Subjective Statistical Methods: A Mathematical, Philosophical, and Historical Introduction
LANGE, RYAN, BILLARD, BRILLINGER, CONQUEST, and GREENHOUSE · Case Studies in Biometry
LAWLESS · Statistical Models and Methods for Lifetime Data
LEBART, MORINEAU., and WARWICK · Multivariate Descriptive Statistical Analysis: Correspondence Analysis and Related Techniques for Large Matrices
LEE · Statistical Methods for Survival Data Analysis, *Second Edition*
LEPAGE and BILLARD · Exploring the Limits of Bootstrap
LEVY and LEMESHOW · Sampling of Populations: Methods and Applications
LINHART and ZUCCHINI · Model Selection
LITTLE and RUBIN · Statistical Analysis with Missing Data
LYBERG · Survey Measurement
MAGNUS and NEUDECKER · Matrix Differential Calculus with Applications in Statistics and Econometrics
MAINDONALD · Statistical Computation
MALLOWS · Design, Data, and Analysis by Some Friends of Cuthbert Daniel
MANN, SCHAFER, and SINGPURWALLA · Methods for Statistical Analysis of Reliability and Life Data
MASON, GUNST, and HESS · Statistical Design and Analysis of Experiments with Applications to Engineering and Science
McLACHLAN and KRISHNAN · The EM Algorithm and Extensions
McLACHLAN · Discriminant Analysis and Statistical Pattern Recognition
McNEIL · Epidemiological Research Methods
MILLER · Survival Analysis
MONTGOMERY and MYERS · Response Surface Methodology: Process and Product in Optimization Using Designed Experiments
MONTGOMERY and PECK · Introduction to Linear Regression Analysis, *Second Edition*
NELSON · Accelerated Testing, Statistical Models, Test Plans, and Data Analyses
NELSON · Applied Life Data Analysis
OCHI · Applied Probability and Stochastic Processes in Engineering and Physical Sciences
OKABE, BOOTS, and SUGIHARA · Spatial Tesselations: Concepts and Applications of Voronoi Diagrams

*Now available in a lower priced paperback edition in the Wiley Classics Library.

Applied Probability and Statistics (Continued)

OSBORNE · Finite Algorithms in Optimization and Data Analysis
PANKRATZ · Forecasting with Dynamic Regression Models
PANKRATZ · Forecasting with Univariate Box-Jenkins Models: Concepts and Cases
PORT · Theoretical Probability for Applications
PUTERMAN · Markov Decision Processes: Discrete Stochastic Dynamic Programming
RACHEV · Probability Metrics and the Stability of Stochastic Models
RÉNYI · A Diary on Information Theory
RIPLEY · Spatial Statistics
RIPLEY · Stochastic Simulation
ROSS · Introduction to Probability and Statistics for Engineers and Scientists
ROUSSEEUW and LEROY · Robust Regression and Outlier Detection
RUBIN · Multiple Imputation for Nonresponse in Surveys
RYAN · Modern Regression Methods
RYAN · Statistical Methods for Quality Improvement
SCHOTT · Matrix Analysis for Statistics
SCHUSS - Theory and Applications of Stochastic Differential Equations
SCOTT · Multivariate Density Estimation: Theory, Practice, and Visualization
SEARLE · Linear Models
SEARLE · Linear Models for Unbalanced Data
SEARLE · Matrix Algebra Useful for Statistics
SEARLE, CASELLA, and McCULLOCH · Variance Components
SKINNER, HOLT, and SMITH · Analysis of Complex Surveys
STOYAN, KENDALL, and MECKE · Stochastic Geometry and Its Applications, *Second Edition*
STOYAN and STOYAN · Fractals, Random Shapes and Point Fields: Methods of Geometrical Statistics
THOMPSON · Empirical Model Building
THOMPSON · Sampling
TIERNEY · LISP-STAT: An Object-Oriented Environment for Statistical Computing and Dynamic Graphics
TIJMS · Stochastic Modeling and Analysis: A Computational Approach
TITTERINGTON, SMITH, and MAKOV · Statistical Analysis of Finite Mixture Distributions
UPTON and FINGLETON · Spatial Data Analysis by Example, Volume 1: Point Pattern and Quantitative Data
UPTON and FINGLETON · Spatial Data Analysis by Example, Volume II: Categorical and Directional Data
VAN RIJCKEVORSEL and DE LEEUW · Component and Correspondence Analysis
WEISBERG · Applied Linear Regression, *Second Edition*
WESTFALL and YOUNG · Resampling-Based Multiple Testing: Examples and Methods for p-Value Adjustment
WHITTLE · Optimization Over Time: Dynamic Programming and Stochastic Control, Volume I and Volume II
WHITTLE · Systems in Stochastic Equilibrium
WONNACOTT and WONNACOTT · Econometrics, *Second Edition*
WONNACOTT and WONNACOTT · Introductory Statistics, *Fifth Edition*
WONNACOTT and WONNACOTT · Introductory Statistics for Business and Economics, *Fourth Edition*
WOODING · Planning Pharmaceutical Clinical Trials: Basic Statistical Principles
WOOLSON · Statistical Methods for the Analysis of Biomedical Data
*ZELLNER · An Introduction to Bayesian Inference in Econometrics

Tracts on Probability and Statistics

BILLINGSLEY · Convergence of Probability Measures
KELLY · Reversability and Stochastic Networks
TOUTENBURG · Prior Information in Linear Models